Plant Strategies, Vegetation Processes, and Ecosystem Properties

Plant Strategies, Vegetation Processes, and Ecosystem Properties

SECOND EDITION

J P GRIME FRS
Unit of Comparative Plant Ecology
Department of Animal and Plant Sciences
University of Sheffield, UK

JOHN WILEY & SONS, LTD
Chichester · New York · Weinheim · Brisbane · Singapore · Toronto

National 01243 779777
International (+44) 1243 779777
e-mail (for orders and customer service enquiries):
cs-books@wiley.co.uk
Visit our Home Page on http://www.wiley.co.uk
or http://www.wiley.com

Reprinted July 2002

Other Wiley Editorial Offices

John Wiley & Sons, Inc., 605 Third Avenue,
New York, NY 10158-0012, USA

WILEY-VCH Verlag GmbH, Pappelallee 3,
D-69469 Weinheim, Germany

John Wiley & Sons Australia Ltd, 33 Park Road, Milton,
Queensland 4064, Australia

John Wiley & Sons (Asia) Pte Ltd, 2 Clementi Loop #02-01,
Jin Xing Distripark, Singapore 129809

John Wiley & Sons (Canada) Ltd, 22 Worcester Road,
Rexdale, Ontario M9W 1L1, Canada

Library of Congress Cataloging-in-Publication Data

Grime, J. Philip (John Philip)
Plant strategies, vegetation processes, and ecosystem properties / J.P. Grime—2nd ed.
p. cm.
Rev. ed. of: Plant strategies and vegetation processes. c1979.
Includes bibliographical references (p.).
ISBN 0-471-49601-4 (cased)
1. Plant ecology. I. Grime, J. Philip (John Philip) Plant strategies and vegetation processes. II. Title.

QK901 .G84 2001
581.7—dc21

00-069342

British Library Cataloguing in Publication Data

A catalogue record for this book is available from the British Library

ISBN 0-471-49601-4

Contents

Preface

As Stearns (1976) observed, it is relatively easy to devise theories of adaptive specialisation in plants and animals but much more difficult to test them. The first edition of this book consisted of a series of hypotheses arising directly from my experience as a field ecologist and laboratory researcher and drew upon a fragmentary literature none of which had been designed with the specific purpose of testing for the existence of primary strategies. In consequence, the objectives of the first edition were restricted to describing the main ideas and their origins. The objectives in writing this second edition have been to re-examine the concept of primary strategies in the context of world literature, to review the results of various attempts at falsification, and to illustrate the expanding role of strategy theories in fundamental and applied ecology. An additional purpose has been to examine the implications of strategy theory for our understanding of ecosystem structure and dynamics. In 1979 the main reason for studying plant strategies was to interpret and predict vegetation processes; in 2000 an equal concern has developed to use this approach to understand the many effects of plants on ecosystem properties.

Immediate reactions to the first edition covered the spectrum from approval to condemnation and, in retrospect, provided a fair reflection of the diversity of opinions current in 1979 with regard to the future conduct of ecological research. Twenty years later, it is possible to explain why the book was controversial in terms of the very different philosophies prevailing at the time. Elsewhere (Grime 1989a), I have commented on the extent to which, in the classical Kuhnian pattern (Kuhn 1962), the debate (e.g. Harper 1982; Grime 1989a) involved arguments about the propriety of words such as 'strategy', 'stress', and 'disturbance'; I submit that these now stand revealed as surrogates for much more important arguments about the scale at which ecological theory should be developed and applied. Initially the notion of primary functional types was often mistakenly regarded as a challenge (or alternative) to demographic and genecological studies of plant populations. In 2000, there is a more general recognition of the need for a diversity of perspectives and methods addressed to different scales of enquiry. Global changes in land use, climate, and biodiversity have achieved changes in outlook that did not emerge from scholarly discussion! During the last 10 years there has been a major expansion of research on large-scale ecological processes, and efforts to recognise plant functional types are now playing an essential part in the task of incorporating vegetation dynamics into models of ecosystem function at a global and regional scale. The need to analyse and predict impacts of changing land-use and climate on vegetation has provided greater impetus to the effort to recognise and validate primary plant strategies.

Inclusion of a section specifically devoted to plant functional types in the Global Change in Terrestrial Ecosystems (GCTE) Core programme of research on global change provides a current example of the importance now attached to this subject. Until recently it has not been unusual to find theories of functional types which drew their inspiration exclusively from population biology (Caswell 1978, 1989; Whittaker and Goodman 1979) and the view was sometimes expressed (e.g. Harper 1982) that demographic studies could provide an adequate basis for the interpretation of vegetation dynamics. However, this school of thought has been notably silent on the topic of climate change where the need for physiological information (e.g. concerning responses to rising temperature and atmospheric carbon dioxide) is indisputable. Similarly, many ecologists, most notably Leps *et al.* (1982), have begun to question whether approaches to the ecosystem that rely exclusively on population and food web models (May 1974; Pimm 1991) can adequately predict ecosystem properties. This book proceeds on the assumption that it is now generally agreed that, wherever possible, functional types should be defined by reference to both demographic criteria *and* those features of life-history, physiology, and resource dynamics which determine the responses of plants to biotic factors, to soils, land-use, and climatic factors, and their role in ecosystems.

The theoretical framework of the original version of this book rested to a substantial extent on the triangular model of primary plant strategies (CSR theory). The passage of 20 years has not diminished my enthusiasm for this model and I make no apology for retaining in this edition some of the original arguments relating to this theory. Support for CSR theory is now available from theoretical studies (Bolker and Pacala 1999) and from various formal tests which are reviewed at the end of Chapter 1. However, equally strong arguments for the model relate to its explanatory power when applied to the properties of communities and ecosystems. There is no mystery about the mechanistic linkages between C-, S-, and R-strategists and their distinctive effects upon the ecosystems in which they occur. Supplementing and even sometimes supplanting the life-history traits which were the focus of attention in early research (e.g. MacArthur and Wilson 1967; Pianka 1970; Whittaker and Goodman 1979), the attributes that have more recently come to dominate our understanding of the primary strategies (resource capture and utilisation, tissue chemistry and life-span, anti-herbivore defence, rates of decomposition) have obvious and direct connections to the functioning of ecosystems. A key objective of this book is to establish the nature and usefulness of these connections and, with this in mind and before attempting a formal review, Chapter 10 begins with two particularly graphic examples where the resource dynamics of dominant plants control specific ecosystem properties. The first example concerns the persistence of ^{137}Cs in pastures affected by the Chernobyl incident whilst the second refers to the complementary roles of grasses, sedges, and forbs in nitrogen retention in a calcareous grassland.

In its core mechanisms the triangular model presented in this book remains essentially the same as that described in 1979 but in one important detail the theory has evolved. Few ecologists doubt the existence of plants throughout the world with the set of traits associated with stress-tolerance. However, as reflected in the first edition, there has been considerable uncertainty about the selection processes promoting this syndrome. In recent years (Grime 1991; Grime *et al.* 1997) I have become convinced that the evolution of stress-tolerance, despite its occurrence in an enormous variety of contrasted habitats, is invariably and causally associated with mineral nutrient stress. The arguments leading to this conclusion are presented in Chapters 1 and 8. There are several fundamental implications that follow from this refinement, the most radical of which is to follow the example of White (1993) and to recognise mineral nutrients as the primary limiting currency of vegetation and ecosystem processes.

Uncertainties and unresolved debates about the mechanisms and role of competitive interactions remain the greatest single obstacle to progress towards general theory in plant ecology. In particular the penetrating insights of Weldon and Slauson (1986) on this subject have been largely ignored. In consequence, a faultline of mutual incomprehension has often separated those ecologists who seek to identify the circumstances where competition dictates the *kind* of plants recruited into communities (and the *kind* of communities they become) from those who are primarily interested in the control of relative abundance *within* communities. In an attempt to reconcile these two very different perspectives a new chapter (Chapter 5) has been added.

There are two further major alterations to the structure of this book. The first is the insertion of additional chapters (Chapters 6 and 7) dealing respectively with extinctions and invasions. The argument for these inclusions was irresistible, first because they provide an obvious counterpoint to dominance, the subject of Chapter 4, and second because both have recently become a subject of universal concern.

The second major structural alteration consists of the addition of Part III dealing with the impacts of plant strategies on ecosystem properties. Here again there is a fortunate relationship to the immediately preceding chapter which is concerned with the control of species richness in plant communities. Since 1994 an interesting international debate has developed concerning the relative importance of dominant plant traits and species richness as controllers of ecosystem properties. To an extraordinary extent, this field has developed without reference to the large volume of research on species richness by plant ecologists. In reviewing this relatively new area of research my primary concern therefore has been to connect studies of the mechanisms controlling species richness to more recent speculations about its consequences for ecosystems. A specific purpose has been to introduce the humped-back model (Grime 1973a) to the debate about species richness and ecosystem function and to remind researchers in this field of the rather precise predictions that arise from this widely-validated model with respect to the narrow range of circumstances where benefits of species richness *might* occur.

As many commentators have recognised (e.g. Watt 1971; Brinck 1980; Southwood 1988; Keddy 1993), the development of ecology as a rigorous predictive science now depends upon our success in recognising principles of wide generality and avoiding submergence in the rising flotsam of case studies, specialist observations, and untested theories. As MacArthur (1972) observed: 'To do science is to search for repeated patterns not accumulate facts.' The challenge in writing this second edition therefore has been to draw upon a massive new literature whilst preserving a concise thread of argument. This has been attempted by providing summaries of each chapter at the front of the book and by adopting a parsimonious approach to the citation of publications and data. As far as possible, attention has been drawn to early and seminal contributions and particular reference has been made to investigations that provide a wider perspective by including comparisons between plants of contrasted ecology. Here, I have recognised a particular responsibility to acknowledge the work of LG Ramenskii, a pioneer of plant strategy theory, who was unknown to me in 1979.

Since 1979, many of the research projects at the Unit of Comparative Plant Ecology in Sheffield have sought either to test hypotheses related to plant strategies or to use them to interpret or predict vegetation dynamics. In these endeavours, I have enjoyed the close support and excellent company of many colleagues and visitors. In the case of Mr SR Band, Dr JG Hodgson, Dr R Hunt, Mr AM Neal, Professor IH Rorison, Mrs N Ruttle, and Dr K Thompson, it is a pleasure to acknowledge collaborations extending over many years. I offer my thanks also to those who have relished/survived the more frenetic UCPE research campaigns and funding traumas of recent times; here I am especially grateful to Mrs RE Booth, Dr SM Buckland, Dr MJW Burke, Dr BD Campbell, Mr RL Colasanti, Dr GAF Hendry, Dr SH Hillier, Dr S Diaz, Dr CM MacGillivray, Dr A Jalili, Dr LH Fraser, Ms S Hubbard, Dr P Wilson, Dr V Greggains, Mr G Burt-Smith, Mrs JML Mackey, Dr ER Rincon, Mrs RE Spencer, Mr F Sutton, and Mr PC Thorpe. It is a particular pleasure to acknowledge the major inputs from Dr MA Davis to the ideas presented in Chapter 7. Val Greggains, David Hollingworth, and Graham Burt-Smith have played a major part in the preparation of the manuscript and figures for this book and I am extremely grateful to them for their skill, dedication, and support.

After some deliberation, I have retained, with slight expansion, the original title of this book, resisting the temptation to replace 'strategies' with the now fashionable 'functional types'. I am pleased to do this as a mark of respect to those ecologists who first used the word. Their achievement was to recognise the value to ecology of classification by function rather than evolutionary affiliation. This may yet prove to have been a crucial step in defining the nature and method of ecological enquiry.

J P Grime
University of Sheffield
August 2000

Preface to the First Edition

In this book I have attempted to analyse in simple terms the processes which control the structure and composition of vegetation. The concepts and the data upon which it is based derive from three types of research which may be described, respectively, as the *correlative*, *direct*, and *comparative* approaches to plant ecology.

The *correlative* approach consists of the attempt to explain variation in the composition of vegetation by reference to associated environmental variation. One of the successes of this approach has been the identification of certain of the climatic and edaphic factors which determine the character of vegetation in severe environments. More recently this type of research has been extended to studies of the effects of pollutants on the distribution of vegetation types, species, and genotypes.

The *direct* approach relies upon detailed observation and recording of the establishment, longevity, and reproduction of individual plants in natural vegetation at specific sites in the field. Data which have been collected in this way allow vegetation to be interpreted as a function of events in the life-histories and population dynamics of the component plants. The strength of the direct approach is related to the opportunity which it provides to observe the process of natural selection at first hand. However, recognition, over the last decade, of the genetic and ecological fluidity of many plant populations, whilst confirming the value of intensive investigations of local populations, has also exposed the need for extreme caution in making extrapolations from such parochial studies.

The *comparative* approach involves the study under standardised experimental conditions of the germination, growth, and reproductive physiology of large numbers of species and populations of contrasted ecology. This broadly-based type of research provides a context for more localised and intensive studies, and enables the ecologist to recognise the main avenues of adaptive specialisation in plants and to identify characteristics of life-history and physiology which determine fitness (or lack of fitness) in particular habitats.

In attempting a synthesis of the main results from these three fields of research it has been convenient to follow the example of many animal ecologists by focussing on 'strategies'. This form of presentation permits a more condensed style of writing and has provided several opportunities to explore ecological and evolutionary parallels between plants and animals.

During the preparation of this book, and in the preceding years of research in Sheffield and New Haven, it has been my good fortune to work with and to learn from many dedicated and talented ecologists. In particular I should like to acknowledge the guidance I have received from Dr PE Waggoner, Professor CD Pigott, the late Dr PS Lloyd, Dr JG Hodgson, Professor TC Hutchinson, Dr BC Jarvis, Dr OL Gilbert, and Dr R Law. I should also like to record my thanks to many past and present colleagues and friends in the Unit of Comparative Plant Ecology, some of whom, including Professor AR Clapham, Dr IH Rorison, Dr R Hunt, Dr AS Mahmoud, Dr K Thompson, Mr M Spray, Dr SB Furness, Dr MM Al-Mufti, Dr CL Sydes, Dr C Sydes, Dr YD Al-Mashhadani, Mr SR Band, Mr AM Neal, and Mr AV Curtis, have kindly allowed reproduction of unpublished data and photographs.

In conclusion, it is a particular pleasure to thank Mrs IN Ruttle for her immaculate typing and unfailing good humour and to record the deep debt of gratitude which I owe to my family for their constant support and encouragement throughout this venture.

J P Grime
University of Sheffield
May 1978

Introduction

A large quantity of data is now available to those who seek to understand how vegetation functions and is caused to vary in composition from place to place and with the passage of time. Although it is generally recognised by plant ecologists that there is a need to deploy this wealth of information within a succinct conceptual framework, opinions differ as to how this may be achieved. The purpose of this book is to examine one approach to the problem by attempting to recognise the major adaptive strategies (functional types) which have evolved in plants, and to relate these to the processes which determine the structure and species composition of vegetation and the properties of ecosystems. Plant strategies may be defined as **groupings of similar or analogous genetic characteristics which recur widely among species or populations and cause them to exhibit similarities in ecology**.

Classifications of functional types differ with respect to the geographical scale at which they are applied, the criteria they use, and the purpose for which they are designed. However, it is often useful to recognise a continuous hierarchy or nesting (Day *et al.* 1988) and Figure 1 classifies functional types arbitrarily into those with *global*, *regional*, *local*, or *within-population* applicability. At the global scale are systems of classification which include all plants and animals and attempt to aggregate them into a small number of primary functional types (MacLeod 1894; Ramenskii 1938; Hutchinson 1951, 1959;

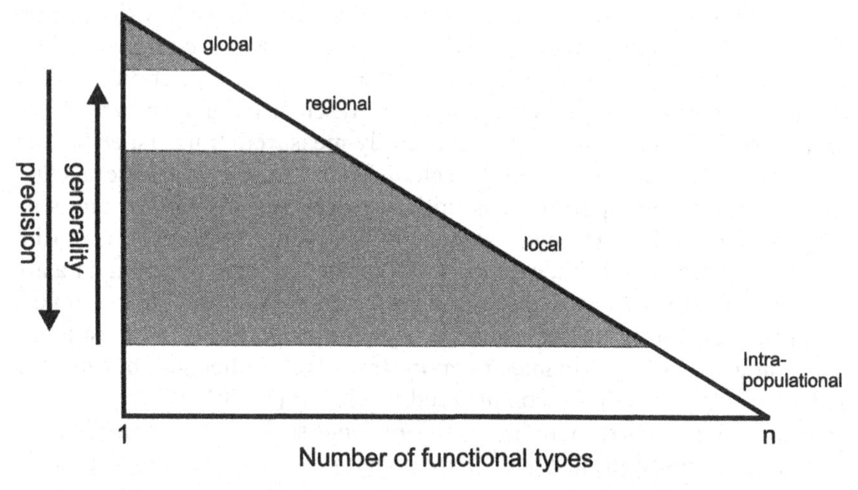

Figure 1. A hierarchy of plant functional types

MacArthur and Wilson 1967; Rabotnov 1985). Here, generality is sought at the expense of precision and the number of useful criteria is restricted by the need to work with attributes common to all organisms; an inevitable consequence of this wide focus is a strong dependence upon life-histories and the capture and utilisation of resources, storage, defence, and reproduction-interrelated subjects, the study of which forms an essential unifying theme in any modern and mechanistic approach to ecology. Success in defining primary functional types of plants will be measurable by the extent to which the structure and dynamics of vegetation are explained, and ecosystem properties at locations anywhere in the world are predicted.

It is perhaps important here to emphasise the focus on 'vegetation' and 'ecosystem properties'. The desire to analyse the detailed ecology of individual species is deeply-rooted in plant ecology and the effort to devise theories applicable to the apex of the hierarchy in Figure 1 continues to attract the critical attention of those who would prefer a finer level of discrimination (e.g. Austin and Gaywood 1994; Grubb 1976, 1980, 1985, 1992, 1998) relevant to the ecology of particular species. This is a recent extension of an old debate. As MacArthur (1968) explained:

> Ecological patterns, about which we construct theories, are only interesting if they are repeated. They may be repeated in space or in time, and they may be repeated from species to species. A pattern which has all of these kinds of repetition is of special interest because of its generality, and yet these very general events are only seen by ecologists with rather blurred vision. The very sharp-sighted always find discrepancies and are able to say that there is no generality, only a spectrum of special cases. This diversity in outlook has proved useful in every science, but it is nowhere more marked than in ecology.

However, the 'blurred vision' of MacArthur implies recognition of that which is general, important, and enduring, and should not be confused with compromise in scientific rigour. Westoby (1998), recognising the continuing difficulties in making theories of primary plant strategies operational, has advocated a pragmatic approach by which all vascular plant species could be rapidly classified with respect to three easily-measured traits (specific leaf area, canopy height, and seed mass), each of which would provide useful clues to the ecology of the species. This initiative deserves support as one step towards a universal functional classification of plants to complement Linnaean classification and is likely to gain support because the required data are easy to collect. However, the three traits selected by Westoby do not reflect many of the vegetation processes and ecosystem properties of greatest interest to ecologists and land managers; more traits will be needed, but not the prohibitively large number recommended by Grubb (1998)!

Functional types corresponding to the *regional* scale depicted in Figure 1 have been recognised primarily by plant geographers and physiologists seeking to explain patchiness in species distributions and vegetation types across the land surface. Much of this patchiness coincides with the world's major climatic zones and has resulted in classifications that focus upon variation in the primary mechanism of photosynthesis (C_3, C_4, or CAM) or which rely upon morphological traits suspected to confer a selective advantage under particular combinations of temperature and moisture supply (Raunkiaer 1934; Holdridge 1947; Box 1981; Woodward 1987; Ehleringer 1993; Smith *et al.* 1996; Grubb 1998; Cunningham *et al.* 1999; Diaz Barrados *et al.* 1999; Diaz *et al.* 1999). At a finer scale, the definitions of regional functional types may often bear a strong imprint of the underlying geology and soil types; particular importance has been attached to functional types associated with saline, calcareous, ultramific, and acidic substrata. A variety of plant strategies are also evident with respect to the capacity of particular taxa to fix atmospheric nitrogen or to capture phosphorus in conditions where this element is in short supply (Sprent 1979; Pate and Dell 1984; Pate 1993; Smith and Read 1997).

At the *local* scale, it is possible to define functional types more narrowly by restricting attention to selected functional aspects in one habitat (e.g. Landsberg *et al.* 1999; Kleyer 1999; Lavorel *et al.* 1999). Quite clearly, as suggested in Figure 1, work at this scale has the potential to generate an infinite number of arrays of functional types and, in particular cases of academic or economic imperative, such a narrow focus can be fully justified. However, despite strong advocacy for research directed at this scale (Harper 1982; Woolhouse 1982), it is difficult to envisage investigations of this specialised type providing an easy or sure path to ecological theories of wide generality. At the base of the hierarchy of functional types we must consider the strong possibility of fine-scale variation within local populations. This subject has very recently become the subject of ecological experiments, and Chapter 9 presents evidence (page 286) that genetic variation within populations may be critical to the persistence of plant species within communities.

The potential value of the concept of primary functional types as a unifying approach to ecology depends upon the extent to which it may be applied to all living organisms. In this book attention will be mainly confined to autotrophic organisms, and most of the evidence which will be considered refers to vascular plants. However, the functional types considered include several which are evident throughout both the plant and the animal kingdoms. In the attempt to define functional types, a complication arises from the need to consider different phases in the life-cycle of the same organism. Even where they experience the same habitat conditions, juvenile and mature stages within the same population may be subject to different forms of natural selection, or alternatively, because of differences in size and function, they may respond in different ways to the same selection force. The resultant 'uncoupling' between juvenile and mature functional types has long been recognised in animals. In many invertebrates the transition between larval and adult phases in the life-history is characterised by radical alterations in structure, physiology, and ecology, and a less dramatic but essentially similar phenomenon is evident in features such as the specialised nutrition, more strictly-defined habitat requirements, and dependence upon parental care which are usually exhibited by the offspring of vertebrate animals.

The same principle may be applied to plants, and from many investigations conducted over the last 25 years (e.g. Grubb 1977; Harper 1977; Grime 1979; Shipley and Parent 1991; Leishman and Westoby 1992; Grime *et al.* 1997) it is clear that in order to understand the basic features of their ecology it is necessary to examine the functional types adopted during two different parts of the life-history—the established (mature) phase and the regenerative (immature) phase. The scheme in Figure 2a describes the regenerative and established phases in the life-cycle of an annual flowering plant. The regenerative phase consists of a series of stages (seed release, dispersal, dormancy, germination, and seedling establishment), each of which varies in duration and in mechanism according to the species or population. The established phase is characterised by a variety of interrelated functions including the capture of resources, the maintenance, replacement, and enlargement of roots and shoots, survival of stress and damage, and the production of seeds.

The ecology of many annual plants may be interpreted as a function of two strategies which are determined, respectively, by the characteristics of the established and regenerative phases of the life-cycle. However, in certain annuals, seeds originating from the same parent may bring about quite

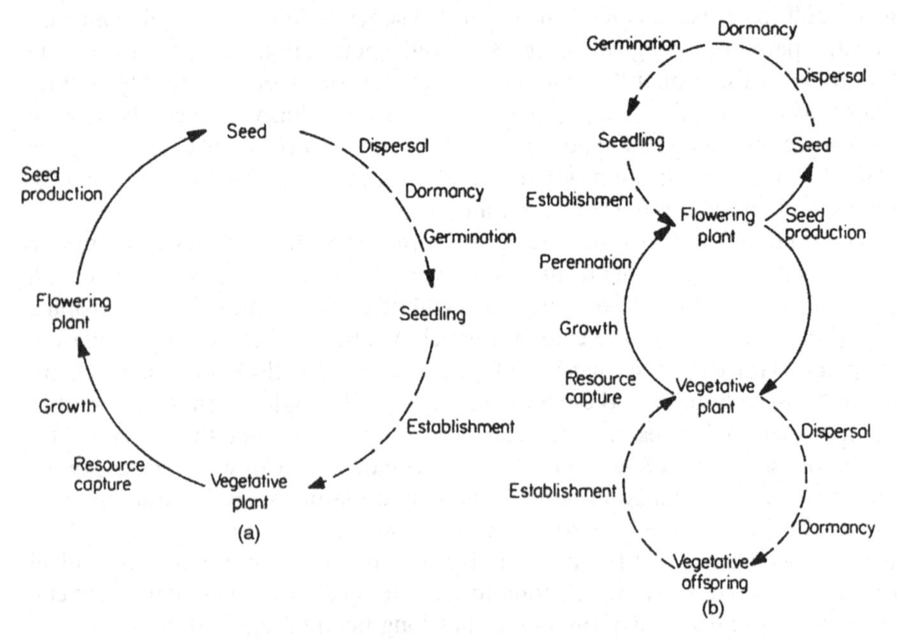

Figure 2. Schemes illustrating the activities associated with the regenerative (- - -) and the established (—) phases in the life-cycles of (a) an annual flowering plant and (b) a perennial plant producing both seeds and vegetative offspring

distinct types of regeneration. Some seeds, for example, may be dispersed considerable distances to new habitats in which they germinate and establish without delay, whilst others become buried *in situ* and after a period of dormancy in the soil established in the place originally occupied by the parent. Such diversity in the regenerative capacity of a single species may arise from genotypic variation within the seed population or from differing experience during maturation on the parent plant, or may be the result of a sophisticated seed physiology. Regardless of its causation, this variety of mechanism exerts a profound effect upon the ecology of the species and must be taken into account in any functional analysis. The importance of multiple forms of regeneration in plants is particularly clear when attention is turned to the life-cycles of perennial species, many of which reproduce both vegetatively and by seed (Figure 2b). In the majority of perennials the two types of offspring differ radically in characteristics such as size, times of production, efficiency of dispersal, and rates of mortality, and for this reason they may be expected to add quite different dimensions to the ecology of a species or population.

The first two chapters of the first section of this book are concerned with functional types of the established phase. Those of the regenerative phase are examined in Chapter 3. The second section (Chapters 4–9) explores the significance of functional types in relation to the processes which control the structure and dynamics of plant communities, whilst the third section (Chapter 10) examines the impacts of plant strategies on ecosystem properties.

In recent years a consensus has begun to develop with regard to the procedures necessary for the recognition of plant functional types and for tests of our ability to use them to interpret and predict the structure and properties of communities and ecosystems. Figure 3 summarises a research protocol advocated (Grime 1993) as a general approach to this problem. The starting point is the accumulation of standardised information on the functional characteristics of large numbers of crops, weeds, and native plant species of contrasted ecology (Clapham 1956). Examples of this approach include Baker (1972), Grime and Hunt (1975), Jurado *et al.* (1991), Hunt *et al.* (1991), Shipley and Parent 1991, Keddy (1992), Ashenden *et al.* (1996), Diaz and Cabido (1997), Diaz *et al.* (1999, 2001), Mabry *et al.* (2000), and Wright and Westoby (2000). The purpose of such screening is to document variation in basic attributes of plant morphology, physiology, and biochemistry. This is followed by multivariate analyses to determine whether there are positive or negative associations between particular traits; when correlations between traits recur widely in screening operations conducted in different floras we may begin to suspect the existence of primary plant strategies. There is growing evidence that certain attributes used singly (Noble and Slatyer 1979; van der Valk 1981) or as sets (Grime *et al.* 1987a; Leishman and Westoby 1992; Gitay *et al.* 1999; White *et al.* 2000a, b) can provide a basis for interpreting individual plant responses to specific changes in climate, soils, or management (Chapters 1–3), or to predict the composition of plant communities and the relative abundance of component populations (Chapters 4 and 5). Where sufficient and appropriate data are available to classify the majority of component species into functional types it may be possible to predict primary and secondary successional changes and vegetation responses to specific scenarios of changed climate or land use (Chapter 8). It is an unfortunate fact, however, that in many investigations, the screening of traits has not been possible and in consequence the interpretation of many observational and experimental studies continues to rely upon classifications that are taxonomic (e.g. grass, sedge, forb) or rely upon single criteria (e.g. C_3, C_4, or CAM photosynthesis) that produce groups of uncertain ecological significance and coherence.

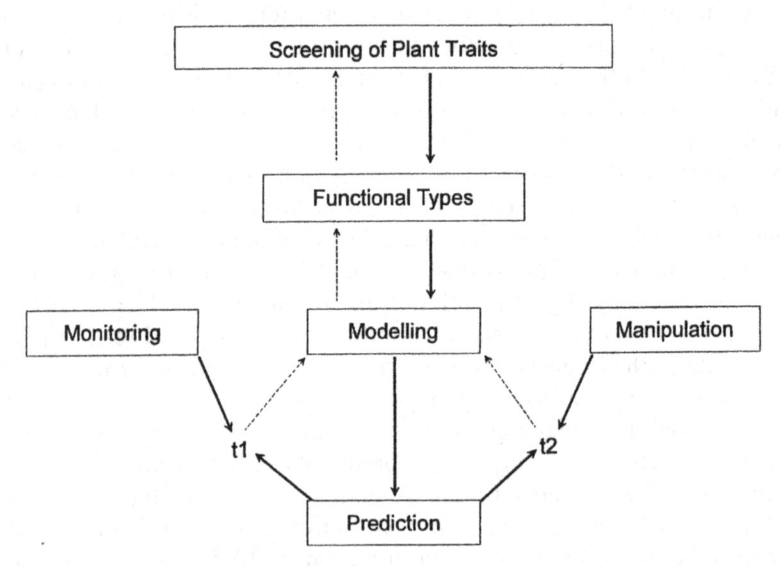

Figure 3. Protocol for testing the predictive value of plant functional types. Discrepancies revealed in the tests at t1 and t2 initiate further modelling cycles, each of which may necessitate refinement of the functional types or even additional screening. (Reproduced from Grime 1993 by permission of *Oikos*)

In Figure 3 two alternative methods are suggested by which predictions of community and ecosystem structure and properties based upon plant functional types may be tested and refined. The first, illustrated in the left-hand side of the figure, involves comparison of model predictions against field data collected by surveys or long-term monitoring of permanent plots. This represents an efficient mechanism for recognising errors and weaknesses in that, as indicated by the arrows in Figure 3, discrepancies can stimulate not only changes in the model but also, where necessary, further data inputs from new screening procedures.

Monitoring studies on communities and ecosystems are few in number, and in consequence alternative mechanisms of hypothesis testing (illustrated on the right-hand side of Figure 3) must be used. These follow a logical pathway similar to that involving surveys and monitoring but rely upon manipulative experiments, some of which can be conducted on a small scale and involve synthesis of vegetation and ecosystems under controlled conditions (e.g. Grime *et al.* 1987b; Diaz *et al.* 1993; Lawton *et al.* 1993; Thompson *et al.* 1993). Often, however, it is necessary to perform replicated manipulations at a comparatively large scale in natural environments and necessitating co-operation between ecologists and engineers (e.g. Thorpe *et al.* 1993; Schulze and Mooney 1993a; MacGillivray and Grime 1995; White *et al.* 2000a, b).

Chapter Summaries

INTRODUCTION

Plant strategies, otherwise described as plant functional types, can be defined as groupings of similar or analogous genetic characteristics which recur widely among species or populations and cause them to exhibit similarities in ecology. A hierarchy can be observed with respect to the geographical scale at which plant strategies are recognised and applied. At each scale, alternative functional classifications can be developed using particular criteria to address different facets of plant biology and ecology, including life-history, capture and utilisation of resources, responsiveness to climate and regenerative biology.

Although the origins of plant strategy theory are to be found in theoretical studies attempting to generalise about the factors controlling the ecology of individual plant species and populations, to an increasing extent, the primary purpose in recognising plant functional types is to understand the assembly of plant communities and ecosystems and to interpret their responses to environmental change and management. In order to pursue these objectives it is necessary to classify large numbers of species, populations, and cultivars with respect to strategy. In recent years a consensus has begun to develop with regard to the procedures appropriate for this task. These include screening of plant traits, multivariate analysis to recognise recurring associations between sets of traits and particular ecologies, and testing of predictions by reference to long-term field studies, manipulative experiments in natural habitats, or synthesis of communities and ecosystems in field plots or sealed microcosms.

PART I PLANT STRATEGIES

Chapter 1 Primary Strategies in the Established Phase

Early attempts to achieve functional classifications of terrestrial plants in the established (mature) phase of the life-history were based upon the architecture of the above-ground parts and through the efforts of Raunkiaer (1934) and his successors it has been possible to identify many consistent correlations between plant form and climate. A key factor explaining the popularity and success of classifications of plants by reference to architecture and climate has been the availability world-wide, as a by-product of taxonomy, of the required information on plant form and structure.

A more difficult task is to recognise functional types that reflect the dynamic relationships of plants with other plants, with resources, with herbivores, carnivores, detritivores, pathogens, and symbionts and with the disruptive effects of climatic extremes and human interference on vegetation development. Attempts to achieve this objective can be broadly classified into two types—those that propose two extremes of adaptive specialisation in plants and those involving three.

The origin of the two-strategy model can be traced to the early speculations of MacLeod (1894) who interpreted vegetation dynamics as an interplay between 'capitalists' that monopolised resource capture and short-lived 'proletarians' that depended upon disturbed conditions to sustain their fugitive existence. This idea received formal recognition in the theory of r- and K-selection (MacArthur and Wilson 1967) and is expressed as a trade-off between competition and dispersal in many subsequent papers (e.g. Levins and Culver 1971; Levin 1974; Slatkin 1974; Hastings 1980; Tilman 1994). A major weakness of the two-strategy models of primary functional types is the failure to differentiate between two completely different types of resource dynamics and dominance mechanisms that occur in different kinds of perennial vegetation. In the first, associated with productive habitats subject to continuous resource replenishment (e.g. alluvial terraces, fertilised meadows), dominance is achieved by rapid rates of resource capture. In the second, coinciding with the low productivity conditions of primary succession in skeletal habitats or mature late-successional vegetation, success does not depend upon a superior ability to capture resources; the key traits are those allowing survival of long-lived tissues and protection of captured resources against herbivores and pathogens. A further deficiency in the two-strategy model is its dependence upon a tradeoff between competition and colonisation; as explained in later chapters empirical studies on many different floras do not support this proposed relationship.

The three-strategy model originates from the ideas of Ramenskii (1938) and Grime (1974,1977), who recognised that much of the variation in the adaptive responses of plants can be predicted and explained by recognising the overwhelming importance of habitat productivity and the frequency and severity of biomass destruction (disturbance) in the evolution and current ecology of plants. When a templet comprising four types of plant habitat is constructed by combining the extremes of high and low productivity with high and low disturbance it is evident that only three of the resulting contingencies are capable of supporting vegetation. Plants are excluded where low productivity coincides with frequent and severe destruction. It is proposed that the three remaining cells of the productivity:disturbance templet correspond to primary plant strategies (competitors (C), stress-tolerators (S), and ruderals (R)) of universal occurrence and with distinctive traits.

The defining characteristic of the competitor is the ability to rapidly monopolise resource capture by the spatially-dynamic foraging of roots and shoots. Stress-tolerators are distinguished by the capacity of their long-lived tissues to resist herbivory and effects of environmental stress in conditions where growth is severely restricted by low rates of mineral nutrient supply. Ruderals are characterised by a short life-history and the tendency to rapidly invest captured resources in the production of offspring.

The existence of C, S, and R as widely-recurrent adaptive extremes implies that fundamental internal constraints operate to restrict the paths of functional specialisation available to plants. The laws of physics and chemistry dictate 'tradeoffs' which restrict the evolutionary responses of plants (and animals) to a limited range of alternative solutions to the problems posed by environments. A tradeoff is defined as an evolutionary dilemma whereby genetic change conferring increased fitness in one circumstance inescapably involves sacrifice of fitness in another. The pervading nature of C, S, and R axes of specialisation in the established phase of plant-life histories is a consequence of the occurrence of unavoidable and identifiable tradeoffs in core aspects of plant function including mechanisms of resource capture, growth, storage, defence, and reproduction. It is possible to reconcile the occurrence of both fundamental tradeoffs and primary plant strategies with the individuality of plant ecologies. The constraints that predetermine channelling into C, S, and R paths of specialisation do not preclude fine-tuning of ecologies (for example through possession of particular types of juveniles (regenerative strategies), distinct phenologies, resistances to particular edaphic or climatic stresses or association with various symbionts).

If recognition of C-, S-, and R-strategists does not exclude the possibility of numerous dimensions of additional ecological variation within each of the three categories it is necessary to define as precisely as possible the ecological phenomena that can be predicted and interpreted from CSR theory. These are explored in subsequent chapters and relate mainly to the role of plants in large-scale processes such as community assembly, vegetation succession, and the functioning of ecosystems.

In addition to its primary role in scaling-up from the traits of individual plant species to the properties of vegetation and ecosystems, the validity and usefulness of CSR theory must be judged also by the extent to which it solves a problem that has haunted ecology throughout its history. This problem concerns the nature of competition for resources and its role in different organisms and circumstances. As Milne (1961) observed, the failure to adopt a mechanistic definition of competition for resources has, since the time of Darwin, generated considerable controversy mainly by allowing confusion between competition for resources, fitness, and dominance of communities. CSR theory has addressed this problem by restricting the definition of competition to the attempt to capture specified resources in the presence of neighbours. The theory identifies a set of traits that is characteristic of the C-strategy and permits relatively high rates of resource capture in crowded productive undisturbed vegetation and, in terms of tradeoffs, explains the selective disadvantage of these same traits in circumstances of low productivity or frequent and severe disturbance. In proposing an association between competition and success in productive, relatively-undisturbed vegetation, CSR theory has attracted the criticism that in other circumstances, in particular on infertile soils, quite different traits may permit superior rates of resource capture. However, evidence strongly supporting the identification of competitive ability for resources with a particular set of genetic traits has emerged from experiments reporting the invariability of hierarchies in yield within sets of ecologically-contrasted plants allowed to compete at high and low soil fertility. When experiments at low soil fertility are conducted in the presence of generalist herbivores, or in circumstances of intermittent (pulsed) mineral nutrient supply, the advantage of the C-strategists is lost suggesting, in accordance with CSR theory, that the fitness of S-strategists in low productivity communities arises mainly from the capacity of their tissues to remain viable for long periods and to resist herbivory rather than the ability to capture resources in competition with the leaves and roots of neighbouring plants.

It is therefore a key assertion of CSR theory that there is a general decline in the importance of competition, relative to other components of fitness, as we approach habitats of chronic infertility or frequent and severe vegetation disturbance. The pattern which is proposed has theoretical support (Bolker and Pacala 1999) and displays startlingly exact and complete parallels with variation in the dynamics of the business-world and human society across the spectrum from rich to poor. CSR theory proposes that competition reaches maximum expression in circumstances where resource supplies are sufficient to permit the rapid construction of large individuals which have the capacity for monopoly through rapid reinvestment and spatial redistribution of captured resources allowing dynamic foraging above and below ground. It is suggested that under resource-limited conditions such high rates of reinvestment run the risk of bankruptcy: selection is more likely to promote conservation and protection of captured resources and low-cost opportunistic pulse interception by long-lived tissues. Where vegetation is frequently destroyed there is clearly little opportunity for construction of extensive and dynamic vegetative structures and competition for resources is predicted to be intermittent and inconclusive due to its rapid curtailment as captured resources are invested in offspring. On this basis, restriction of the advantage of high competitive ability to productive, relatively undisturbed conditions can be explained by the declining opportunity for interaction between neighbouring individuals as vegetation development and the selective advantage of large dynamic plant individuals are constrained by low resource supply or frequent disturbance or some combination of the two.

On first inspection a radical alternative to the CSR approach to competition is presented in the R* hypothesis of Tilman (1982). Here it is suggested that competition for resources changes in character but remains equally important under conditions of low soil fertility. It is proposed that the long-lived, slow-growing plants that occupy infertile soils owe their success to an ability to draw down critically-limiting mineral nutrients to levels at which more demanding species cannot obtain them in sufficient quantities. However, the R* hypothesis is not supported by experiments that demonstrate that it is species of fertile soils that possess a superior ability to extract mineral nutrients from low external concentrations. A reconciliation between the R* hypothesis and CSR theory is possible by close examination of the theoretical formula used to calculate R*. Essentially this formula refers to the total budget of the plant for a specific limiting nutrient and contains both gain and loss components. When slow-growing plants are found to be unusually capable of surviving on infertile soil it may be erroneous to assume that their success arises from superior rates of nutrient absorption. It is much more consistent with the empirical data to conclude that low values of R* arise from the low loss rates of slow-growing plants due to internal recycling, low rates of tissue-turnover, and the higher resistances to herbivory predicted by CSR theory.

A potential source of confusion arises from the fact that stress-tolerant plants are not restricted to vegetation of low biomass. Although S-strategists of small stature are strongly associated with skeletal habitats with shallow soil and low mineral nutrient supply, S-strategists also occur as large, long-lived woody species in the late stages of succession. Here the stress-tolerant strategy prevails not because the total stock of mineral nutrients in the ecosystem is low but because mineral nutrients are scarcely available due to tight recycling and their sequestration in living and dead components.

Various methods have been applied in attempts to falsify the CSR model. These include comparative studies within particular plant and animal taxa to seek tradeoffs and patterns of ecological specialisation similar or distinct from those proposed in CSR theory. Mathematical models have been developed to examine the consequences of different life-histories and mechanisms of resource capture and utilisation. Strong support for the CSR model has also become available from (1) individualistic dynamic simulation models of plants and communities using cellular automata, (2) recognition of tradeoffs and recurring sets of plant traits by screening procedures and multivariate analysis, and (3) vegetation synthesis under controlled conditions of productivity and disturbance.

Chapter 2 Secondary Strategies in the Established Phase

The habitats exploited by C-, S-, and R-strategists represent extremes in the range of conditions available to plants; secondary strategies have evolved in habitats experiencing intermediate productivities and intensities of disturbance. In addition to the three primary functional types, four secondary strategies corresponding to particular equilibria between productivity, intensity of disturbance, and competition for resources can be recognised within a triangular model.

Various criteria are available with which to locate species and populations within the triangular model. In cool temperate regions with a well-defined growing season, measurement of shoot phenology provides a valuable diagnostic tool. For precise classification, laboratory screening of plant traits (the so-called 'hard' tests) followed by multivariate analysis is desirable but for projects requiring more rapid but less exact attributions in large numbers of species 'soft' test techniques can be developed and calibrated against results from hard tests.

Consistent differences are apparent when comparisons are made between major life-forms and taxa with regard to their range of occurrence within the triangular model. These differences have immediate implications for the role of life-forms in vegetation structure and successional processes. It is also apparent that phylogenetic constraints have restricted the strategic ranges of particular plant families and some larger taxonomic units such as gymnosperms, bryophytes, and lichens. Within angiosperms there is evidence that plant families of more recent origin have stronger representation among ruderal and competitive functional types and that this may explain the increasing abundance of these families in floras heavily impacted by man.

Chapter 3 Regenerative Strategies

When multivariate analyses are conducted on sets of attributes measured on large numbers of species it is usually observed that there is weak coupling between juvenile and mature plant traits. This confirms that the selection forces operating upon the regenerative phase are often concerned with the exploitation of 'safe sites' by juveniles (Harper *et al.* 1965) and are often quite different from those occurring later in the life-history. Perhaps the most fundamental basis of the uncoupling is the widespread occurrence in the regenerative phase of mechanisms (production of numerous wind-dispersed propagules (W), accumulation of persistent seed or spore banks (B_s)) that disperse individuals through either space or time. Both of these regenerative strategies appear to represent devices whereby juveniles may escape or reduce the deleterious effects of established plants. A similar explanation can be applied to the strategy of seasonal regeneration (S) in which cohorts of seedlings or vegetative propagules appear at times in the year when predictable disturbances of the vegetation may improve the chance of successful establishment by independent offspring. The dispersal of offspring also occurs through the activities of animal vectors and may be particularly important where the seeds are heavy.

In marked contrast to W, B_s, and S, two other regenerative strategies (V and B_{sd}) do not have prominent dispersal connotations and may in fact rely critically upon continued connection with the parent or derive benefits from close physical contact with established vegetation.

It is argued that the differentiation of regenerative strategies arises mainly from the inflexibility of reproductive characters rather than the existence of inevitable tradeoffs. Consistent with this conclusion it is not unusual for a species or population to exhibit several regenerative strategies and for each of these to add a recognisable dimension to the ecology. This is particularly obvious in the case of the combinations V + W and V + B_s where the capacity for local consolidation of populations is combined with the potential for dispersal in space or time. Species with multiple forms of regeneration are included among those that have achieved high abundance in the world flora. Changes in the abundance and ecology of a species may arise from the progressive failure of regenerative strategies as geographical limits are approached. The metapopulation dynamics of species at their climatic limits appear to be strongly determined by the number and type of functioning regenerative strategies.

PART II PLANT STRATEGIES AND VEGETATION PROCESSES

Chapter 4 Dominance

In natural vegetation and in experiments with plant communities examples of the impact of large plant size on the fate of smaller species and individuals are ubiquitous but they vary in detailed mechanisms. In analysing the nature of these effects it is necessary to distinguish between mechanisms of competition (which focus exclusively upon resource capture) and those involved in dominance (which include competition and several other processes including beneficial effects). It is helpful to recognise with Boysen-Jensen (1929) two components which engage in a positive feedback to drive the dominance mechanism: (1) the mechanism whereby the dominant plant achieves a size larger than that of its neighbours—this varies according to strategy; and (2) the +ve and –ve effects of large plants on smaller neighbours. This definition allows recognition of competitive, stress-tolerant, and ruderal dominants which achieve their status in communities by fundamentally different routes with profoundly different implications for the flux of resources and the dynamics of component organisms in the ecosystem concerned. This model of dominance corresponds closely to that described for sessile marine plants and animals by Paine (1984).

The impact of a dominant plant may be exerted upon neighbours at various stages of their life-cycles but the most widespread effects are those operating upon seedlings. Role-reversal is frequently observed during vegetation succession; seedlings of woody species may be subject to dominance by herbs which are later or simultaneously dominated by established specimens of the same shrubs or trees.

Graphic examples of the need for the study of dominance to embrace more than the struggle to capture resources are available from field studies and experiments on the effects of litter in woodlands and herbaceous vegetation. Both seedling establishment and the species composition of mature herbaceous vegetation appear to be directly controlled in many situations by the depth, persistence, and physical structure of the litter deposited by dominant species

Dominant plant species differ considerably in their effects on plant species richness and it is frequently observed that diversity is higher in communities dominated by stress-tolerators. The explanation for this relationship is complex (see Chapter 9) but a key factor appears to be the slow-dynamics of stress-tolerant dominants. This attribute causes the microhabitats created by the dominants including the surfaces of living leaves and their litter derivatives to be of sufficient duration to permit exploitation by populations of subordinate species of plants. This is in marked contrast to communities dominated by competitors and ruderals where the dynamic growth and rapid turnover of tissues creates an extremely unproductive and hazardous environment for smaller plant species.

Chapter 5 Assembling of Communities

For more than a century plant ecologists have recognised that within geographical regions plants aggregate into assemblages of consistent structure and species composition. The populations in particular communities are a subset of those present in the local flora and it is necessary therefore to identify the filters that control admission from the local species pool into each community. A separate challenge arises from the need to explain what controls the relative abundance of species within communities. Are the consistent hierarchies observed within plant communities controlled by specific genetic traits and, if so, what are the selective advantages associated with playing a consistently subordinate role? In seeking answers to these problems ecologists have gathered insights by field observation, manipulation of natural communities, and by synthesising communities from seed under controlled conditions.

Following Whittaker (1965), it is instructive to rank the plant species present in a community according to their biomass. In the resulting dominance–diversity curve there are three conspicuous components—the dominants, subordinates, and transients. The first two of these are consistent members of the community but they differ in size and total contribution to the biomass. The transients occupy the tail of the dominance–diversity curve and consist of a heterogeneous assortment of juveniles originating from the seed rain or seed banks and including species that play a dominant role in other, neighbouring communities. It is suggested that both subordinates and transients may play a vital but intermittent role in the assembly of plant communities. Subordinates can act as a filter controlling the recruitment of dominants and transients may be transformed into dominants should changes in management or successional events encourage the assembly of a different type of ecosystem.

It is concluded that investigations of the changes that take place in structure and composition following natural or experimental manipulations of plant communities, such as the addition of fertilisers or the exclusion of grazing animals, have been informative, particularly in exposing the role of dominant plants. In contrast, the large number of experiments that have involved species removal or introduction of seedlings or small plants (phytometers) have not provided a useful basis for interpreting community structure. It is suggested that this may be due to predication of such experiments (Goldberg and Werner 1983) on the false assumption that the reactions of a community to manipulation are a reliable indicator of the sequence of events through which the community was assembled and is currently maintained.

An exciting alternative approach is to assemble communities from a large pool of plant functional types and species in experimental plots or sealed microcosms with controlled factorial manipulations of environmental factors and trophic structure. It is concluded that this approach provides insights into mechanisms that are not currently accessible by direct observation in the field.

Chapter 6 Rarification and Extinction

Under the pressures generated by rising human populations and changing patterns of land use many plant species are experiencing decline and, in some cases, extinction. For the purposes of long-term conservation and management it is necessary, wherever possible, to interpret and predict rarification and extinction in terms of measurable functional shifts within regional or national floras and within individual plant communities.

Comparisons between various Western European countries with respect to the CSR profiles of expanding and declining plant species have revealed consistent differences correlated with human population density. In countries with relatively few people no strategic differences can be detected between increasing and decreasing components of the national flora but where human density exceeds 100 Km^{-2} a polarisation is evident between declining stress-tolerators and expanding C- and R-strategists. It would appear therefore that human population density can provide a convenient surrogate for a set of landscape changes that above a critical threshold have the combined effect of causing a predictable functional shift.

By monitoring changes in the abundances of primary and secondary CSR strategies in herbaceous vegetation in permanent plots it is possible to detect functional shifts within plant communities and to discriminate between effects of eutrophication, dereliction, and increased disturbance, The same technique also provides an early-warning system for changes likely to threaten the survival of local populations of rare plant species.

Chapter 7 Colonisation and Invasion

In a tradition originating from Elton (1958), investigations of the spread of alien species have been dissociated from many relevant areas of ecological theory and research and it is suggested that this estrangement explains to a major extent our failure to predict plant invasions with any reliability.

When invasions (outbreaks beyond former boundaries) are compared with colonisations (expansions or redistributions within existing ranges) the two processes appear to be functionally indistinguishable. Both require conjunction between a large output of propagules and phenomena that increase the receptivity of vegetation to invasion.

Based upon insights from experiments and long-term monitoring studies it is suggested that fluctuations in resource availability are the key factor controlling plant community invasibility. It is proposed that vegetation becomes invasible following episodes in which the supply of unused resources, particularly mineral nutrients, rises due to increased supply, decreased consumption by the resident species, or the combined effect of both. Plant invaders and invaded communities cover a very wide range of plant functional types and environmental circumstances. Many different factors are capable of affecting resource supply and consumption making predictions of invasibility extremely difficult. The elusive nature of the invasion process appears to arise therefore from the fact that it depends upon conditions of resource enrichment and release that have a variety of causes but occur only intermittently and, to result in invasion, must coincide with the availability of invader propagules.

Chapter 8 Succession

All vegetation is subject to temporal changes in species composition and in the relative importance of constituent life-forms. In successional change there is a progressive alteration in vegetation structure and in the species represented whereas in cyclical change similar plant communities recur at intervals in the same place. A further distinction can be drawn between primary succession where a new habitat, often initially lacking in soil and vegetation, is colonised and secondary succession in which recolonisation of a disturbed habitat occurs.

It is proposed that both successional and cyclical changes in vegetation can be described within the framework provided by the CSR model and can succinctly accommodate seminal contributions to successional theory such as the initial floristic composition hypothesis (Egler 1954) and the concept of tolerance, inhibition, and facilitation (Connell and Slatyer 1977). Key features of the perspective that emerges from the application of CSR theory are: (1) identification of the circumstances where competition for resources drives succession; (2) clarification of the role of herbivores in inhibiting or accelerating succession; and (3) a focus in investigations of primary succession and the later stages of secondary succession on the size of mineral nutrient stocks and the extent of their sequestration in living and dead components.

Successional processes cannot be fully interpreted without reference to the juvenile phase in plant life-histories and distinctive roles can be recognised for the regenerative strategies identified in Chapter 3. Whereas small wind-dispersed offspring (W) are conspicuous during the early stages of succession, persistent seeds and spores (B_s) are more commonly associated with cyclical vegetation processes and persistent seedlings (B_{sd}) are more likely to be effective during primary succession or in the later stages of secondary succession.

As described in Chapter 3, in landscapes heavily impacted by modern forms of landuse some of the regenerative strategies of plants are developing a new ecological significance. In circumstances where late successional plant species are now confined to small geographically-isolated fragments of vegetation a dependence upon persistent seedlings or dispersal by particular animal vectors may leave these plants vulnerable to extinction. In contrast in landscapes where there is increasing transport of soil the capacity of seeds to remain ungerminated in the soil appears to be assuming an important role in the dispersal of colonising species.

Chapter 9 Co-existence

Although variation in species richness occurs in all types of terrestrial vegetation research on the mechanisms that permit coexistence of different species has been concentrated on communities of herbaceous plants. From field surveys and experiments synthesising plant assemblages from seed it is apparent that there are two quite distinct circumstances in which species richness falls close to zero. The first occurs where conditions permit dominance by a small number of robust species. The second is associated with extremely low biomass in environments where low productivity or frequent disturbance or some combination of these two factors excludes all except a few highly specialised species. Between these two contrasted conditions there is a corridor of higher potential richness occupied by species which are neither potential dominants nor capable of tolerating extreme conditions. This pattern resolves into the humped-back model (Grime 1973a) in which the 'hump' corresponds approximately to the range 250 to 750 g m^{-2} in maximum standing crop and litter. Patterns in species richness conforming with the model have been detected in a large number of investigations in which sampling was conducted along gradients of productivity and disturbance. Such unimodal relationships between species richness and standing crop + litter are not observed where the investigated transect does not extend sufficiently into one or both extremes of the environmental gradient. Only weakly-defined patterns occur where biomass fluctuates from year to year due to climatic variation or inconsistent management.

Two additional mechanisms appear to influence the level of species richness attained in the 'hump'. The first is related to niche differentiation and the second to recruitment from the local reservoir of plant species. Various axes of spatial variation have been recognised within herbaceous vegetation and implicated in mechanisms of species richness. Examples include differences in soil depth, horizontal and vertical gradients in soil moisture or acidity, variation in litter depth and quality, in bryophyte density, and in small-scale patterns corresponding to vegetation and soil disturbance at scales ranging from earthworm casts to ant-hills. However, high species richness is not confined to circumstances with such obvious spatial heterogeneity. In communities exploiting more uniform soil conditions coexistence mechanisms involving temporal niche differentiation may be suspected in some cases from the presence of C_3 and C_4 photosynthetic mechanisms and large differences in genome size within the community. It is not uncommon for species-rich plant communities to contain species that are widely-contrasted in seed size, seed persistence, and germination requirements, suggesting that coexistence may depend upon exploitation of long-term variation in the seasonal distribution of opportunities for seedling regeneration.

Another source of variation recently implicated in coexistence mechanisms within species-rich vegetation is genetic variation within component species; experimental evidence supports the hypothesis (Antonovics 1976) that such variation may reduce species losses within communities by diversifying the outcome of competitive interactions in local neighbourhoods and reducing the risk of species extinctions by pathogens.

Following the arguments developed in Chapter 5, species richness can be identified as one of the community attributes likely to be strongly affected by the size and composition of the pool of species available to an assembling plant community. The effectiveness of the seed rain and seed banks as a source of species is determined by large-scale historical events during the development of regional floras and by local and recent patterns of land management. It is particularly important to recognise that even a large species pool may contain only a small number of 'genuine' candidates for admission to a community. This is particularly evident in the case of communities on acidic soils where the numbers of suitable species in the local species pool are small and reflect the global sparsity of calcifuge species.

PART III PLANT STRATEGIES AND ECOSYSTEM PROPERTIES

Chapter 10 Trophic Structure, Productivity, and Stability

In the second half of the 20th century, ecologists of many different persuasions began to converge upon terrestrial ecosystems as an object of study. Differences in method and interpretation have been evident between, on the one hand, theoreticians and many animal ecologists, for example MacArthur (1955), Elton (1958), May (1972, 1974), and Pimm (1991), who have relied upon insights from population biology, and, on the other hand, many plant ecologists, soil biologists, and limnologists, for example Chapin (1980), Leps *et al.* (1982), Grime (1987), Carpenter (1988), and Reynolds (1998), who have focused upon resource dynamics. In consequence, a lively debate has developed between those seeking to explain ecosystem properties as a function of the number and interaction of component species and those suggesting that ecosystem functioning, productivity, resistance, resilience, and sustainability are determined to a major extent by the physics and chemistry of the dominant plants. A particularly good example of the control of ecosystem properties by plant functional types relates to the contrasted effects of C- and S-strategists on the sequestration of pollutants in vegetation and soils. The difficulties experienced in predicting the persistence of ^{137}Cs deposited in upland areas of the British Isles following the Chernobyl Incident provides a graphic example of this phenomenon.

In comparison with the well-defined trophic structures encountered in most freshwater lakes, those of terrestrial ecosystems are less amenable to quantitative study and experiment. Complications arise from the long life-span of many of the dominant plant species and their localised effects on the spatial structure and resource dynamics both above and below the soil surface. It is often extremely difficult to recognise the boundaries of a terrestrial ecosystem particularly where there are herbivores and carnivores with extensive territories.

Following the initial speculations of Hairston *et al.* (1960), a functional classification of terrestrial ecosystems has been proposed by Fretwell (1977) and elaborated by Oksanen *et al.* (1981). The essential feature of this scheme is the hypothesis that the influence of animals on vegetation depends on productivity. At the lowest productivity the plant biomass is too small or unpalatable to support large populations of herbivores. At intermediate productivity herbivore numbers rise and have important controlling effects on vegetation quantity and quality. As productivity rises still further it is proposed that herbivores become a consistent food source for their predators which are sufficiently numerous to provide 'top-down protection' to a large, relatively palatable plant biomass.

In its emphasis upon productivity and coincident variation in plant defence, CSR theory shows strong convergence with the Fretwell–Oksanen hypothesis and suggests that trophic structures, including those associated with decomposer organisms, may be predictable from the primary functional types of the dominant plants. Support for this proposition has been obtained in field experiments involving the removal of invertebrates from sites along natural productivity gradients and in microcosm studies in which ecosystems have been allowed to assemble at contrasted levels of soil fertility.

Although the productivity of a terrestrial ecosystem depends to a considerable extent upon the supplies of resources available to the vegetation, modifying effects occur according to the functional biology of the dominant plants which can exercise controls on losses to herbivores and can promote or inhibit the recycling of mineral nutrients through differences in the quality and quantity of their litter.

During the period 1993–1999, several experimental studies involving the synthesis of plant assemblages from species-poor and species-rich seed mixtures have purported to demonstrate benefits of high plant species diversity to productivity. These effects appear to have been artefacts associated with the higher frequencies of unusually productive species such as nitrogen-fixing legumes or larger, faster-growing species in the species-rich assemblages. Further research is required to determine the extent to which the positive relationship between species richness and yield in the rising curve (Zone B) of the humped-back model (Chapter 9) arises from a complementary exploitation that drives up productivity or is merely a reflection of the admission of more species under increasingly hospitable conditions.

Following the definitions of Westman (1978) resistance to perturbation is recognised as the ability of an individual, community, or ecosystem to resist displacement of its biomass and resilience as the speed and completeness of the subsequent return to control levels. To date most studies of the resistance and resilience of ecosystems have focused on the above-ground component of the plant biomass, mainly because suitable data of this kind are widely available from monitoring studies.

Predictions of resistance and resilience in vegetation arise directly from CSR theory and propose: (1) a tradeoff between the two very different sets of traits associated with resistance and resilience; and (2) a predictable relationship between resistance and the traits of stress-tolerators and between resilience and the traits of ruderals and competitors. In some instances more refined predictions can be developed by reference to additional plant traits; persistent seed banks, for example, have been used as a predictor of resilience following severe vegetation disturbance.

Validations of the use of CSR-related traits as predictors of resistance and resilience have become available from studies of natural extreme events and from experimental manipulations of climatic factors in contrasted types of vegetation.

It is concluded that a primary need in seeking a scientific basis for the study of sustainability is a unified functional classification of ecosystems that reflects variable attributes such as productivity, physical structure, trophic structure, and the resistance and resilience of main components. In addressing this objective, it is suggested that recognition of the adaptive strategies of the established and regenerative phases in the life-histories of the dominant plants is a useful and attainable first step. In particular, this would allow predictions of sustainability to be based in part upon the resistance and resilience of the vegetation to the changes in land-use and climate that are the main drivers of global change and ecosystem deterioration. It is suggested that a full understanding of sustainability also requires that account is taken of the importance of declining plant diversity at the landscape scale on the recruitment of dominant plant species during the reassembly of ecosystems.

I Plant Strategies

1 Primary Strategies in the Established Phase

INTRODUCTION

In many early attempts to provide a functional classification of terrestrial plants, reliance was placed upon the distinction between trees, shrubs and herbs, bryophytes and lichens, an approach which provides scope for finer distinctions (broad-leaved and needle-leaved trees, evergreen and deciduous trees, grasses, sedges and forbs etc.). Particular extensions of this approach have been the dichotomy between guerrilla and phalanx strategies in the architecture of clonal herbs (Harper 1977) and the elaboration of theories based upon the geometry of leaf canopies (Horn 1971; Givnish 1982). A strong factor favouring such reliance on morphology and taxonomy was the relative ease with which the resulting classification systems could be applied to vegetation description. Undoubtedly also a factor encouraging ecology in this direction was the availability, world-wide, as a by-product of taxonomy, of the required information on plant morphology. The persistent appeal of descriptions based on morphology is that they are operational; but are they adequate and reliable predictors of the functional responses of plant populations, communities, and ecosystems?

CLASSIFICATION BY ARCHITECTURE

Morphologically-based classifications have been explored most thoroughly in relation to climatic control of plant distribution. The system devised by Raunkiaer (1934) and later elaborated by Holdridge (1947), Hallè and Oldeman (1975), and Box (1981, 1996) relied upon correlation of plant architecture, phenology, and leaf form with the World's Climatic Zones. The use of such a structural approach has continued to the present day and has been fortified by three developments:

1 It has been found possible to predict the distribution of structurally-defined plant types over the land surface of the world using mechanistic models driven by mathematical equations describing plant–climate interactions and based upon readily-available data on rainfall and temperature (Monteith 1973; Jarvis and McNaughton 1986; Woodward 1987; Waring 1988).

2 Some features of plant architecture are detectable by remote sensing and conventional aerial photography, producing obvious opportunities for model validation and implementation for various applied purposes.
3 Models of present and future climates increasingly rely upon calculations of feedbacks from vegetation to the global and regional climate. Structural features of plants are highly relevant to estimates of the modifying effects of vegetation on energy exchange, surface roughness, albedo, and evapotranspiration (Nemani and Running 1996).

It is clear, therefore, that plant functional types based upon architecture have a continuing role to play in plant ecology and have a particular relevance to studies relating vegetation to climate. However, plant structure cannot by itself provide us with all the information required to understand vegetation processes and ecosystems. There are a wide range of phenomena, mainly relating to the dynamics of resources, plants, communities, and ecosystems which, at best, can be addressed only superficially by reference to plant architecture. It is also important to note that many extensive areas of the world are dominated almost exclusively by one morphological type, e.g. forest trees or turf-grasses. There is a clear requirement, therefore, for an additional approach to plant functional types which reflects dimensions of specialisation that are not adequately represented in an architectural classification.

TWO- AND THREE-STRATEGY MODELS

The attempt to recognise a small number of functional types of universal applicability has fundamental importance and a surprisingly long history. In Figure 4, this activity had been reduced to a scheme describing landmarks in this search during which theories have tended to polarise between two- and three-strategy models. Before commenting on the relative merits of these two paths it is worth noting that with respect to both there have been contributions from many sources. Some have been Darwinian in the sense that direct observations of patterns in nature or in experiment have been distilled into a general scheme (e.g. Ramenskii 1938; Grime 1974, 1988c; Southwood 1977; Pugh 1980; Greenslade 1983); others have emerged from a more theoretical, mathematical approach with strong emphasis on population biology (MacArthur and Wilson 1967; Levins and Culver 1971; Tilman 1994; Bolker and Pacala 1999); and in a third group attention has focused on the capture and utilisation of resources (Grime 1979; Chapin 1980; Tilman 1990; Lambers and Poorter 1992; Reiche *et al.* 1992; Grime *et al.* 1997).

The theory of r- and K-selection proposed by MacArthur and Wilson (1967) has been highly influential in the development of two-strategy models and there is no doubt that many differences in life-history that distinguish between organisms appearing early and late in successional sequences in

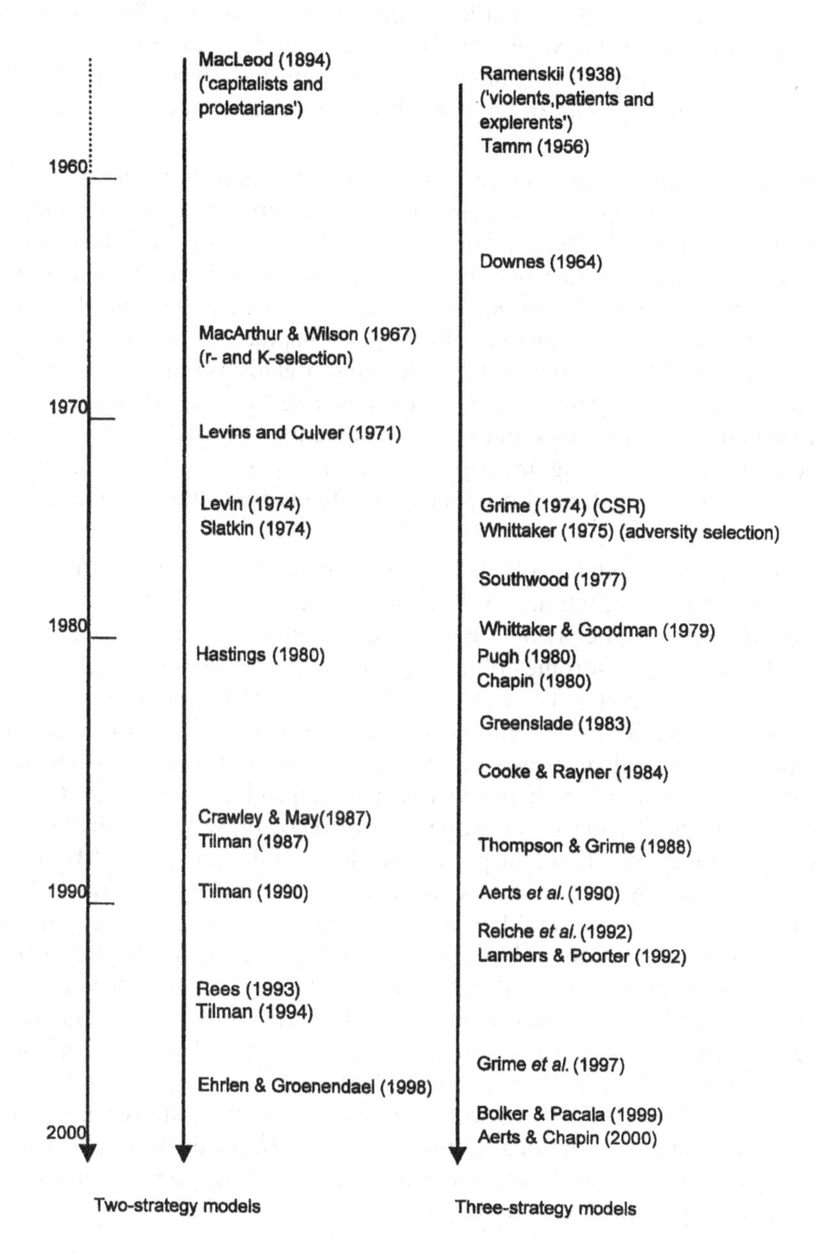

Figure 4. Some contributions to the development of general theories of functional types in plants and animals

productive, unexploited habitats are captured by this model. The legacy of the theory of r- and K-selection is still evident in the frequent references to a tradeoff (see page 10) between colonisation and competition in recent publications (e.g. Hastings 1980; Tilman 1990; Ehrlen and Groenendael 1998). The persistent inadequacies of the two-strategy model are two-fold:

1 Whilst the combination of rapid growth of short-lived individuals with early reproduction envisaged in r-selected organisms clearly conforms to a widely-recurrent functional type exploiting temporary but productive habitats, no provision is made for the existence of highly-effective colonising abilities in a wide range of quite different organisms including long-lived perennial species of productive and unproductive habitats (see Chapters 3, 7, and 8).
2 The functional traits expected for K-selected organisms remain a source of uncertainty and controversy. If we characterise them as organisms with conservative life-histories and resource dynamics how are we to accommodate the fast-growing, monopolistic perennial species that are capable of long-term occupation of productive, relatively-undisturbed habitats?

The three-strategy model, explained in more detail in the next section, deals with both of these limitations. With respect to 1, a closer approach to the realities of colonising behaviours is provided by observing a stricter separation of the adaptive syndromes of the regenerative and established phases of life-histories. The advantages of this approach are explained in the review of regenerative strategies presented in Chapter 3 and in the discussion of colonising and successional processes in Chapters 7 and 8. The need to include long-lived organisms of both productive and unproductive habitats (see 2 above) can be met by the simple expedient of recognising two distinct strategies of common occurrence in perennial plants and characterised by very different resource dynamics—the acquisitive and the retentive (*sensu* Grime *et al.* 1997). In Figure 4 several key contributions concerned with recognising and validating this dichotomy in perennial plants are identified (Grime 1977; Chapin 1980; Aerts *et al.* 1990; Reiche *et al.* 1992; Lambers and Poorter 1992; Grime *et al.* 1997; Aerts and Chapin 2000). Until recently plant population biologists have tended to develop models emphasising colonisation–competition tradeoffs (e.g. Levins and Culver 1971; Hastings 1980; Crawley and May 1987; Tilman 1994). However, a recent theoretical study which employs spatial moment equations to investigate the strategies by which plants occupy space and achieve fitness under different conditions finds support for the three-strategy model, concluding that:

> specializations for colonization, exploitation and tolerance are all possible, and these are the only possible spatial strategies; among them, they partition all of the endogenous spatial structure in the environment
>
> Bolker and Pacala (1999)

It is interesting to consider why the two alternative approaches to recognising primary strategies have persisted to the present day. Until the publication of Bolker and Pacala, the continuing dichotomy was perhaps largely a symptom of the lack of interaction between population biologists and ecophysiologists. The difference in perspective between Tilman and Grime, however, requires more explanation since both have concerned themselves with plant adaptation to different conditions of resource supply. Careful comparison reveals that the substantive differences arise from the reluctance of Tilman to recognise as a recurring strategy the 'competitors', i.e. fast-growing, perennial herbs, shrubs, and trees that rapidly monopolise productive habitats by local consolidation and 'active foraging' by shoots and roots (see page 23). In seeking an explanation for this difference it may be significant that in the two systems which have been the main subjects of Tilman's work (unicellular phytoplankton and the nitrogen-deficient sandplain vegetation at Cedar Creek, Minnesota) monopolistic competitors are scarcely represented and it is perhaps for this reason that their involvement in vegetation processes has been dismissed as a contribution to 'transient dynamics' (Tilman 1988; Gleeson and Tilman 1990). In marked contrast, large fast-growing monopolists are prominent in the area of Northern England investigated by Grime; here there has been a long history of eutrophication, disturbance, and dereliction in which plants illustrating strategies of colonisation, exploitation, and tolerance (*sensu* Bolker and Pacala 1999) are unmistakable as a recurring fine-grained patchwork across the landscape.

THE CSR CLASSIFICATION

By definition, primary strategies, if they exist, must encompass all living matter (plants *and* animals) and must be fashioned by selection mechanisms operating in all habitats. With respect to plants, this appears to rule out of contention any of the individual selective factors (e.g. shade, low phosphorus, salinity, drought, frost, fire) known to be restricted in their occurrence to particular biomes or ecosystems. However, this would not apply if, for the very 'broad brush' purposes associated with the recognition of primary strategies, these individual factors could be shown to be functionally-equivalent and capable of aggregation into broader groups which *are* universally represented. In the argument which follows it is suggested that aggregation is possible by recognising a fundamental dichotomy in ecological factors.

The external factors which limit the amount of living and dead plant material present in any habitat may be classified into two categories. The first, which we may describe as *stress*, consists of the phenomena which restrict photosynthetic production such as shortages of light, water, and mineral nutrients, or sub-optimal temperatures. The second, here referred to as *disturbance*, is associated with the partial or total destruction of the plant biomass and arises from the activities of herbivores, pathogens, man (trampling, mowing, and ploughing), and from phenomena such as wind-damage, frosting, droughting, soil erosion, and fire.

In the spectrum of plant habitats provided by the world's surface the intensities of both stress and disturbance vary enormously. However, when the four permutations of high and low stress with high and low disturbance are examined (Table 1) it is apparent that only three of these are viable as plant habitats. This is because, in highly disturbed habitats, the effect of continuous and severe stress is to prevent a sufficiently rapid recovery or re-establishment of the vegetation.

Table 1. Suggested basis for the evolution of three strategies in plants. (Reproduced from Grime 1977 by permission of *Am Nat.*)

Intensity of disturbance	Productivity	
	High	Low
Low	Competitors	Stress-tolerators
High	Ruderals	No viable strategy

It is suggested that the three remaining contingencies in Table 1 have been associated with the evolution of established strategies conforming to three distinct types. These are the *competitors*, which exploit conditions of low stress and low disturbance, the *stress-tolerators* (high stress–low disturbance), and the *ruderals* (low stress high–disturbance). The three strategies are, of course, extremes of evolutionary specialisation and in Chapter 2 plants adapted to habitats experiencing intermediate intensities of stress and disturbance will be examined.

The Untenable Triangle

In order to begin to dissect more carefully the logic behind the three-strategy model of Table 1, it is useful to follow the method of Southwood (1977) and Greenslade (1983), who classified habitats by ordinating them within a rectangular space representing variation in productivity and rate of biomass destruction (Figure 5). When the co-ordinate representing rate of construction (plant growth) is plotted against that describing rates of destruction it is immediately apparent that only a triangular area remains as viable plant habitat. The remaining half of the square ('the untenable triangle') corresponds to the very wide range of combinations of productivities and rates of destruction which are sufficient to eliminate vegetation completely. Here it is important to recognise that:

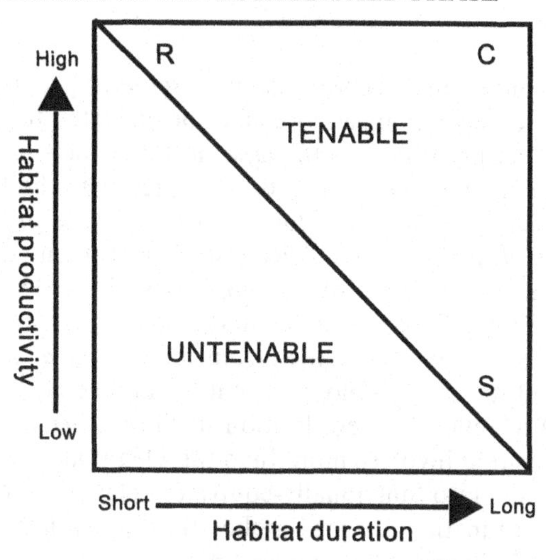

Figure 5. The tenable and untenable triangles

1 Figure 5 represents contingencies in which productivity and rate of destruction remain constant and interact frequently in relation to plant generation time.

2 The spatial scale at which productivity interacts with destruction in Figure 5 is set to exceed the space exploited by individual plants. In the real world, of course, productivity and destruction can vary considerably even within local habitats. This means that questions of sample size are a vital consideration in efforts to falsify the empty cell of Table 1 and the untenable triangle of Figure 5. Several authors including Southwood (1977), Grubb (1976, 1980, 1985, 1998), and Westoby (1998) have considered quite large units of landscape in their searches for recurring functional types, and it is not surprising that these authors have contested one of the central tenets of CSR theory that no organisms can tolerate continuously low resource supply and frequent destruction. Both Southwood and Grubb claimed to have identified organisms and circumstances where plants survived simultaneous exposure to low productivity and frequent destruction. However, the landscape units examined by Southwood and Grubb and found to contain local patches of vegetation are large enough to contain microhabitats in which the productivity and incidence of disturbance depart significantly from the levels estimated by averaging for the units as a whole. Figure 5 differs from the habitat templet (Southwood 1977) and reaches a different conclusion by recognising that rates of plant biomass construction and destruction show small-scale variation and operate locally upon individuals and not at scales and mean habitat values convenient for the landscape ecologist.

Tradeoffs

Although Table 1 and Figure 5 serve a useful purpose in proposing a range of circumstances where vegetation can or cannot be expected to persist, a much more challenging feature of both is the assertion that there are recognisable and predictable functional types of plants associated with the three extremes corresponding to *high productivity—low disturbance*, *low productivity—low disturbance*, and *high productivity—high disturbance*. It might be supposed that a classification founded upon such a simple basis is hardly consistent with the rich diversity of plant life represented in the plant kingdom. However, it is the main contention of this book that from an *ecological* perspective this diversity in detailed structure, biology, evolution, and ecology belies remarkably stereotypical avenues of specialisation in life-history and resource dynamics. This proposition involves more than the idea that selection processes in various habitats fall into functionally-equivalent classes. The recurrence of the same few basic functional types implies the occurrence of design constraints (Grime 1965; Stearns 1976; Wedin 1995) in the organisms themselves. This is to suggest that the laws of physics and chemistry dictate 'tradeoffs' which restrict the evolutionary responses of plants (and animals) to a limited range of alternative pathways. **A tradeoff is here defined as an evolutionary dilemma whereby genetic change conferring increased fitness in one circumstance inescapably involves sacrifice of fitness in another**. Quite clearly, the most important tradeoffs are those which involve core activities such as resource capture, growth, reproduction, storage and defence. Tradeoffs suspected to be involved in the CSR model are described later in this chapter (pages 101–103).

A useful first step in testing the validity of the CSR classification is to consider each of the three primary strategies in turn and to examine the evidence for its existence. Our survey begins with the strategists that have proved to be the most contentious—the competitors.

COMPETITORS

As discussed later in this chapter (page 40), some of the difficulties in analysing the role of competition in ecological processes have arisen from confusion between competition for resources and other aspects of 'the struggle for existence' (Darwin 1859) such as the competition for mates, for pollinators, or for dispersal agents or tolerance of unfavourable conditions. However, even where, as here, we are concerned exclusively with the contest between neighbours for the capture of the resource units required for plant growth, difficulties remain. One reason for this is that early text-books of plant ecology tended to rely upon correlative methods and placed strong emphasis upon soils, climate, and management as direct controls upon the distribution of plant species and vegetation types. References to competition were usually rather guarded and speculative and often coincided with circumstances where simpler explanations of spatial pattern had failed. Competition remained a kind of last resort for descriptive ecology—mechanistically ill-defined and occupying a shadowy place in the background of ecological theory. Although contemporary studies in plant ecology recognise the need for a more analytical approach to competition, the preconditioning effect of this early literature cannot be ignored. Without exposure to rigorous tests several ideas achieved wide currency in ecological thinking. Some persist to the present day. In particular:

1 There is often a residual reluctance to place competition in the foreground of debates concerning primary vegetation mechanisms.
2 Competition is frequently assumed to be a factor increasing in importance under conditions where one or more resources are limiting productivity (Andrewartha and Birch 1954; Weins 1977; Fowler 1986).
3 Competition is regarded by some ecologists as a process which operates through radically-different mechanisms in different habitats (Tilman 1985; Goldberg and Barton 1992).
4 Although competitive exclusion is frequently observed to be responsible for reductions in species diversity in communities many evolutionary biologists regard competition as a process that promotes niche-differentiation and species co-existence, i.e. present-day examples of species co-existence may be a 'ghost of competition past' (e.g. Connell 1980).

A Definition of Competition Between Plants

Wherever plants grow in close proximity to each other, whether they are of the same or different species, differences in vegetative growth, seed production, and mortality are observed. It would be a mistake, however, to attribute all such differences to the process of competition. The dangers in such an assumption are clear if it is recognised that disparities in the performance of neighbouring plants may arise from independent responses to the prevailing physical and biotic environment. It follows that if the term competition is to be useful to the analysis of vegetation mechanisms, a definition must be found which effectively distinguishes it from other processes which influence vegetation composition and species distribution.

In a most revealing paper, Milne (1961) traced the history of biologists' attempts to define competition. He concluded that, far from achieving the necessary distinctions, a majority of authors have not sought to make them and have often used the word as a synonym for 'the struggle for existence' (Darwin 1859). It is perhaps because of the resulting confusion that Harper (1961) proposed that use of the term should be abandoned. However, it may be argued that competition as a process is too important, and as a term too useful, to be allowed to suffer such a fate. An alternative course it to apply a strict definition and to use the term as precisely as possible.

Here, competition is defined as **the tendency of neighbouring plants to utilise the same quantum of light, ion of mineral nutrient, molecule of water, or volume of space** (Grime 1979). This choice of words allows competition to be defined in relation to its mechanism rather than its effects, and the risk is avoided of confusion with mechanisms operating through the direct impact of the physiochemical environment or through biotic effects such as selective predation. The need to make such distinctions has often been recognised in investigations of animal interactions where the term 'apparent competition' has been used to describe the impacts of a shared predator on two prey populations (Holt 1977). According to this definition, competition refers exclusively to the capture of resources and is only part of the mechanism whereby a plant may suppress the fitness of a neighbour by modifying its environment. This dissection of the phenomena which are often 'lumped together' in the more traditional usage of the term has distinct advantages. We are able to classify the variety of mechanisms whereby plants are successful in crowded environments and, in particular, as explained in Chapter 4, we are able to analyse more satisfactorily situations in which dominance of vegetation is achieved by plants which exhibit relatively slow rates of resource capture.

Some ecologists have been reluctant to accept a definition which places such exclusive emphasis upon resource capture. Goldberg and Fleetwood (1987), for example, have advocated an operationally-based approach in which the competitive ability of an individual is assessed in terms of its effects on neighbours. Tilman (1982, 1988) has adopted a more extreme stance in which the strongest competitors are defined as those most tolerant of low resource levels. It has been argued (Keddy 1989) that the failure of ecologists to agree upon a definition of competition and a mechanistic approach to its study has been a significant limiting factor in the development of ecological theory. The definition adopted in this book narrows attention to the struggle between neighbours to capture resource units, and as we shall see later (page 23) attaches particular importance to the foraging activities of shoots and roots. The advantage of this approach is that it leads to testable hypotheses concerning the habitats in which competition is important and the genetic attributes associated with high competitive ability.

Some ecologists have objected to the idea of naming plants of high competitive ability 'competitors'. Rabotnov (1983), for example, maintains that 'any plant growing together with other plants competes with them'. However, it has never been the contention in CSR theory that competition disappears completely under conditions of low productivity or intense disturbance. It does seem logical, however, to use a terminology which reminds us of the circumstance where competition for resources is a dominant (but not exclusive) influence on the membership of plant communities.

A more serious objection to the CSR conception of competition arises from the argument (Newman 1973; Huston and Smith 1987; Tilman 1988) that it is misguided to associate maximal intensity of competition with habitats of high productivity and low disturbance. According to these authors the nature of competition itself changes according to the conditions such that different resources and plant attributes become important in different habitats. This divergence of view has generated experimental tests and a long-running controversy which is now approaching resolution as more definitive experimental evidence becomes available (pages 33–48).

Competition Above and Below Ground

Having recognised that plants may compete for several different resources, it is necessary to consider whether any generalisations can be made with regard to the way in which the focus of competition varies according to vegetation type. In the first place, it is clear that competition for light may become a major influence upon vegetation composition only in circumstances in which the canopy is sufficiently dense for an overlap of leaves to occur. In the early stages of colonisation of a fertile disturbed habitat, such as an arable field, shoots of the invading plants scarcely impinge on each other and competitive interactions, where they occur at all, are likely to be mainly confined to those operating within the soil. It is interesting to note, however, that there is now strong experimental evidence (Ballaré *et al.* 1987; Novoplansky *et al.* 1990) that some plants 'anticipate' competition for light by responding to changes in the quality of light reflected from the encroaching canopies of neighbours.

In the case of the arable field, the relative importance of competition above and below the soil surface is a function of the maturity of the vegetation. As vegetation development continues and the leaf canopy closes there is opportunity for competition to occur simultaneously above and below ground. It seems likely, however, that this relationship is characteristic only of situations in which plant colonisation is allowed to proceed undisturbed in conditions of moderate to high productivity. Where vegetation is developing in a habitat of low potential productivity, such as a rock outcrop with shallow soil, or where there is continuous and severe damage to the vegetation such as on a heavily trampled path, the leaf canopy remains sparse and competitive interactions will be mainly confined to the root environment.

Many descriptive studies of natural vegetation and some competition experiments contain assumptions about the importance of root and shoot competition based upon visual inspection of the above-ground parts. There are two major dangers in such an approach:

1 It cannot be assumed that a large shoot mass is always associated with high competitive ability. Success in competition for light frequently relies upon leaf dynamics rather than shoot architecture.

2 The presence of a conspicuous shaded zone beneath a leaf canopy is not a reliable indication that competition is dominated by above-ground interactions. Dense canopies and shading frequently coincide with an extensive root mat and with nutrient and water depletion (Grubb 1994; Coomes and Grubb 2000).

Characteristics which Determine the Competitive Ability of Established Plants

Competitive ability, as defined for the purposes of CSR theory, is a function of the area, the activity, and the distribution in space and time of the plant surfaces through which resources are intercepted and as such it depends upon a *combination* of plant characteristics. It is important to bear in mind therefore that each of the attributes considered in the review which follows is not by itself diagnostic of high or low competitive ability. As we shall see later, many of the characteristics dealt with under the following headings are capable, in other contexts, of assuming a different strategic significance.

Storage Organs

Just as the competitive ability of a young seedling may be influenced by seed size (Black 1958), so that of an established plant may be affected by the quantity of reserves stored in perennating organs (Al-Mufti *et al.* 1977; Chapin *et al.* 1990). The extremely rapid expansion of the leaf canopy characteristic of many of the larger perennial herbs such as *Chamerion angustifolium*, *Petasites hybridus*, and *Pteridium aquilinum* (Figure 6) is the result of the mobilisation of large reserves of energy and structural materials accumulated in underground storage organs during the later stages of the previous growing season (Bradbury and Hofstra 1976). An obvious competitive advantage arising from such a rapid expansion of foliage is the preemption of space in the leaf canopy.

Height

In the early studies of Boysen-Jensen (1929) and more recent investigations of Grubb *et al.* (1982) and Givnish (1982) it was recognised that where perennial plants are competing for light, small differences in stature may exercise a critical effect on survival and relative abundance. A comprehensive validation of this fact is available from the work of Gaudet and Keddy (1988), who conducted a standardised competition experiment comparing the success of 44 species of herbaceous plants, each matched individually against the same phytometer species (*Lythrum salicaria*). Within closed herbaceous vegetation small differences in height are associated with large changes in the intensity, direction, and quality of radiation, and the ability of a seedling or established

Figure 6. Herbaceous C-strategists of Western Europe. (a) *Fallopia japonica*, (b) *Urtica dioica*, (c) *Petasites hybridus*, (d) *Pteridium aquilinum*, (e) *Calystegia sepium*, (f) *Elytrigia repens*, (g) *Phragmites australis*, (h) *Arrhenatherum elatius*

plant to compete successfully for light may depend upon the extent to which the leaves can rapidly penetrate to superior positions in the canopy. Height growth in the shoots of established plants is determined, first, by the supply of energy and structural materials available from storage organs or current photosynthesis and, second, by the morphology of the shoot. Among herbaceous plants, the highest canopies occur in species in which the growing points are carried aloft at the apex of tall shoots, e.g. *Fallopia japonica*, *Urtica dioica* (Figure 6a, b). In the majority of those species in which the growing points remain close to the ground surface there is a limited capacity for height growth. However, in certain species, e.g. *Petasites hybridus* (Figure 6c), and the rhizomatous fern, *Pteridium aquilinum* (Figure 6d), an elevated canopy is achieved through the production of a small number of massive leaves (fronds in the case of *P. aquilinum*). In a third variant exemplified by *Calystegia sepium* (Figure 6e) the canopy consists of long scrambling shoots which form a dense blanket over the surface of the vegetation.

Similar arguments apply to the role of rapid height growth in woody species of high competitive ability. Extremely large annual increments in height are characteristic of the species which rapidly monopolise disturbed fertile soils in tropical (e.g. *Cecropia obtusifolia*, Figure 7a), temperate (*Sambucus nigra*, Figure 7b), or boreal (*Populus grandidentata*, Figure 7c) forests.

Lateral Spread

It is possible for a plant to reach a considerable height without capturing a major share of the resources present in its habitat. In both herbs and woody species, effective competition for light, water, mineral nutrients, and space is characteristic of species in which tall stature is allied to a growth form in which lateral expansion results in a high density of shoots and roots. The main advantage of clonal expansion in a C-strategist appears to be the pre-emptive occupation of space and capture of resources. This provides a route to dominance in many plant communities (see page 181) and contrasts strongly with species and circumstances where lateral spread is not associated with the potential for monopoly but instead confers an ability to explore patchy environments and to exploit them by translocating captured resources within a diffuse network (Tissue and Nobel 1988; see page 131). Among herbaceous species the growth forms most clearly conforming to this pattern are the branching rhizomes of dicotyledonous species such as *Urtica dioica*, *Chamerion angustifolium*, and *Solidago canadensis*, and of invasive grasses such as *Elytrigia repens* (Figure 6f) and *Phragmites australis* (Figure 6g), and the dense expanded tussocks of grasses such as *Arrhenatherum elatius* (Figure 6h).

The same principle applies to those woody species, e.g. *Populus grandidentata* (Figure 7c), and bamboos (Numata 1979), which combine rapid height growth with an ability to spread into large clonal patches by root suckers.

Figure 7. Woody C-strategists. (a) *Cecropia obtusifolia* in secondary tropical rainforest, Los Tuxtlas, Mexico,. (b) *Sambucus nigra* from derelict, productive farmland, South Yorkshire, England. The annual increment to a branch of this fast-growing shrub (left) is compared to that on a branch of the stress-tolerant competitor, *Quercus petraea* (right). The scale is in centimetres. (c) A clonal population of *Populus grandidentata* on an alluvial terrace, Fairbanks, Alaska

Shoot Thrust

The ability of the above-ground parts of a plant to capture resources is strongly determined by a combination of traits that allow rapid development of a large canopy. Where several species are simultaneously extending their shoots into the same volume of space physical contact is inevitable and it is reasonable to propose that in some cases the capacity either to push aside the foliage of neighbours or to resist physical displacement may be a significant component of competitive interactions. To test this hypothesis Campbell *et al.* (1992) measured the development of shoot thrust in eight contrasted herbaceous species using a conical framework of weighted windows (Figure 8) that could be opened by an expanding shoot. The experiment showed that the greatest ability to generate thrust occurred in species such as *Arrhenatherum elatius*, *Bromopsis erectus*, and *Urtica dioica* that are frequent vegetation dominants. It was also found that shoot thrust was a good predictor of the hierarchy in biomass (Figure 9) that developed when the eight species were grown together in an equiproportional mixture for 16 weeks in a competition experiment.

From these results it appears that when combined with other traits such as tall stature, lateral spread, and rapid dry matter production, shoot thrust can influence competitive interactions. However, as explained in Chapter 4, shoot thrust also plays a part in the mechanisms whereby dominance is eventually achieved late in succession by long-lived slow-growing herbs, shrubs, and trees which owe their status to an ability to retain captured resources rather than to a capacity for pre-emptive foraging.

Phenology

Development of attributes such as a high density of tall leafy shoots involves the production and deployment of a large quantity of photosynthate. This in turn depends upon an extended period of photosynthetic activity under climatic conditions conducive to high productivity.

In temperate climates like that of the British Isles, productive herbs such as *Chamerion angustifolium*, *Petasites hybridus*, and *Pteridium aquilinum* (Figure 10) tend to attain full leaf expansion over the period June–August when daylengths, light intensities, and temperatures are favourable to high rates of photosynthesis. In some of these species the development of flowers and fruits is delayed until the late summer. This phenomenon is consistent with maximum leafiness during the period most favourable to high rates of photosynthesis, and in certain species, e.g. *Fallopia japonica*, flowering may be so delayed that, at latitudes close to the northern limit of distribution, flowering coincides with the onset of winter.

Leaf Longevity

In the trees, shrubs and herbaceous plants which exploit productive, relatively undisturbed habitats, there is a strong tendency for the leaf canopy to be

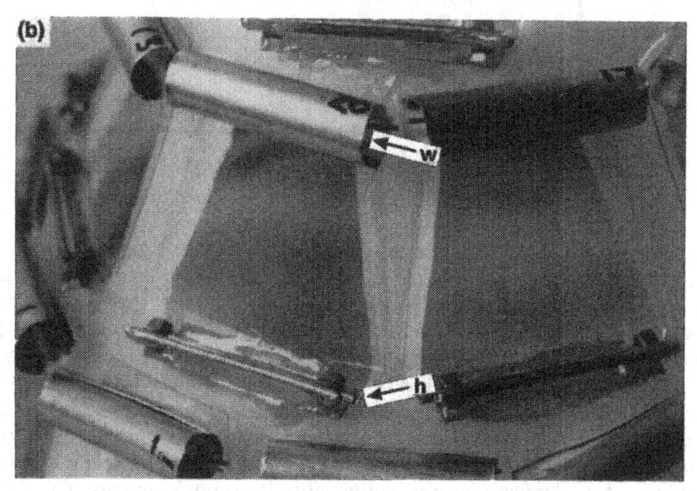

Figure 8. (a) The cone apparatus used to measure shoot thrust; (b) detail of window construction showing hinge (h) and compartment holding lead weights (w)

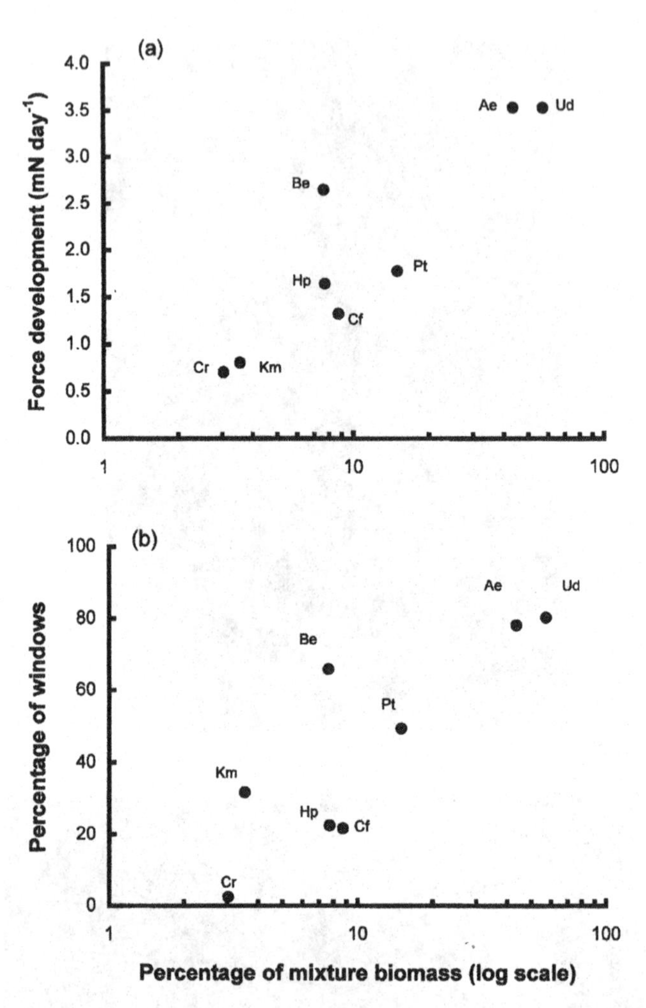

Figure 9. Shoot thrust and canopy architecture measurements from the cone apparatus plotted against species ranking in an experimental plant community synthesised by growing the eight species together in an equiproportional mixture for 16 weeks. (a) Rate of force development against a window with a resistance of 22.7 mN plotted against percentage contribution to mixture biomass ($r = 0.89$, $P < 0.01$). (b) Proportion of opened windows occurring in the top tier of the cone with windows offering 1.5 mN of resistance plotted against percentage contribution to mixture biomass ($r = 0.80$, $P < 0.05$). Calculations refer to the opening of a target number of windows; this target was the number of 1.5 mN windows opened 20 days after window opening commenced. The species are: Ae, *Arrhenatherum elatius*; Be, *Bromopsis erectus*; Cf, *Cerastium fontanum*; Cr, *Campanula rotundifolia*; Hp, *Hypericum perforatum*; Km, *Koeleria macrantha*; Pt, *Poa trivialis*; Ud, *Urtica dioica*

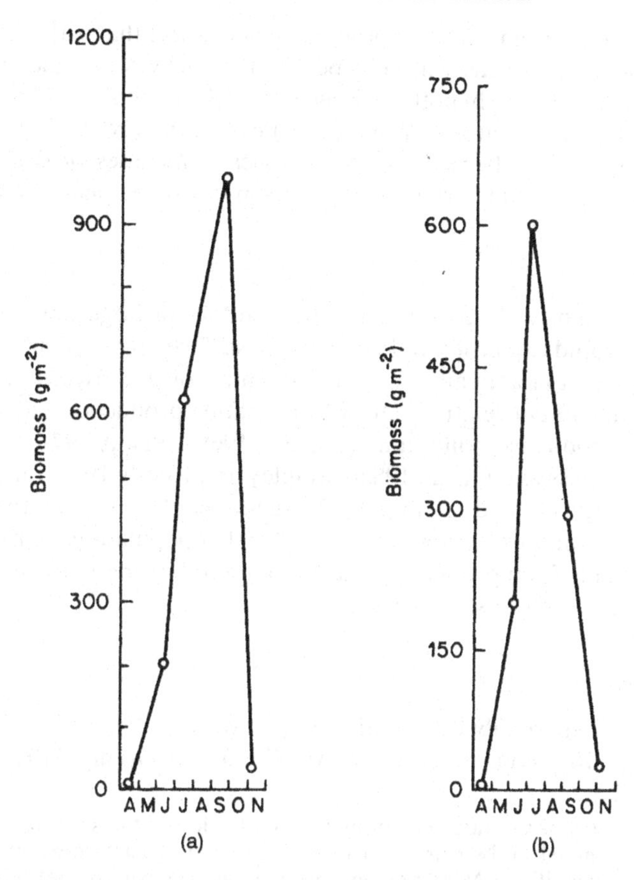

Figure 10. Seasonal changes in the shoot biomass of two perennial competitors growing in fertile, relatively undisturbed vegetation in northern England. (a) *Pteridium aquilinum*, (b) *Chamerion angustifolium*. (Reproduced from Al-Mufti *et al.* 1977 by permission of Blackwell Scientific Publications Ltd.)

exceedingly dynamic as a consequence of the continuous production of leaves at the shoot apices. However, this phenomenon also arises from the short life-span of individual leaves (Chabot and Hicks 1982). As we shall see later (page 30), the high turnover of leaves is often due, in part, to herbivory (Coley 1983), but demographic studies on marked leaves (Williamson 1976; Sydes 1984; Reich *et al.* 1991; Reich *et al.* 1992; Rogers and Clifford 1993) have confirmed that leaf life-span remains short in such dynamic canopies even in the absence of herbivore attack.

Root Longevity

Later in this review of competitive attributes (page 29) it is concluded that there is strong functional integration between the resource foraging capacities of root and shoot systems. On this basis we might expect that within a species a strong parallel in life-spans will exist between the leaves and the roots. Evidence consistent with this hypothesis is available (Ryser and Lambers 1995) but as explained by Eissenstat and Yanai (1997) conclusive proof of this relationship is lacking, mainly because of the technical difficulties of distinguishing between living, dead, and partially-dead roots in long-lived plant species.

Growth Rate

Tall stature, extensive lateral spread, the build-up of large perennating organs, and the rapid expansion of leaf and root surface areas are all dependent upon the production annually of a large quantity of photosynthate. Estimations of the maximum relative rates of dry matter production (R_{max}) under standardised productive conditions (Parsons 1968a; Pigott 1971; Grime and Hunt 1975; Furness and Grime 1982b; Shipley and Keddy 1988; Lambers and Poorter 1992; Rincon and Huante 1994; Huante *et al.* 1995; Saverimutta and Westoby 1996; Cornelissen *et al.* 1998) suggest that populations of herbaceous and woody plants from productive habitats tend to be among the most rapid-growing perennial species (Table 2).

Leaf Nutrients

Following the remarkably early and prescient work of Pearsall (1950, see his Table 7, page 109) comparative surveys (Band and Grime 1981; Field and

Table 2. Estimates of the rates of dry matter production in the seeding phase in six competitive perennial herbs compared to those attained by two crop plants grown under the same conditions. Measurements were conducted over the period 2–5 weeks after germination in a controlled environment. (Temperature: 20°C day, 15°C night; daylength: 18 h; visible radiation: 38.0 W m^{-2}; root medium: sand + Hewitt nutrient solution.) Rates of dry matter production are expressed as maximum (R_{max}) and mean (R) relative growth rates and as a function of leaf area (E_{max}). (Reproduced from Grime and Hunt 1975 by permission of *Journal of Ecology*.)

Species	R_{max} week^{-1}	R week^{-1}	E_{max} g m^{-2} week^{-1}
Competitive herbaceous perennials			
Alopecurus pratensis	1.3	1.3	69
Arrhenatherum elatius	1.3	1.3	115
Chamerion angustifolium	1.4	1.4	80
Epilobium hirsutum	1.8	1.8	126
Phalaris arundinacea	1.2	1.2	91
Urtica dioica	2.4	2.2	89
Annual crop plants			
Lycopersicon esculentum (Tomato)	1.6	1.0	93
Hordeum vulgare (Barley)	1.6	0.9	64

Mooney 1986; Poorter and Bergkotte 1992; Reich *et al.* 1992; Thompson *et al.* 1997) have revealed that the leaves of perennial plants capable of rapid rates of growth have a distinctive chemical composition. High leaf nitrogen coincides with high concentrations of the enzyme ribulose 1,5-bisphosphate carboxylase/oxygenase (Rubisco) which appear to confer the potential for rapid rates of carbon fixation. In comparison with slow-growing species, competitors also exhibit relatively high foliar concentrations of phosphorus, a feature consistent with the capacity to achieve rapid rates of photosynthesis.

Specific Leaf Area (SLA)

As Lambers and Poorter (1992) observed, the high concentrations of nitrogen and phosphorus in the leaves of fast-growing perennials (and ephemerals) are also explicable in terms of the low investment in cell wall material and in other chemical constituents conferring leaf thickness and toughness. The possibility that natural selection for leaf defence is reduced under conditions of intense competition for resources will be discussed later (page 30). Here it will suffice to record that several authors (Garnier and Laurent 1994; Huante *et al.* 1995; Westoby 1998; Wilson *et al.* 1999) have noted that the canopies of fast-growing perennials of productive habitats are often characterised by high specific leaf areas. However, caution must be applied in any attempt such as that of Westoby (1998) to use SLA as an easily-measured predictor in plant ecology. In a survey of 769 herbaceous species of the British flora (Wilson *et al.* 1999) it was found that this trait exhibited considerable variation both within populations and between leaves from the same individuals. Perhaps a more serious difficulty (see page 53) arises from evidence that high SLA values are not peculiar to plants competing for resources in the dense and dynamic canopies of productive vegetation. Thin leaves and high specific leaf area are also characteristic of the comparatively sparse, long-lived shoots of slow-growing shade plants such as *Oxalis acetosella.*

Rapid Growth Responses to Localised Resource Depletions (Active Foraging)

Although high R_{max} appears to be directly advantageous in competition, its significance may extend beyond the capacity for rapid dry matter production and it may be more helpful to regard it as one of the more easily quantified expressions of a group of physiological attributes which facilitate the capture of resources. These may include, in particular, a rapid rate of response to environmental variation, especially in the extension growth of stems, petioles, and fine roots, and in the expansion of leaf area. It is well known that there are rapidly-changing and complex gradients in light quantity and quality within developing leaf canopies (Holmes and Smith 1975) and equally dynamic spatial patterns of nutrient depletion and local enrichment are observed in the rhizosphere (Bhat and Nye 1973; Figure 11). It is not difficult to foresee the extent to which rapid adjustments in morphology in response to local depletions in resources arising during competition could facilitate the capture of mineral nutrients, water, light, or space. There is now widespread recognition that such rapid adjustments are an integral part of the resource-foraging mechanisms of plants (Smith 1982; Grime *et al.* 1986; Hutchings and de Kroon 1994). Various techniques have been developed to examine the responses of shoots and roots to spatial patterns in resource supply in the laboratory (Drew 1975; Campbell and Grime 1989a; Granato and Raper 1989; Hendry and Grime 1993) (Figures 12 and 13) and in the field (Caldwell and Pearcy 1994; Jackson *et al.* 1990; Robinson 1994; Robinson *et al.* 1998). In order for these responses to be effective it seems likely that they will involve early detection of changes in environment arising from encroachment by neighbours. It is particularly interesting therefore to find that plants have been shown (see page 13) to be capable of responding to changes in the spectral quality of light reflected from adjacent canopies.

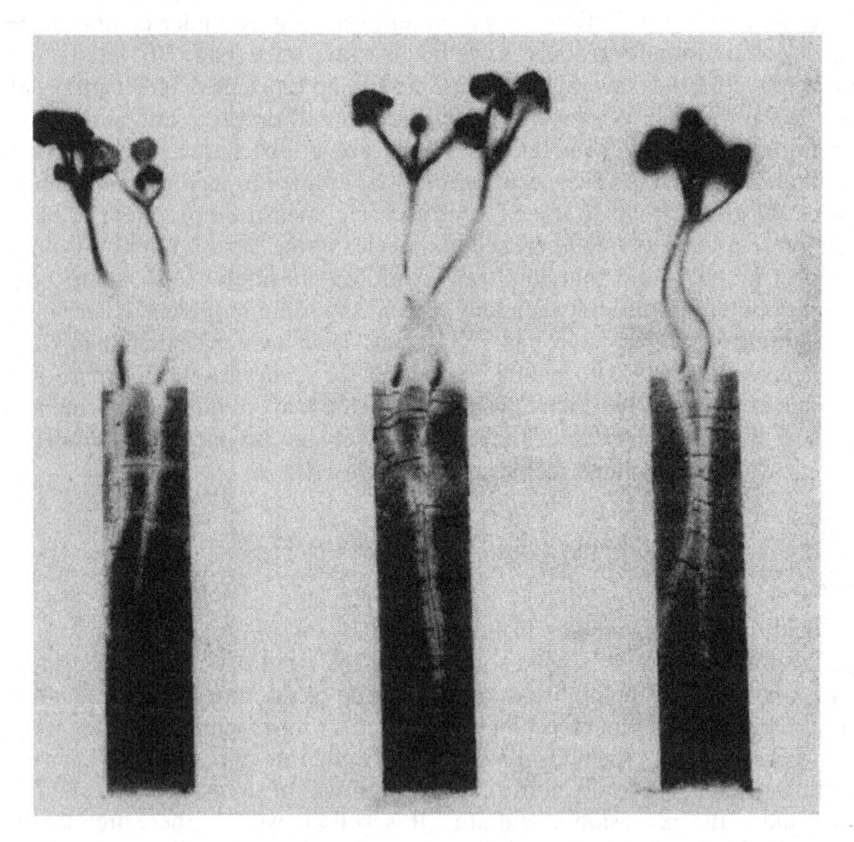

Figure 11. Autoradiographs showing the zones of ^{32}P depletion around the roots of mustard seedlings exploiting uniformly labelled soil. (Reproduced from Bhat and Nye 1973 by permission of *Plant and Soil.*)

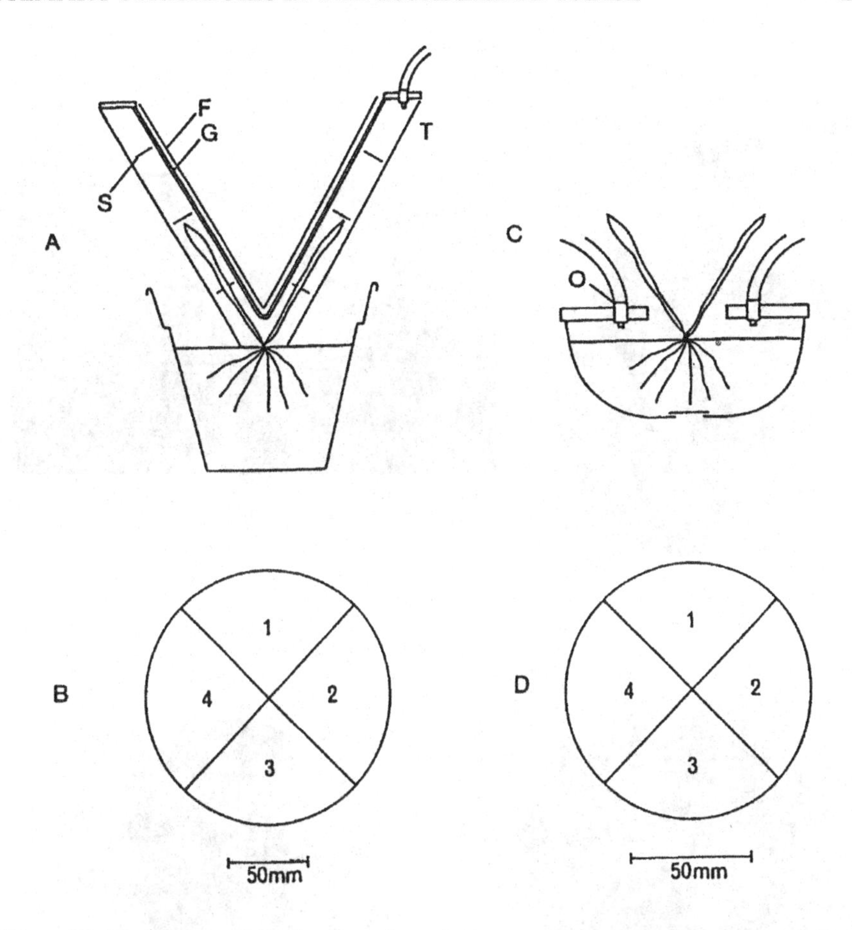

Figure 12. Imposing standardised patches of resource depletion on growing shoot and root systems. Two techniques are used to assay the 'resource-foraging' attributes of the leaf canopies and root systems of individual plants grown in isolation under standardised conditions simulating those experienced during competition. An important feature of both techniques is that plants are presented with standardised patches of resource depletion created without the use of partitions or barriers that could impede growth between patches.

Light patches (A and B). (A) Section through cone-shaped chamber for imposing partial shading on developing shoot system. A transparent glass upper surface (G) is covered with filters (F) to produce standardised patches of shade. Fine struts (S) support leaves and the chamber is supplied with compressed air (T). (B) View from above of shading pattern. Quadrants 1 and 3 are fully illuminated and quadrants 2 and 4 are shaded by filters.

Nutrient patches (C and D). (C) Section through bowl used to impose patches of nutrient depletion on developing root system. Nutrient solution fed by peristaltic pumps is dripped continuously onto the surface at symmetrically-arranged outlets (O). (D) View from above of nutrient distribution pattern. Quadrants 1 and 3 are supplied with nutrient-rich solution, quadrants 2 and 4 with nutrient-poor solution

(a)

(b)

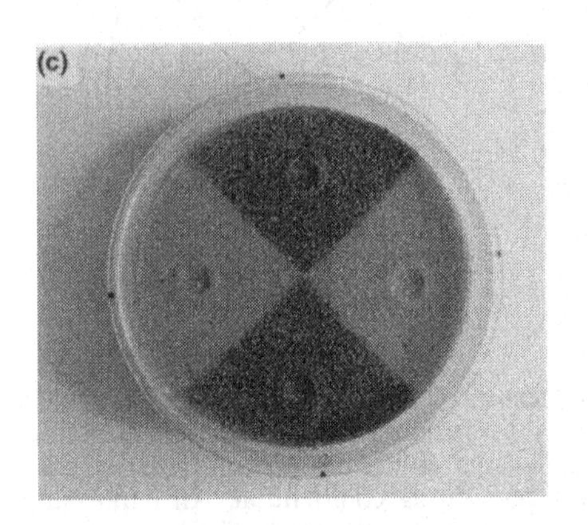
(c)

(d)

(e)

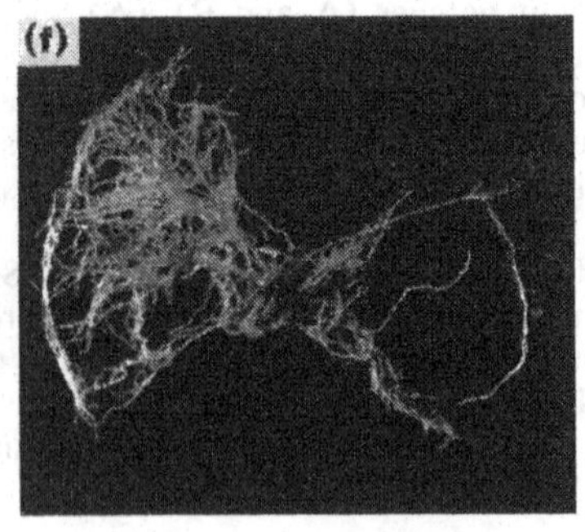
(f)

To serve the interests of ecology, research on resource-foraging by plants must address a number of related questions. Do plants of productive and unproductive habitats differ in their ability to respond to resource patchiness above and below ground? Are differences in foraging mode and success due to genetic differences in either sensitivity to environmental cues or in architectural flexibility? Do plants distinguish between patches and pulses in resource supply? How is foraging for light related to foraging for mineral nutrients and water? In evolutionary terms, has a tradeoff in allocation occurred between foraging for light and foraging for below-ground resources as suggested by Newman (1973), Huston and Smith (1987), and Tilman (1988), or is it more likely, as proposed by Donald (1958), Grime (1973b, 1994), and Colasanti and Hunt (1997), that there will be a functional integration and close equivalence between foraging above and below ground?

These questions are so fundamental to ecology and to the objectives of this book that evidence relating to them is introduced at several places in succeeding chapters (e.g. pages 97 and 201). Here, for the purposes of this brief review of competitive traits, it is perhaps sufficient to refer to Figure 14, which presents measurements of the ability of eight common herbaceous plants to exploit patchiness above and below ground (Campbell *et al.* 1991a). For each species, assays (Figures 12 and 13) were conducted to examine the capacity of the leaf canopy and root system to exploit standardised resource patchiness created without using partitions or barriers which could impede growth between patches. In both root and shoot assays, measurement was made of the partitioning of dry matter allocation between depleted and undepleted sectors imposed after an initial growth period in uniform productive conditions.

In order to test the predictive value of data obtained using the two procedures, a comparison was made between performance of the eight species in the foraging assays and the status achieved in a conventional competition experiment in which the eight species were grown together in an equiproportional mixture under productive glasshouse conditions for 16 weeks. The

---

Figure 13. (*opposite*) Equipment used to subject plants to controlled patchiness in light intensity and quality (a, b) and in mineral nutrient supply (c–f). (a) Inserting the glass cone which forms the roof of the shoot compartment and includes two opposite sectors with three gelatine filters mimicking the light-absorbing properties of a cucumber leaf; (b) a shoot-foraging experiment in progress showing the compressed air supply to the plant chambers; (c) vertical view of the patchiness in soil chemistry applied in the root-foraging experiments. The pattern has been revealed by including a dye in the solution supplied to two of the four symmetrically-distributed delivery points on the sand surface; (d) side view of (c) to show that the geometrical patchiness is maintained down to the base of the container; (e) a root-foraging experiment in progress showing the four tubes connecting nipples to peristaltic pumps and two enriched sectors indicated by a blue-green alga on the sand surface; (f) a harvested root system of *Urtica dioica* with roots concentrated in the two enriched sectors. (Reproduced from Campbell *et al.* 1991a by permission of *Oecologia.*)

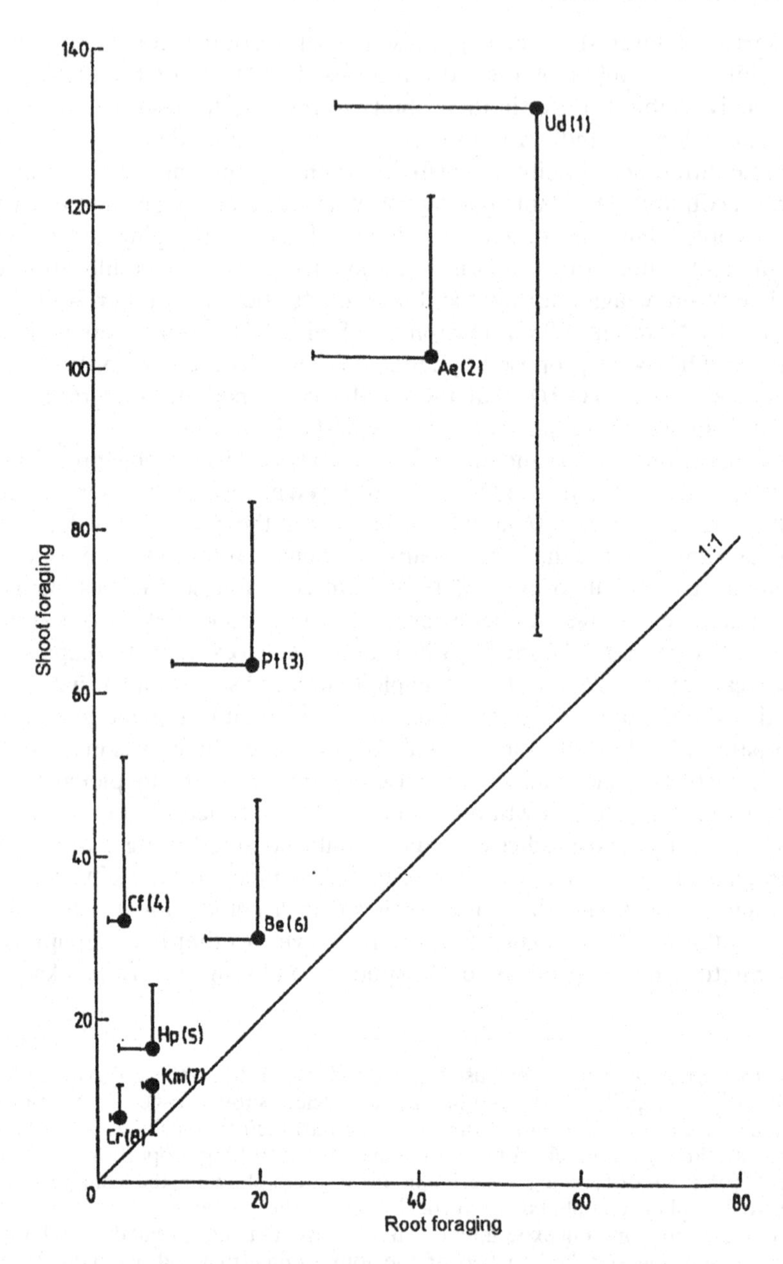

Figure 14. An examination of the relationship between root and shoot responses to resource heterogeneity in eight herbaceous species of contrasted ecology. A description of the methods used to expose the plants to resource patchiness is provided in Figure 12. Scales of foraging by the roots and shoots in the foraging assays are expressed as the respective increments of biomass (mg) to two undepleted quadrants, which in both assays constitute 50% of the available volume. The numbers in brackets

results (Figure 14) reveal a consistent relationship between the increment of dry matter in the undepleted sectors in both assays and the ranking of the species in the competition experiment. Covariance between roots and shoots in their responses to resource patchiness is also apparent. This relationship is maintained despite the consistent tendency for the scale of leaf canopy adjustment to exceed that of the root system. This, of course, arises from the freedom of movement of leaves in air and the encasement of roots in soil.

These results suggest that the status of plants in perennial herbaceous communities of high productivity may be predictable from specific, measurable, plant attributes. They also tend to confirm that, at least under productive conditions, there is a strong interdependence of foraging for light and mineral nutrients as originally proposed by Donald (1958).

Response to Damage

The competitive ability of a plant may be considerably reduced where the leaf area or root surface area is subject to predation or other forms of damage. Although there have been no extensive comparative studies of plant responses to damage, field observations and the results of experiments (Milton 1940; Mahmoud 1973) indicate that there are various morphogenetic responses to defoliation which may be characteristic of highly-competitive plants. These are particularly obvious in meadow grasses such as *Lolium perenne*, *Arrhenatherum elatius*, and *Alopecurus pratensis*, which are capable of rapid regrowth of the leaf canopy after defoliation. It would appear that the success of these plants in vegetation subjected to infrequent but severe damage by mowing (meadows and road verges) is related to their ability to respond to defoliation either by renewed growth of severed leaves or by expansion of new shoots, processes which may involve the diversion of an increased proportion of the photosynthate into shoot growth with a concurrent check in root development and capacity to exploit nutrient-rich patches in the soil (Mackie-Dawson 1999). It seems reasonable to suggest that such responses are specifically adapted to competition in that they rapidly cause the plant to re-establish a tall, dense leaf canopy.

Under conditions in which productive vegetation is subject to frequent mowing or grazing a slightly different phenotypic response appears to characterise strongly competitive species. In productive turf-grasses such as *Agrostis*

---

refer to the species ranking in a conventional competition experiment in which all eight species were grown together in an equiproportional mixture on fertile soil for 16 weeks. The vertical and horizontal lines refer to 95% confidence limits. The species are: Ae, *Arrhenatherum elatius*; Be, *Bromopsis erecta*; Cf, *Cerastium fontanum*; Cr, *Campanula rotundifolia*; Hp, *Hypericum perforatum*; Km, *Koeleria macrantha*; Pt, *Poa trivialis*; Ud, *Urtica dioica*. (Reproduced from Campbell *et al.* 1991a by permission of *Oecologia*.)

capillaris or pasture genotypes of *Arrhenatherum elatius* (Mahmoud *et al.* 1975), the effect of repeated defoliation is to stimulate the development of a very large number of small tillers with the result that a dense and rapidly-repaired leaf canopy is formed close to the ground surface.

Palatability

Comparative experiments examining the consumption of leaves of a wide range of plant species by generalist invertebrate herbivores (e.g. Grime *et al.* 1968, 1996; Coley 1983) suggest that foliage of many of the fast-growing perennial plants of productive habitats are relatively palatable. In some habitats such as second growth tropical rainforest (Figure 15) and monocultures of palatable perennial herbs in temperate regions (Figure 16) vulnerability to herbivory is evident from the large number of holes in leaves. Low expenditure of captured resources on anti-herbivore defences in productive vegetation is predictable from an evolutionary perspective. Species or genotypes which divert a high proportion of their resources into physical or chemical defence are likely to become vulnerable to competitive exclusion by neighbours that continue to allocate to new leaves and roots.

However, when this phenomenon of low expenditure on defence in fast-growing perennials is reviewed in a broader and long-term perspective it is clear that there are further issues that need to be addressed. In particular there is a need to consider why large stands of productive herbaceous or woody vegetation on productive soils do not become the subject of devastating attack by large populations of generalist herbivores. A possible answer to this question, discussed in Chapter 10 (page 313), is that competitors are protected from herbivores by the continuous presence of substantial populations of carnivores and parasitoids. In addition there are situations in which particular species can combine high productivity with effective defence. A remarkable example of this is provided by the experimental study of Clay and Holah (1999) in which it has been demonstrated that enhanced defence against herbivores resulting from the presence of a fungal endophyte in the leaves of *Festuca arundinacea* not only improved the performance of this species but also suppressed the species richness of communities dominated by the species.

Intraspecific Variation in Competitive Ability

A fundamental problem underlying any attempt to analyse the impact of competition in vegetation arises from the proposition, scarcely disputed among ecologists, that the competitive ability of a plant species varies according to the conditions in which it is growing.

In the first place, it is clear that plant characteristics which affect the competitive ability of the species may be subject to genetic variation. The literature provides numerous examples of intra-specific variation in competitive

Figure 15. Insect damage to the canopy of secondary tropical rainforest, Los Tuxtlas, Mexico

Figure 16. Gastropod damage to the canopy of a stand of the tall herb, *Filipendula ulmaria*, Coombsdale, North Derbyshire, England

attributes (e.g. Clausen *et al.* 1940; Bradshaw 1959; Cook *et al.* 1972; Gadgil and Solbrig 1972; Mahmoud *et al.* 1975).

A second source of variation in competitive ability arises from the fact that environments differ in the extent to which they allow the competitive characteristics of a plant to be expressed. It is clear that the development of large and dynamic leaf canopies and root systems may be restricted by forms of stress and damage. An example of modification of the competitive ability of a species by environment is illustrated in Figure 17, which compares the seasonal pattern of shoot production by bracken (*Pteridium aquilinum*) in two contrasted habitats. The results show that, at an unshaded productive site, the vegetation consisted of a vigorous single-species stand of *P. aquilinum* in which the density of living shoot material reached a summer maximum of approximately 1000 g m^{-2}. Although the second site, situated on an acidic woodland soil, contained a high density of fronds, the dry weight of living shoot material of *P. aquilinum* at the summer maximum was restricted (presumably by mineral nutrient stress and shading) to a value corresponding to 4% of that achieved at the productive site.

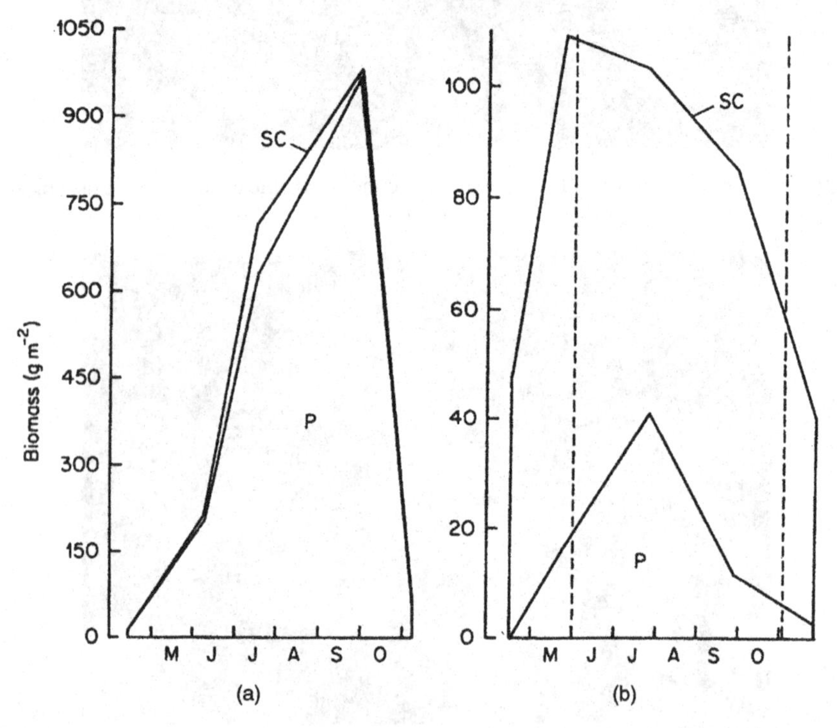

Figure 17. Comparison of the shoot biomass of *Pteridium aquilinum* (P) and its contribution to the standing crop (SC) in two contrasted habitats (Northern England, 1975): (a) in an unshaded situation on a deep brown earth soil; (b) in an oakwood on an incipient podzol. Limits of the period of maximum shading by the tree canopy in (b) are indicated by the broken lines. (Reproduced from Al-Mufti *et al.* 1977 by permission of *Journal of Ecology*.)

Further evidence of the modifying effect of environment and management upon competitive ability is available from a number of field experiments in which fertiliser additions have been made to nutrient-deficient meadows (e.g. Brenchley and Warington 1958), pastures (Smith *et al.* 1971), and sand-dunes (Willis 1963). A characteristic of the responses to fertiliser applications described in these investigations is the rapid expansion of grasses such as *Arrhenatherum elatius*, *Holcus lanatus*, *Agrostis stolonifera*, and *Festuca rubra* which were present, at low frequency, in the untreated vegetation. It would appear, therefore, that in these various types of vegetation, prior to the addition of fertilisers, the competitive ability of certain component species was held in check by mineral nutrient stress.

Both phenotypic and genetic variation in competitive ability create problems for the ecologist who is attempting to analyse the role of competition in natural vegetation. It is clear, however, that, with care, both types of variation may be recognised and taken into account. What is potentially a more serious difficulty arises from the suggestion, implicit in many ecological writings, that the nature of competition itself may vary fundamentally from one field situation to another so that, relative to other species, a particular species or genotype may be a strong competitor in one site but a weak competitor in another. Inspection of the data which have been quoted in support of this view (e.g. Newman 1973; Ellenberg and Mueller-Dombois 1974) shows that, often, this concept has evolved in tandem with loose definitions of competition. This is to say that, in certain cases, shifts in the fortunes of 'competitors' coincident with changes in environment may be attributed more correctly to non-competitive effects (e.g. differential predation) rather than to any alteration in the relative abilities of the plants to compete for resources.

When non-competitive effects have been discounted, the remaining objection to a unified concept of competitive ability arises from the fact that competition occurs with respect to several different resources including light, water, various mineral nutrients, and space. Hence it might be supposed either that the ability to compete for a given resource varies independently from the ability to compete for each of the others or that different competitive abilities trade-off against each other. There are grounds, both theoretical and empirical, for rejecting these possibilities as major sources of variation in competitive ability. It seems more logical to predict that the effect of natural selection will be to cause the abilities to compete for light, water, mineral nutrients, and space to vary in concert and to be developed to a comparable extent in any particular genotype.

In circumstances which allow the rapid development of a large standing crop, the most obvious competition is that occurring above ground with respect to space and light, and it may not be immediately obvious why success should also depend on the effective capture of water and mineral nutrients. However, as pointed out by Mahmoud and Grime (1976) and Chapin (1980), it is clear that rapid production of a large biomass of shoot material, a prerequisite for effective above-ground competition, is dependent upon high rates of uptake of water and mineral nutrients, characteristics which are themselves dependent upon a considerable expenditure of photosynthate on root development. It would appear, therefore, that the abilities to compete for light, mineral nutrients, water, and space are closely interdependent. We may suspect that, although in productive habitats competition above ground for space and light is more conspicuous, the outcome may be strongly influenced by competition below ground. Positive interaction between competitive abilities above and below ground is strongly suggested by the results of resource-foraging studies (Figure 14; page 28) and the consequences of such interaction are illustrated in Figure 18 which depicts diagrammatically the predicament of a seedling undergoing the process of competitive suppression in dense productive vegetation. A popular account of this situation (e.g. Tilman 1988) suggests that the seedling is being 'shaded out' under conditions of ample nutrient supply. A more satisfactory, if more complicated, explanation must take account of the morphogenetic response of the plant to shade; this will dictate the retention of photosynthate in the shoot, causing starvation and restricted development of the root, confining it to the zones of nutrient depletion which, even in fertile soils, surround roots (Bhat and Nye 1973). In these circumstances, competitive suppression is more accurately described as a dynamic process in which resource limitations above and below ground interact progressively to enfeeble the seedling.

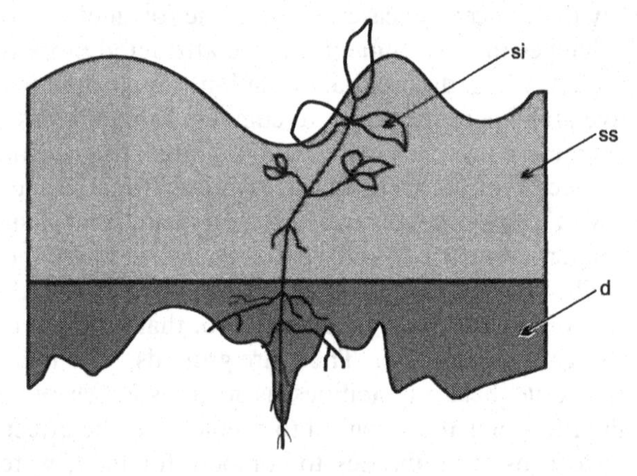

Figure 18. Competitive suppression in dense productive vegetation. si, suppressed individual; ss, shaded stratum; d, zone of mineral nutrient depletion

The models of competition for resources in terrestrial vegetation devised by Tilman (1982, 1988) have been intellectually appealing and highly influential but throughout their development they relied upon simplifying assumptions that do not match in important features the physics and chemistry of resource supply, depletion, and release above and below ground in terrestrial ecosystems. Some of these assumptions allow underestimations of the resource loss rates (to herbivores and pathogens) incurred by fast-growing species on resource-poor soils (see Figure 22) but the most serious weaknesses of the Tilman models are their failure to account for the consequences and costs of resource capture by roots growing in a solid medium (soil). The assumption of a uniform 'draw-down' of nutrient levels is justified for phytoplankton populations in well-mixed solutions (Titman 1976) but does not capture the realities of localised resource depletion experienced by plants exploiting fertile soils (Bhat and Nye 1973; Caldwell and Pearcy 1984; Huston and De Angelis 1994; Jackson and Caldwell 1996). Recently an additional dimension has been added to this argument with the discovery (Gordon and Jackson 2000) that the finer roots which are relatively short-lived and suffer the highest levels of herbivore damage contain higher levels of nitrogen, phosphorus, and magnesium. These same authors also report that the concentrations of nitrogen in living and dead fine roots are identical, suggesting that there is little retranslocation and, therefore, high nitrogen loss, associated with dynamic root foraging.

The importance of below-ground interactions in competition in productive environments is well illustrated by a classic experiment conducted by Donald (1958) involving competition between the perennial grasses, *Lolium perenne* and *Phalaris tuberosa.* In this pot experiment, a system of partitions, above and below ground, was devised so that it was possible to compare the yield of each of the two species under four treatments, i.e. competition above ground only, competition below ground only, competition above and below ground, and no competition (control). The experiment was carried out at high and low levels of nitrogen supply, but here it is only the former which is relevant to the immediate point at issue. The results of this experiment are presented in Table 3 and show that the clear advantage of *Lolium perenne* over *Phalaris tuberosa* under productive conditions derived from the superior competitive ability of the species both above and below ground.

Table 3. Measurements of the effects of root and shoot competition between two perennial grasses, *Lolium perenne* and *Phalaris tuberosa*, at high and low rates of nitrogen supply. The values tabulated refer to the dry weight of the shoot (g) after 105 days. (Reproduced from Donald 1958 by permission of *Australian Journal of Agricultural Research.*)

	No interspecific competition		Interspecific competition					
			Between shoots		Between roots		Between roots and shoots	
	Lolium	*Phalaris*	*Lolium*	*Phalaris*	*Lolium*	*Phalaris*	*Lolium*	*Phalaris*
High nitrogen	4.71	4.67	4.19	3.19	4.31	1.17	4.72	0.32
Low nitrogen	2.45	2.00	2.71	1.63	2.12	0.35	2.77	0.18

Competition in Productive and Unproductive Conditions

Resistance to the idea that competition has greater impact in productive environments also originates from animal ecologists (e.g. Andrewartha and Birch 1954) who have maintained the view that the intensity of competition between animals reaches greatest intensity under conditions of low resource supply. Kadmon and Shmida (1990) conducted an experimental test of the applicability of this hypothesis to plants by examining the effects of competing neighbours on the yield of individuals of the desert annual grass, *Stipa capensis*, under the contrasted conditions of resource supply (water and mineral nutrients) associated with slopes, depressions, and wadis. The results (Figure 19) provide convincing falsification of the notion of competition increasing at lower rates of resource supply. Not only is the intensity of competition greater (wadis>depressions>slopes) in the more productive habitats but within the two least productive habitats supplementary watering significantly increased the intensity of competition.

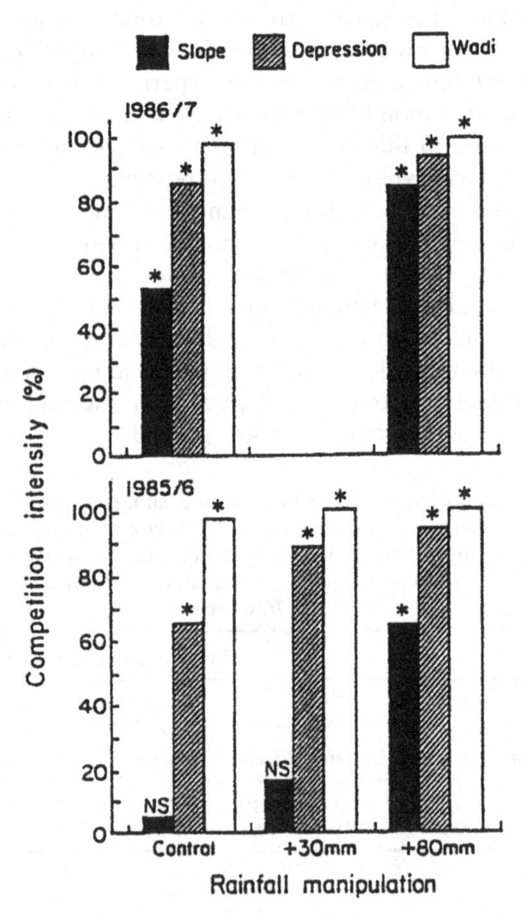

Figure 19. Competition intensity experienced in the desert annual, *Stipa capensis* ($100(Y_{max}-Y)/Y_{max}$, where Y_{max} denotes mean seed yields of experimentally-isolated plants, and Y denotes mean seed yield of control plants), as it varies between habitat types, seasons, and rainfall manipulation treatments. Asterisks indicate statistically significant differences between control and experimentally-isolated plants. (Reproduced from Kadmon and Shmida 1990 by permission of *Israel Journal of Botany*.)

Some ecologists are extremely reluctant to recognise the declining importance of competition for resources in unproductive habitats. A remarkable example of this is provided in Rees *et al.* (1996) in which demographic records over the period 1979–1988 were used to explore the population dynamics of four diminutive coexisting winter annuals exploiting an area of unproductive stabilised sand-dune subject to summer drought and rabbit damage. The authors concluded that 'interspecific interactions are extremely weak' and that 'there was no evidence of interspecific competition for any of the species in any year', conclusions entirely consistent with the proposition from CSR theory that interactions between component species will be strongly reduced in communities in which potential monopolists (C-strategists and their close allies) are excluded by resource shortage and the habitat is occupied by small stress-tolerators and by ephemerals exploiting local gaps in the turf. Undeterred by this negative result, Rees *et al.* (1996) noted that the cohorts of seedlings of the tiny annuals in the patches of disturbed vegetation exhibited self-thinning and intra-specific competition and on this basis they were able to conclude that 'we find little to support Grime's (1979) verbal theory, which asserts that competition is unimportant in "stressful environments".' As Welden and Slauson (1986) recognised, there is an obvious and continuing conceptual gulf dividing those ecologists who are measuring the role of competition in modulating the relative abundance of species co-existing in plant communities and those pursuing a larger predictive framework that identifies the circumstances where competition determines the kind of plant species admitted to communities.

Rejection of the Resource-ratio Model of Competitive Ability

Until recently a popular alternative explanation for the role of competition in vegetation was that attributing a primary role to evolutionary tradeoffs between root and shoot allocation. Support for the model was drawn from experiments (Brouwer 1962a, b; Corré 1983; Hunt and Nicholls 1986) in which predictable alterations in partitioning of dry matter between shoots and roots were induced by exposing plants to either low mineral nutrient concentrations or to shade treatments. In a wide range of plants, these manipulations were shown to be capable of modifying root–shoot ratios to a profound extent and this has led to the proposition that there are effective homoeostatic mechanisms that, through modifying allocation between root and shoot, preserve a balance between photosynthesis and mineral nutrient capture. Such results based upon the phenotypic plasticity of plants encouraged the prediction that genetically-determined patterns of variation in competitive ability would occur and would follow a similar pattern. In particular it was predicted that plants of fertile and infertile soils or shaded and unshaded habitats will differ in root-shoot ratio. This hypothesis has been developed with particular reference to evolutionary tradeoffs in allocation of captured resources between root and shoot by Iwasa and Roughgarden (1984) and Huston and Smith

(1987) and is explored in greatest detail by the model ALLOCATE devised by Tilman (1988).

The idea of evolutionary tradeoffs between root and shoot allocation has immediate and persistent intellectual appeal and has wide currency in ecological textbooks (e.g. Begon *et al.* 1996). However, as explained by Robinson and Van Vuuren (1998), the ultimate value of attempted generalisations in ecology must depend upon the extent to which, regardless of their elegance, they can predict field reality:

> A generalisation that seems, intuitively, as though it should be true, but is not, is particularly dangerous. Genuine scientific progress is stifled if an intuitively obvious, but wrong, generalisation remains unchallenged and becomes embedded in the literature. Conversely, when a generalisation survives comparison with a large body of independent data, its usefulness is strengthened.

Experimental evidence in support of the resource-ratio hypothesis is lacking; in fact consistent falsification (Berendse and Elberse 1989; Gleeson and Tilman 1990; Olff *et al.* 1990; Shipley and Peters 1990; Aerts *et al.* 1990; Campbell *et al.* 1991a; van der Werf *et al.* 1993; Brown and Parker 1994; Canham *et al.* 1994; Ryser and Lambers 1995; Huante *et al.* 1995; Denslow *et al.* 1998; Coomes and Grubb 2000) has been a feature of the recent literature.

Until recently many reviews (e.g. Grace 1991) on the subject of plant competition at high and low productivity have presented this subject area as an unresolved ongoing debate. However, in the recent literature a consensus appears to be developing from experiments conducted under carefully controlled conditions and recognising the functional interplay between roots and shoots. The following quotations are not unrepresentative of the recent literature:

> Based on the present analysis, we favour the suggestion of Grime (1979) and Chapin (1980), that mechanisms enabling plants to conserve nutrients, whether via biomass longevity or via remobilisation of previously stored nutrients, have been the subject of natural selection in slow-growing species and are among the key-factors explaining the high abundance of inherently slow-growing species in unproductive habitats. Contrary to the suggestions of Grime and Chapin, Tilman (1988) stressed the significance of maximising nutrient uptake by increasing allocation of carbon and nitrogen to roots at low nutrient supply. Indeed, our model (Van der Werf *et al.* 1993) showed that optimal root weight ratio is increased with decreasing nitrogen supply. However, these optimal ratios differ among species and depend on the functional equilibrium between roots and shoots. For instance, if the slow-growing *Briza media* (which has inherently a lower root weight ratio than *Dactylis glomerata*) has the same inherent high root weight ratio as the fast-growing *Dactylis*, it would have a lower carbon gain per plant (cf. Van der Werf *et al.* 1993). With time this would lead to a strongly decreased ability to compete for nutrients. Therefore we reject the possibility that inherently slow-growing species have a superior competitive ability due to more efficient nutrient acquisition, as put forward by Tilman (1988, 1990).
>
> Van der Werf *et al.* (1993)

> We conclude that, although there must be a trade-off in allocation of plant biomass either to above- or below-ground organs (Tilman 1988) and although there may be considerable plasticity in this allocation pattern, this does not necessarily mean that a plant is unable to maintain a high capacity for resource capture simultaneously above and below ground. It is possible for a species to be a superior competitor for above- as well as below-ground resources over a wide range of conditions, as already suggested by Grime (1979).
>
> Ryser and Lambers (1995)

Why has the resource-ratio hypothesis failed to match reality? Two explanations can be suggested:

1 The majority of habitats cannot be simply classified with respect to single limiting factors. Although shading, for example, is often a conspicuous feature of grasslands and woodlands, mineral nutrients also frequently limit plant production at particular sites (Coomes and Grubb 2000).
2 Models such as those of Huston and Smith (1987) and Tilman (1988) and experiments such as those of Brouwer (1962a, b) and Hunt and Nicholls (1986) involve circumstances in which resource depletion is imposed uniformly within the aerial or the rooting environment. Although this scenario may be applicable to models and experiments where phytoplankton exploit well-mixed cultures (e.g. Tilman 1981) it does not hold true for the environments experienced by terrestrial plants. As already explained in relation to Figure 18, it is essential for models of competing vascular plants to take account of the dynamic and heterogeneous nature of resource supply both above and below ground.

An alternative explanation for the fitness of plants adapted to conditions of low soil fertility will be presented later (page 55). The essence of this theory is the suggestion that although competition, especially that for water and mineral nutrients, is not restricted to productive habitats, its importance in unproductive habitats is small relative to the ability to conserve the resources which have been captured and to resist the severe hazards to survival (e.g. herbivory, extreme climatic events) which characterise many infertile environments. This argument may be extended to the full range of unproductive habitats including those in which light and water supply are the most obvious limiting factors. Further evidence of the decline in importance of competition for resources in unproductive vegetation is available from a wide range of comparative studies (Ashton 1958; Pigott and Taylor 1964; Grime and Jeffrey 1965; Grime 1965, 1966; Clarkson 1967; Hutchinson 1967; Loach 1967, 1970; Rorison 1968; Parsons 1968a, b; Higgs and James 1969). As explained later (page 91), it would appear that competitive characteristics such as high potential growth-rate and rapid phenotypic adjustments in the deployment of photosynthate within the root and shoot systems become disadvantageous in circumstances of extreme and more or less continuous environmental stress.

An Alternative View; the R* Model of Competition of Tilman

One of the key assertions of the CSR model of primary plant strategies is that the importance of competition for resources as a determinant of the admission or exclusion of species from communities reaches a maximum under productive conditions and declines on highly-infertile soils. An apparent direct challenge to this interpretation is presented by the R* model of competition developed by Tilman (1982, 1988) in which it is maintained that under infertile conditions competition remains important but occurs through a different mechanism in which success is achieved through the capacity of certain species to reduce limiting nutrients in the soil to levels below which other species cannot maintain their populations.

Before considering the evidence for or against this hypothesis it is important to point out that in several important respects the mechanisms proposed by Tilman (1982, 1988) are not substantially different from those proposed in CSR theory (Grime 1974, 1977, 1979). Both theories recognise that as we move down the soil fertility gradient there is a transition from mechanisms of swiftly-resolved competition through pre-emptive resource capture (the 'transient dynamics' of Tilman (1988)) to less dynamic community processes that rely upon the relative abilities of plants to survive chronically-unproductive conditions. There is also unlikely to be a major difference between Tilman and Grime with regard to mechanisms structuring the most severely nutrient-constrained vegetation, such as where lichens, mosses, and rosette plants grow as isolated specimens (Figure 29a, b, page 66); here without direct contact with other plants fitness cannot, by definition, involve competition for resources and will depend on tolerance of low resource supply and endurance of severe conditions. On this basis the main difference between the CSR model and the theories proposed by Tilman (1982, 1988) appears to refer to the mechanisms operating where plants are growing in close association on infertile soil (Figure 29c–g). The different predictions of the two theories with respect to community processes under such conditions were summarised by Wedin and Tilman (1993) as follows:

> Grime (1979) proposed that the abilities to compete for light and nutrients are positively correlated because the greater size or growth rate of a superior competitor allows it to preempt both above- and below-ground resources. In contrast, Tilman (1982, 1988) proposed that the levels to which a species reduces light and nutrients depends on its allocation pattern, and that there is an inherent tradeoff between competitive ability for above- and below-ground resources. Consequently, Grime (1979) predicts that the competitive rankings of species should remain constant along productivity gradients, whereas Tilman (1988) predicts they should change.

The extensive literature falsifying the notion of tradeoffs between competitive abilities for above- and below-ground resources has been presented and discussed elsewhere (page 37) and need not be repeated here. This brings the long debate involving Tilman and Grime and many others within sight of a final resolution. Quite clearly the last and decisive step must involve experimental tests of the apparently diametrically-opposed predictions summarised in the last sentence of the quotation from Wedin and Tilman (1993):

> Consequently, Grime (1979) predicts that the competitive rankings of species should remain constant along productivity gradients whereas Tilman (1988) predicts that they should change.

In order to examine the extent to which experiments have resolved this argument it is essential to clarify exactly what the two sides are predicting. First it should be noted that there is no difference between Tilman and Grime with respect to the changes predicted to occur in the species composition and rankings in abundance of species along productivity gradients *in the field*. As Wedin and Tilman (1993) acknowledge, the debate is about the contribution of changes in relative competitive abilities to the shifts in relative abundance that self-evidently occur along natural productivity gradients. Our task in designing experiments, therefore, is to determine whether the observed changes in species composition coinciding with the transition from fertile to infertile soils arise from a switch in the mechanism of competition for resources (Tilman) or are due to the declining importance of competition and to an increased impact of other selective mechanisms such as drought or herbivory (Grime).

If we are to differentiate effectively between competition for resources and the other mechanisms that determine success and failure on infertile soils it is impracticable to conduct experiments under natural field conditions. The very revealing experiment of Brown and Gange (1992) involving the removal of above- and below-ground herbivores by pesticide treatments provides a graphic illustration of the confusion between competition for resources and 'apparent competition' (*sensu* Holt 1977) due to differential herbivory that could arise if competition studies were to be conducted in natural habitats.

In fact, quite a large number of competition experiments involving plant species from productive and unproductive habitats have been conducted under laboratory and glasshouse conditions on fertile and infertile soils. In one such (Mahmoud and Grime 1976), three perennial grasses of contrasted ecology were allowed to compete under productive and unproductive (low nitrogen) conditions. One of the species, *Festuca ovina*, is restricted to unproductive sites; another, *Agrostis capillaris*, is associated with habitats of intermediate fertility; whilst the third, *Arrhenatherum elatius*, is a frequent dominant of productive meadows. Each species was grown in monoculture and in separate 1 : 1 mixtures with each of the other two grasses. The results obtained in the productive treatment are included in Figure 20 and show that the yield of A. *elatius* in monoculture was twice that of A. *capillaris* and approximately seven times that of *F. ovina*. In the mixtures grown under productive conditions, the interactions between the species were in each case conclusive in that *F. ovina* was totally eliminated by both of the other grasses and there were no survivors of *A. capillaris* when this species was grown with A. *elatius.* In the low-nitrogen treatment all three species showed a marked reduction in yield, most pronounced in *A. elatius* and *A. capillaris*. However, there is no evidence in Figure 20 to suggest that the competitive ability of *F. ovina* increased under conditions of nitrogen stress and no impact of *F. ovina* was detected on the yields of either of the other two species.

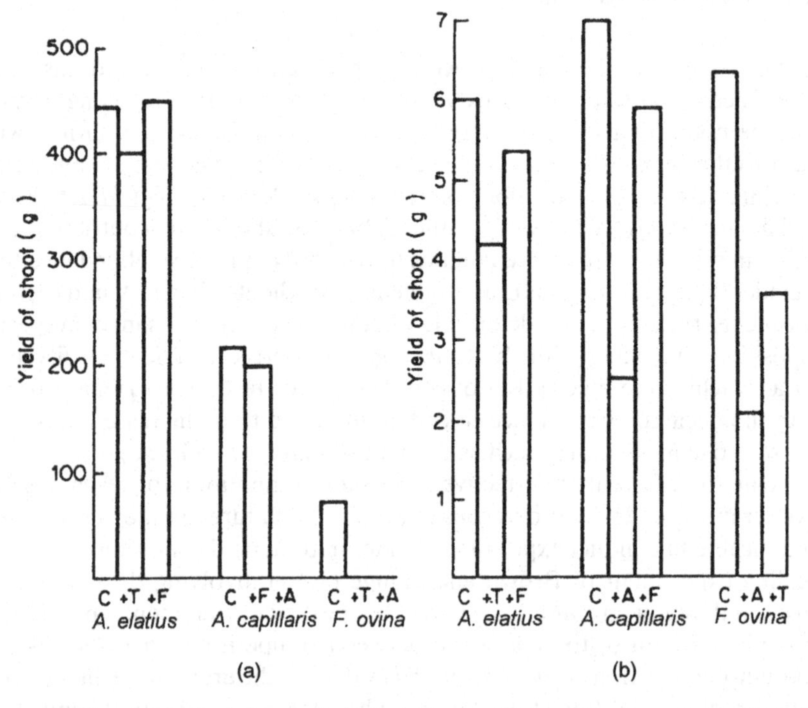

Figure 20. The yield (g) of shoots of *Arrhenatherum elatius* (A), *Agrostis capillaris* (T), and *Festuca ovina* (F) in mixtures (+) and controls (C): (a) under productive conditions (N 176 mg/l), (b) under nitrogen stress (N 5 mg/l). (Reproduced from Mahmoud and Grime 1976 by permission of Blackwell Scientific Publications Ltd.)

From this result, it would appear that the ability to compete for nitrogen plays little part in the mechanism whereby *F. ovina* is adapted to survive conditions of low soil fertility. A similar conclusion may he drawn with respect not only to nitrogen but also to other essential mineral nutrients from a number of laboratory experiments (e.g. Bradshaw *et al.* 1964; Hackett 1965; Clarkson 1967; Rorison, 1968; Higgs and James 1969; Fisher *et al.* 1974; Whelan and Edwards 1975; McLachlan 1976; Scaife 1976) in which plants of infertile habitats have been grown on nutrient-deficient soils and solution-cultures. These studies provide no convincing evidence that plants indigenous to poor soils are more efficient in the uptake of mineral nutrients when these are present at low concentrations. Further evidence of the competitive superiority of fast-growing species across a wide range of soil fertilities is available from more recent experiments (Campbell and Grime 1992; Keddy *et al.* 1994, 2000; Weiher and Keddy 1995; Fraser and Grime 1999).

If fast-growing species from fertile soils outcompete species of infertile soils under low nutrient regimes in the laboratory it is necessary to identify the additional factors that prevent this competitive superiority from expression in unproductive communities under natural conditions. Several hypotheses can be advanced on the basis of comparative investigations of species exploiting fertile and infertile soils. It is possible that strong competitors for resources at continuously high or continuously low concentrations lose their advantage under conditions where mineral nutrients are available only as brief pulses of mineralisation (Campbell and Grime 1989; Jonasson and Chapin 1991). Fast-growing species with high turnover of tissues may be more vulnerable to root pathogens and herbivores and less able to benefit from mycorrhizal colonisations. Comparative experiments confirm that fast-growing annuals and perennials are more palatable to generalist herbivores (Grime *et al.* 1968, 1996), and in two recent experiments involving the synthesis of plant communities at high, moderate, and low fertility in outdoor microcosms (Figures 21 and 22) strong selective suppression of fast-growing species by generalist invertebrate herbivores has been demonstrated (Fraser and Grime 1999b; Buckland and Grime 2000). A remarkable feature of both of these experiments was the persistent dominance of the communities developed in the absence of herbivores on highly infertile soil by plants such as *Lolium perenne* and *Poa annua* (Figure 23), species which under natural field conditions are restricted to fertile soils.

Against this background a critical reassessment is necessary for evidence of strong competitive effects on infertile soils that is based upon measurements of R* in field conditions. Experiments such as those of Wedin and Tilman (1993) in which plants of species associated with fertile and infertile soils were grown in field plots on soil of low nitrogen status over extended periods of time do not provide an adequate test of the two theories. They are clearly capable of identifying the capacity of certain species to survive under conditions in which the nutrient status of the soil has been reduced to a low level but they are unable to discriminate between the alternative mechanisms proposed to explain such effects because they have been conducted in field conditions in which not only differences in rates of resource capture but many other selective mechanisms likely to cause large differences in rates of resource loss including herbivory are operating. From soil nitrogen analyses in the experiments of Wedin and Tilman (1993) there is evidence of the ability of slow-growing plants of infertile conditions to survive and to reduce the levels of available nitrogen to low levels. Unfortunately, it is not possible to assess to what extent their success is due to a superior ability to capture nitrogen at low external concentrations or is the result of the capacity to retain nitrogen more efficiently by greater resistance to herbivores and pathogens and through

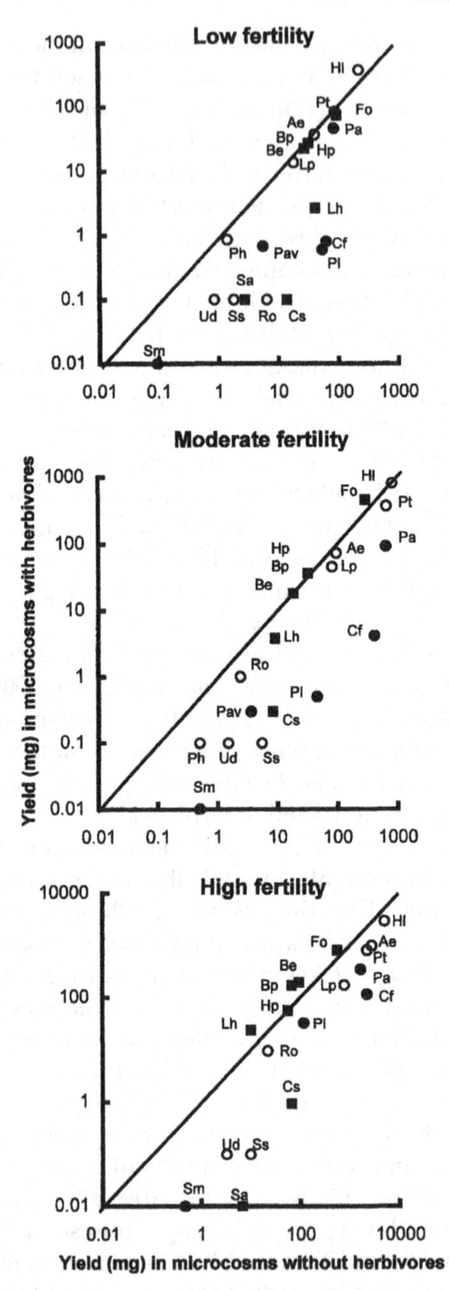

Figure 21. Effects of herbivory (snails and grass aphids) on various species grown together for two growing seasons in outdoor microcosms at (a) low soil fertility, (b) moderate soil fertility, and (c) high soil fertility.(●), ruderals; (○), competitors; (■), stress-tolerators. The statistical significance of changes in weight associated with

other components of the syndrome of nutrient-conserving attributes associated with stress-tolerance (see page 89).

There is an additional source of information that suggests that the capacity of slow-growing plant species to dominate vegetation on infertile soils is related more to the ability to protect nutrient capital than to capture nutrients at low external concentrations. In a recent investigation (Burt-Smith *et al.* 2001), a comprehensive screening of plant traits was conducted on 24 prairie perennials that had been sown together in field plots under the severely nitrogen-deficient conditions of the savannah grasslands at Cedar Creek, Minnesota (Tilman 1996; Tilman *et al.* 1996, 1997a). The resulting database allows recognition of the traits that are the strongest predictors of the relative abundance of species in the synthesised communities over the course of the experiment. In the first year, relatively fast-growing early-successional species such as *Rudbeckia hirta* were prominent but subsequently the vegetation became dominated by slow-growing C_4 grasses such as *Andropogon geradii* and *Schizachyrium scoparium*. The species trait that is the strongest predictor of the dominance hierarchy in the vegetation developed after the first growing season is the palatability of the foliage to the generalist herbivore, *Acheta domestica*, in a cafeteria-style feeding trial conducted on intact growing plants over the period 6 to 9 weeks after germination. Insect herbivores are of common occurrence at Cedar Creek (Ritchie and Tilman 1995) and in experiments with grasshoppers it has been demonstrated that the low digestibility of the bundle sheath cells of C_4 grasses can provide an effective deterrent to attack (Caswell and Reed 1976; Caswell *et al.* 1973). This provides circumstantial evidence that the control of vegetation composition under nutrient-limited conditions such as those prevailing at Cedar Creek depends primarily upon reduction of the loss components in the total nutrient budget of the plants. It would not be surprising if the species which suffer the lowest losses of nutrients had a greater capacity to survive at low values of R*. It would be entirely misleading, however, to ascribe this capacity to high competitive ability. More research is required but the balance of evidence currently points to nitrogen conservation and apparent competition (Holt 1977) as the dominant mechanisms structuring the grasslands at Cedar Creek.

---

herbivory is indicated as follows: * $P < 0.05$; ** $P < 0.01$; *** $P < 0.001$. The species are: Ae, *Arrhenatherum elatius*; At, *Arabidopsis thaliana*; Be, *Bromopsis erectus*; Bp, *Brachypodium pinnatum*; Ca, *Chenopodium album*; Cs, *Centaurea scabiosa*; Ct, *Cerastium fontanum*; Fo, *Festuca ovina*; Ga, *Galium aparine*; Hl, *Holcus lanatus*; Hn, *Helianthemum nummularium*; Hp, *Helictotrichon pratense*; Lh, *Leontodon hispidus*; Lp, *Lolium perenne*; Pa, *Poa annua*; Pav, *Polygonum aviculare*; Ph, *Petasites hybridus*; Pl, *Plantago lanceolata*; Pt, *Poa trivialis*; Ro, *Rumex obtusifolius*; Sa, *Sedum acre*; Sm, *Stellaria media*; Ss, *Stachys sylvatica*; Ud, *Urtica dioica*; Hp and Bp in (b) overlap and therefore share the same symbol. (Reproduced from Fraser and Grime 1999b by permission of *Journal of Ecology*.)

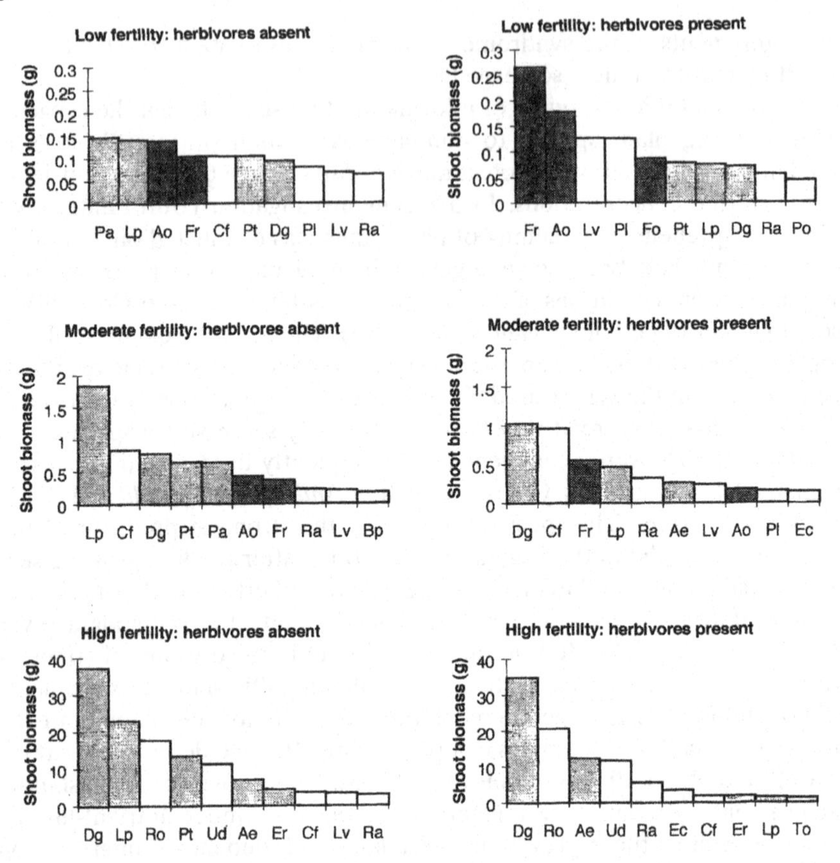

Figure 22. The 10 highest contributors to the shoot biomass in vegetation developed over two growing seasons in closed outdoor microcosms sown with a standardised seed mixture of 48 species of a wide range of plant functional types and provided with three levels of fertility in the presence and absence of slugs and aphids. ■ = slow-growing grasses, ▒ = fast-growing grasses, □ = forbs. The species are: Ae, *Arrhenatherum elatius*; Ao, *Anthoxanthum odoratum*; Cf, *Cerastium fontanum*; Dg, *Dactylis glomerata*; Ec, *Epilobium ciliatum*; Er, *Elytrigia repens*; Fo, *Festuca ovina*; Fr, *Festuca rubra*; Lp, *Lolium perenne*; Lv, *Leucanthemum vulgare*; Po, *Pilosella officinarum*; Pa, *Poa annua*; Pt, *Poa trivialis*; Pl, *Plantago lanceolata*; Ra, *Rumex acetosa*; Ro, *Rumex obtusifolius*; To, *Taraxacum officinale*; Ud, *Urtica dioica*. (Reproduced from Buckland and Grime 2000 by permission of *Oikos*.)

General Features of Competitors

Phenology and Phenotypic Plasticity

From the evidence and arguments which have been presented in this chapter it seems reasonable to conclude that high competitive ability is recognisable as a family of genetic characteristics which permit a high rate of acquisition of

Figure 23. Vegetation synthesised from a mixture of 48 herbaceous species allowed to develop for two years in highly nutrient-deficient conditions in the absence of herbivores in outdoor microcosms. Individuals of *Poa annua* are arrowed

resources in productive, crowded vegetation. Under these conditions, natural selection appears to favour those plants which are best equipped both to tap the surplus of resources above and below ground and to maximise dry matter production. In this respect two competitive characteristics are of particular importance. The first consists of the potential to produce a dense canopy of leaves and a large root surface area during the period in the year when conditions are most favourable to high productivity. The second is the capacity for rapid morphogenetic adjustments both in the apportionment of photosynthate between root and shoot and in the size, morphology, and distribution of individual leaves (Blackman and Wilson 1951a, b; Blackman and Black 1959; Grime and Jeffrey 1965; Smith 1982) and roots (Drew *et al.* 1973; Rincon and Huante 1994; Robinson 1994), a characteristic which involves a high rate of reinvestment of captured resources in growth and in respiration (Poorter *et al.* 1991; van der Werf *et al.* 1993). The effect of such responses, coupled with the rapid turnover of the leaves and roots, is to bring about, during the growing season, a constant readjustment of the spatial distribution of the absorptive surfaces (i.e. the leaf canopy and root surface area) of the plant. The advantage which the competitor appears to derive from this high flexibility is the potential to respond rapidly to changes in the distribution of resources within the habitat. On page 90 this foraging mechanism of the competitor is contrasted with the very different mechanism of resource-capture which appears to be characteristic of the stress-tolerator.

Life-form and Ecology

Until growth analysis experiments and competition studies have been conducted on a wide variety of species from contrasted habitats it may not be possible to recognise with certainty the life-form classes and vegetation types which include species of high competitive ability. However, some tentative conclusions may be drawn with regard to the relationship between competitive ability and life-form. In particular it is evident, from the limited amount of data which is available, that no broad distinction with respect to competitive ability may be drawn between herbaceous and woody species. Perennial herbs, shrubs, and trees each appear to encompass a wide range of competitive abilities. As explained in Chapter 6, this observation is of considerable significance in relation to the process of vegetation succession.

STRESS-TOLERATORS

A Definition of Stress

Dry matter production in vegetation is subject to a variety of environmental constraints, the most frequent of which are related to shortages and excesses in the supply of solar energy, water, and mineral nutrients. Plant species and even different genotypes of the same species may differ in susceptibility to particular forms of stress and, in consequence, each may exercise a different effect upon vegetation composition. Because, over the course of a year, several stresses may operate intermittently in the same habitat, analyses of the impact of stress may become quite complex (Mooney *et al.* 1991).

A further complication arises from the fact that certain stresses either originate from or are intensified by the vegetation itself. Among the most important types of plant-induced stress are those arising from shading and depletion of the levels of mineral nutrients in the soil following their accumulation in the plant biomass. In addition, growth-inhibitors may be released into the soil either by direct secretion or as a result of microbial decay of plant residues.

In order to accommodate its diverse forms, stress will be defined simply as **the external constraints which limit the rate of dry matter production of all or part of the vegetation.**

Stress in Productive and Unproductive Habitats

Identification of the forms of stress characteristic of particular habitats is but one aspect of the analysis which is necessary in order to determine the influence of stress upon vegetation. Another requirement is to estimate the extent to which stress is limiting primary production in different types of vegetation. The need for such studies arises from the fact that the role of stress changes according to the productivity of the environment.

In productive, undisturbed habitats the vegetation is composed of potentially-large, fast-growing plants, and stress arises mainly as a consequence of local or temporary depletion of resources by competitors. In these circumstances, as we have seen earlier (pages 23 and 34), natural selection is likely to favour plants in which exposure to stress, whether plant-induced or imposed directly by the environment, induces rapid morphogenetic responses which tend to maximise the capture of resources and the production of dry matter. However, we may suspect that such responses to stress may have little survival value under conditions in which severe stress in one form or another is a constant feature of the environment.

In chronically-unproductive habitats, therefore, there is little opportunity for the phenology or morphogenetic responses of the plant to provide mechanisms of stress-avoidance and the most conspicuous effect of severe and continuous stress is to eliminate or to debilitate species of high competitive ability and to cause them to be replaced by plants which are capable of tolerating the prevailing forms of stress. The stress-tolerators comprise an extremely diverse assortment of plants which on first inspection appear to be far too varied in life-form and in ecology to be included in the same category. However, the morphological and taxonomic diversity among the stress-tolerators belies the conformity of life-history and physiology whereby these plants are adapted to survive in continuously-unproductive conditions. This is to suggest that although the various types of stress-tolerators differ through the possession of mechanisms adapted to the specific forms of stress operating in their habitats, all exhibit a suite of characteristics necessary for survival in conditions in which the level of production remains consistently low. The recurrence of the same set of traits in very different kinds of unproductive habitats immediately poses the question: 'Is there a common factor operating in these contrasted circumstances or do different environmental stresses have convergent evolutionary effects?' Some of the evidence which can be used to examine this problem will now be reviewed by considering briefly adaptation to severe stress in four contrasted types of habitat. In two of these (arctic–alpine and arid habitats) the plant biomass is small and stress is mainly imposed by the environment. In the third (shaded habitats) stress is plant-induced, whilst in the fourth (nutrient-deficient habitats) stress may be due to the low fertility of the habitat or to sequestration of the mineral nutrients in the vegetation or to a combination of the two.

Tolerance of Severe Stress in Various Types of Habitat

Stress-tolerance in Arctic and Alpine Habitats

Without doubt, the dominant environmental stress in arctic and alpine habitats is low temperature. In these habitats, the opportunity for growth is limited to a short summer season. For the remainder of the year growth is prevented by low temperature and the vegetation is either covered by snow or, where it remains exposed on ridges, is subjected to extreme cold coupled with the desiccating effect of dry winds. During the growing season itself, production is often severely restricted not only by low temperatures but also by desiccation and mineral nutrient stress (Haag 1974, Sonesson and Callaghan 1991), the latter arising largely as a result of the low microbial activity of the soil. Alpine vegetation types are subject to additional stresses peculiar to high altitudes, including strong winds and intense solar radiation.

The adaptive characteristics of the terrestrial plants of arctic and alpine habitats have been examined in excellent reviews by Billings and Mooney (1968), Bliss (1962, 1971, 1985), Savile (1972), Callaghan and Emanuelsson (1985), Chapin and Shaver (1985), Körner and Larcher (1988), and Körner *et al.* (1989), and here comment will be restricted to certain of the most generally-occurring adaptations.

The predominant life-forms in arctic and alpine vegetation are very low-growing evergreen shrubs, small perennial herbs, bryophytes, and lichens. The adaptive significance of the small stature of all of these plants appears to be related, in part, to the observation that during the winter when the ground is frozen and no water is available to the roots the aerial parts of low-growing plants tend to be insulated from desiccation by a covering of snow. However, as suggested by Boysen-Jensen (1932), quite apart from the risk of winter-kill of shoots projecting through the snow cover, the absence of larger plants and, in particular, erect shrubs and trees is also related to the fact that the productivity of tundra and alpine habitats is so low that it is unlikely that there will be a surplus of photosynthate sufficient to sustain either wood production or the annual turnover of dry matter characteristic of deciduous trees and tall herbs.

A conspicuous feature of arctic and alpine floras is the preponderance of evergreen species (Polunin 1948). Annual plants are extremely rare and the majority of bryophytes, lichens herbs, and small shrubs remain green throughout the year. The advantage of the evergreen habitat is particularly clear in long-lived prostrate shrubs. Here, Billings and Mooney (1968) suggest that the main advantage of the evergreen habitat is that it obviates the necessity to spend food resources on a wholly new photosynthetic apparatus each year. In this connection it is interesting to note that Hadley and Bliss (1964) found that evergreen tundra shrubs had lower photosynthetic rates than the deciduous shrub, *Vaccinium uliginosum*, and various herbaceous species. The ability to survive, despite these lower rates, is apparently due to the longer functional life of individual leaves. Hadley and Bliss (1964) suggest that the older leaves may act as winter food storage organs since lipids and proteins are mobilised and translocated from old to new leaves during the growing season. It is also well-established that long-lived arctic perennials store and recycle a high proportion of the mineral nutrients that are mobilised with the tissues each year (Jonasson and Chapin 1991) and have the capacity to absorb mineral nutrients at low soil temperatures (Chapin 1983). Billings and Mooney (1968) note that evergreen shoots in alpine conditions tend to break bud dormancy later than deciduous species exposed to the same conditions and suggest that evergreens can afford the apparent waste of these days of uncertain weather early in the growing season since their older leaves are already in photosynthetic operation.

The importance of the evergreen habit in arctic vegetation has been exemplified by a detailed analysis (Callaghan 1976) of the life-cycle of the sedge, *Carex bigelowii*, in which it has been found that the life-span of individual tillers in arctic populations frequently extends to four years.

It would appear, from the relatively small amount of information which is available, that seed production in arctic and alpine habitats is erratic and hazardous. However, as Billings and Mooney (1968) point out, this is compensated by the fact that the majority of species are long-lived perennials, many of which are capable of vegetative reproduction (see page 140).

Stress-tolerance in Arid Habitats

The comprehensive reviews by authors such as Walter (1973), Slatyer (1967), and Levitt (1975) make clear the pitfalls in any attempt to generalise about the ways in which plants are adapted to exploit conditions of low annual rainfall. However, if attention is confined to those species which, in their natural habitats, experience long periods of desiccating conditions without access to underground reservoirs of soil moisture, common adaptive features are apparent. These xerophytes are perennials in which the vegetative plant is adapted to survive for extended periods during which little water is available. Several types of xerophytes may be distinguished according to the severity of moisture stress which they can survive.

In habitats which experience a short annual wet season the most commonly occurring xerophytes are the sclerophylls (Grieve 1956; Monk 1966). This group includes both small shrubs and trees, such as the evergreen oaks and the olive, all of which are distinguished by the possession of small, hard leaves which are retained throughout the dry season. It has been established (Gates 1968) that under conditions of high radiation load and restriction of transpiration by stomatal closure, small leaves dissipate heat more efficiently than large ones. Perhaps even more important is the suggestion of Walter (1973) that the ecological advantage of sclerophylly is related to the ability of sclerophyllous species to conduct active gaseous exchange in the presence of an adequate water supply but to cut it down radically by shutting the stomata when water is scarce, a mechanism which enables these plants to survive months of drought with neither alteration in plasma hydrature nor reduction in leaf area and, when rains occur, to take up production again immediately.

Under severely desiccating conditions, the most persistent of the xerophytes are the succulents. These are plants in which water is stored in swollen leaves, stems, or roots. During drought periods no water absorption occurs, but following rain, small, short-lived roots may be produced extremely rapidly. Succulents are also distinguished by peculiarities in stomatal physiology and metabolism.

The stomata open at night and remain tightly closed during the day. Gaseous exchange occurs during the period of stomatal opening, at which time carbon dioxide is incorporated into organic acids. These are decarboxylated in the daylight to release carbon dioxide for photosynthesis. This mechanism, known as crassulacean acid-metabolism (Thomas 1949), appears to represent a mechanism whereby low levels of photosynthesis may be maintained with minimal transpiration. A well-known feature of many xerophytes and, in particular, the succulents is the rarity or erratic nature of flowering.

Stress-tolerance in Shaded Habitats

On superficial acquaintance, the stresses associated with shade appear to differ from those considered under the two previous headings in that they are not directly attributable to gross features of climate. However, although shade itself is not imposed by the physical environment it becomes important only in climatic regimes which are conducive to the development of dense canopies. It may be important to bear in mind, therefore, that shade at its greatest intensities frequently coincides (1) with the high temperatures and humidities of tropical and subtropical climates or with the warm summer conditions of temperate regions, and (2) with conditions of mineral nutrient depletion associated with the development of a large root mass (Grubb 1994; Coomes and Grubb 2000). Hence, although the discussion which follows refers to adaptations to shade, there is a strong probability that some of the plant characteristics described are related to co-tolerance of shade, warm temperatures, and mineral nutrient stress.

The intensity of shade experienced near the ground surface depends upon the number of layers of foliage present and upon the light absorbing and reflecting characteristics of the canopy. Although the amount of light intercepted by a dense community of herbaceous species may be comparable with that intercepted by forest (Monsi and Saeki 1953), there is, of course, a major difference with respect to the height of the shaded stratum. Within herbaceous vegetation, the shaded stratum is low and all or part of it is renewed annually by extension of shoots and individual leaves from positions near the ground. In forest, however, the shaded stratum is high and arises by expansion of foliage *in situ*.

In dense herbaceous vegetation, whether in the open or in forest clearings, small differences in height above ground are associated with large changes in intensity, direction, and quality of radiation, and success among the component herbs and tree seedlings may depend to a considerable extent upon the ability to compete for light and to project leaves into the higher light intensities above the herb layer. In contrast, beneath dense tree canopies (and, more especially, those composed of evergreen species), herbaceous vegetation is usually sparse and, in consequence, vertical gradients in light intensity are less pronounced near the forest floor. Here, the ability to compete for light is likely to be secondary in importance to the capacity to tolerate shade conditions. The need to distinguish between competition for light and tolerance of shade becomes clear when an attempt is made to analyse the results of shade experiments.

The effect of shading upon growth-rate and morphogenesis has been examined in a large number of experiments by growing plants, usually as seedlings, under screens of cotton, plastic, or metal (e.g. Burns 1923; Blackman and Rutter 1948; Blackman and Wilson 1951a, b; Bordeau and Laverick 1958; Blackman and Black 1959; Grime and Jeffrey 1965; Björkman and Holmgren 1963, 1966; Loach 1970; Corré 1983; Rincon and Huante 1994). The general conclusions which may be drawn from these experiments are: (1) that, in response to shade, the majority of plants produce less dry matter, retain photosynthate in the shoot at the expense of root growth, develop longer internodes and petioles, and produce larger, thinner leaves; and (2), that species differ very considerably both with respect to the magnitude and to the rate of these various responses. However, when the responses of different ecological groups of plants are compared, a paradox becomes apparent. This arises from the fact that the capacity to maximise dry matter production in shade through modification of the phenotype is most apparent in species characteristic of unshaded or lightly-shaded environments, whilst plants associated with deep shade tend to grow slowly and to show much less pronounced morphogenetic responses to shade treatment. An early examination of this difference is available from the investigations of Loach (1967, 1970), who used screens to compare the shade responses of several North American trees, and a similar approach has been applied recently by Rincon and Huante (1994) to compare shade-tolerant and gap-exploiting trees in tropical deciduous forest in Mexico. The results of these investigations establish quite clearly that morphological plasticity in shade is most apparent in species such as *Liriodendron tulipifera* and *Heliocarpus pallidus* which normally colonise unshaded habitats or woodland clearings, whilst shade-tolerant species, such as *Fagus grandifolia* and *Caesalpinus platyloba*, exhibit a much smaller degree of phenotypic modification. It is reassuring to observe that the same pattern has been found in a recent review (Veneklaas and Poorter 1998) comparing leaf area responsiveness to shade in seedlings of 194 species of tropical trees.

From these results it would appear that the rapidity of phenotypic response and the comparatively high growth rates in shade of species such as *L. tulipifera* are attributes which allow seedlings of these species to compete for light in rapidly-expanding herbaceous vegetation such as that which occurs in forest clearings. The low rates of growth and the small extent of phenotypic response to shading in the shade-tolerant trees suggest that adaptation to shade in these plants may be concerned more with the ability to survive for extended periods in deep shade than with the capacity to maximise light interception and dry matter production. A similar conclusion is prompted by the observation that many of the most shade-tolerant herbaceous plants, e.g. *Pachysandra* spp. and *Deschampsia flexuosa*, have morphologies which allow considerable self-shading.

In a number of comparative experiments (Grime 1965; Loach 1970; Grime and Hunt 1975; Rincon and Huante 1994) shade-tolerant herbs and tree seedlings have been found to exhibit consistently slow relative growth-rates under conditions in which plants characteristic of productive habitats grew extremely rapidly. The results of shade-screen experiments with tree seedlings (Loach 1970) and with the shade-tolerant grass, *Deschampsia flexuosa* (Mahmoud and Grime 1974), indicate that these low relative growth-rates are genetically determined in that they are maintained at both high and low light intensities.

Measurements in the dark on leaves from a variety of species (Bordeau and Laverick 1958; Björkman and Holmgren 1963, 1966; Grime 1965; Loach 1970; Taylor and Pearcy 1976) suggest that shade-tolerant species have comparatively low respiratory rates and that these tend to remain low in shade--grown plants. Experiments by Woods and Turner (1971) showed that stomatal responses to changes in light intensity in a range of North American trees are consistently more rapid in shade-tolerant species. This observation suggests that rapid opening of stomata may allow shaded leaves to exploit brief periods of illumination due to sun flecks (Tinoco-Ojanguren and Pearcy 1992). In a review of the long-term effects of sunflecks on forest understorey plants Pearcy *et al.* (1994) concluded that their utilisation was critically important to the carbon balance of these species and could account for up to 60% of carbon gain.

An additional mechanism whereby carbon and energy may be conserved in shade-tolerant plants was detected through the early work of Björkman (1968a, b), who demonstrated that the lower photosynthetic rates of shade-adapted ecotypes of the herbaceous perennial, *Solidago virgaurea*, were related to the lower Rubisco content of the leaves. This evidence prompted Mooney (1972) to observe that since a major portion of the leaf protein is Rubisco, there can be a considerable conservation of carbon by making less of this enzyme in light-limited habitats, where it would be of little advantage. Consistent with this hypothesis investigations by Taylor and Pearcy (1976) showed that in a number North American shade-tolerant herbs, including *Trillium grandiflorum*, *Podophyllum peltatum*, and *Solidago flexicaulis*, shoots developed under shaded conditions exhibited marked reductions in carboxylating capacity whereas species such as *Erythronium americanum* and *Allium tricoccum*, which are vernal shade-intolerant species, showed no evidence of such flexibility.

Among the trees, climbing plants, and epiphytes which occupy the shaded stratum of tropical forests, evergreen species predominate. In temperate woodlands also, the most shade-tolerant herbaceous plants tend to remain green throughout the year (Kubicek and Brechtl 1970; Hughes 1975). In the British Isles, for example, deciduous woodlands contain a variety of slow-growing evergreen species such as *Lamiastrum galeobdolon*, *Milium effusum*, *Deschampsia flexuosa*, and *Veronica montana.* A characteristic of many shade-tolerant species is the paucity of flowering and seed production under heavily-shaded conditions. This phenomenon is particularly obvious in British woodlands where flowering in many common shade plants such as *Hedera helix*, *Lonicera periclymenum*, and *Rubus fruticosus* agg. is usually restricted to plants exposed to sunlight at the margins of woods or beneath gaps in the tree canopy.

Stress-tolerance in Nutrient-deficient Habitats

Under this heading attention will be confined to habitats in which mineral nutrient stress arises from the impoverished nature of the habitat. This is because it is in these habitats that mechanisms of adaptation to mineral nutrient stress have been most closely studied. However, it is vital to the arguments developed in Chapter 8 to recognise that severe stress may also arise under conditions in which mineral nutrients are sequestered in the living or non-living parts of the biomass.

Since the incursion of experimental methods into agricultural and ecological research, considerable effort has been devoted to the task of identifying the stress factors which cause particular soils to be infertile. The types of naturally-occurring infertile soils which have been intensively studied include those which are acidic, calcareous, or derived from serpentine rock. Industrial spoils have been investigated also and particular attention has been paid to coal mine waste and spoil contaminated by heavy metals such as lead, copper, and zinc.

In the majority of habitats which have been examined, low soil fertility has been found to be associated with several different forms of stress operating simultaneously or in seasonal succession (Pate 1993). The complexity possible in an analysis of the causes of soil infertility may be illustrated by reference to the list (Table 4) composed by Hewitt (1952) of the stresses which may contribute to low productivity on highly acidic soils. A similar degree of complexity is suspected in the mechanism inhibiting vegetation development on serpentine soils. Walker (1954), for example, states that, in order to survive in serpentine habitats, plants must be tolerant not only of low calcium levels but also of one or more of the following: high concentrations of chromium and nickel, high magnesium, low levels of major nutrients, low availability of molybdenum, drought, and other undesirable aspects of shallow stony ground.

Before any progress can be made in an attempt to generalise about the way in which vegetation is attuned to conditions of low soil fertility it is necessary to distinguish between those soil characteristics which are common to most

Table 4. List of the main factors contributing to low productivity on highly acidic soils. (Reproduced from Hewitt 1952 by permission of International Society of Soil Science.)

(1) Direct injury by hydrogen ions (low pH)

(2) Indirect effects of low pH
  (a) Physiologically-impaired absorption of calcium, magnesium, and phosphorus
  (b) Increased solubility, to toxic extent, of aluminium, manganese, and possibly iron and heavy metals
  (c) Reduced availability of phosphorus partly by interaction with aluminium or iron, possibly after absorption
  (d) Reduced availability of molybdenum

(3) Low base status
  (a) Calcium deficiency
  (b) Deficiencies of magnesium, potassium, or possibly sodium

(4) Abnormal biotic factors
  (a) Impaired nitrogen cycle and nitrogen fixation
  (b) Impaired mycorrhizal activity
  (c) Increased attack by certain soil pathogens

(5) Accumulation of soil organic acids or other toxic compounds due to unfavourable oxidation-reduction conditions

infertile habitats and those which are peculiar to specific types of soils. When a survey is made of the extensive literature concerning various forms of soil infertility, the most consistent components of the 'infertility complex' are major nutrient deficiencies and, in particular, those of phosphorus and nitrogen. Accordingly, the remainder of this section is concerned with an examination of the adaptations of vegetation to conditions of major nutrient deficiency.

When the range of vegetation types characteristic of severely nutrient-deficient soils are surveyed, certain common features are immediately apparent. Although the identity of the species involved varies in different parts of the world and according to local factors such as soil type and vegetation management, similar plant morphologies may be recognised. The herbaceous component shows a marked reduction in growth form. Among the grasses, narrow-leaved tussock forms predominate and a high proportion of the dicotyledons are creeping or rosette species. Under management conditions which allow the development of woody vegetation, nutrient-deficient soils are usually colonised by small (often coniferous) trees or sclerophyllous shrubs (Mason 1946; Gardner and Bradshaw 1954; McMillan 1956; Monk 1966; van Steenis 1972; Vogl 1973). In Europe and North America, members of the Ericaceae such as *Calluna vulgaris*, *Vaccinium myrtillus*, and *Erica cinerea* are particularly common on highly acidic soils, whilst on shallow nutrient-deficient calcareous soils species such as *Helianthemum nummularium* and *Thymus polytrichus* occur. The xeromorphy evident in all these shrubs could be interpreted to be an adaptation to winter or summer desiccation. However, this explanation cannot be applied in the case of the xerophytic shrubs which are known to occur on infertile soils in tropical and subtropical regions. In New Caledonia, for example, Birrel and Wright (1945) described xerophyllous shrub vegetation varying between 1–2 m in height on serpentine soil and commented upon its 'unusual appearance in this region of high rainfall in which tropical forest is the ordinary plant cover'.

The foliage of plants associated with nutrient-poor soils is usually characterised by low concentrations of mineral nutrient elements (Pearsall 1950; Band and Grime 1981; Field and Mooney 1986; Coley 1988; Reich *et al.* 1991; Foulds 1993; Thompson *et al.* 1997). The most likely explanation for this phenomenon is the large investment in cell wall material and in other carbon-based structural components (Coley 1988; Reich *et al.* 1991) which appear to be involved in the defence of potentially long-lived leaves against herbivores.

Another feature which is characteristic of the vegetation of nutrient-deficient soils is the high frequency of species of inherently slow growth-rate. Among the first to recognise this was Kruckeberg (1954), who noted that, even when grown on fertile soils, certain ecotypes of herbaceous plants adapted to survive on serpentine soils in North America grew slowly in comparison with species and ecotypes from fertile habitats. Another important early contribution was that of Beadle (1954, 1962), who recognised that slow growth-rates were characteristic of species of *Eucalyptus* growing on Australian soils with low phosphorus availability. Slow rates of growth were also found to occur in *Festuca ovina* and *Nardus stricta*, two grasses of widespread occurrence on infertile soils in the British Isles (Bradshaw *et al.* 1964), and Jowett (1964) concluded from his investigation with the grass, *Agrostis capillaris*, that the effect of natural selection under conditions of severe mineral nutrient-deficiency had been to reduce the potential growth-rate of local populations established on mine-waste in Wales. Subsequent investigations (e.g. Hackett 1965; Clarkson 1967; Higgs and James 1969; Grime and Hunt 1975) have confirmed that there is a strong correlation between low potential growth-rate and tolerance of mineral nutrient deficiencies.

It would appear, therefore, that adaptation for survival on infertile soils has involved, in both woody and herbaceous species, reductions in stature, in leaf form, and in potential growth-rate. The explanation which has been put forward, with varying degrees of elaboration, by a number of authors (e.g. Kruckeberg 1954; Loveless 1961; Jowett 1964) to account for this phenomenon is that it is primarily an adaptation for survival under conditions of low mineral nutrient supply. This is to suggest that, under conditions in which elements such as phosphorus and nitrogen are scarcely available, natural selection has led to the evolution of plant species and ecotypes which grow slowly and are capable of surviving in conditions where external concentrations of nutrients are exceedingly low. Consistent with this hypothesis are the results of a number of experiments in which species from infertile habitats have been grown under various levels of supply of major mineral nutrients (Bradshaw *et al.* 1964; Hackett 1965; Clarkson 1967). From these studies there is no convincing evidence that species normally restricted to infertile soils are better adapted than species from fertile habitats to maintain dry matter production under conditions in which mineral nutrients are provided at low rates of supply. In Figure 24, for example, the results of an experiment by Bradshaw *et al.* (1964) illustrate clearly that, under low levels of nitrate nitrogen supply, *Festuca ovina* and *Nardus stricta*, both grasses of infertile pastures, are outyielded to a considerable extent by *Lolium perenne* and *Cynosurus cristatus*, species which are normally restricted to fertile soils. The pattern observed in flowering plants is also evident in bryophytes. In a standardised screening programme involving species of widely-contrasted ecology (Furness and Grime 1982b) low rates of growth were characteristic of mosses exploiting bare rock surfaces and infertile soils.

If plants which are restricted to infertile soils are not differentially capable of capturing mineral nutrients from low external concentrations it is pertinent

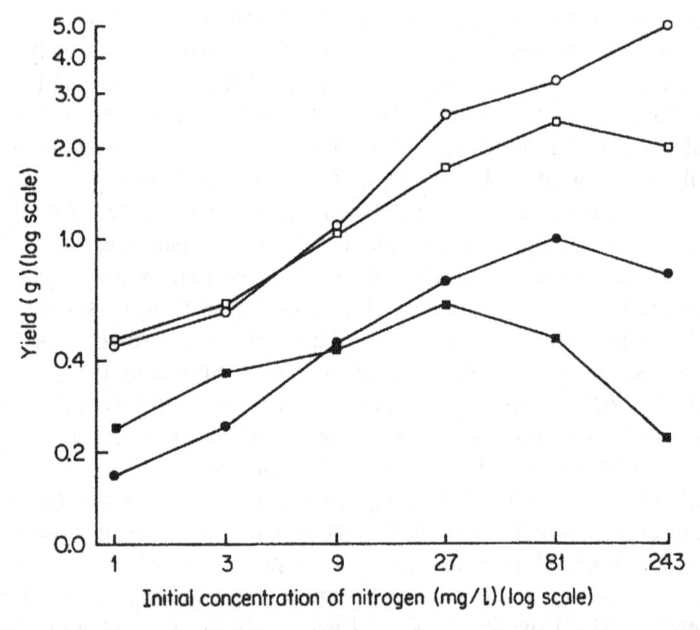

Figure 24. The yield of four grasses grown for eight weeks in unreplenished sand cultures provided with various initial concentrations of nitrogen. Two of the species, *Lolium perenne* (o-o) and *Cynosurus cristatus* (□-□), are characteristic of fertile pastures, whilst the remaining pair, *Festuca ovina* (●-●) and *Nardus stricta* (■-■), are species of infertile grassland. (Reproduced from Bradshaw *et al.* 1964 by permission of Blackwell Scientific Publications Ltd.)

to ask how they derive a selective advantage under such conditions. In seeking an answer to this question it is necessary to recognise that for long periods in many infertile soils mineral nutrients may be sequestered in the plant and microbial biomass and become available for only brief periods following climatic events such as drought or frost (Shields *et al.* 1973; Taylor *et al.* 1982; Marimoto *et al.* 1982a, b). There is insufficient time for rapid exploitation of such brief enrichments to be achieved by growing new roots capable of high specific absorption rates. Instead an advantage may be expected for those plants which maintain a viable if relatively inactive root system throughout the year. Campbell and Grime (1989a) tested this hypothesis by comparing the yield and nitrogen capture of *Festuca ovina* and *Arrhenatherum elatius* grown under conditions in which opportunities for nutrient capture were restricted to pulses of enrichment varying in duration. The results (Figure 25) confirmed that whereas *A. elatius*, the species usually associated with fertile soils, was superior in yield and nitrogen capture when the duration of pulses exceeded 10 hours, below this threshold the advantage switched to *F. ovina*, a species which exploits infertile conditions, and is capable of maintaining a viable root system under severely nutrient-limited conditions.

Further evidence of the ability of slow-growing plants of unproductive vegetation to capture nutrients from short pulses is available from field experiments in which isotopic enrichments have been applied to infertile soils. A particularly good example is that of Jonasson and Chapin (1991), in which it was shown that following an application of ^{32}P to a tundra sedge community there was incorporation into the roots and shoots of the tussock species, *Eriophorum vaginatum*, followed by a rapid decline to negligible rates as the added phosphorus became fixed in the soil and unavailable to the plants.

An analogous but mechanistically-distinct form of pulse interception in slow-growing plants of nutrient-impoverished habitats is exhibited by species that capture and digest arthropods. Investigations of *Drosera* species in south-western Australia (Dixon *et al.* 1980; Schulze *et al.* 1991; Karlson and Pate 1992) have demonstrated that carnivorous species are strongly dependent upon animal prey as a source of both nitrogen and phosphorus.

An advantage which a low rate of dry matter production confers upon a plant growing on an infertile soil is that, during periods of the year when mineral nutrients are more readily available, uptake is likely to exceed the rate of utilisation in growth, allowing reserves to accumulate. An example of this 'luxury uptake' which presumably may benefit the plant during subsequent periods of nutrient shortage is provided in Figure 26, which is taken from an experiment by Clarkson (1967) and shows the concentration of phosphorus accumulated in the plant tissues by two species of *Agrostis* grown at various levels of phosphorus supply. The data show that in *A. curtisii*, a slow-growing species of infertile acidic grasslands, there was a progressive accumulation of phosphorus within the plant as the concentration of the element increased in the external medium. In contrast, *A. stolonifera*, because it

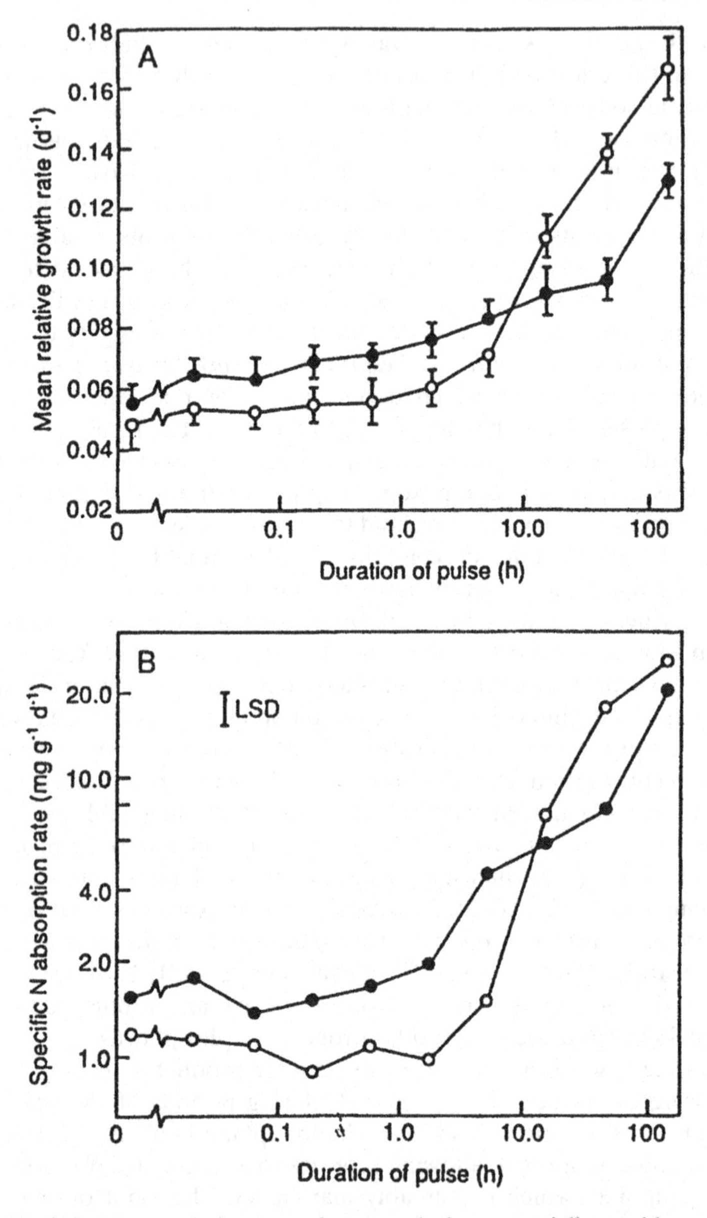

Figure 25. (A) Mean relative growth rate of the potentially rapid-growing *Arrhenatherum elatius* (○) and slow-growing *Festuca ovina* (●) plants exposed once every six days to pulses of nutrient enrichment of differing durations. Vertical bar in (B) is L.S.D. ($P < 0.05$) for comparing means on logarithmic scale. Means in (A) are shown ±95% confidence limits. (B) Mean specific nitrogen absorption rate of *A. elatius* (○) and *F. ovina* (●) plants exposed once every six days to pulses of nutrient enrichment of differing duration. Vertical bar is L.S.D. ($P < 0.05$) for comparing means on logarithmic scale. (Reproduced from Campbell and Grime 1989a by permission of *New Phytologist*.)

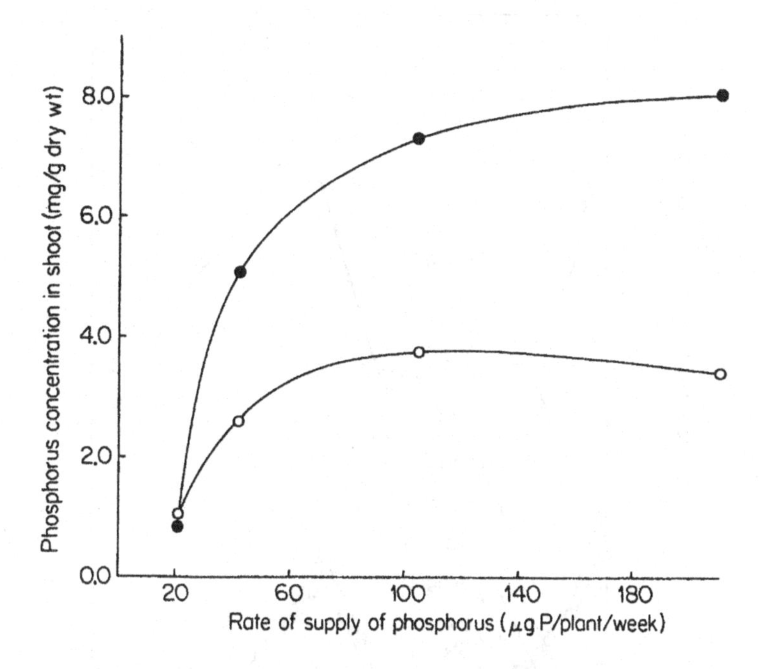

Figure 26. Measurements of the concentration of phosphorus in two perennial grasses grown for a period of six weeks at various rates of phosphorus supply. One species, *Agrostis stolonifera* (○-○), is associated with soils of high fertility whilst the other, *A. curtisii* (●-●), is restricted to infertile acidic soils. (Reproduced from Clarkson 1967 by permission of Blackwell Scientific Publications Ltd.)

responded to increasing levels of phosphorus by growing more rapidly, did not accumulate phosphorus to a level surplus to the growth requirements of the plant.

Whilst the capacity of slow-growing plants of infertile soils to accumulate reserves of nutrient elements may be of considerable survival value in the natural habitats of these species, there is evidence which suggests that this characteristic may be disadvantageous when these same plants colonise more fertile soils. In a number of experiments (Bradshaw *et al.* 1964; Jeffries and Willis 1964; Ingestad 1973; Jones 1974) in which plants associated with nutrient-deficient habitats were supplied with high rates of mineral nutrients, elements appear to have been accumulated in quantities which were detrimental to the growth of the plants.

An additional feature of many of the slow-growing species characteristic of infertile soils is the lack of a sharply-defined seasonal variation in shoot biomass (Figure 27). The majority of both the woody and the herbaceous species characteristic of infertile habitats are evergreen plants in which the leaves have a comparatively long life-span. Measurements by Williamson (1976), for example, on grasses of calcareous pastures in southern England have shown

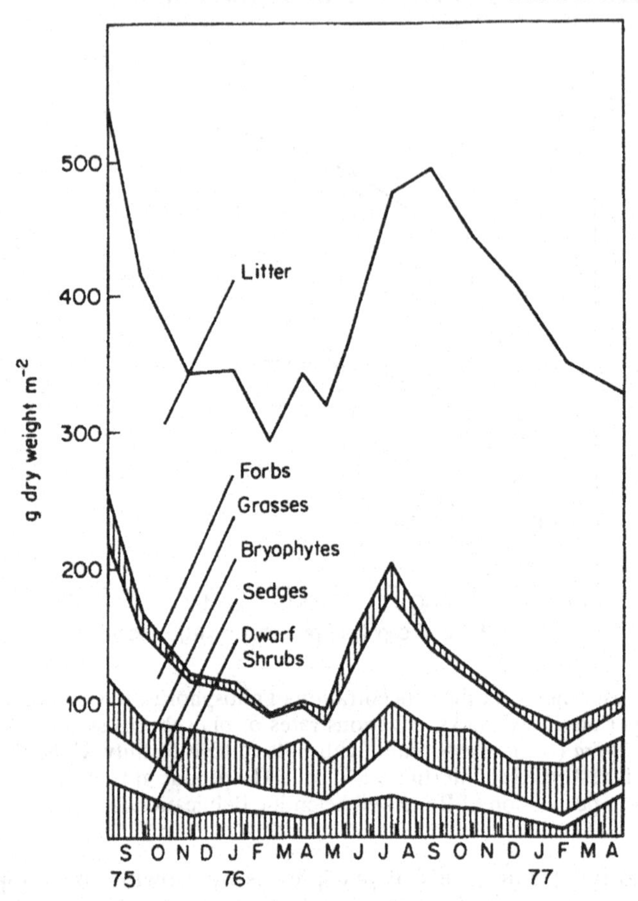

Figure 27. Seasonal variation in the composition of the standing crop in an unfertilised limestone pasture in northern England. A feature of the figure is the lack of pronounced seasonal changes in the contribution made by any of the five life-form classes, a consequence of the high frequency of slow-growing evergreen species. (Reproduced by permisson of Furness 1978.)

that in the evergreen species of nutrient-deficient soils, *Helictotrichon pratense* and *H. pubescens*, the functional life of the leaf is considerably longer than in *Arrhenatherum elatius* and *Dactylis glomerata* species which are usually associated with more fertile soil conditions and show a well-defined summer peak in shoot biomass. A consequence of the greater longevity and slower replacement of leaves in plants of infertile habitats is a reduction in the rate of nutrient cycling between plant and soil and a smaller risk of loss of mineral nutrients either by leaching or by incorporation into other organisms exploiting the habitat (Monk 1966, 1971; Thomas and Grigal 1976). This observation is consistent with the general conclusion that plants of infertile habitats are adapted to conserve mineral nutrients rather than to maximise the rate of capture. Consistent with this interpretation there is also evidence that some plants exploiting infertile soils are better able to withdraw nutrient elements from senescing leaves (Pate and Dell 1984; Aerts 1996).

Stress-tolerance in Urban Habitats

In areas of high human population density disturbance of vegetation is inevitable and in Chapter 6 the predictable functional shifts that arise from man's activities are examined in some detail and confirm that it is in stress-tolerators that we see the most severe losses of species as a consequence of industrialisation, urbanisation, and intensive agriculture. On this basis we might suppose that when we examine the centres of cities and towns, the impact of man would exclude completely long-lived, slow-growing plant species. However, as Gilbert (1984) recognised, a consistent effect of urban building is rapidly to enlarge the range of skeletal habitats that hitherto had been represented mainly by rock outcrops and cliffs by creating a large surface area of hard materials such as stonework, brickwork, concrete, and cinders. In industrial cities colonisation of these habitats is restricted by air pollution but careful searching usually reveals local populations of bryophytes, small ferns, and perennial herbs such as those illustrated in Figure 28.

General Features of Stress-tolerance

From the preceding survey and from other reviews of the traits common to plants of low-resource environments (Levitt 1956; Chapin 1991; Grubb 1998) it is apparent that vascular plants exploiting various types of chronically-unproductive habitats are comparatively long-lived (Tamm 1956, 1972; Currey 1965; Inghe and Tamm 1985) and exhibit a range of features which, although varying in detailed mechanism, represent basically similar adaptations for endurance of such unfavourable conditions. These features, which are listed in Table 6, include inherently slow rates of growth, the evergreen habit, long-lived organs, sequestration, and slow turnover of carbon, mineral nutrients, and water, infrequent flowering, and the presence of mechanisms which allow the intake of resources during temporarily favourable conditions. The latter consist not only of the presence throughout the year of functional leaves and, probably, roots also (Jeffrey 1967; Chapin and Bloom 1976; Thomas and Grigal 1976; Aerts 1996; Eissenstat and Yanai 1997), but, in addition, special mechanisms such as the rapid activating of stomata in scelerophylls and shade plants and the rapid sprouting of roots by succulents.

A feature consistently associated with the stress-tolerator is low morphogenetic plasticity. In terms of the growth physiology of these plants this characteristic is not difficult to understand. Growth in stress-tolerators occurs

Figure 28. Stress-tolerators of urban environments. (a) *Sedum acre*, (b) *Asplenium ruta-muraria*, (c) *Polypodium vulgare*, and (d) graveyard mosses

intermittently, and for most of the time, therefore, differentiating (i.e. potentially-responsive) tissue forms a very small proportion of the biomass.

In stress-tolerators the most important responses to environmental variation are physiological rather than morphogenetic. It is now apparent that the long functional life of individual shoots and roots in many stress-tolerators is characterised by seasonal or short-term changes which maintain their viability and functional efficiency under changing environmental conditions. Reference has been made already (page 54) to evidence of the capacity of some woodland plants to modify the size of the carboxylating system in response to shade conditions. Changes in the photosynthetic apparatus are also apparent in the temperature acclimation and seasonal variation in efficiency of carbon assimilation described for vascular plants in desert and temperate localities (Mooney and West 1964; Strain and Chase 1966; Mooney and Shropshire 1967; Mooney and Harrison 1970) and in arctic, alpine, and subarctic species (Billings *et al.* 1971; Grace and Woolhouse 1970; Billings and Godfrey 1968). Physiological changes corresponding to fluctuations in seasonal conditions have been reported also for temperate forest species (Cannell and Sheppard 1982), bryophytes (Hosakawa *et al.* 1964; Miyata and Hosakawa 1961; Oechel and Collins 1973; Hicklenton and Oechel 1976), and arctic lichens (Larsen and Kershaw 1975). It is interesting to note that acclimation to temperature may be extremely rapid in certain species. Mooney and Shropshire (1967) have shown that in coastal populations of the Californian desert shrub, *Encelia californica*, changes in photosynthetic response to temperature may be induced within a period of 24 hours, and a similar phenomenon has been described (Hicklenton and Oechel 1976) in the subarctic moss, *Dicranum fuscescens*. As in the case of the competitive strategy, no simple generalisation can be made with regard to the stature and life-form of stress-tolerant plants. Although, in extremely unproductive environments stress-tolerance is associated with trees, shrubs, and herbs of reduced stature, account must be taken of the fact that many of the shade-tolerant trees in temperate and tropical habitats are long-lived and attain a comparatively large size at maturity. This wide diversity in size and growth form in the stress-tolerators is illustrated in Figure 29.

Stress-tolerance and Mineral Nutrition: a Unifying Hypothesis

In the preceding survey of vegetation characteristics in cold, dry, shaded, and nutrient-limited habitats the assembled evidence supports the conclusion that, in all of these very different conditions, low productivity coincides with a set of plant traits including long life-span, slow turnover of tissues, defence against herbivory, and several others listed in Table 6 (page 89). The occurrence of such close similarities in plants occupying habitats that are so different, apart from the common feature of low productivity, raises a fundamental issue concerning the selection processes promoting stress-tolerance. This issue can be clarified into two alternative hypotheses:

1 The sets of traits associated with stress-tolerance and conferring fitness at low productivity are selected regardless of the environmental factors constraining production.

Figure 29. Structural, taxonomic, and ecological variety in stress-tolerators. (a) Alpine lichens, bryophytes, and rosette plants, Mount Washington, USA. (b) The epiphyte, *Tillandsia recurvata*, in the Pedrigal, Mexico City. (c) Monoculture of *Deschampsia flexuosa*, on a podzolic railway embankment at Arkwright Town, North Derbyshire, England. (d) Sclerophyllous shrubs (including mulberries and olives), Delphi, Greece. (e) The palm, *Chamaedorea tepejilote*, in the understorey of primary rainforest, Los Tuxtlas, Mexico. (f) Remnant of the Caledonian Forest dominated by *Pinus sylvestris* and heather (*Calluna vulgaris*), Sutherland, Scotland. (g) Arborescent cacti in dry tropical forest, Chamela, Mexico

2 The sets of traits associated with stress-tolerance and conferring fitness at low productivity are selected only in circumstances where productivity is constrained by the same underlying limiting factor.

When we examine the merits and weaknesses of these two hypotheses it is helpful to rid ourselves of anthropomorphic attitudes to the ecology and evolution of native plants. In particular it may be helpful to set aside agricultural concepts of plant productivity (e.g. Liebig 1840) that have reviewed limiting factors within the strict confines of our efforts to maximise returns from short-lived crop species exploiting relatively-productive habitats. For the purpose of agriculture it has been analytically correct and rewarding to recognise the various stresses operating within the growing season as additive and, at least in certain respects, functionally-equivalent as depressors of yield. If the control of crop yield by various stresses operating in arable systems provided a good model of the ecological and evolutionary consequences of the operation of multiple stresses in more natural habitats it would be necessary to accept hypothesis 1. However, crop plants, their environments, and their associated research literature are of limited value as we seek to identify the selection factors responsible for the widespread recurrence of the stress-tolerant syndrome of traits. In the absence of the disturbance and fertiliser applications of cultivated systems, perennial life-histories prevail and stresses interact in a complex way to fashion the potential niches for different life-histories and physiologies.

In order to test hypotheses 1 and 2 it is useful to ask the question: Can single stresses such as low temperature, aridity, shade, low mineral nutrients *by themselves* promote vegetation with stress-tolerant traits? In an attempt to answer this question it is necessary to return to the earlier review of stress-tolerance in cold, arid, and shaded habitats (pages 49–63). Whilst reaffirming that the set of traits associated with stress-tolerance recurs widely in these three habitat-types, it is necessary to recognise that within each type, other primary plant strategies (R or C) occur locally in circumstances of mineral nutrient enrichment. In arctic habitats, fast-growing trees, shrubs, and herbs are not uncommon on alluvial river terraces or in disturbed, eutrophic conditions associated with human settlement; this provides clear evidence that low temperatures and shortgrowing seasons do not by themselves confer a decisive advantage upon stress-tolerant traits. Low temperatures have played a major role in determining the low microbial activity and infertility of the majority of soils of polar and alpine areas of the world and in this sense they provide the ultimate explanation for the prevalence of the stress-tolerance syndrome in these regions. However, it would appear (Jonasson 1989) that the proximal determinant of their ecology and productivity is likely to be low soil fertility.

A similar situation can be proposed for many deserts and dryland areas of unproductive grassland and scrub, where low rainfall and soil infertility coincide. Although the dominant plant strategy in these regions is that of the stress-tolerator it is most revealing to observe the shift to ruderals (Pate and Dell 1984) or competitors in local circumstances (fire-prone areas, wadis, drainage ditches, oases, human settlements, roadsides, ant-hills) where mineral nutrient enrichment has taken place. Again, as in the cold regions, we can identify climatic factors (in this case, low rainfall and high temperatures) as the ultimate causes of low soil fertility and, hence, stress-tolerance, but the immediate selective factor promoting stress-tolerance can be recognised as mineral nutrient stress.

The suggestion that mineral nutrient status determines whether dryland habitats are occupied by stress-tolerant perennials or ephemerals is likely to remain controversial for some time. As the following quotation illustrates, there is a strong predisposition to explain the ecology of desert vegetation exclusively by reference to water supply.

> Water is the environmental parameter most affecting plant life in warm deserts. Growth, productivity and phenological activity are tightly linked to the brief periods in which adequate soil moisture is available. In response to these selective pressures, warm desert plants have evolved a diversity of physiological, morphological and life-cycle adaptations that allow plants to maximise net carbon gain during periods of high soil moisture availability, to enhance carbon gain during early drought, and to maximise survival through extended drought periods.
>
> Ehleringer (1985)

The importance of water supply in controlling opportunities for resource capture and growth in dry habitats is unquestionable and it is perhaps for this reason that little work has been conducted to examine the consequences of mineral nutrient applications on the relative success of different plant functional types in arid zone habitats. However, it is well-established that deserts are extremely patchy with respect to mineral nutrient supply with both plants and animals providing mechanisms that concentrate nutrients locally. There is now an urgent requirement for experimental manipulations of mineral nutrient supply in dryland ecosystems such as that conducted by James and Jurinak (1978) to test the validity of the null hypothesis that the ecology of annuals and perennials in warm deserts can be explained without reference to mineral nutrient supply.

Recently an attempt has been made (Cunningham *et al.* 1999) to determine whether the set of traits associated with stress-tolerance can arise in circumstances where rainfall but not mineral nutrients are limiting. The method was to examine leaf structure and chemistry in perennial species distributed along natural gradients in rainfall and total soil phosphorus status in New South Wales, Australia. The results showed a strong convergence in that, with some minor differences in chemical composition and defensive structure, similar sets of stress-tolerant traits were observed under low rainfall and on soils of low total phosphorus. However, it may by unwise to regard this interesting investigation as a definitive proof that moisture stress alone is capable of driving plant evolution along the path towards the stress-tolerance syndrome. First, it is necessary to recognise that restricted nutrient uptake is unavoidable in perennials exploiting habitats in low rainfall conditions (Dunham and Nye 1976; Mackay and Barber 1985). Secondly, a bias in favour of the convergence hypothesis may have been introduced by restricting the comparison to perennial species; it can be argued that successful exploitation of habitats combining high soil fertility with low rainfall rests almost exclusively with ephemeral species.

A further parallel is apparent in the distribution of plant functional types in shaded habitats. Long-lived, slow-growing herbs are of widespread occurrence throughout the world beneath evergreen and deciduous tree canopies but their abundance declines sharply in circumstances of high soil fertility. In cool temperate regions this phenomenon is particularly obvious during the spring in deciduous woodlands (Al-Mufti *et al.* 1977). On infertile soils this vernal phase is not characterised by major phenological events; the spring period simply coincides with flowering and modest expansions of the biomass of the evergreen stress-tolerators that dominate the herbaceous layer. A very different sequence of events occurs in neighbouring areas of deciduous forest on fertile soils; here the vernal phase commences with an eruption of ephemerals and ephemeroids and, following closure of the tree canopy, terminates with an abrupt decline and senescence of the herbaceous canopy. This remarkable contrast in the species composition and seasonal dynamics of the herbaceous layer indicates a controlling effect of soil fertility on the strategic composition beneath deciduous forest. Although stress-tolerators of the woodland herb layer are distinguished by physiological traits that confer shade-tolerance (page 52), it would appear that tolerance of mineral nutrient stress is fundamental to their ecology. The dependence of these plants upon infertile soils is related to the fact that where the constraint of mineral nutrient stress is removed the vernal phase becomes a 'window of opportunity' exploitable by more productive, mineral nutrient-demanding species.

Further evidence of the importance of low soil fertility in determining the presence of stress-tolerant herbs beneath forest canopies is available from field experiments. Grubb (1994) and Coomes and Grubb (2000) have assembled the literature relating to experiments in which nutrient enrichment of the herb layer on infertile woodland soils has been brought about by trenching procedures that sever the feeder roots of trees. These experiments confirm that in a wide range of circumstances, and particularly on infertile or seasonally-dry soils, the productivity of herbaceous vegetation in woodland herb layers is frequently constrained by mineral nutrient limitations rather than shade; in root-trenching experiments conducted in shaded habitats, Coomes and Grubb (2000) reported positive responses in the ground flora in 40 out of 47 seven cases.

Experimental proofs are still lacking but the weight of evidence from field investigations in unproductive ecosystems points to acceptance of hypothesis 2. Tentatively, we may conclude therefore that although a large number of stresses, most notably low temperatures, moisture stress, and shade, are capable of inhibiting plant growth, these factors alone do not provide the conditions and selection processes responsible for the evolution and widespread success of the stress-tolerant strategy. Underlying the very wide range of circumstances where a great diversity of stress-tolerant life-forms have evolved, a common factor can be identified in the form of mineral nutrient stress. Why has this played such a pervasive and distinctive role in the evolution of the stress-tolerant syndrome of plant traits? Two main factors must be taken into account in seeking the answer:

1 With the exception of fertilised croplands, horticultural areas, and gardens, most habitats contain finite quantities of available nutrients. In many parts of the world this contrasts strongly with the continuous refreshment of supplies of solar energy and water. Even where the stocks of mineral nutrients in the ecosystem are large it is common for their availability to be low because elements such as nitrogen and phosphorus are sequestered in the plant biomass and in the litter and surface layers of the soil. In both mature forest and old grassland ecosystems it is not unusual for mineral nutrients to be tightly held, recycled, and recaptured by the dominant plants and soil organisms with consequences for both annual production and resistance to invasion by new plant individuals.
2 As explained above, at a superficial level we can attribute the success of stress-tolerators to the widespread occurrence of mineral nutrient limitations in ecosystems. However, at an ecophysiological level it is also necessary to explain why the particular set of plant traits associated with stress-tolerance recurs with such universal fidelity as an evolutionary solution to conditions of severe mineral nutrient stress. In particular, we must ask why limitations imposed by other factors do not appear to evoke the same adaptive response. The answer to this question almost certainly lies in the distinctive nature of mineral nutrient limitation and release. In natural ecosystems on infertile soils mineral nutrients tend to be released in brief mineralisation pulses that are rapidly captured by the plant roots and micro-organisms that live in intimate relationship with the sites of temporary enrichment. In these circumstances, as we have seen from the result of foraging experiments (page 59), effective mineral nutrient capture, retention, and recapture is more likely where the long-lived root system of a long-lived plant remains in place throughout the year. Could the same opportunistic strategy of pulse interception by long-lived structures facilitate the capture of solar energy and water where these resources, rather than mineral nutrients, are limiting productivity? Undoubtedly, such mechanisms exist in the activating of stomata in sunflecks (page 54) and in

desert succulents (page 52). Here the problem is to determine whether these mechanisms have evolved directly in response to shade and water stress or are a secondary consequence of the adoption of stress-tolerant traits imposed by chronic mineral nutrient stress.

Sophisticated, comparative, multivariate experiments with carefully-selected species will be required to resolve these issues but even in our present state of knowledge there are indications that the syndrome of stress-tolerant traits is invariably associated with mineral nutrient-limited conditions. In circumstances where mineral nutrients are not severely limiting there is a tendency for evolutionary specialisations to low rainfall and shade to follow paths of evasion rather than tolerance. Investigations of the ecology of woodland annual plants such as *Galeopsis tetrahit* (Slavikova 1958) and *Impatiens parviflora* (Peace and Grubb 1982) reveal the extent to which under shade conditions selection may favour an ephemeral life-history provided mineral nutrient supplies remain sufficient for rapid exploitation of short periods when the intensity of shade or drought is relaxed.

Stress-tolerance and Palatability

The possibility that the reduced stature and slow growth-rates of stress-tolerant plants may cause them to be particularly vulnerable to physical damage has been recognised by several authors (e.g. Whittaker 1975; Grime 1979; Reader and Southwood 1981; Southwood *et al.* 1986; Coley 1983, 1988; Coley *et al.* 1985). Plants growing under conditions of severe stress are likely to exhibit slow rates of recovery from defoliation by predators, and during their long phase of establishment they will be particularly susceptible to the activity of herbivores. To species which exist for an extended period as small slow-growing plants, grazing, even by small invertebrates, presents a major threat to survival. It may be predicted, therefore, that many stress-tolerant plants will have experienced intensive natural selection for resistance to predation. Chabot and Hicks expressed the same ideas succinctly by focussing in particular on the penalties likely to follow herbivore damage to the potentially long-lived leaves of many slow-growing species:

> Herbivory should always decrease the ability of a leaf to gain carbon and thus increase the amount of time necessary for the start-up costs to be paid back. The effect may be differential, with greater impact on the always apparent leaves of evergreens. The costs arise in three ways: (a) both the tissue and the future productivity of the leaf are lost; (b) the low nutrient content and high proportion of support tissue found in evergreen leaves may in part make the leaves less palatable to herbivores but they also decrease photosynthetic rates; and (c) chemical protection by secondary compounds presumably necessitates diversion of carbon from the other functions and is metabolically expensive.
>
> Chabot and Hicks (1982)

A comprehensive analysis of the frequency of unpalatable plants in stressed environments is not yet possible because of the shortage of reliable data and because it is often difficult to allow for the fact that palatability is a variable attribute depending upon the characteristics of plant and herbivore and the circumstances in which they interact (Tribe 1950). Despite these complications, however, it is possible to refer to a large amount of circumstantial evidence (see Levin 1971; Janzen 1973; Feeny 1975; Rhoades and Cates 1976; Cooper-Driver 1985; Choong *et al.* 1992) which supports the hypothesis that in habitats subject to severe environmental stress there is a general decline in palatability.

In unproductive habitats, both physical and chemical deterrents to herbivory are conspicuous. These include the spines of many cacti and succulents, the hard or leathery texture of the foliage of many arid-zone shrubs and shade-plants, and the coarsely siliceous or needle-like leaves of many of the perennial herbs present in arid, arctic–alpine, or nutrient-deficient habitats. Evidence of low palatability and low rates of consumption by mammals is available for desert scrub communities (Chew and Chew 1970), whilst the limited scope for herbivory in montane rain-forests has been commented upon by Leigh (1975), who describes the leaves of the dominant trees as 'built to last, and filled with poisons to keep off hungry insects'. In both temperate and tropical environments gastropods often provide a major hazard to survival, particularly at the seedling stage, and it is interesting to note that for slugs (Cates and Orians 1975), snails (Grime *et al.* 1968, 1996), crickets (Grime *et al.* 1996), and aphids (Fraser and Grime 1999a) there is evidence that the palatability of vascular plants from unproductive or late-successional vegetation is lower than that of ruderal species.

With respect to both vertebrate and invertebrate herbivores circumstances have been described (e.g. Nicholson *et al.* 1970; Feeny 1968, 1969, 1970; Rhoades and Cates 1976) in which foliage is vulnerable to attack only during the phase of leaf expansion. It would seem quite possible, therefore, that the slow turnover of leaves in many stress-tolerators may contribute to the mechanisms resisting predation.

Examples of chemical defence against herbivores (e.g. Harris 1960; Smith 1966; Rhoades 1976) are evident in the strongly aromatic or resinous compounds which, either as leaf constituents or when released into the atmosphere, deter predators in a range of herbaceous vegetation types including the chaparral of North America, the garigue of Southern Europe, and the mulga of Central Australia. As several authors such as Levin (1971), Mooney (1972), and Whittaker (1975) have suggested, synthesis of chemical deterrents to predation or microbial attack, in some species, accounts for a considerable fraction of the photosynthate. The possibility may be recognised that defence expenditure involving either physical or chemical mechanisms is a factor contributing to the low potential growth-rates of some stress-tolerant plants.

Defence against invertebrate herbivores also arises in some grasses of infertile soils such as species of *Danthonia* through the presence of fungal endophytes in the above-ground tissues (Clay 1990; Clay *et al.* 1995).

It is interesting to note that many lichens are not subject to severe predation. In this connection, it may be significant that the biochemistry of lichens is characterised by the production of a wide range of compounds (described collectively as the 'lichen substances') of no known metabolic function.

From a wide range of sources, including Davis (1928), Muller and Muller (1956), Naveh (1961), and Del Moral and Muller (1969), there have been reports of phytotoxic effects of chemicals originating from the living parts or from the litter of plants adapted to stressed environments. Whilst these phenomena deserve consideration as evidence of allelopathic mechanisms (see page 195), there is also the possibility that such effects could arise from the release and persistence of compounds which have evolved primarily as a defence against predation.

Stress-tolerance and Decomposition

An additional significance has been attached to the low palatability of many stress tolerators by the hypothesis (Grime and Anderson 1986) that the physical defences which protect the living foliage often remain operational after senescence, retarding the breakdown of litter by decomposing organisms. Recently this hypothesis has been tested by comparing the rates of decomposition of naturally-shed leaves from a large number of British and Argentine trees, shrubs, and herbs of widely-contrasted ecology (Cornelissen *et al.* 1996, 1997, 1999). The results (Figure 30) confirm that rates of decomposition are relatively slow in the litter of slow-growing evergreens of unproductive vegetation.

These results are consistent with the high organic content of the surface layers of many infertile soils in cool temperate regions (Kubiena 1953) and may also explain the deep accumulations of litter under many slow-growing trees, shrubs, and herbs (e.g. *Quercus petrea*, *Fagus sylvatica*, *Rhododendron ponticum*, *Calluna vulgaris*, *Brachypodium pinnatum*).

Stress-tolerance and Carbon Isotope Fractionation

Since the pioneering studies of Nier and Gulbransen (1939) and Murphy and Nier (1941) it has been known that in comparison with inorganic carbonaceous materials, plant tissues are slightly depleted in the heavy isotope of carbon, ^{13}C and show interspecific variation in this respect. Subsequent work has established that the lowered concentrations of ^{13}C arise, first, from the slower rate of diffusion of the heavier isotopic species during its penetration into the leaf and to the site of photosynthesis, and, second (and more importantly) from isotopic fractionation during the carboxylation process itself.

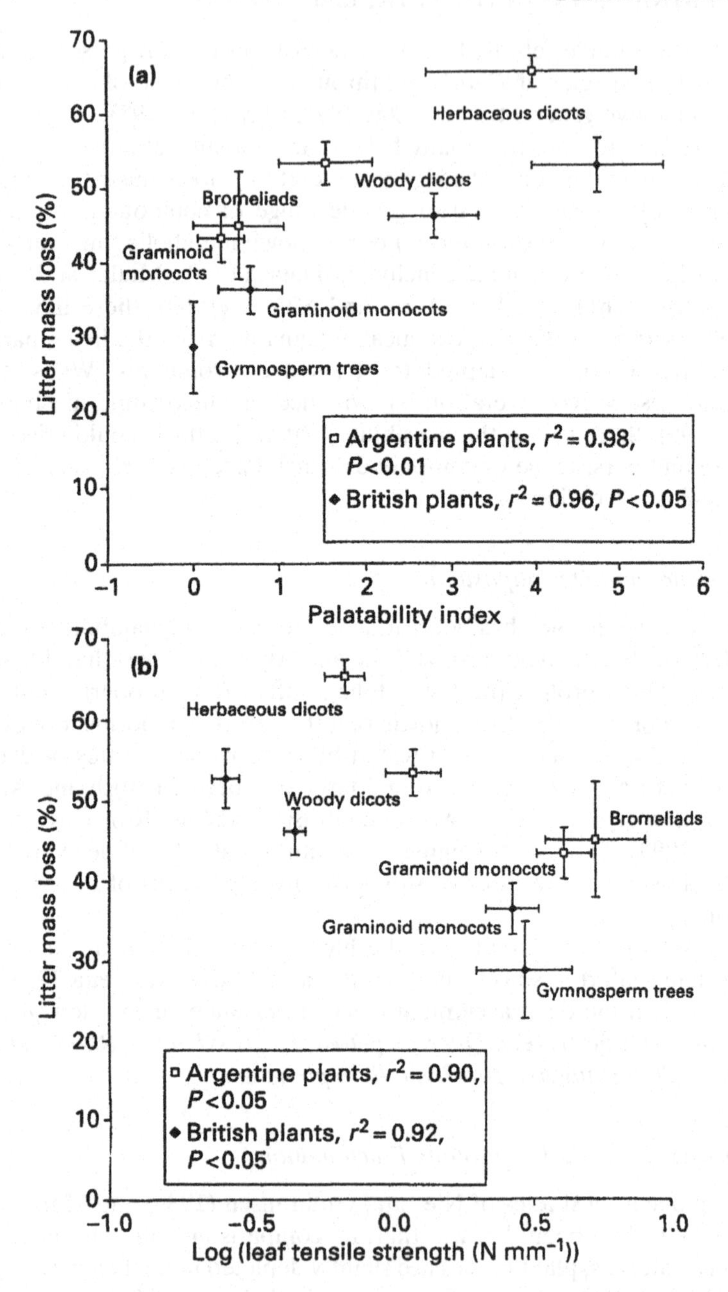

Figure 30. Percentage leaf litter mass loss of different species assemblages in terms of life-form and higher taxonomy as a function of traits of fresh leaves: (a) leaf palatablility index; (b) leaf tensile strength. Linear regressions on the means (SE bars shown) were performed separately for the Argentine and British data sets. (Reproduced from Cornelissen *et al.* 1999 by permission of *New Phytologist*.)

Comparative investigations of the extent of ^{13}C depletion in plants have tended to focus on the differences between C_3 and C_4 species (e.g. Troughton 1972; Vogel 1980) and the use of the $\delta^{13}C$ values (relative deviation of the $^{13}C/^{12}C$ ratio from that in a reference standard expressed in parts per thousand) as a predictor of water use efficiency (dry matter produced per unit of transpiration) in crop plants and forage species. However, there is an additional possibility that $\delta^{13}C$ values may vary in accordance with plant functional type. Comparative studies reported by Ehleringer (1993) have revealed that in deserts there is a clear tendency for short-lived plants to have carbon isotope values that are lighter (i.e. more depleted in ^{13}C) than those in longer-lived species. It is possible that these differences reflect rooting depths and leaf water relations but the recent observation that similar differences associated with life-history and leaf life-span occur in the cool, damp conditions of northern England (Figure 31) suggests that differences may also be related to the variations in leaf life-span and chemistry that coincide with gradients in habitat productivity and plant functional type (pages 101–105).

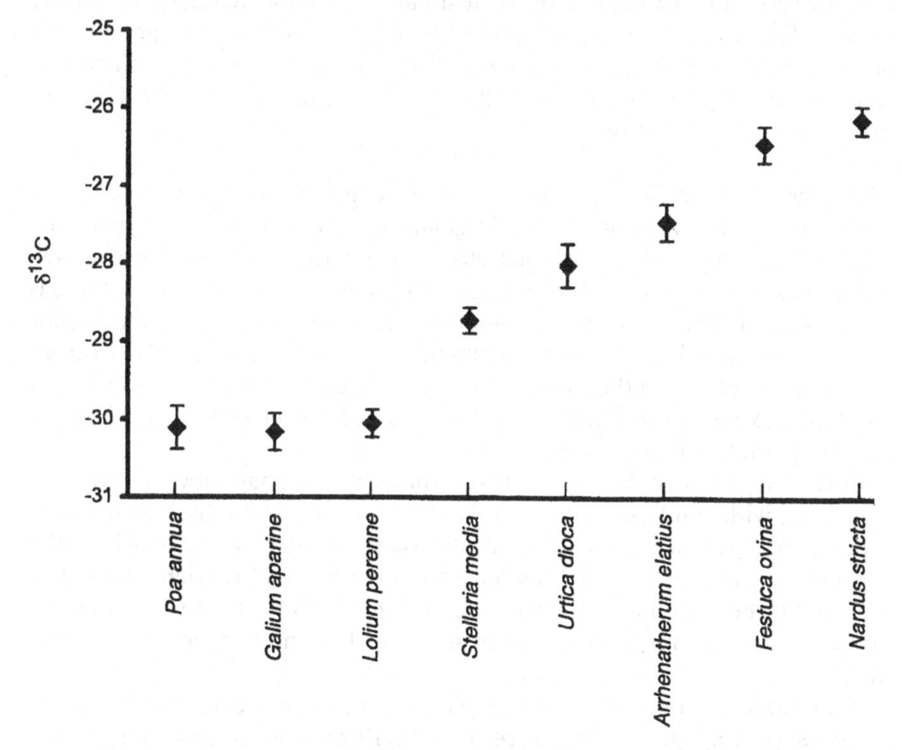

Figure 31. The relationship between carbon isotope discrimination ($\delta^{13}C$) and leaf life-span in eight common herbaceous plants sampled widely throughout North Central England. The species have been ranked in order of increasing life-span. The vertical lines are standard errors. (Reproduced by permisson of Grime and Handley.)

Stress-tolerance and Symbiosis

Lichens. Four features which, under the last heading, have been associated with stress-tolerance in flowering plants (slow growth-rate, longevity, opportunism, and physiological acclimation) are expressed in a most extreme form in lichens. This is perhaps hardly surprising in view of their ecology. Lichens are able to survive in extremely harsh environments under conditions in which vascular plants may be totally excluded and in which they experience extremes of temperature and moisture supply and are subject to low availability in mineral nutrients. Although there is some diversity of opinion with regard to the longevity of lichens (see, for example, the review by Billings and Mooney 1968) there is general agreement that many are exceedingly long-lived. From experimental studies, such as those of Farrar (1976a, b, c), there is abundant evidence of the tendency of lichens to sequester a high proportion of the photosynthate rather than to expend it in growth. Much of the assimilate appears to be stored as sugar-alcohols (polyols) in the fungal component. Lichens are able both to remain alive during prolonged periods of desiccation and, on rewetting, to resume nutrient uptake and photosynthesis extremely rapidly. The studies of Kershaw and his colleagues on Canadian populations of *Peltigera canina* and *P. polydactyla* (Kershaw 1977a, b; MacFarlane and Kershaw 1977) provide an ample illustration of the capacity of lichens for rapid seasonal acclimation.

Mycorrhizas. Interesting parallels may be explored between lichens and other types of symbiosis involving fungi and root systems. It is now generally accepted that mycorrhizal associations of one form or another are represented in all but the most fertile and waterlogged conditions. However, there is considerable variation in structure and function between and within the major types of mycorrhizas (vesicular-arbuscular mycorrhizas, ectomycorrhizas, ericoid mycorrhizas, and orchidaceous mycorrhizas) and caution must be applied in seeking to establish any general framework linking mycorrhizas to primary plant functional types.

Review of the distribution of mycorrhizas shows that they are strongly associated with conditions of mineral nutrient-stress. These include not only vegetation types such as heathland and sclerophyllous scrub in which the vegetation density is severely limited by the low mineral nutrient content of the habitat, but also mature temperate and tropical forests in which mineral nutrients' scarcity in the soil arises from the scale of nutrient absorption into the plant biomass.

A considerable amount of physiological work has been carried out in order to assess the ecological significance of mycorrhizal associations, and the results have been reviewed extensively (Harley 1969, 1970, 1971; Smith and Read 1997). It has been clearly established that the uptake of mineral nutrients such as phosphorus and nitrogen may be facilitated, and the yield of the host plant increased, by the presence of mycorrhizal infections. Mycorrhizas are abundant near the soil surface and it seems likely that they enable nutrient elements mineralised during the decay of litter to be efficiently re-absorbed into the plants. In the case of ecto- and ericoid mycorrhizas there is accumulating evidence that the fungi are able to obtain mineral nutrients from complex organic molecules within plant litter (e.g. Nemeth *et al.* 1987; Finlay *et al.* 1992; Leake 1994). This suggests that involvement in these types of mycorrhizal association can be regarded as a stress-tolerant trait in the sense that a consequence of the association is an extrapolation of the internal recycling and retention mechanism to include the litter component.

In comparison with the roots of crop plants and species restricted to fertile soils, many mycorrhizal roots appear to be long-lived, and the possibility may be considered that this attribute is itself a considerable advantage to a host plant growing under conditions of severe mineral nutrient stress. In competitive herbs adapted to exploit soils in which there is a large reservoir of available mineral nutrients, high rates of uptake are attained, firstly through the production of a very large absorptive surface composed of fine roots and root-hairs and, secondly, through continuous and rapid morphogenetic responses in root : shoot ratio and in the extension growth of individual parts of the root system. There is, however, continuous decay and replacement of roots, a process which is analogous to the rapid turnover of leaves which occurs simultaneously above ground in the same species. It is clear that this system of nutrient absorption, whilst effective in absorbing mineral nutrients, even from infertile soils, is achieved at the cost of a high expenditure of photosynthate, and high rate of reinvestment and risk of loss of captured mineral nutrients (Gordon and Jackson 2000). The main benefit of a mycorrhizal association to a host plant growing under conditions of mineral nutrient stress may therefore lie in the provision of an absorptive system which, because it remains functional over a long time and under varying conditions, allows exploitation of temporary periods of increased mineral nutrient availability, at relatively low cost to the synthetic resources of the plant.

Phosphorus scavengers. In the preceding section it has been emphasised that mycorrhizal associations function as a vital component of the stress-tolerant syndrome in many plant species that exploit infertile soils. Phosphorus capture and recapture is particularly important in ancient sandplains and calcareous systems where this element is frequently the controlling factor in plant productivity (Beadle 1954; Lloyd and Pigott 1967). There is now a large volume of experimental research implicating mycorrhizal associations in the phosphorus economy of both herbaceous and woody species and this linkage is particularly evident in vesicular-arbuscular mycorrhizas (Sanders *et al.* 1977; Smith *et al.* 1994).

It is interesting to note that mycorrhizas are not exclusively successful as a solution to plant survival on phosphorus-deficient soils. Plant communities exploiting conditions of extreme phosphorus stress frequently contain mixtures of mycorrhizal and non-mycorrhizal species, in which many of the latter are characterised by the presence of what appear to be uninfected cluster-roots that mobilise phosphorus by secreting organic acids (Davies *et al.* 1973; Lamont 1984; Bowen and Pate 1991); see Figure 32. From experiments such as those of Jalili (1991) and Booth and Grime (2001b; Figure 120, page 333) there is strong evidence of the capacity of the cluster-roots to stimulate yield in phosphorus-limited conditions. There is now a need to identify the factors that determine the relative abundance of mycorrhizal species and cluster-root species in plant communities.

Nitrogen fixers. In the mycorrhizal species and in phosphorus-scavengers (whether mycorrhizal or not) it is relatively easy to incorporate discussion within the framework of the stress-tolerator strategy. A more difficult task is

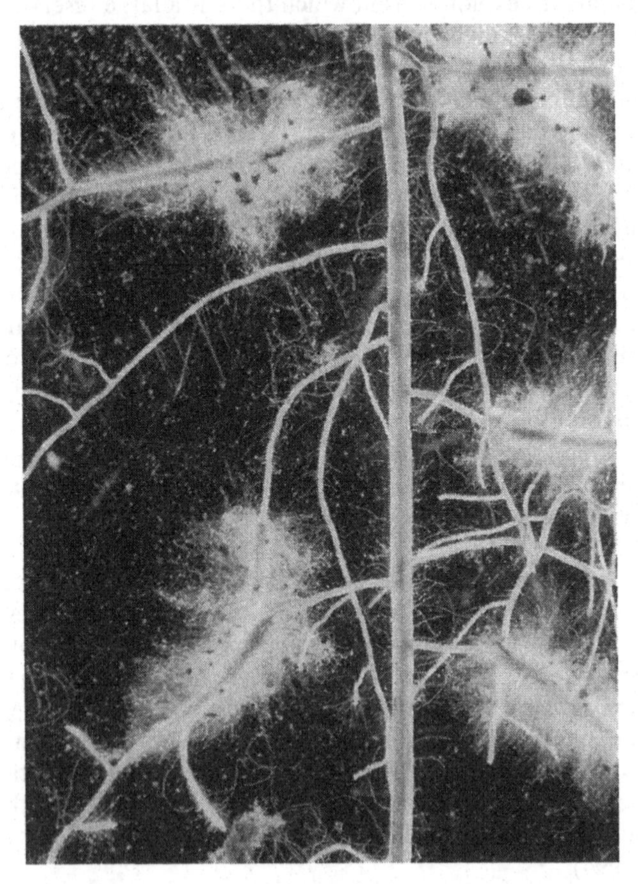

Figure 32. The dauciform cluster-roots of *Carex flacca*. (Reproduced by permission of S Ballard.).

encountered when attention is turned to the very large number of plants that by association with nitrogen-fixing organisms capture nitrogen from the atmosphere. Whereas phosphorus-scavengers are mainly involved in determining the 'market-share' of individuals and species exploiting a finite small pool of resource, the nitrogen-fixers are tapping a very large external source and are capable of rapidly enlarging the pool size (Crocker and Major 1955; Jeffries *et al.* 1981). This fact has profound implications for the role of nitrogen-fixing symbioses in ecosystem productivity and as controller of vegetation succession (page 240). For the same reason it is clear that only a restricted subset of nitrogen-fixers (lichens, small herbs, and shrubs) are stress-tolerators. In particular, many of the leguminous herbs and shrubs exploit moderately-productive ecosystems in which nitrogen is the major limiting factor and nitrogen-fixing species are under strong selection pressures to match in growth-rate and competitive ability co-existing non-fixers that respond to the rising nitrogen status of the soil (Leith 1960; Willis 1963). It is interesting to note that some plant genera of extremely impoverished soils in Australia (Macrozamia, Casuarina, and legumes of the Fabaceae and Mimosaceae) have root systems that combine nitrogen-fixation with phosphorus-scavenging by cluster-roots or mycorrhizas. It is not surprising that this formidable combination of traits allows such species to achieve dominance of vegetation in post-fire conditions in which phosphorus release coincides with nitrogen impoverishment (Monk *et al.* 1981; Hansen and Pate 1987).

Stress-tolerance and Facilitation

Lichens and mycorrhizas appear to represent adaptive solutions to the problems posed by combinations of stresses and threats to survival in unproductive habitats. If such organic connections between two or more distinct organisms can confer fitness upon the participants it would not be surprising to find other examples where individual or mutual benefits from close physical associations arise on an *ad hoc* basis. Where plants are exposed to extreme stress such as in desiccated open habitats it is frequently observed that seedling establishment is promoted under the canopies of established vegetation (Franco and Nobel 1989; Carlsson and Callaghan 1991; Aguiar and Sala 1994). The mechanisms implicated in these various examples of facilitation include reduced water stress in the shade of 'nurse' plants, avoidance of heat stress, and protection against herbivory. However, as recognised by the authors of these studies and many commentators (e.g. Callaway and Walker 1997; Holmgren *et al.* 1997; Brooker and Callaghan 1998), facilitation frequently coincides or alternates with negative effects of competition for resources and in consequence fitness benefits can only be assessed as the net balance of counterveiling forces measured over the life-history of each individual. Facilitation occurs widely but is more frequently documented in unproductive habitats and appears to attain its greatest significance in the earliest stages of primary succession (page 239). The pattern emerging from the recent explosion of interest in facilitation is neatly summarised by Tielborger and Kadmon (2000) as follows:

> There is now a broad consensus that the balance between negative and positive interactions should shift along environmental gradients, with competition prevailing under environmentally benign conditions and positive interactions dominating under harsh conditions.

RUDERALS

A Definition of Disturbance

Reference has been made in the preceding section to habitats in which the density of living and dead plant material remains low because production is severely restricted by environmental stress. It is quite clear, however, that low vegetation densities are not confined to unproductive habitats. Some of the world's most productive terrestrial environments, including many arable fields and pastures, are characterised by a rather sparse vegetation cover. Here the low densities arise from the fact that the vegetation is subject to partial or total destruction.

Although they are more obvious in particular habitats such as arable land, the effects of damage upon vegetation are ubiquitous. The amount of vegetation and the ratio of living to dead plant material in any habitat at any point in time depend upon the balance obtaining between the processes of production and destruction. Even within one environment there may be considerable variety in the mechanisms which bring about the destruction of living or dead vegetation components. In addition to natural catastrophies (e.g. floods and windstorms) and the more drastic forms of human impact (e.g. ploughing, mowing, trampling, and burning), account must be taken of more subtle effects such as those due to climatic fluctuations and the activities of herbivores, decomposing organisms, and pathogens. For the purpose of analysing the primary mechanism of vegetation, therefore, a term is required which encompasses this wide range of phenomena yet is capable of a simple definition. The term which will be used here is *disturbance*, which may be said to consist of **the mechanisms which limit the plant biomass by causing its partial or total destruction**.

Forms of disturbance differ with respect to their selectivity. Whilst the effects of herbivores and decomposing organisms tend to be restricted to the living or the dead material, respectively, certain phenomena, e.g. fire, may affect both components of the vegetation. In general, there is an inverse correlation between degree of selectivity and intensity of disturbance, a relationship which may be exemplified at one extreme by molecular discrimination between litter constituents by microbial decomposing organisms and at the other by the total vegetation destruction associated with phenomena such as severe soil erosion.

With respect to the intensity of damage experienced by the vegetation, a continuous range of plant habitats may be recognised. One end of the spectrum is represented by relatively undisturbed habitats such as mature temperate forests in which the loss of plant material is a more or less continuous process associated with a comparatively low rate of predation and with senescence and the decomposition of litter (Odum 1969). At the other extreme there are habitats such as arable fields in which, at frequent intervals, a large proportion of the plant biomass is summarily destroyed. Among the forms of disturbance which affect living components, a distinction may be drawn between mechanisms which involve the immediate removal of plant structures from the habitat (e.g. grazing, mowing) and those in which plant material is killed but remains *in situ* (e.g. frost, drought, application of herbicides). In the latter, destruction proceeds in two stages. An initial rapid loss of solutes is followed by a rather longer phase in which the residue of plant structures is attacked by decomposing organisms.

Vegetation Disturbance by Climatic Fluctuations

Some the most important forms of severe disturbance leading to the development of ruderal vegetation are effects of climate. Since climatic factors such as low temperature and low rainfall have been associated already with the quite different phenomenon of stress (pages 49–52), it is necessary to draw a distinction between the circumstance in which the main effect of an unfavourable climate is merely to reduce productivity and that in which the influence is disruptive. Whether factors such as low rainfall result in disturbance or in stress depends not so much upon their severity as upon the constancy of their occurrence. Vegetation disturbance by climate is prevalent where conditions encouraging the establishment (but not the uninterrupted growth) of competitors alternate abruptly with seasons imposing severe stress. In these circumstances neither competitors nor stress-tolerators can gain a secure advantage, and natural selection favours potentially fast-growing ephemeral plants. Ecologists are unlikely to challenge the suggestion that marked fluctuations in water supply and temperature can act as agents of vegetation disturbance. Perhaps rather more controversial is the assertion (Grime 1979) that in temperate woodlands on fertile soils the seasonal reduction in light intensity resulting from the expansion of a deciduous tree canopy functions primarily as an agent of disturbance to the underlying populations of herbs, shrubs, and tree seedlings by inducing respiratory losses and pathogen attack. Here an important piece of evidence is the established fact (e.g. Scurfield 1953; Al-Mufti *et al.* 1977) that in deciduous woodlands on very infertile soils no marked floristic changes or major disturbance coincides with canopy expansion, and the ground floras tend to consist of stable communities of slow-growing evergreen herbs. This observation suggests that the onset of shading by deciduous trees has a disruptive effect on the herb layer only in systems which are sufficiently fertile to permit the establishment of shade-intolerant competitive or ruderal plants in the light phase during the early spring. On infertile soils, the conditions remain unfavourable to these plants throughout the year, and the effect of canopy expansion in the spring is merely to add the limiting effect of shade to that of mineral nutrient stress.

Adaptation to Frequent and Severe Disturbance in Various Types of Habitat

Just as the impact of stress upon vegetation changes according to its intensity (page 49), so also does that of disturbance. As described on page 29, in potentially productive environments, low intensities of disturbance function as a modifier to competition by favouring those species which tend to maintain their competitive ability either by avoidance of damage or by rapid recovery from its effects. However, when we turn to productive habitats exposed to repeated and severe disturbance it is apparent that competitors are excluded and that a quite different strategy, that of the ruderal, prevails.

In order to recognise the basic characteristics of ruderal plants, the vegetation in six types of habitat subject to regular and severe disturbance will now be examined.

Ruderals of the Sea-shore

Where sea coasts and inlets are composed of unstable sand or shingle, the effect of wave action and daily inundation is to prevent the establishment of vascular plants on the lower parts of the beach. However, at a position just above the normal upper limit of the tide—the so-called drift-line—a sparse vegetation commonly occurs. On the coasts of Britain, the most characteristic species of the drift-line are *Salsola kali*, *Cakile maritima*, and *Tripleurospermum maritimum*, all three of which are annual plants and are apparently resistant to salt-spray. Drift-line species are often rooted in organic debris, including the decaying remains of seaweed, and in this habitat they are frequently associated with potentially fast-growing annual plants such as *Galium aparine*, *Stellaria media*, and *Senecio vulgaris*, which are familiar in a variety of other disturbed habitats.

The drift-line vegetation is subject to frequent disturbance at high tides and during storms, and the colonising species suffer high rates of mortality but an outstanding adaptive feature appears to be the ability of the survivors to grow and to produce seeds rapidly during the relatively short intervals between disturbances.

Ruderals of Marshland

In areas marginal to open water such as the sides of rivers, lakes, reservoirs, ponds, and ditches, considerable areas of bare mud and silt may become available for colonisation when the water level falls during dry periods. These substrata are usually moist and rich in available mineral nutrients and often support extremely rapid plant growth (van Dobben 1967). As in the case of the drift-line, the period available for growth may be relatively short and it is therefore no surprise to find that the species which exploit disturbed marshland are annuals of high potential growth-rate. This point may be illustrated by reference to Table 5, which includes estimates of the potential maximum relative growth-rates measured in populations of a number of annual species which are frequent colonists of bare mud and silt in the British Isles.

Table 5. Estimates of dry matter production in the seedling phase in six ruderal plants compared to those attained by two crop plants grown under the same conditions. Measurements were conducted over the period 2–5 weeks after germination in a controlled environment. (Temperature: 20°C day, 15°C night; daylength: 18 h; visible radiation: 38.0 W m^{-2}; root medium: sand + Hewitt nutrient solution.) Rates of dry matter production are expressed as maximum (R_{max}) and mean (R) relative growth rates and as a function of leaf area (E_{max}). (Reproduced from Grime and Hunt 1975 by permission of *Journal of Ecology*.)

Species	Seed weight (mg)	R_{max} $week^{-1}$	R $week^{-1}$	E_{max} g m^{-2} $week^{-1}$
Ruderals				
Poa annua	0.26	2.7	1.7	120
Polygonum aviculare	1.5	1.4	1.4	134
Persicaria maculosa	2.1	1.3	1.3	177
Stellaria media	0.35	2.4	2.1	101
Veronica persica	0.52	1.7	1.3	75
Rorippa islandica	0.07	2.3	16	60
Crop plants				
Lycopersicon esculentum (Tomato)	3.1	1.6	1.0	64
Hordeum vulgare (Barley)	37.8	1.6	0.9	64

Arable Weeds

In relation to the time-scale for the evolution of modern flowering plants, agriculture and the attendant expansion in the populations of arable weeds are comparatively recent events (Godwin 1956). It seems reasonable to conclude therefore that, prior to the advent of agriculture, the majority of arable weeds had already evolved in habitats subject to more 'natural' forms of disturbance. This conclusion is supported by the fact that many arable weeds are of common occurrence in disturbed habitats in marshes and on the seashore.

The majority of arable weeds are annuals of high potential growth-rate (Table 5) and under favourable conditions each plant is usually capable of producing a very large number of seeds (Salisbury 1942), many of which tend to become buried and remain dormant in the soil for long periods. The significance of these and other characteristics of the reproductive biology of annual weeds is examined in Chapter 3.

Ruderals of Trampled Ground

Another group of habitats in which plants have evolved in relation to severe and persistent disturbance comprises the grasslands exploited by herds of wild or domesticated grazing animals. Particularly where damage by animals coincides with seasonal effects of drought, pastures may be composed almost exclusively of annual grasses, e.g. *Poa annua* and *Bromus hordeaceus*, and annual legumes, e.g. *Trifolium dubium* and *Medicago lupulina.* Conditions favourable to ruderals arise in grassland not only from excessive trampling by the larger mammals including man, but also as a result of local effects such as uprooting of plants, scorching by urine, and burrowing and scraping by smaller animals. In farmland the most severe damage arises on trackways, and here again the most successful species are annual plants, many of which (e.g. *Poa annua* and *Polygonum aviculare*) are represented in trampled habitats by prostrate genotypes.

Desert Annuals

In arid climates the vegetation is usually composed almost exclusively of stress-tolerant perennials (page 51). However, in areas where non-saline and relatively fertile soils are exposed to a wet season, circumstances favourable to rapid growth may occur for a relatively short period. Under such conditions competitive species are excluded by drought, but annual plants may contribute significantly to the flora. Numerous studies of the ecology of desert annuals (e.g. Went 1948, 1949; Koller 1969; Aronson *et al.* 1992) have drawn attention to the ability of these plants to persist as dormant seeds in the dry season and to germinate, grow, and to produce seeds extremely rapidly during periods in which moisture is temporarily available.

Fire Ephemerals

In many ecosystems where fires are of frequent occurrence a distinctive component of the flora consists of ephemerals that germinate and briefly exploit the post-fire conditions. The life-histories and reproductive biology of fire ephemerals have been well-documented in the sandplains of south-west Australia (Pate *et al.* 1985). These studies have shown that these short-lived plants are distinguished by extremely high relative growth rates, high foliar concentrations of mineral nutrients, and marked plasticity in size and seed output. There is a clear implication from the investigations of Pate *et al.* (1985) that the dominant factor controlling the ecology and abundance of these ephemerals is the temporary release of limiting nutrients in the immediate aftermath of fire.

General Features of Ruderals

Life-cycle

It would appear that flowering plants adapted to persistent and severe disturbance have several features in common. The most consistent among these is the tendency for the life-cycle to be that of the annual or short-lived perennial, a specialisation clearly adapted to exploit environments intermittently favourable for rapid plant growth. A related characteristic of many ruderals is the capacity for high rates of dry matter production (Baker 1965; Grime and Hunt 1975; Table 5), a feature which appears to facilitate rapid completion of the life-cycle and maximises seed production.

In many ruderals flowering commences at a very early stage of development. The process of seed ripening may be extremely rapid and it is not uncommon in genera such as *Polygonum*, *Atriplex*, and *Chenopodium* for flowers and ripe seed to occur in the same inflorescence. Such features appear to be fully consistent with the habitat conditions experienced by ruderals; particularly where the effect of repeated disturbance is to cause a high rate of mortality it may be expected that natural selection will favour the early production and maturation of seeds.

Even in the absence of disturbance, ruderal plants are short-lived and, in the majority of annual species, seed production is followed immediately by the death of the parent. In this respect the ruderal differs consistently from both the competitor and the stress-tolerator. The significance of this difference is examined on page 91.

Response to Stress

In preceding sections (pages 47 and 63) it has been concluded that differences in the rate and extent of morphogenetic responses to stress are of crucial importance in distinguishing the physiology of competitors and stress-tolerators. Experimental data are available which suggest that the stress responses of ruderal plants are in certain respects quite different from those which are characteristic of either of the other two primary strategies. Before considering this evidence, it may be helpful to remember that two forms of stress, i.e. shortages of water and mineral nutrients, which ruderals, in common with other plants, experience in nature, may arise either from the direct impact of an unfavourable environment or may be due to the depletion of resources by neighbouring plants. There is no reason to suspect that a plant may differentiate in its response according to whether shortage of water or mineral nutrients is plant-induced or not. This is a fact of considerable importance since, as explained in Table 7, the consequences of particular forms of response vary according to ecological context.

An informative method of studying the response of ruderal species to stress is to compare performance, and in particular seed production, of monospecific cultures sown at high and low densities. One of the first experiments of this type was that of Salisbury (1942), who compared capsule production in populations of a cornfield poppy (*Papaver argemone*) sown at high and low density on the same soil. The results of this experiment revealed that at the high sowing density the population showed a roughly normal distribution with respect to capsule number and included some plants with more than one hundred capsules. The most important fact to emerge from the experiment was that, although at the high sowing density mineral nutrient stress (perhaps associated with some mutual shading) caused severe reductions in vegetative development and seed production, *each plant produced at least one capsule.* More refined experiments (e.g. Hodgson and Blackman 1956; Hickman 1975; Raynal and Bazzaz 1975; van Andel and Vera 1977; Boot *et al.* 1986; Kadmon and Shmida 1990) which have been conducted on other species subsequently have confirmed the tendency of annual crops and weeds to sustain seed production, at however reduced a level (Figure 33), when subjected to severe stress.

Figure 33. An illustration of the maintenance of reproductive effort under stress in the annual weeds, *Matricaria discoidea* (left) and *Polygonum aviculare* (right). In both species large and small individuals bearing flowers are shown. All specimens were removed from the entrance to a cattle pasture at Wardlow Mires, North Derbyshire, England

One of the most interesting and thorough investigations of the stress physiology of a ruderal plant is that of Kingsbury *et al.* (1976), who examined the response of the annual halophyte, *Lasthenia glabrata*, to various periods of osmotic stress of the type normally experienced in the natural habitat of the species. Salt-water treatments applied at different stages of the life-cycle were found to induce a pulse of increased reproductive activity within a few weeks after stress began and the authors concluded that 'osmotic stress from high salt concentrations causes a shift in the hormone balance to favour reproductive development over vegetative growth'. It was observed that prolonged exposure to salt stress conditions induced 'a high ratio of flowers to biomass, a relatively low vegetative yield, a higher probability of early senescence, and a life-cycle which was 'accelerated and condensed.'

Information relating to the life-cycles and stress-responses of ruderal plants is highly relevant to studies of the interactions between annual crops, and in a large number of experiments interactions between annual plants have been examined by sowing two species together in 1 : 1 mixtures and in various unequal proportions (e.g. DeWit 1960, 1961). One of the most interesting features of the results of these experiments is the extent to which, even when employing high total densities of seed, the species composition of the original seed mixture has remained the overriding determinant of the contribution to the total yield made by each of the two species. It seems reasonable to presume that this is due to the short-term and inconclusive nature of competition between ruderal plants. This in turn may be related in part to the limited capacity of the majority of annual plants for vegetative expansion, continuing resource capture, and monopolisation of the environment. It seems likely also that an important contributory factor here is the tendency of ruderal plants, when impinging upon each other, to exhibit growth-responses to the resulting stresses which maximise seed production at the expense of a rapid curtailment of vegetative development.

PREDICTIONS ARISING FROM THE CSR MODEL

Three Sets of Traits

The information reviewed in this chapter suggests that, during the evolution of plants, the established phase of the life-cycle has experienced, in different habitats, three fundamentally different forms of natural selection. The first of these (C-selection) has involved selection for high competitive ability which depends upon plant characteristics that maximise the capture of resources in productive, relatively undisturbed conditions. The second (S-selection) has brought about reductions in both vegetative and reproductive vigour, adaptations which allow endurance of continuously unproductive environments. The third (R-selection) is associated with a short life-span and with high seed production and has evolved in severely disturbed but potentially productive environments.

It is also concluded that R-selected (ruderal), S-selected (stress-tolerant), and C-selected (competitive) plants each possess a distinct family of genetic characteristics and an attempt has been made in Table 6 to list some of these.

Three Types of Plant Response to Stress

It would appear that a crucial generic difference between competitive, stress-tolerant, and ruderal plants concerns the form and extent of phenotypic response to stress. Many of the stresses which are a persistent feature of the environments of stress-tolerant plants are experienced, although less frequently and in different contexts, by competitive and ruderal plants, although, from the evidence reviewed in this chapter, it would appear that the response in each case is quite distinct. We may suspect that such differences constitute one of the more fundamental criteria whereby the three strategies may be distinguished.

In Table 7 an attempt has been made, firstly, to define the circumstances in which competitive, stress-tolerant, and ruderal plants arc normally exposed to major forms of stress; secondly, to describe the stress response which appears to be characteristic of each strategy; and, thirdly, to predict the different consequences attending each type of response in different ecological situations.

It is predicted that the slow rate and relatively small extent of morphogenetic response and high physiological adaptability which is associated with endurance of protracted and severe stress in stress-tolerant plants will be of low survival value in environments where stress is a prelude either to competitive exclusion or to disturbance by phenomena such as drought. Similarly, the rapid and highly plastic growth-responses to stress of competitive plants (tending to maximise resource capture and vegetative growth) and of ruderals (tending to curtail vegetative growth and maximise seed production) are predicted to be advantageous only in the specific circumstances associated, respectively, with competition and disturbance.

To summarise, therefore: we can predict that the stress-response of the ruderal ensures the production of seeds, those of the competitor maximise the capture of resources, whilst those of the stress-tolerator allow the conservation of captured resources. This, of course, is an extremely simplified account of the immediate consequences of the three types of response. In the context of the entire life-history the three types of stress-response can be regarded as components of the rather different mechanisms promoting fitness in different types of environment. A most interesting implication concerns the difference in method of resource-capture exhibited by the competitors and the stress-tolerators. As explained on page 23, the effect of the stress-responses of the competitors, when coupled with the rapid turnover of leaves and roots, is to

Table 6. Some characteristics of competitive, stress-tolerant, and ruderal plants

		Competitive	Stress-tolerant	Ruderal
(i)	**Morphology**			
1.	Life-forms	Herbs, shrubs, and trees	Lichens, herbs, shrubs, and trees	Herbs
2.	Morphology of shoot	High dense canopy of leaves. Extensive lateral spread above and below ground	Extremely wide range of growth forms	Small stature, limited lateral spread
3.	Leaf form	Robust, often mesomorphic	Often small or leathery, or needle-like	Various, often mesomorphic
(ii)	**Life-history**			
4.	Longevity of established phase	Long or relatively short	Long—very long	Very short
5.	Longevity of leaves and roots	Relatively short	Long	Short
6.	Leaf phenology	Well-defined peaks of leaf production coinciding with period(s) of maximum potential productivity	Evergreens, with various patterns of leaf production	Short phase of leaf production in period of high potential productivity
7.	Phenology of flowering	Flowers produced after (or, more rarely, before) periods of maximum potential productivity	No general relationship between time of flowering and season	Flowers produced early in the life-history
8.	Frequency of flowering	Established plants usually flower each year	Intermittent flowering over a long life-history	High frequency of flowering
9.	Proportion of annual production devoted to seeds	Small	Small	Large
10.	Perennation	Dormant buds and seeds	Stress-tolerant leaves and roots	Dormant seeds
11.	Regenerative* strategies	V, S, W, B_s	V, B_{sd}, W	S, W, B_s
(iii)	**Physiology**			
12.	Maximum potential relative growth-rate	Rapid	Slow	Rapid
13.	Response to stress	Rapid morphogenetic responses (root-shoot ratio, leaf area, root surface area) maximising vegetative growth	Morphogenetic responses slow and small in magnitude	Rapid curtailment of vegetative growth, diversion of resources into flowering
14.	Photosynthesis and uptake of mineral nutrients	Strongly seasonal, coinciding with long continuous period of vegetative growth	Opportunistic, often uncoupled from vegetative growth	Opportunistic, coinciding with vegetative growth

(*continued over*)

Table 6. (*continued*)

		Competitive	Stress-tolerant	Ruderal
15.	Acclimation of photosynthesis, mineral nutrition and tissue hardiness to seasonal change in temperature, light and moisture supply	Weakly developed	Strongly developed	Weakly developed
16.	Storage of photosynthate and mineral nutrients	Most photosynthate and mineral nutrients are rapidly incorporated into vegetative structure but a proportion is stored and forms the capital for expansion of growth in the following growing season	Storage systems in leaves, stems and/ or roots	Confined to seeds
(iv)	**Miscellaneous**			
17.	Litter	Copious, not usually persistent	Sparse, often persistent	Sparse, not usually persistent
18.	Palatability to unspecialised herbivores	Often high	Low	Usually high

*Key to regenerative strategies (considered in full detail in Chapter 3): V—vegetative expansion, S—seasonal regeneration in vegetation gaps, W—numerous small wind-dispersed seeds or spores, B_s—persistent seed bank, B_{sd}—persistent seedling bank.

bring about a continuous spatial re-arrangement of the absorptive surfaces which causes the plant to adjust to changes in the distribution of resources during the growing season. This type of growth response, although highly effective in resource-capture, involves high rates of reinvestment of these resources and is clearly advantageous in productive but crowded environments where the effect of localised resource depletion by the rapidly-growing vegetation is to create within the habitat severe and continuously changing gradients in the distribution of light, water, and mineral nutrients. In contrast with the 'foraging' growth responses of the roots and shoots of the competitor, resource-capture in the stress-tolerator appears to be a more conservative activity primarily adapted to exploit temporal variation in the availability of resources in chronically unproductive habitats, i.e. the absorptive surfaces of the plant are predicted to be long-lasting physiologically-adaptive structures which, at least on an annual basis, tend to remain in the same location and to exploit temporary periods during which resources become available.

Table 7. Morphogenetic responses to desiccation, shading, or mineral nutrient stress of competitive, stress-tolerant, and ruderal plants and their ecological consequences in three types of habitat

		Consequences		
Strategy	Response to stress	Habitat 1*	Habitat 2†	Habitat 3‡
Competitive	Large and rapid changes in root : shoot ratio, leaf area, and root surface area	Tendency to sustain high rates of uptake of water and mineral nutrients to maintain dry matter production under stress and to succeed in competition	Tendency to exhaust reserves of water and/or mineral nutrients both in the rhizosphere and within the plant: etiolation in response to shade increases susceptibility to fungal attack	Failure rapidly to produce seeds reduces chance of rehabilitation after disturbance
Stress-tolerant	Changes in morphology slow and often small in magnitude	Overgrown by competitors	Conservative utilisation of water, mineral nutrients, and photosynthate allows survival over long periods in which little dry matter production is possible	
Ruderal	Rapid curtailment of vegetative growth and diversion of resources into seed production		Chronically low seed production fails to compensate for high rate of mortality	Rapid production of seeds ensures rehabilitation after disturbance

* In the early successional stages of productive, undisturbed habitats (stresses mainly plant-induced and coinciding with competition).

† In either continuously unproductive habitats (stresses more or less constant and due to unfavourable climate and/or soil) or the late stages of succession in productive habitats.

‡ In severely disturbed, potentially-productive habitats (stresses either a prelude to disturbance, e.g. moisture stress preceding drought fatalities, or plant-induced, between periods of disturbance.

Until recently, these predictions of differences in foraging behaviour between competitors and stress-tolerators would have remained in the realm of speculation. However, since 1985, there has been a remarkable expansion in the number of studies designed to test quantitatively our understanding of how the roots and shoots of plants exploit spatial and temporal patchiness in resource supply. Comprehensive reviews are available with which to examine progress in this field of research (Caldwell and Pearcy 1994; Hutchings and de Kroon 1994).

Unfortunately, many of the experiments are not comparative and cannot be used to test for the existence of the differences in foraging behaviour predicted to distinguish between plants of productive and unproductive habitats. However, results suitable for this purpose are available from two experiments.

The first experiment (Crick and Grime 1987) involved partitioned containers in which each plant was exposed to controlled patchiness in mineral nutrient supply. The main conclusion drawn from the results was that high rates of nitrogen-capture from a stable patchy rooting environment were achieved by *Agrostis stolonifera*, a fast-growing species with a small but morphologically dynamic root system. Under the same conditions, nitrogen-capture was much inferior in *Scirpus sylvaticus*, a slow-growing species with a massive root system but a relatively low nitrogen-specific absorption rate and slow root adjustment to mineral nutrient patchiness. This result is consistent with the hypothesis that rapid morphogenetic changes ('active foraging') are an integral part of the mechanism whereby fast-growing plants sustain high rates of resource-capture in the rapidly changing resource mosaics created by actively-competing plants.

In the second experiment, already described on page 59 (Campbell and Grime 1989a), plants characteristic of fertile and infertile soils were compared with respect to their ability to capture nitrogen from pulses of various durations. The results (Figure 25) confirm that, where pulses were short, a condition likely to occur on infertile soils, an advantage was enjoyed by the slow-growing species, *Festuca ovina*, which appeared to have low rates of tissue turnover and roots which remained functional despite exposure to long periods of mineral nutrient stress.

Three Types of Life-history

It has been noted already (page 85) that, in the majority of ruderals, seed production is followed by the death of the parent plant. This phenomenon impinges upon an important principle which has interested several evolutionary theorists (e.g. Cole 1954; Williams 1966; Stearns 1976; Ricklefs 1977; Sibly and Calow 1983; Promislow and Harvey 1990) and which concerns the partitioning of captured resources between parent and offspring and the optimising of life-histories by natural selection in various environments.

It is clear that when resources are expended upon reproduction at an early stage of the life-history there is an increased risk of parental mortality. However, in environments as uncertain as those of the ruderal, high rates of mortality are inevitable and the cost of a marginally-increased rate of parental fatality is outweighed by the benefit of high fecundity. As we might expect, therefore, the result of natural selection in most ruderals has been the development of early and 'lethal' (Harper 1977) reproduction.

In comparison with the habitats of ruderals, the environments colonised by competitors and stress-tolerators are characterised by a low intensity of disturbance. This has two main effects: the first is to reduce the risks of mortality in long-lived plants, whilst the second is to limit drastically the opportunities for seedling establishment. A high risk of mortality to juvenile members of the plant population is characteristic of the environments of both the competitor and the stress-tolerator but, despite this point of similarity, the evolution of life-histories has taken a rather different course in the two types of habitat. Whereas in the competitors, reproduction occurs at a relatively early stage in the life-history and usually involves the expenditure each year of a considerable proportion of the captured resources, the stress-tolerators commence reproduction later and tend to show intermittent reproductive activity over a long life-history. Analysis of this difference requires some reference to the regenerative strategies of plants (Chapter 3) and to the process of vegetation succession (Chapter 8). Here, therefore, only a preliminary explanation will be attempted.

In the crowded but productive environments colonised by competitive plants, seedling establishment is restricted because the habitat is occupied by a vigorously-expanding mass of established vegetation. Hence, although successful competitors accumulate resources at rates sufficient to sustain abundant seed production, there is little opportunity for the population to expand its immediate frontiers by seedling establishment. As explained in Chapter 3, this dilemma has evoked several types of evolutionary response in competitive herbs, shrubs, and trees. One is evident in the high incidence of vegetative expansion, a form of asexual regeneration which is viable in dense vegetation and is compatible with the maintenance of high competitive ability (see page 140). A second is the production annually of numerous wind-dispersed seeds, which facilitate the colonisation of new habitats. This form of regeneration is particularly common among the perennial herbs, shrubs, and trees which appear in the early and intermediate stages of vegetation succession in fertile, undisturbed habitats. Seed production in these plants commences relatively early in the life-history and appears to be related to the fact that as vegetation development proceeds and resource depletion occurs the environment becomes progressively less hospitable to the competitor. Early and continuous expenditure on wind-dispersed seeds in many competitors may be interpreted, therefore, as an indication that, in common with many ruderals, competitors tend to lead a 'fugitive' (Hutchinson 1951) existence.

Stress-tolerators occur in environments in which mineral nutrients are severely restricted by absolute shortage or by sequestration in the vegetation itself. In contrast to the populations of many competitors it seems likely that those of the stress-tolerators often remain in continuous occupation of the same habitat for many hundreds, perhaps thousands, of years. The biomass remains fairly constant and opportunities for reproduction may be exceedingly rare and dependent upon senescence or occasional damage to established members of the populations. In these circumstances, low parental mortality and a conservative but sustained reproductive effort can make a major contribution to the maintenance and expansion of population size. It is not surprising, therefore, that we should find that in stress-tolerators, such as *Pinus aristata*, *Pinus sembra*, *Pseudotsuga menziesii*, *Tsuga canadensis*, and *Northofagus pumilo* (Currey 1965; Smith 1970; Janzen 1971; Barrera *et al.* 2000; Woods 2000), the onset of reproduction is delayed and flowering tends to occur only intermittently during a long life-history.

TESTS OF THE CSR MODEL

Until recently the most compelling evidence supporting the CSR model has been the large number of very different ecological phenomena which appear to be explained by the theory; many of these are described in later chapters of this book. However, to an increasing extent in future the fate of the model will rest upon the results of various independent tests. These vary enormously in scope and rigour (Grime 1988a) and may be usefully classified into informal and formal tests.

Informal Tests

Despite its general implications for ecological theory, the CSR model drew inspiration mainly from comparative studies of vascular plants in Europe (Ramenskii 1938; Grime 1974). In retrospect this origin is explicable in terms of the large fund of botanical information available following the long history of field and laboratory investigations in Europe and, in particular, on the small and well-defined flora of the British Isles. Efforts to falsify the model by matching its predictions against patterns of ecological specialisation observed in other plants and in other parts of the world and in diverse plant and animal taxa have been limited in many instances by lack of information. In consequence, many such extensions or applications of CSR theory cannot be regarded as definitive tests of the model. They are valuable, however, as informal tests, and, in many cases, e.g. Shepherd (1981), Raven (1981), Dring (1982), Menges and Waller (1983), Kautsky (1988), Rogers (1988, 1990), Silvertown *et al.* (1992), Brzezieki and Kienast (1994), Hills *et al.* (1994), Rincon and Huante (1994), Topham (1997), and Onipchenko *et al.* (1998), they have identified organisms likely to provide excellent subjects for more critical work. In particular, several authors (Lawrence 1990; Andersen 1991, 1995; Hodgson 1991, 1993; Convey 1997; Ohtonen *et al.* 1997; Lawrence and Baghim 1998) have attempted direct comparisons of primary strategies between plant and animal taxa; this constitutes a difficult but essential step in the effort to use CSR theory to analyse ecosystem structure and dynamics (see page 304).

Formal Tests

Mathematical Models

There is an urgent need for models which not only examine the influence of life history and reproductive schedules on fitness in defined circumstances, e.g. Caswell (1989), but also take account of the crucial role of resource-capture and utilisation. A leading role in the effort to devise mathematical models incorporating life-history theory and resource dynamics has been played by Tilman (1982, 1988) and it is singularly unfortunate that his models of plant strategies have sometimes (Tilman 1988) relied upon tradeoffs between root and shoot allocation and function that have been falsified by comparison with experimental data (Shipley and Peters 1990, 1991). In order to devise models which adequately represent the wealth of information now available concerning the resource dynamics and growth of plants in various environments it is necessary to incorporate:

1 'Patch' and 'pulse' supply of resources within the leaf canopy *and* the rhizosphere.
2 The interdependence of root and shoot function (i.e. the nutrient demand of leaves and dependence of roots upon a supply of photosynthate).
3 The tradeoff between high rates of resource-capture (and loss) in potentially fast-growing plants and efficient retention of captured resources by inherently slow-growers.

Several authors have attempted to meet these requirements and models have been devised which define the very different circumstances conducive to the success of fast-growing perennials as compared to plants with extended life-histories, slow growth rates, and conservative patterns of resource use (Sibly and Grime 1986; Aerts and van der Peijl 1993; Colasanti and Grime 1993).

Another approach that can be used to seek evidence of a fundamental trichotomy in plant functional types of the established phase is to develop mathematical models that examine the way in which plants capture space. Using spatial moment equations Bolker and Pacala (1999) have recently demonstrated that there are only three fundamentally different strategies whereby plant functional types can exploit space. The first, clearly equivalent to the R-strategist, involves rapid colonisation and vacation of local areas of unpopulated ground; the second depends upon long-term occupation of space and tolerance of resource depletion (S-strategist); and the third relies upon rapid local consolidation to exclude competitors (C-strategist). The rigorous mathematical proof by Bolker and Pacala (1999) is a welcome development although historians of ecology may wonder at the delay in its appearance some 25 years after Grime (1974) and 61 years after Ramenskii (1938)! As explained already in relation to the parallel histories of two- and three-strategy models (Figure 4), the main obstacle to progress had been controversy surrounding recognition of the C-strategist which as pointed out by Bolker and Pacala (1999) has been consistently overlooked by many previous modellers (e.g. MacArthur and Wilson 1967; Levins and Culver 1971; Hastings 1980; Tilman 1994). Why then was the C-strategist so underestimated? Patently it was not for lack of illustrative botanical examples from the real world: clonal monopolistic species (e.g. *Pteridium aquilinum*, *Urtica dioica*, *Fallopia japonica*; see Figure 6) occur widely, and as described in Chapters 6 and 7 they are familiar beneficiaries of the cycles of disturbance and dereliction that characterise so many modern landscapes. Perhaps the most likely explanation for the delayed entry of the C-strategist upon the stage lies in the roots of the modelling activity itself that has drawn its inspiration mainly from animal populations where the mechanisms and consequences of competition have been more difficult subjects for investigation.

Cellular Automaton Models

In addition to models that deal with the occupation of space, more detailed approaches have been developed to examine the contrasted ways in which different plant functional types capture, utilise, and release units of resource. An important step in the development of this form of modelling was the self-assembling model published by Colasanti and Hunt (1997) in which each individual plant was represented as a two-dimensional structure, displayed in vertical section as above- and below-ground parts (Figure 34a). In this model, the plant has a branching modular structure and each module has the capacity to sense the resources present in its immediate environment and also in the modules to which it is immediately attached. The rulebase controlling its behaviour specifies the local conditions necessary for resource uptake, transfer of resources in or out of the module, maintenance requirements, and vegetative and reproductive growth. Before its application to tests of CSR theory, an extended series of simulations was conducted to examine its approximation to field reality. In view of the extremely local nature of the rulebase (i.e. restricted to individual modules) it was reassuring to observe that several classical physiological and ecological phenomena were reproduced successfully. These included concentration of roots and shoots in resource-rich patches, competitive exclusion of individuals by their confinement to resource-depleted zones above and below ground, reversible shifts in root to shoot biomass induced by manipulating resource concentrations above or below ground, and a precise reproduction of the –3/2 thinning law on overcrowded populations (Yoda *et al.* 1963).

Following the completion of these proving trials, the two-dimensional cellular automata have been applied to a wide range of vegetation processes (Colasanti 2000), involving the assembly of 'virtual vegetation' from mixtures of C-, S-, and R-strategists subjected to various combinations of resource

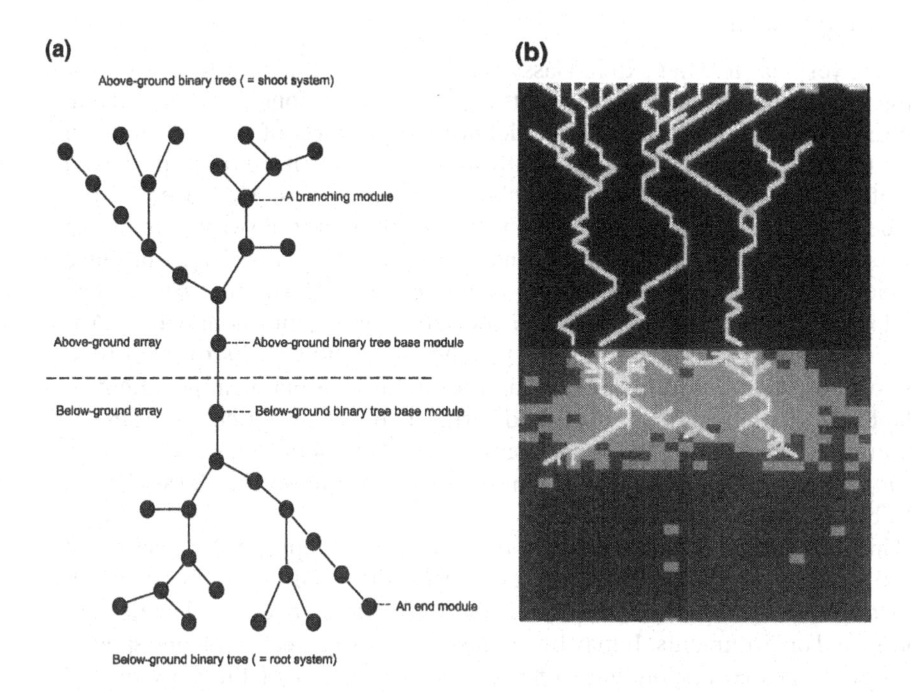

Figure 34a. The modular basis of the self-assembling cellular automata plant. Each module can 'sense' the resources present in its immediate environment and also in the modules to which it is immediately attached. The rulebase controlling its behaviour sets requirements for resource uptake, transfer of resources in or out, maintenance requirements, and vegetative and reproductive growth. All the rules are 'tuned' to create different possibilities for behaviour according to CSR theory and observations, but the outcome in terms of plant growth, plant-to-plant competition, and community ecology results only from the interaction of the resource regime and these 'bottom-up' rules. (Reproduced from Colasanti and Hunt by permission of *Functional Ecology*.)

Figure 34b. A cellular automata simulation of competing plants. Note the depletion of mineral nutrients around each root system. (Reproduced by permission of Colasanti and Hunt.)

supply, climate, and management impacts (Figure 34b). Because the rules governing the resource transactions, growth, reproduction, and survival of each plant are extremely simple and reside entirely at the level of the individual modules, the simulations permit an examination of the consequences for individuals or the vegetation of very specific changes in the size or resource processing characteristics of root or shoot modules. This promises to provide a mechanism for measuring the importance of particular combinations of traits not only at the level of CSR strategies but also in more detailed simulations at or below the level of the population.

Intraspecific Studies

It is not yet clear to what extent classical genecological methods will provide a basis for testing the CSR model. Plants exploiting conditions corresponding to the extremities of the triangular model are the products of natural selection over many generations in highly contrasted environments and are predicted to differ with respect to a large number of genetic traits. Moreover, as described later (page 132), according to their evolutionary history, families of flowering plants appear to differ considerably in their capacity to produce species with ruderal or competitive characteristics (Hodgson 1986c). Such phylogenetic constraints on adaptive specialisation are not confined to Angiosperms; Bond (1989) has pointed out the incorrigibly stress-tolerant biology of most Gymnosperms and in the Pteridophyta a major impediment to adaptive radiation into the disturbed fertile habitats of agricultural land is discernible from the fact that only one ephemeral species of fern (*Anogramma leptophylla*) has been recorded in the world flora (Grime 1985b).

This background is highly relevant to the choice of appropriate species and the development of realistic objectives in tests of the CSR model by selection experiments or comparisons of populations of the same species collected from contrasted environments. It may be unreasonable to expect to find one species with genotypes corresponding to more than one corner of the CSR model. However, species which are predominantly C, S, or R do not offer the best prospects for experimental tests. Among herbaceous plants, the most promising subjects appear to be species of relatively wide ecological amplitude and with characters intermediate between those associated with C, S, and R (e.g. *Agrostis capillaris*, *Poa pratensis*, *Holcus lanatus*, *Deschampsia cespitosa*, *Trifolium repens*, *Plantago lanceolata*). All of these species are known to be genetically variable and would provide excellent experimental material. One insight into the potential of common wide-ranging herbaceous plants to allow tests of the CSR theory is available from studies of the common pasture grass, *Poa annua* (Law *et al.* 1977; Law 1979). These investigations have revealed variation between individuals with respect to the onset of flowering and have enabled an analysis of the tradeoff between early allocation to reproduction and sustained vegetative vigour, which is clearly an important component of the R–C dimension of the triangular model. There also exists a small number of pioneer studies which appear to explore intraspecific variation corresponding to the C–S axis of the model (Kruckeberg 1954; Böcher 1949, 1961).

Screening of Individual Traits

Standardised measurements of particular attributes on large numbers of species of contrasted ecology (Salisbury 1942; Baker 1972; Grime and Hunt 1975; Grime *et al.* 1981) were involved in the development of the CSR theory. As explained with great clarity by Keddy (1992a) there is scope for independent tests of aspects of the model by screening programmes involving other strategy-related characteristics measured on plants and animals. Unfortunately, screening (in common with monitoring) has an undeserved reputation for robotic activity, lack of hypothesis-testing, and an uncertain return for effort. However, there appears to be a growing awareness that the broader perspectives resulting from screening frequently lead to discoveries and generate hypotheses. As Clutton-Brock and Harvey (1979) recognised: 'as more species are considered it becomes progressively more difficult to fit several adaptive hypotheses to the empirical facts'. Screening experiments of specific relevance to tests of CSR theory include those conducted by Shipley and Keddy (1988), Hunt *et al.* (1991), Jurado *et al.* (1991), Boutin and Keddy (1992), Lambers and Poorter (1992), Reader *et al.* (1992), Reiling and Davison (1992), Bennett and Leitch (1995), Ashenden *et al.* (1996), Cornelissen *et al.* (1996), and Castro-Diez *et al.* (1998).

Screening and Multivariate Analysis of Many Traits

Although measurement of individual traits provide limited tests of parts of the model, a more critical approach is to bring together information on diverse aspects of the biology of large numbers of plants drawn from contrasted habitats to seek evidence for or against the co-occurrence of attributes in the sets (Table 6) predicted by the CSR model.

In two early attempts to follow this approach (Grime *et al.* 1987a; Grime *et al.* 1988) multivariate analyses were conducted on large data-sets of species attributes, assembled from existing published sources of common British herbaceous plants. Apart from the obvious and expected distinction between ephemerals and perennials these analyses failed to reveal clear patterns of ecological specialisation in the established phase. In retrospect, this outcome was inevitable. The data-bases were heavily biased towards morphological attributes which, although universally available as a by-product of taxonomy, are of uncertain relevance to ecological prediction. Quite clearly, a desirable alternative to *ad hoc* analyses of data-sets abstracted from the published literature would be one conducted on the results of a standardised screening programme designed specifically to test for the existence of primary functional types. With this objective in mind the Integrated Screening Programme (ISP) was initiated in Sheffield in 1987.

The ISP applied standardised experimental procedures and measurements for the ecological, physiological, and biochemical characterisation of a plant species, population, or cultivar. Procedures, some illustrated in Figure 35, were devised to cover many aspects of the established and regenerative phase including those predicted to be affected by the tradeoffs at the core of the CSR model. A laboratory manual was produced (Hendry and Grime 1993) describing each procedure and illustrating its use in ecological prediction.

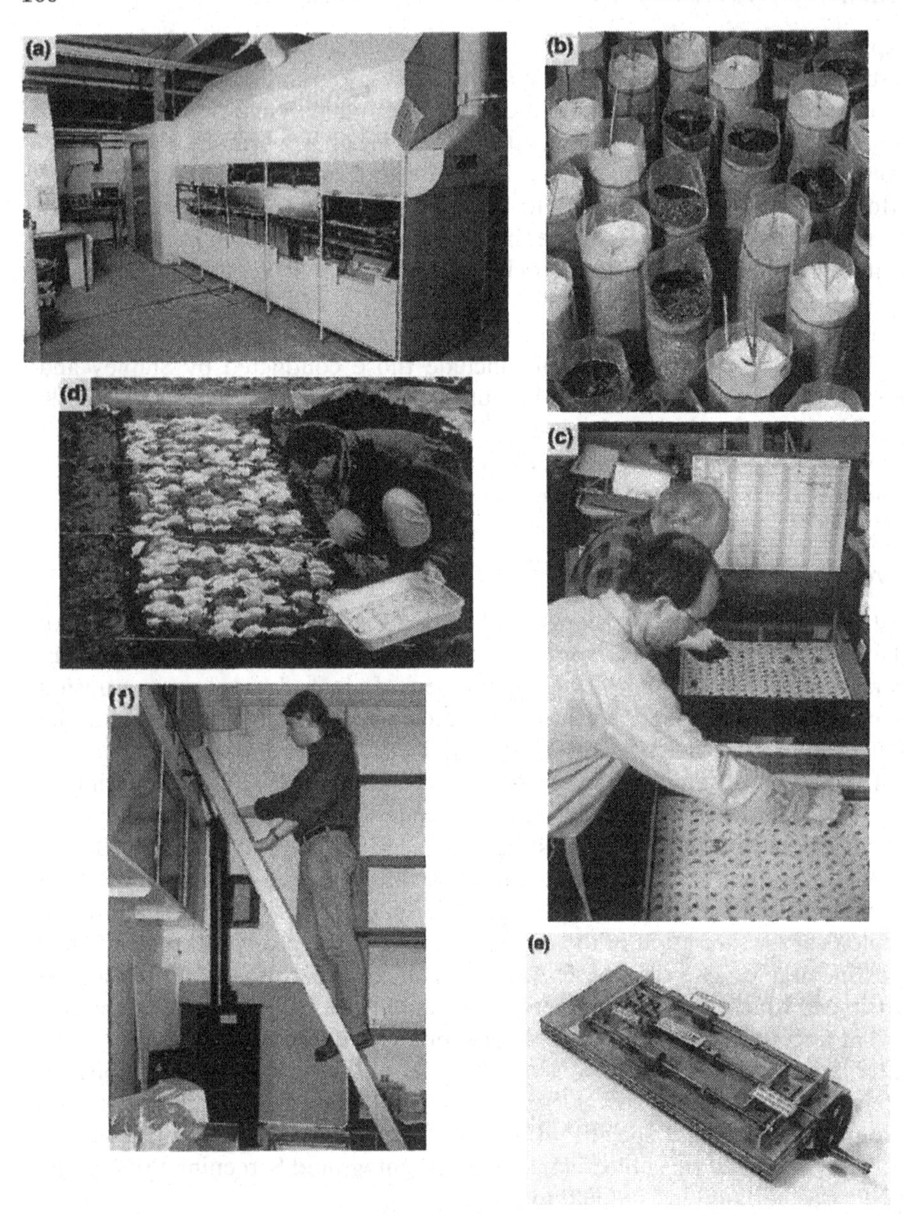

Figure 35. Some procedures used in the Integrated Screening Programme: (a) the temperature gradient tunnel employed to measure growth responses to temperature and their interaction with light intensity and drought; (b) metre length tubes of sand designed to record seedling root penetration; (c) palatability investigations presenting leaves to the garden snail, *Helix aspersa*; (d) standardised decomposition studies involving burial of senescent material in mesh bags; (e) tensioning device for determination of leaf tensile strength; (f) measurement of seed buoyancy by timing the descent of dispersules through horizontal laser fans

Some of the tests examined basic features of anatomy, morphology, physiology, and biochemistry at different stages of development (seed, seedling, established plant), others measured survival and growth responses under particular environmental stresses. Further sets of procedures were designed with specific relevance to effects of mineral nutrient supply and climatic factors (temperature, moisture supply, CO_2 concentration) and efforts were made to measure attributes relevant to competitive interactions and the potential to dominate plant communities. The ISP was applied to 43 species (one seed source per species), of which two were crop plants whilst the remainder were native British herbs and sub-shrubs from a wide range of habitats. The majority of the procedures in the ISP were conducted over a relatively short period of time, under standard conditions on material of known and consistent genetic origin. However, some important attributes were not amenable to this approach and it was necessary to rely upon data collected from the field or reported in the literature. Examples in this category include seed persistence in the soil and the nature and extent of mycorrhizal infections. Caution has been applied in using these latter sources of data, since they allow contamination of the data-base by less precise information.

When the complete set of results from the ISP were analysed (Grime *et al.* 1997) patterns of variation in individual traits across the 43 species were detected and multivariate analysis allowed primary axes of variation involving sets of traits to be recognised. The most significant feature of the data was the emergence from a principle component analysis of a first axis accounting for 22% of the total variation occurring in all measured attributes of the established phase. In view of the fact that measurements were made on so many different aspects of plant functioning (growth, morphology, resource acquisition and use, defence, decomposition, etc.) and on such widely-contrasted ecologies, it is perhaps surprising that almost a quarter of the variation could be related to one axis. From this fact one may infer many strong interrelationships and incompatibilities (tradeoffs) and these are indeed evident when rank correlation coefficients between individual traits are reviewed (Table 8).

Mineral nutrition is obviously implicated in the pattern of specialisation detected in Axis 1. A consistent correlation occurred between foliar concentrations of N, P, K, Ca, and Mg, high concentrations of which coincided with a capacity for rapid growth in productive conditions and a failure to sustain yield under limiting supplies of nutrients. Other traits entrained in Axis 1 included life-history, rapid foraging responses in both root and shoot in resource-patchy conditions, the morphology, longevity, tensile strength and palatability of leaves to generalist herbivores, and the decomposition rate of leaf litter. These patterns were observed in both monocotyledons and dicotyledons, and appear to confirm the occurrence across the species investigated of the tradeoff, predicted from CSR theory (Grime 1974), between a set of attributes conferring an ability for high rates of resource acquisition

Table 8. Correlations between pairs of attributes with correlation coefficients of 0.65 or above in the Integrated Screening Programme (Reproduced from Grime *et al.* 1997 by permission of *Oikos*.)

Rank	*r*	Description	Description
1.	0.93	Specific leaf area	Leaf area ratio
2.	0.84	Leaf Mg content	Leaf Ca content
3.	0.83	Leaf P content	Leaf N content
4.	0.83	Shoot foraging increment to unshaded quadrants (mg)	Root foraging increment to undepleted quadrants (mg)
5.	−0.81	Leaf tensile strength	Leaf Ca content
6.	0.79	Yield in low nutrients	Yield in low N
7.	−0.76	Leaf tensile strength	Leaf Mg content
8.	−0.75	Leaf width	Leaf tensile strength
9.	0.74	Life-history	Lateral spread
10.	−0.72	Leaf tensile strength	Palatability index
11.	0.72	Leaf K content	Leaf P content
12.	0.71	Abaxial epidermal cell size	Spongy mesophyll cell size
13.	0.69	Palatability index	Leaf Ca content
14.	0.69	Palatability index	Decomposition % wt loss
15.	0.67	Leaf max surface area	Root increment to undepleted quadrants (mg)
16.	0.66	Leaf K content	Leaf N content
17.	−0.66	Leaf tensile strength	Leaf N content
18.	0.66	Adaxial epidermal cell size	Leaf Mg content
19.	−0.65	Leaf width	Yield in low nutrients
20.	0.65	Palatability index	Leaf Mg content

(and loss) in productive habitats and another set conducive to retention of resource capital in unproductive conditions.

As pointed out by Tessier *et al.* (2000) this tradeoff has surfaced repeatedly in the ecological literature in association with a wide variety of organisms including bacteria (Fredrickson and Stephanopoulos 1981), algae (Sommer 1989; Turpin 1988), invertebrates (Schmitt 1996), vertebrates (Brown 1989), fish (Winemiller and Rose 1992), and in more general reviews (e.g. Arendt 1997).

It might be argued that the prominence of mineral nutrients in Axis 1 arose from their strong representation in the tests conducted in the ISP. However, when rankings of the 43 species were compared in two independent multivariate analyses conducted on mineral nutrient-related attributes and on attributes with no immediate and obvious involvement in mineral nutrition (Figure 36) the Axis emerged from both and there was little difference in the ranking of species. This finding strongly supports the hypothesis that mineral nutrition is involved in a primary axis of ecological specialisation that has repercussions on many other aspects of the plants' biology.

The second axis of specialisation to emerge from the principle-component analysis of the ISP data accounted for 11.4% of the total variation and was

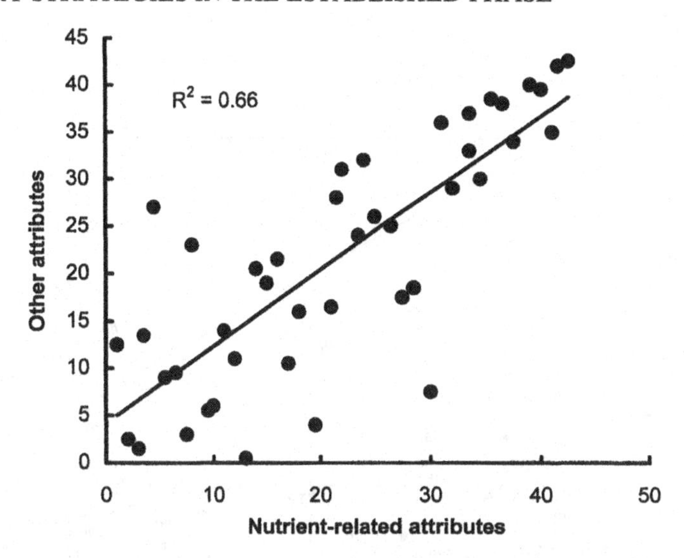

Figure 36. Ranking of 43 herbaceous species on Axis 1 of two separate DECORANA ordinations conducted respectively on mineral nutrient-related attributes and on other attributes. The two rankings are significantly ($P < 0.001$) correlated. (Reproduced from Grime *et al.* 1997 by permission of *Oikos*.)

clearly related to phylogeny in that it effected a consistent separation of monocotyledons from dicotyledons. Much more interesting in the effort to test CSR theory was the association of 7.7% of the variation with a third axis reproducing the distinction between ephemerals and perennials noted in earlier multivariate analyses on *ad hoc* data sets (e.g. Grime *et al.* 1988). When a scatter-diagram is constructed by plotting species positions on Axes 1 and 3 (Figure 37) it is immediately apparent that groups of species with sets of traits and ecologies consistent with CSR theory occupy characteristic areas within a triangular space.

The distribution of plant species in Figure 37 strongly resembles patterns obtained much earlier using simpler ordinations of species and relying upon a small number of morphological and physiological traits (e.g. Grime 1974). On first inspection, therefore, Figure 37 hardly justifies the resources and commitments that were required to complete the ISP. However, the significance of the main results extends beyond validation of a particular typology of plants useful to plant sociologists. Ultimately the value of the ISP will depend upon the ecological significance we can attach to the traits (and tradeoffs between traits) represented by Axes 1 and 3. It is of considerable interest, therefore, to observe in Table 8 that the traits (growth rates, resource capture and retention, anti-herbivore defence, decomposition) and their tradeoffs not only involve aspects of the core functioning of plants but also relate directly to plant community and ecosystem properties. Returning to the key objective identified at the beginning of this book (page xix, Figure 1) we can see the main value

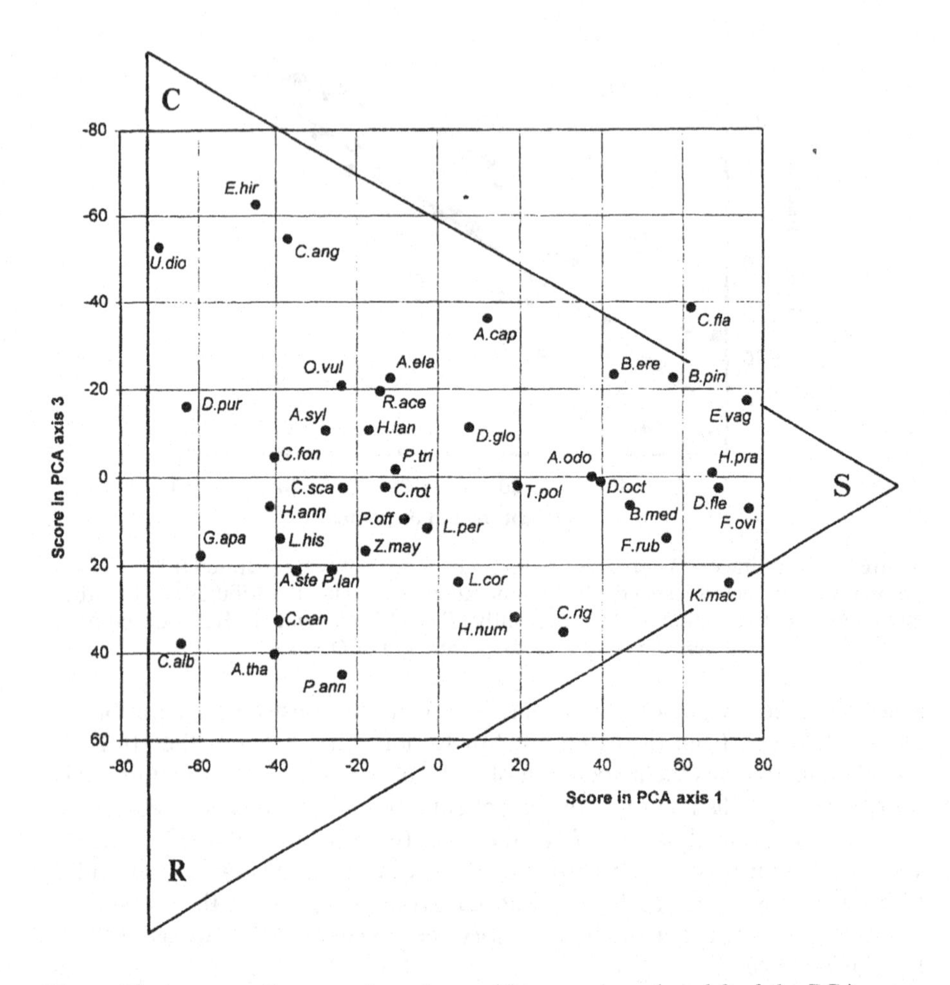

Figure 37. A scatter-diagram of species positions on Axes 1 and 3 of the PCA conducted on the established phase in the Integrated Screening Programme. The superimposed equilateral triangle has been inserted to indicate the consistency of the distribution of species with the three poles of the CSR system (Grime 1974, 1977, 1979). The species are: *Agrostis capillaris, Anisantha sterilis, Anthoxanthum odoratum, Anthriscus sylvestris, Arabidopsis thaliana, Arrhenatherum elatius, Brachypodium pinnatum, Briza media, Bromopsis erecta, Campanula rotundifolia, Carex flacca, Catapodium rigidum, Centaurea scabiosa, Cerastium fontanum, Chamerion angustifolium, Chenopodium album, Conyza canadensis, Dactylis glomerata, Dryas octopetala, Epilobium hirsutum, Eriophorum vaginatum, Festuca ovina, Festuca rubra, Galium aparine, Helianthemum nummularium, Helianthus annuus, Helictotrichon pratense, Holcus lanatus, Koeleria macrantha, Leontodon hispidus, Lolium perenne, Lotus corniculatus, Origanum vulgare, Pilosella officinarum, Plantago lanceolata, Poa annua, Poa trivialis, Rumex acetosella, Thymus polytrichus, Urtica dioica, Zea mays.* (Reproduced from Grime *et al.* 1997 by permission of *Oikos*.)

of the ISP and allied screening operations (e.g. Field and Mooney 1986; Boutin and Keddy 1992; Lambers and Poorter 1992; Reich *et al.* 1992; Diaz and Cabido 1997) as a mechanism by which to discriminate between plant traits of fundamental importance in the functioning of all plant communities and ecosystems and those responsible for the fine-tuning of individual plant ecologies.

Vegetation Synthesis under Controlled Conditions

As explained on page 8, a central assertion of the triangular model is that competition for resources reaches maximum importance under conditions of high productivity and declines as a vegetation determinant at sites experiencing severe mineral nutritional constraints on production or frequent vegetation disturbance. In order to test this hypothesis it is essential to measure the importance of competition under specified conditions of productivity and disturbance using methods which are independent of organism responses (van der Steen and Scholten 1985). In natural habitats this is difficult because there are no easy and reliable methods of measuring productivity and disturbance in the field. Exposure of bare soil in arctic fellfields (Fox 1981), soil depth in grassland (Reader and Best 1989), and sediment organic content on lakeshores (Wilson and Keddy 1986) have been used as indirect measures but these cannot be regarded as fully satisfactory since, in some cases at least, they do not always discriminate adequately between impacts of low productivity and effects of disturbance.

A more effective approach to testing these hypotheses is to synthesise artificial communities under simple experimental conditions in which productivity and disturbance are under direct control. Usually, studies of species distribution and competitive displacement on synthetic environmental gradients use conventional, randomised pot designs for the experiments, where each level of the gradient is assigned to a separate small container. Examples of this method are the nutrient gradient experiments of Austin and Austin (1980), Austin (1982), Parrish and Bazzaz (1982), Snow and Vince (1984), and Austin *et al.* (1988). This approach has several drawbacks. In particular: (1) there is a need to guard against edge effects in pots (see Austin and Austin 1980); (2) there is a risk that chance establishment failures may predetermine the final outcome (see Austin *et al.* 1988); and (3) the many permutations that arise when multiple factors or factor levels are considered make the experiment logistically prohibitive. A simpler approach, which addresses these difficulties, is to construct *continuous* gradients within single containers. This approach has been used to study simple soil depth gradients (Sharitz and McCormick 1973) and water-table depth gradients using either incomplete partitions between factor levels (Pickett and Bazzaz 1978) or continuous gradients (Ellenberg 1963, 1974; Mueller-Dombois and Sims 1966; Grace and Wetzel 1981).

An extension of these approaches was developed by Campbell and Grime (1992) in an attempt to measure the role of competition under a wide range of known combinations of productivity and disturbance. The objective was to subject artificial communities, either single-species stands or mixed-species stands, to two continuous gradients, one of nutrient supply, one of simulated grazing and trampling intensity. The gradients were superimposed at right angles to each other in square plots to produce a factorial matrix of productivity × disturbance intensities. An additive experimental design was favoured over a substitutive design (de Wit 1960) because it enables the absolute magnitude of competitive effects to be determined (Mahmoud and Grime 1976; Underwood 1986; Goldberg and Fleetwood 1987; Austin *et al.* 1988).

The experiment was applied to grass species that had been classified in advance with respect to primary strategy. Comparison after two years of the vegetative and reproductive vigour of the species in pure stands and in additive mixture allowed examination of variation in the impact of competition across the matrix. It was also possible to examine the extent to which the species became arranged in the spatial patterns predicted by CSR theory when grown in mixed stands on the stress-disturbance matrix.

The patterns in phytomass and flowering in the monospecific stands in Figure 38 illustrate that species distribution and performance can be influenced by direct impacts of soil fertility and disturbance quite apart from any effects of interspecific competition. Suppression of vegetative and reproductive vigour by the combined effect of low soil fertility and severe disturbance was evident in the robust fast-growing perennial, *Arrhenatherum elatius*, in the small slow-growing tussock perennial, *Festuca ovina*, and in an ephemeral population of the ruderal species, *Poa annua*. All three grasses showed maximum performance in the productive, undisturbed corner of the matrix.

In Table 9 two quite different indices of competition have been calculated for each of the three species at 25 positions on the productivity/disturbance matrix. In the left-hand side of the table values are presented for the proportional effect (CP_{ij}) of interspecific competition at the ith level of fertility and j^{th} level of vegetation damage:

$$CP_{ij} = (YP_{ij} - YM_{ij}) \times 100/(Y_{max} - YM_{ij})$$

where YP is the yield in the pure stand, YM_{ij} is the yield in the mixed stand and Y_{max} = max (max $[YP_{ij}]$, max $[YM_{ij}]$), i.e. the maximum yield observed over all i,j. CP_{ij} is thus the percentage contribution of interspecific competition to the reduction in this yield potential (relative to the direct impacts of stress and disturbance). It is similar to the importance of competition (Welden and Slauson 1986). The results show that, under conditions of low soil fertility or severe vegetation disturbance, competition remained operative but declined in importance relative to the direct effects of low fertility and disturbance on plant yield. A second set of values of the right-hand side of Table 9 refer to the proximal effect *(CX)* of interspecific competition.

$$CX_{ij} = (Yp_{ij} - YM_{ij}) \times 100/Y_{ij}$$

CX_{ij} is the percentage reduction in local yield potential (i.e. at a particular level of stress and disturbance) due to interspecific competition. It is a relative-yield measure of competitive ability.

This index permits an assessment of the influence of soil fertility and disturbance upon the susceptibility of each species to interspecific competition in each section of the fertility-disturbance matrix. The proximal effect of interspecific competition was relatively severe on *P. annua* but, as we might expect in a small ephemeral, the impact of competition was reduced in those parts of the matrix where low fertility or severe disturbance created bare ground and opportunities for seedling regeneration. The results show quite clearly that even in parts of the matrix providing low fertility, *F. ovina* remained more susceptible to competition than *A. elatius*, a species which is normally associated with fertile habitats and, in this experiment, dominated the productive, undisturbed parts of the matrix. This provides evidence of remarkable constancy in the relative competitive abilities of *A. elatius* and *F. ovina* across a very wide range of soil fertility and does not support the hypothesis (Huston and Smith 1987; Tilman 1987) that plants of infertile soils become superior competitors for mineral nutrients under nutrient-limited conditions.

Figure 38a. (*over*) Above-ground phytomass (g m^{-2}) of three species grown for 24 months in pure and mixed stands on an experimental stress-disturbance matrix. The density of seed sown for each species was the same in pure and mixed stands (i.e. the design of the experiment was additive). Although the results presented here refer only to three species, the full experiment involved four other species, *Dactylis glomerata*, *Bromopsis erecta*, *Lolium perenne*, and *Catapodium rigida*. All species were sown at the same density in the mixed stand. An analysis of variance of logarithm transformed data indicated that there was a significant ($P < 0.001$) effect of competition on the phytomass of *Festuca ovina* (competition x stress interaction effect) and *Poa annua* (competition x stress x disturbance interaction effect) but not on the phytomass of *Arrhenatherum elatius* ($P < 0.05$). There was a significant stress x disturbance interaction effect for all three species ($P < 0.001$). Contour lines fitted by eye show changes in phytomass in 10-fold steps from 0.01 g m^{-2} to 1000 g m^{-2}. (Reproduced from Campbell and Grime 1992 by permission of *Ecology*.)

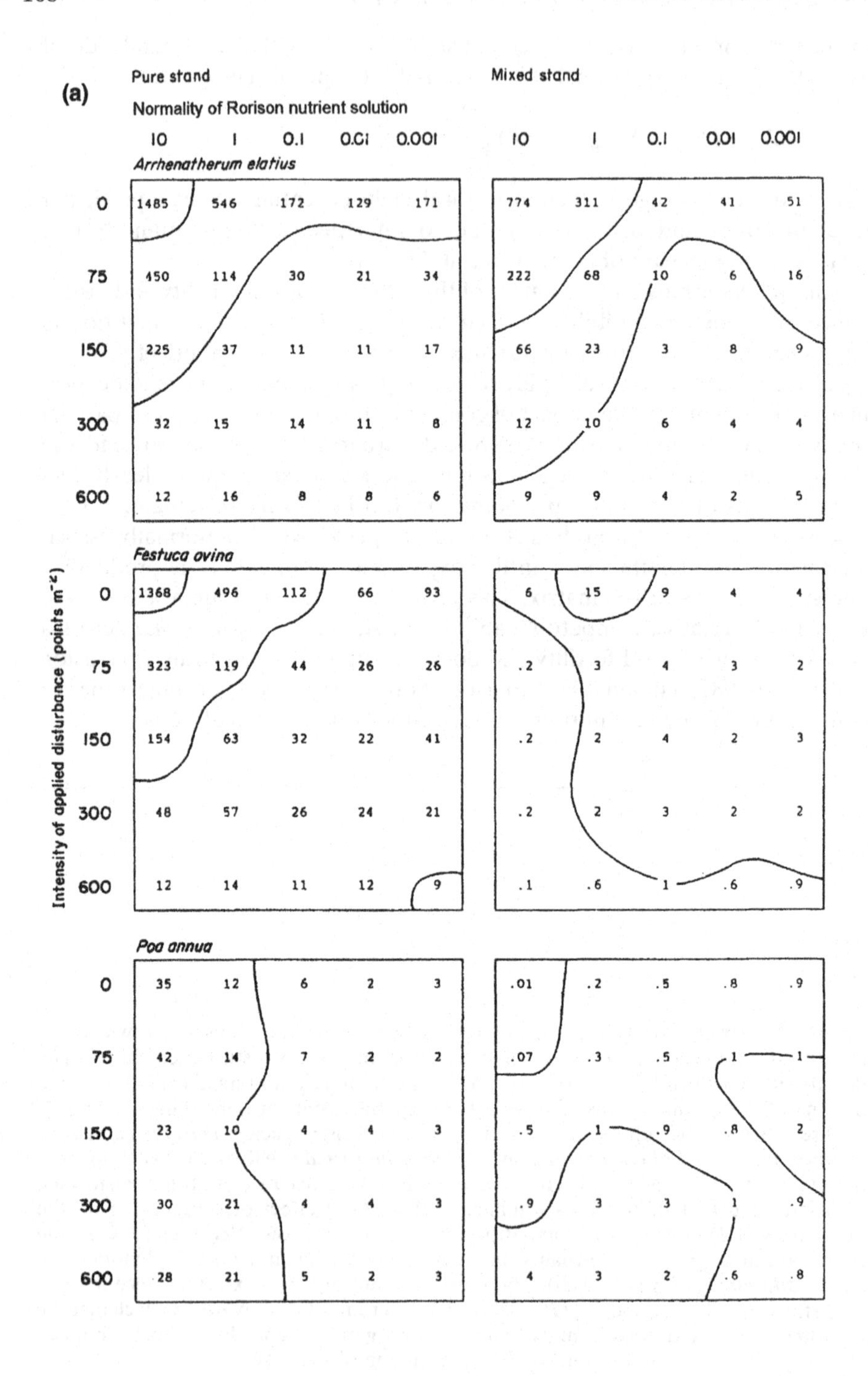
(a)
Pure stand
Mixed stand
Normality of Rorison nutrient solution
10 1 0.1 0.01 0.001
10 1 0.1 0.01 0.001
Arrhenatherum elatius
0
75
150
300
600
1485 546 172 129 171
450 114 30 21 34
225 37 11 11 17
32 15 14 11 8
12 16 8 8 6
774 311 42 41 51
222 68 10 6 16
66 23 3 8 9
12 10 6 4 4
9 9 4 2 5
Festuca ovina
1368 496 112 66 93
323 119 44 26 26
154 63 32 22 41
48 57 26 24 21
12 14 11 12 9
6 15 9 4 4
.2 3 4 3 2
.2 2 4 2 3
.2 2 3 2 2
.1 .6 1 .6 .9
Intensity of applied disturbance (points m⁻²)
Poa annua
35 12 6 2 3
42 14 7 2 2
23 10 4 4 3
30 21 4 4 3
28 21 5 2 3
.01 .2 .5 .8 .9
.07 .3 .5 1 1
.5 1 .9 .8 2
.9 3 3 1 .9
4 3 2 .6 .8

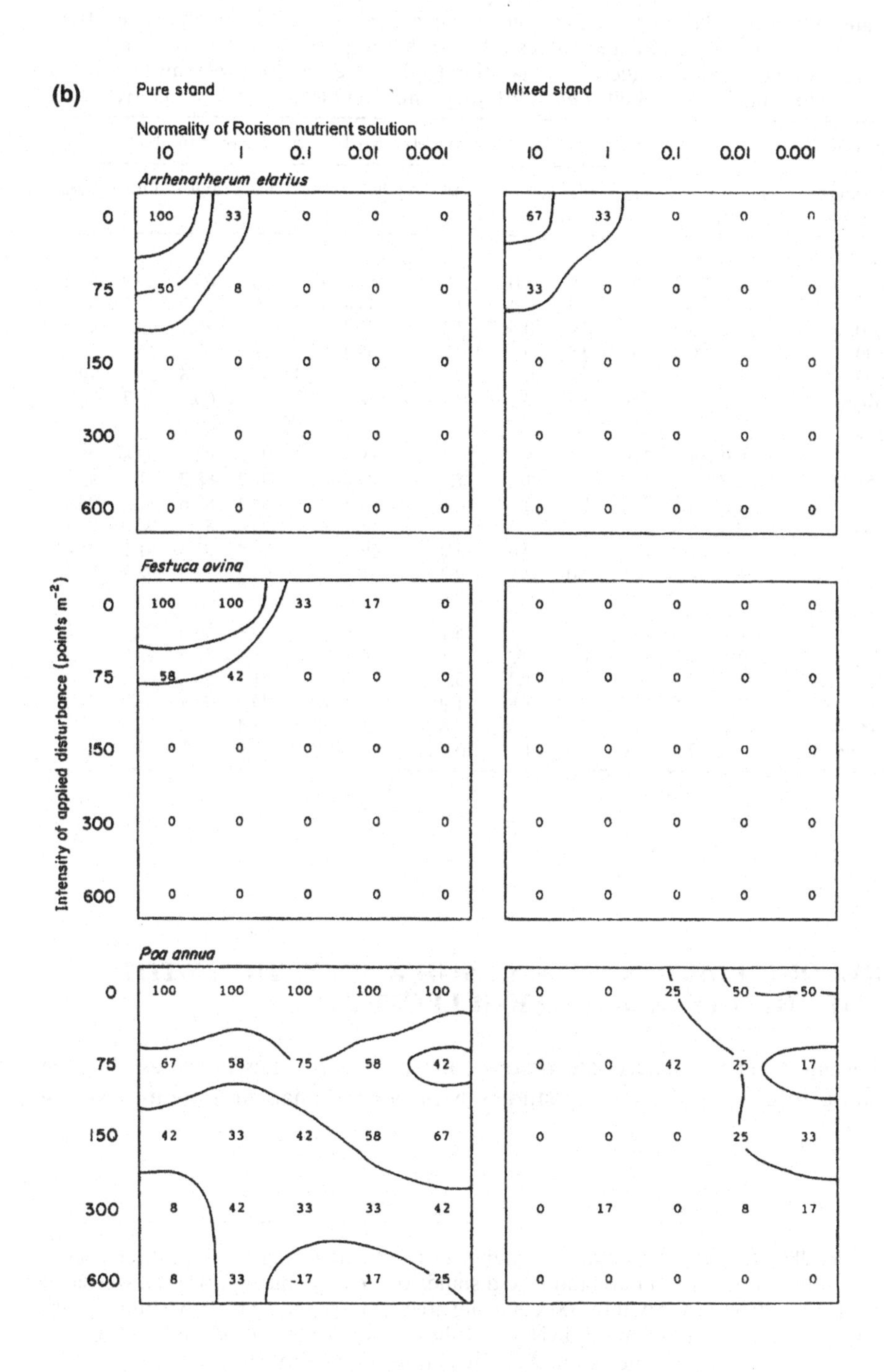
(b)
Pure stand
Mixed stand
Normality of Rorison nutrient solution
10
1
0.1
0.01
0.001
Arrhenatherum elatius
Festuca ovina
Poa annua
Intensity of applied disturbance (points m^{-2})
0
75
150
300
600

Table 9. Two indices of competition on an experimental mineral nutrient stress-disturbance matrix. Tabulated values refer to the proportional effect of competition (left) and the proximal effect of competition (right). See text for explanation. (Reproduced from Campbell *et al.* 1991b by permission of *Functional Ecology*.)

Intensity of applied disturbance (points m^{-2})	Intensity of applied mineral nutrient stress (relative nutrient concentration)											
	10	1	0.1	0.01	0.001	Mean	10	1	0.1	0.01	0.001	Mean
Arrhenatherum elatius												
0	110.0	20.2	8.7	6.6	8.0	28.7	47.4	43.7	72.1	68.7	66.9	59.8
75	17.7	2.7	1.4	1.2	1.4	4.9	45.8	33.5	65.7	74.2	55.3	54.9
150	13.1	0.6	0.7	0.2	0.6	3.1	72.4	23.8	70.1	20.7	50.9	47.6
300	2.2	0.5	0.6	0.5	0.3	0.8	68.4	45.3	59.9	62.3	46.5	56.5
600	0.3	0.5	0.3	0.4	0.1	0.3	27.9	43.0	48.2	73.8	21.7	49.2
Mean	26.7	4.9	2.3	1.8	2.1	7.5	52.4	37.9	63.2	60.0	48.3	52.3
Festuca ovina												
0	100.0	34.8	7.5	4.5	6.9	30.7	99.4	97.0	92.1	93.7	95.8	95.6
75	23.7	8.2	3.0	1.7	1.8	7.7	99.9	97.3	91.7	88.7	91.2	93.8
150	11.5	4.3	2.1	1.4	2.7	4.4	99.8	96.5	88.4	89.6	92.8	93.4
300	3.4	3.9	1.7	1.5	1.4	2.4	98.8	96.2	90.6	89.9	90.0	93.1
600	1.1	1.2	0.7	0.9	0.6	0.9	98.7	96.0	88.8	95.3	91.1	94.0
Mean	27.9	10.5	3.0	2.0	2.7	9.2	99.3	96.6	90.3	91.5	92.2	94.0
Poa annua												
0	100.0	16.8	6.4	1.0	1.5	25.1	100.0	97.4	90.2	50.0	56.7	78.8
75	51.5	16.2	6.7	1.0	0.2	15.1	99.8	97.4	91.7	52.2	12.5	70.7
150	56.1	14.3	3.4	3.2	1.4	15.7	98.6	90.6	81.3	80.4	48.5	79.9
300	31.2	21.5	2.2	2.8	1.8	11.9	96.8	84.5	49.0	72.7	67.7	74.2
600	27.9	18.7	3.1	1.5	2.6	10.8	89.7	86.5	57.4	68.0	76.9	75.7
Mean	53.3	17.5	4.4	1.9	1.5	15.7	97.0	91.2	73.9	64.7	52.5	75.9

RECONCILIATION OF C-, S-, AND R-SELECTION WITH THE THEORY OF r- AND K-SELECTION

The most generally accepted theory concerning strategies in the established phase to emerge from early studies with both animals and plants was the

---

Figure 38b. (*p. 109*) Percentage frequency of rooted flowering stems of three species grown for 24 months in pure and mixed stands on an experimental stress-disturbance matrix was determined within 200 x 200 mm areas corresponding to 25 positions on the matrix using 100 x 100 mm subdivisions of the plot surface. (Reproduced from Campbell and Grime 1992 by permission of *Ecology*.)

concept of r- and K-selection which was originally proposed by MacArthur and Wilson (1967) and expanded by Pianka (1970). This theory prompted many ecologists to recognise as opposite poles in the evolutionary spectrum two types of organism. The first, said to be K-selected, consist of organisms in which the life expectancy of the individual was long and the proportion of the energy and other captured resources which was devoted to reproduction was small. The second, or r-selected, type was made up of organisms with a short life-expectancy and large reproductive effort. It is now widely accepted that the majority of organisms fall between the extremes of r- and K-selection and the results of studies such as those of Gadgil and Solbrig (1972), Abrahamson and Gadgil (1973), and McNaughton (1975) suggested that genetic variation may cause populations of the same species to occupy different positions along the r–K continuum.

Figure 39 attempts to reconcile the concept of the three primary plant strategies with that of r- and K-selection. It is suggested that the ruderal and stress-tolerant strategies correspond, respectively, to the extremes of r- and K-selection and that competitors occupy an intermediate position. This relationship is consistent with the model proposed later (page 243) to explain the involvement of the three strategies in the process of secondary vegetation succession.

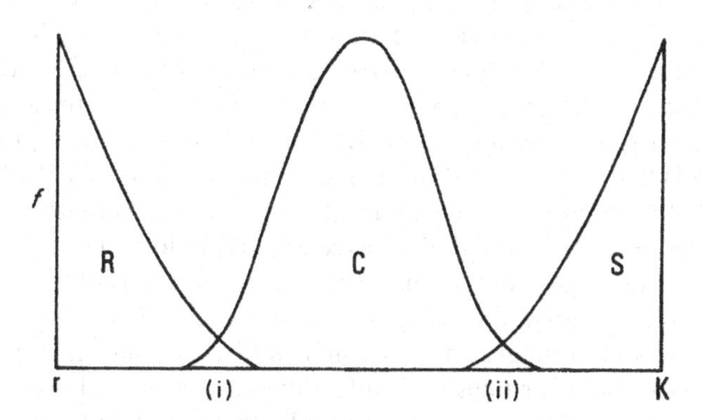

Figure 39. Diagram describing the frequency (*f*) of ruderal (R), competitive (C), and stress-tolerant (S) strategies along the r–K continuum. An explanation of the significance of critical points (i) and (ii) is included in the text. (Reproduced from Grime 1977, *American Naturalist* **111**, by permission of the University of Chicago Press. © 1977 the University of Chicago Press.)

The most substantial way in which the three-strategy model differs from that of the r–K continuum lies in recognition of stress-tolerance as a distinct strategy evolved in intrinsically unproductive habitats or under conditions of extreme resource depletion induced by the vegetation itself. Inspection of Figure 39 suggests that there are two critical points, (i) and (ii), along the r–K continuum. At (i), the intensity of disturbance becomes insufficient to prevent the exclusion of ruderals by competitors, whilst at (ii) the level of supply of resources is depleted below the level required to sustain the high rates of reinvestment of captured resources characteristic of competitors, and selection begins to favour the more conservative physiologies of the stress-tolerators.

It is desirable that any general model of strategies in the established phase should describe also the secondary strategies which have evolved in the various conditions which fall between those associated with the primary strategies. In this respect the scheme represented in Figure 39 is both incomplete and misleading. This is because the linear arrangement does not correspond to the full range of conditions to which strategies of the established phase may be adapted. In order to accommodate the range of equilibria which may exist between stress, disturbance, and competition it is necessary to arrange the primary strategies within the triangular model considered in the next chapter.

ANALOGOUS STRATEGIES IN FUNGI AND IN ANIMALS

In view of the evidence suggesting that in the established phase autotrophic plants conform to three basic strategies and to various compromises between them, it is interesting to consider whether the same pattern obtains in heterotrophic organisms. In fungi, there is strong evidence of the existence of strategies corresponding to those recognised in green plants and these have been explored with explicit reference to CSR theory by Pugh (1980), Cooke and Rayner (1984), and Grime (1988b). 'Ruderal' life-styles are particularly characteristic of the *Mucorales* in which most species are ephemeral colonists of organic substrates. These fungi grow exceedingly rapidly and exploit the initial abundance of sugars, but as the supply of soluble carbohydrate declines they cease mycelial growth and sporulate profusely. 'Competitive' characteristics are evident in fungi such as *Serpula lachrymans* and *Armillaria mellea* which are responsible for long-term infections of timber and produce a consolidated mycelium which may extend rapidly through the production of rhizomorphs (Dowson *et al.* 1986; Rayner *et al.* 1987). Examples of 'stress-tolerant' fungi appear to include the slow-growing basidiomycetes which form the terminal stages of fungal succession on decaying matter and the various fungi which occur in lichens and ectotrophic mycorrhizas. All of the 'stress-tolerant' fungi are characterised by slow-growing, relatively persistent mycelia and low reproductive effort (e.g. Burgeff 1961; Reed 1987).

The majority of the attempts which have been made to recognise strategies in animals have adhered closely to the concept of the r–K continuum. However, Wilbur *et al.* (1974), Whittaker (1975), Nichols *et al.* (1976), and Southwood (1977a) have recognised the need for an extended theoretical framework and have presented information and ideas consistent with the concept of three primary strategies. In the animal kingdom, 'ruderal' life-histories occur in both herbivores and carnivores. They are particularly common in insects (e.g. aphids, blow-flies) where, as in ruderal plants, populations expand and contract rapidly in response to conditions of temporary abundance in food supply (Brown 1962; Southwood 1977b). As pointed out by Southwood *et al.* (1974), similar characteristics are apparent in short-lived birds including not only grain-eaters, such as the zebra finch (*Taeniopygia castanotis*) and the budgerigar (*Melopsittacus undulatus*), but also the insectivorous bluetit (*Parus caeruleus*).

Earlier it has been pointed out that a characteristic of competitive plants is the possession of mechanisms of phenotypic response which maximise the capture of resources, and it has been argued that these mechanisms are advantageous only in environments in which high rates of capture of energy, water, and mineral nutrients can be sustained. It has been suggested (Grime 1977) that the same principle applies in animals and that a direct analogy can be drawn between the 'foraging' responses of competitive plants (page 23) and the liberal expenditure of energy and other captured resources which characterises the very active methods of food-gathering (Newton 1964; Nitsan *et al.* 1981; Gadallah and Jeffries 1995a,b) observed in the species of mammals, birds, and fish which depend for their survival upon continuous access to high rates of food supply. Support for this hypothesis is evident in the theoretical analysis of Norberg (1977), who concluded that 'the most energy-consuming, but also most efficient, search methods should be employed by a predator at the highest prey densities.'

There is also a small amount of evidence of a very direct parallel in foraging biology between some fast-growing animals and plants of productive habitats. Experimental manipulations (e.g. Savory and Gentle 1976) have revealed that reductions in food quality in birds such as Japanese quail induce a reversible increase in the absorptive surface area of the gut, a phenomenum remarkably similar to the rapid changes in leaf and root areas that are observed when C-strategists are exposed, respectively, to shade or low nutrient treatments.

It is interesting to note that Norberg (1977) also concludes that 'when prey density decreases a predator should shift to progressively less energy-consuming search methods although these are connected with low search efficiency.' This same principle has been stated with specific relevance to thermally-inhospitable environments by Morhardt and Gates (1974) as follows: 'Regulation of energy exchange may require a considerable portion of an animal's time and metabolic energy reserves and, in energetically severe habitats, may seriously limit activities which are necessary for the normal biology of the animal, such as seeking food and defending territory.' From these theoretical studies, and also by analogy with plant strategies, therefore, we may predict that in extreme habitats where food is scarce, the level of food capture may be too low and the risks of environmental stress (e.g. desiccation, hypothermia) and predation too great to permit domination of the fauna by species which indulge in very active methods of food gathering. Studies of certain of the animals which occur in habitats such as polar regions (Downes 1964; Andreev 1991), deserts (Chew and Butterworth 1964; French *et al.* 1966; Pianka 1966; Tinkle 1969), coral reefs (Odum and Odum 1955; Muscatine and Cernichiari 1969; Goreau and Yonge 1971), and tropical forests (Montgomery and Sunquist 1974) indicate that, in species adapted to low rates of food supply, strategies of active food gathering give way to feeding mechanisms which are more conservative and appear to be analogous to the assimilatory specialisations observed in stress tolerant plants (Greenslade 1983). Moreover, the similarities between plants and animals exploiting conditions of low resource-availability extend to the life-histories and reproductive biology. Some of the most remarkable parallels with stress-tolerant plants are to be found in the long-life histories and exceedingly delayed and intermittent reproduction in reptiles such as the sphenodon (*Sphenodon punctatum*), and the Aldabran giant tortoise (*Geochelone gigantea*) (Swingland 1977) and birds such as the wandering albatross *(Diomedea exulans)*.

As Nichols *et al.* (1976) have pointed out, there is a scattered but extensive body of evidence which suggests that, in environments with low and uncertain rates of food supply, there is a tendency for animal life-histories to be extended and for reproductive activity to be capable of suspension during exceptionally unfavourable years. In desert environments in North America, for example, it is not unusual, during years of very low rainfall and reduced plant production, for reproduction to be completely inhibited in rodents belonging to genera such as *Perognathus* and *Dipodomys* (French 1967; French *et al.* 1966, 1974). Suspension of reproduction also occurs in various types of Australian animals of desert environments including birds (Keast 1959) and frogs (Bragg 1945; Main *et al.* 1959; Bentley 1966), and a similar phenomenon has been recorded in lizards (Turner *et al.* 1970) and in waterfowl of unpredictable habitats (Frith 1973).

Support for extension of the CSR framework to invertebrates is apparent in a comparative review of species attributes and food plants of British butterflies (Hodgson 1993). In a pattern strongly resembling the resource quality gradient (Axis 1) detected across the 43 plant species examined in the ISP (Grime *et al.* 1997; page 101), it was discovered that variation in the fecundity and rates of development in British butterflies is predictable from the growth rates and nutritional quality of the plants exploited by the caterpillars. Further parallels between the strategies of plants and invertebrates are apparent from several studies in which measurements have been made of the effect of food shortage on longevity and reproductive activity. Of particular relevance are the comparative investigations which have been conducted upon freshwater triclads (Reynoldson 1961, 1968; Calow 1977) and gastropods (Grime and Blythe 1968) which show differences in response which closely parallel those distinguishing ruderal and stress-tolerant plants. In the studies of Calow and Woolhead (1977), for example, it has been shown that, in the annual triclad

Dendrocoelum lacteum, the response to food shortage is to sustain reproductive effort with the result that increased rates of mortalities are experienced in the adult population. In marked contrast, under the same conditions, the perennial species, *Dugesia lugubris*, responds promptly with a cessation of reproduction, a mechanism which, as in the case of competitive and stress-tolerant plants, appears to safeguard the survival of the established population.

ANALOGOUS STRATEGIES IN ECONOMIC SYSTEMS

In 1985 a review was published exploring some of the parallels that can be recognised between resource limitations in plants and the dynamics of currency movement in economic systems (Bloom *et al.* 1985). Although these ideas were not developed with explicit reference to CSR theory it was immediately apparent that a close analogy exists between ecological and economic systems when resource dynamics and, in particular, mineral nutrients are allowed to become the focus of study. At a superficial level, for example, it is possible to recognise a similarity between differences in resource capture and utilisation exhibited by C-, S-, and R-strategists and those business enterprises that represent contrasted policies of financial investment and risk management. It is not the purpose of this book to enlarge upon these rather obvious parallels but rather to comment on one particular issue.

As we have seen in earlier pages of this book a recurring debate among plant ecologists over a considerable period of time has been the question of how the intensity of competition varies in relation to resource availability. In this chapter the conclusion has been drawn that competition between plants achieves greatest intensity in circumstances where interactions are taking place between large fast-growing perennials. In Grime (1987) it is pointed out that under such conditions the very rapid exclusion of weaker competitors is not a simple consequence of resource capture by larger plants. A further consequence of dominance by fast-growing perennials is the creation of an unpredictable and hazardous environment for smaller plants arising from the dynamic foraging, resource depletion and litter deposition of the shoots. In this respect it is tempting to draw a direct parallel with the intense and rapidly-resolved competitions in those parts of the local or global marketplace where business instability and failure coincide not with chronic poverty but with the interaction of large enterprises and the rapid spatial redistribution of capital.

2 Secondary Strategies in the Established Phase

INTRODUCTION

The environments which, in Chapter 1, have been associated with the occurrence of competitors, stress-tolerators, and ruderals form only part of the spectrum of habitats available to plants. It seems reasonable to suppose, therefore, that in addition to the three extremes of evolutionary specialisation there will be various secondary strategies which have evolved in habitats experiencing intermediate intensities of competition, stress, and disturbance.

Figure 40 illustrates the conditions in which various types of secondary strategies may be expected to occur. The model consists of an equilateral triangle in which variation in the relative importance of competition, stress, and disturbance as determinants of the vegetation is indicated by three sets of contours. At their respective corners of the triangle, competitors, stress-tolerators, and ruderals become the exclusive constituents of the vegetation and the remaining areas of the triangle correspond to the various equilibria which are possible between competition, stress, and disturbance. Four main types of secondary strategy are proposed. These consist of:

1. *Competitive ruderal* (C-R)—adapted to circumstances in which there is a low impact of stress and competition is restricted to a moderate intensity by disturbance.
2. *Stress-tolerant ruderals* (S-R)—adapted to lightly-disturbed, unproductive habitats.
3. *Stress-tolerant competitors* (C-S)—adapted to relatively-undisturbed conditions experiencing moderate intensities of stress.
4. *'C-S-R strategists'*—adapted to habitats in which the level of competition is restricted by moderate intensities of both stress and disturbance.

From both field and laboratory investigations, there is evidence confirming the existence of these strategies and it would appear that the criteria which have been used to define the primary strategies (Table 6) also provide a basis for recognition of the secondary strategies. Until comparative studies have been conducted on the ecology, life-histories, and physiology of a wider range of plants, a comprehensive account of the four secondary strategies cannot be attempted. However, mainly by reference to herbaceous plants of temperate

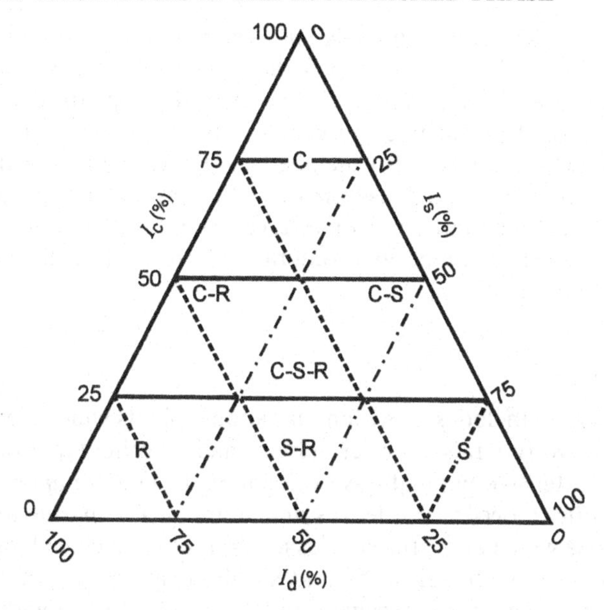

Figure 40. Model describing the various equilibria between competition, stress, and disturbance in vegetation and the location of primary and secondary strategies. *Ic*, relative importance of competition (—); *Is*, relative importance of stress (- - -); *Id*, relative importance of disturbance (-·-·-). A key to the symbols for the strategies is included in the text. (Reproduced from Grime 1977, *American Naturalist*, **111**, by permission of the University of Chicago Press. © 1977 the University of Chicago Press.)

environments, an attempt can be made to illustrate some of their key characteristics. Although, in the descriptions which follow, attention is confined to the established phase of the life-cycle, it is worth noting that certain of the secondary strategies usually occur in combination with particular regenerative strategies, and an assessment of the significance of these associations is attempted at the end of Chapter 3.

COMPETITIVE-RUDERALS

The competitive-ruderals occur in habitats of high productivity in which dominance of the vegetation by competitors is prevented by disturbance. In comparison with habitats populated exclusively by ruderals, *sensu stricto*, the sites colonised by competitive-ruderals experience a smaller effect of disturbance. The reduced impact may be due to the fact that although disturbance is severe it occurs infrequently as, for example, in grasslands which are ploughed and re-sown at intervals greater than two years. Conditions favourable to the persistence of competitive-ruderals may also arise in habitats where damage to the vegetation occurs once annually and is sufficient to check the vigour of competitive species but does not reach the degree of severity necessary to eliminate them from the vegetation. Examples of this type of habitat include fertile meadows and productive grasslands subject to seasonal damage by drought or grazing animals. Also included in this category are the various types of vegetation which are associated with seasonal flood damage, silt deposition, and soil erosion on river terraces and at the margins of ponds, lakes, and ditches. The competitive-ruderals fall into three classes: annuals, biennials, and perennials.

Annual Herbs

The European flora includes many familiar annual plants which may be classified as competitive-ruderals. At river margins and on other types of disturbed fertile ground extensive populations of the annuals, *Galium aparine* and *Impatiens glandulifera*, occur. The leaves and stems of *G. aparine* are covered with small hooks which allow the species to scramble to a considerable height over the shoots of perennial herbs and shrubs (Figure 41). In contrast, *I. glandulifera* often occurs in extensive stands in which each individual is composed of an erect unbranched stem bearing several whorls of large leaves (Figure 42).

In North America, competitive-ruderals may be recognised among the large summer annuals (e.g. *Ambrosia artemisifolia*, *Polygonum pensylvanicum*, and *Abutilon theophrasti*) which colonise abandoned arable fields. Flowering in these plants is preceded by a comparatively long vegetative phase (Raynal and Bazzaz 1975). At maturity, the shoots may be several metres in height and where many individuals have established in close proximity a dense leaf canopy may be developed.

In these potentially large summer annuals, the degree of dominance (page 180) depends not only upon the density of individuals but also upon the duration and vigour of the vegetative phase, termination of which is usually determined by the induction of flowering by increasing daylength. It is clear that in the life-cycles of these species there is a delicate balance between the initial competitive phase and the later reproductive phase. Optimisation of resource capture and seed production in the two phases of the life-history results in some ecotypic variation, particularly with respect to latitudinal races of these plants; this has been considered with respect to *Chenopodium rubrum* by Cook (1976). We may also anticipate that the response to stress in these relatively long-lived annuals is likely to vary according to the stage of development, but little work appears to have been carried out on this subject.

Grasses also provide species which may be described as competitive-ruderals. Examples are *Hordeum murinum*, *Anisantha sterilis*, and *Lolium multiflorum*, all species which are of common occurrence in grasslands in which the perennial component is subject to severe seasonal damage by

Figure 41. The scrambling shoots of *Galium aparine* at Wardlow Mires, North Derbyshire, England

drought, grazing, and trampling. As in all the other species cited under this heading, each is capable of rapid dry matter production and under favourable conditions may reach a large size at maturity. It is interesting to note from Table 10 that there is some variation between these annuals with respect to the mechanism by which large size may be achieved during the vegetative phase. In certain species, e.g. *Fallopia convolvulus*, dry matter production depends upon the combination of a large seed reserve with a modest relative growth rate. In *Chenopodium rubrum*, however, the seed is small and large size may be attained only by a sustained period of rapid photosynthesis.

Many of the species employed in agriculture including *Hordeum vulgare* (barley), *Secale cereale* (rye), *Fagopyron esculentum* (buckwheat), *Zea mays* (maize), *Helianthus annuus* (sunflower), and *Panicum mileaceum* (millet) have characteristics (annual life-cycle, potentially rapid relative-growth rates,

Figure 42. A stand of individuals of *Impatiens glandulifera* on a river terrace at Anston Stones Wood, South Yorkshire, England

capacity to produce a high leaf area index) which allow them to be classified as competitive-ruderals.

Biennial Herbs

A characteristic component of many types of productive, intermittently-disturbed vegetation is made up of large biennial herbs. In temperate regions one family, the *Umbelliferae*, provides a number of good examples including such familiar European species as *Anthriscus sylvestris*, *Heracleum sphondylium*, *Angelica sylvestris*, and *Conium maculatum.* Another family which contributes to this group is the *Compositae* which includes several large biennial species, e.g. *Cirsium vulgare*, *Carduus nutans*, and *Arctium spp.*, which are of frequent occurrence in disturbed vegetation in many parts of the world.

The habitats of these large biennials resemble closely those exploited by the annuals described under the previous heading and it is not unusual for species

Table 10. Estimates of the rates of dry matter production in the seedling phase in ten annual competitive-ruderals (Grime and Hunt 1975). Measurements were conducted over the period 2–5 weeks after germination in a controlled environment (temperature: 20°C (day), 15°C (night); daylength: 18 h; visible radiation: 38.0 W m^{-2}; rooting medium: sand + Hewitt solution). Rates of dry matter production are expressed as maximum (R_{max}) and mean ($\overline{R}$) relative growth rates and as a function of leaf area (E_{max}). The seed weights quoted refer to the seed samples used

Species	Seed weight (mg)	R_{max} week^{-1}	$\overline{R}$ week^{-1}	E_{max}g m^{-2} week^{-1}
Bromus sterilis	8.4	2.3	1.4	134
Chenopodium rubrum	0.09	2.0		116
Galium aparine	7.3	1.5	1.2	96
Helianthus annuus	66.0	0.8	0.8	91
Hordeum murinum	6.6	1.8	1.2	102
Hordeum vulgare	37.8	1.6	0.9	64
Lolium multiflorum	2.5	2.0	1.3	100
Lycopersicon esculentum	3.1	1.6	1.0	93
Polygonum convolvulus	1.3	1.9	1.4	68
Zea mays	273.0	0.9	0.9	107

of both groups to grow together. An obvious point of difference from the annuals lies in the fact that the life-history usually extends over two growing seasons. During the first, a rosette of leaves is formed and photosynthate is accumulated in a swollen rootstock. In the second season the storage organ forms the capital for expansion of flowering shoots which under favourable circumstances may become large structures bearing a considerable number of seeds.

In terms of the established strategy, therefore, it seems reasonable to regard these biennials as close relatives of the large annuals. The main difference is in the more formalised separation of the vegetative and reproductive phases of the life-history. A further distinction, which identifies the biennials as less ruderal in character, is the tendency under unfavourable conditions for the vegetative phase to become extended over several years. A good example of this phenomenon is provided by the study of Werner (1975) in which it was shown that in a North American population of *Dipsacus fullonum* there was a high probability that flowering would be delayed in circumstances where the rosette failed to achieve a diameter of 300 mm by the end of the first season.

Ruderal-perennial Herbs

A third group which may be recognised within the competitive-ruderals consists of a variety of plants which may be conveniently described as ruderal-perennials. These plants often occur as seedlings or small plants during the initial colonisation of bare ground but they are most abundant in circumstances in which the impact of disturbance is less immediate or catastrophic, such as those which occur during the second and third year after colonisation of bare soil in abandoned arable fields, derelict gardens, and construction sites. Species which belong in this category include forbs, e.g. *Ranunculus repens*, *Achillea millefolium*, *Cirsium arvense*, *Tussilago farfara*, and *Trifolium repens*, and grasses, e.g. *Poa trivialis*, *P. pratensis*, *Agrostis stolonifera*, and *Elytrigia repens*.

The majority of these plants are strongly rhizomatous or stoloniferous and show a capacity for rapid vegetative spread. Life-history studies (Ogden 1974; Sarukhan 1974; Watt 1976) of certain of these species have confirmed that, once established in fertile conditions, there is a rapid lateral extension of vegetative shoots either through the soil (e.g. *T. farfara*) or over the ground surface (e.g. *R. repens*). In marked contrast to the offspring orginating from seed, ramets derive considerable benefit from parental plants (Slade and Hutchings 1987; Evans 1988; Alpert 1991; Caraco and Kelly 1991; Turkington and Klein 1991). In many of the ruderal perennials the life-span of an individual shoot is usually less than one year and this characteristic, coupled with the vigorous and often complex proliferation and fragmentation of rhizomes or stolons, brings about a new spatial distribution of shoots in each successive growing season. It is not surprising, therefore, to find that where the effect of seasonal disturbance is to produce a discontinuous vegetation cover, species of this type are efficient colonisers of temporary gaps (Lovett-Doust 1981).

Where the process of colonisation of fertile sites by herbaceous vegetation is allowed to proceed undisturbed for several years the ruderal perennials are progressively excluded by the presence of perennial herbs with taller and more consolidated growth forms, which are capable of producing a dense cover of foliage and plant litter in which there are few gaps. Scrutiny of the vegetation types in which ruderal perennials are a persistent component (e.g. heavily disturbed pastures, meadows, and marshes) suggests that these plants are dependent upon conditions in which the intensity of damage sustained each year by the vegetation is sufficient to debilitate more strongly competitive perennial herbs.

The General Characteristics of Competitive-ruderals

From the preceding review of the characteristics of the annuals, biennials, and perennials included within the competitive-ruderals it is possible to recognise several points of difference from the ruderals *sensu stricto*. In general, competitive-ruderals exhibit a longer period of vegetative growth, and a considerable biomass may be attained before the onset of flowering. A consequence of the delay in seed production is that populations of competitive ruderals are more susceptible than those of ephemerals to habitat disturbance during the growing season. A further distinction lies in the ability of some of the competitive-ruderals to exploit environments already occupied by perennial species. However, it is clear that co-existence with vigorously-growing perennial species occurs only under specific conditions. Phenological studies by Al-Mufti *et al.* (1977), for example, illustrate that in Britain the competitive-ruderals, *Galium aparine* and *Poa trivialis*, persist in vegetation dominated by larger perennials (either herbs or trees) by exploiting periods in the year when the potential impact of the dominant species is at a minimum. An example of this phenomenon is illustrated in Figure 43 which describes the relationship between the shoot phenology of *P. trivialis* and the seasonal change in standing crop and tree litter on an alluvial terrace which in summer is shaded by a dense canopy of deciduous trees. From this figure, it is apparent that *P. trivialis* exhibits a truncated phenology in which rapid growth and flowering occur in the short period in the spring preceding full expansion of

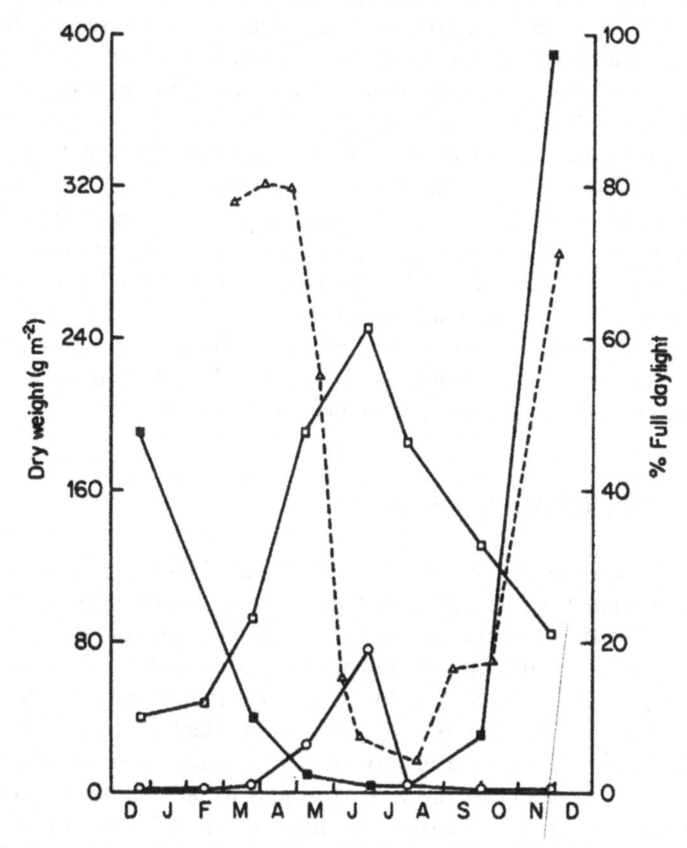

Figure 43. Seasonal variation in the shoot biomass of the perennial grass, *Poa trivialis* (○-○), in a deciduous woodland in northern England, and associated changes in the total biomass of the herb layer (□-□) and in the density of tree litter (■-■). Light intensity reaching the herb layer (Δ-Δ) is expressed as a percentage of that measured simultaneously at an adjacent unshaded site. (Reproduced from Al-Mufti *et al* 1977 by permission of *Journal of Ecology*.)

both tree and herbaceous canopies and coinciding with the annual minimum in density of tree litter. In roadsides, meadows, and river banks a similar phenology occurs in the biennial, *Anthriscus sylvestris*, and in the annual, *Impatiens glandulifera*, in each of which leaf expansion usually occurs in advance of that of the neighbouring shoots of established perennial herbs.

Although the annuals, biennials, and perennials are distinguishable in terms of life-history and reproduction there can be little doubt of the strong ecological affinities between the three classes of competitive-ruderals. There are many vegetation types in which representatives of two or all three of the classes occur together. Reference has been made already to the frequent association of the annual species *Galium aparine* with the short-lived perennial, *P. trivialis*. In damp meadows and marshes it is not unusual to find biennials such as *Heracleum sphondylium* and *Angelica sylvestris* in close proximity to perennial species such as *Ranunculus repens* and *Agrostis stolonifera*, whilst in semi-derelict habitats in urban areas all three classes of competitive ruderals are often represented.

In view of this evidence of close similarities in field distribution it is pertinent to consider why the different life-histories exhibited by the three classes of competitive ruderals do not give rise to greater disparities in ecology. In particular it is necessary to explain why species totally dependent upon seedling establishment for maintenance of their populations (the annuals and biennials) frequently co-exist with perennial ruderals in which the annual replacement of individuals is mainly by vegetative reproduction. However, as we shall see in Chapter 3, the regenerative strategies of the annuals and biennials are not as dissimilar from the perennials as they appear on first inspection.

STRESS-TOLERANT RUDERALS

Stress-tolerant ruderals occur in habitats in which the character of the vegetation is determined by the coincidence of moderate intensities of stress and disturbance. The stress-tolerant ruderals resemble the ruderals *sensu stricto* in that opportunities for growth and reproduction are usually restricted to relatively short periods. However, a distinguishing characteristic of the stress-tolerant ruderals is that they occupy habitats in which stress conditions are experienced during the period of growth. The severity of stress often varies from one growing season to the next, but it is usually sufficient to restrict annual production to well below that achieved in habitats dominated by ruderals or competitive-ruderals and in exceptional years both vegetative growth and reproduction may be very severely restricted.

Two types of plants are strongly represented among the stress-tolerant ruderals. The first consists of small herbs whilst the second is made up of bryophytes.

Herbaceous Plants

The herbaceous plants which may be classified as stress-tolerant ruderals consist of a rather heterogeneous assemblage drawn from widely different habitats encompassing the complete range of world climatic zones. Two main groups may be recognised. The first consists of small annuals and short-lived perennials whilst the second is composed of small geophytes.

Small Annuals and Short-lived Perennials

Terrestrial habitats which experience severe levels of both stress and disturbance tend to be devoid of vegetation; the effect of low productivity and disturbance is such that even short-lived species are unable to complete their life-cycles. However, as we approach conditions in which the effects of stress and disturbance are slightly less severe, environments are encountered in which certain specialised plants can survive. Among the latter are a variety of small annual plants. These include the tiny arctic annual, *Koenigia islandica*, the diminutive desert ephemerals which emerge following rain-showers in some deserts (Went 1948, 1949, 1955), and the wide variety of annuals which exploit shallow unproductive soils in temperate regions. By far the largest contribution to this last group is that composed of the various types of small winter annuals, many of which have been the subjects of intensive studies in Europe (Ratcliffe 1961; Newman 1963; Pemadesa and Lovell 1974; Verkaar and Schenkeveld 1984; Rees *et al.* 1996), North America (Tevis 1958; Beatley 1967), and Australia (Mott 1972).

In temperate regions these plants are restricted to shallow or sandy soils which experience desiccation during the summer or to local habitats such as ant-hills (Grubb *et al.* 1969; King 1977a, b, c) where the vigour of perennial herbs is restricted by animal activity. Growth and flowering occur during the cool wet season and the seeds remain ungerminated throughout the dry summer period. In comparison with summer annuals or the winter annuals which occur on deep fertile soils (e.g. *Galium aparine*) the stress-tolerant winter annuals are small in stature, have lower potential relative growth-rates, and produce smaller seeds. Such characteristics are consistent with the fact that these plants occur in habitats in which production is often severely restricted by mineral nutrient deficiencies and grow at times of the year when photosynthesis is restricted by low temperature.

It is not uncommon, especially in more northern latitudes, for the life-cycles of certain winter-annuals, e.g. *Erigeron acer*, *Clinopodium acinos*, *Linum catharticum*, to be extended over two years while some of the rather less ephemeral monocarpic species of disturbed unproductive habitats, e.g. *Carlina vulgaris* and *Gentianella amarella*, have life-histories which frequently extend for longer periods. Like the small winter annuals, these biennials and short-lived perennials are particularly common in unproductive habitats in which the perennial component is subject to local damage by drought or by biotic factors, e.g. hoof-marks and rabbit-scrapes. Again, in common with the winter annuals, the short-lived perennials show considerable plasticity in growth form and under unusually favourable conditions numerous seeds may be produced.

Small Geophytes

Another distinctive component of disturbed but unproductive vegetation is composed of what Noy-Meir (1973) has described as the 'perennial ephemeroids'. Many of these plants resemble the small winter annuals in that they are abundant on shallow unproductive soils in regions such as southern Europe where herbaceous vegetation is subject to severe desiccation during the summer. Growth in these plants tends to be confined to the moist, cool season and survival of the dry season is in the form of an underground storage organ such as a bulb (e.g. *Scilla verna*), tuber (e.g. *Orchis mascula*), or rhizome (*Anemone nemorosa*, *Primula veris*).

It is evident that in several respects the geophytes resemble the group of annual and biennial species considered under the last heading. Traits common to both types include small stature, rather slow relative growth-rates, and seeds which vary in size from small to minute. All three of these characteristics appear to be related to the low productivity of the habitat. A possible point of difference between the annuals, short-lived perennials, and geophytes relates to the year-to-year fluctuations in population density. Following years of exceptional disturbance by factors such as drought, trampling, or fire, annual and biennial species often show very rapid expansion in numbers (Lloyd 1968). Populations of geophytes appear to be relatively more stable, although in some species, including many orchids in which the juvenile plants are saprophytic and remain below ground, it is necessary to conduct long-term population studies (e.g. Tamm 1972; Wells 1967) in order to obtain accurate estimations of population size.

Bryophytes

The least conspicuous of the stress-tolerant ruderals are mosses and liverworts. This is not to suggest that all bryophytes conform to this strategy. Many mosses, e.g. *Racomitrium lanuginosum* and *Dicranum fuscescens*, occur in extremely unproductive habitats and studies of their ecology, life-history, and physiology (Tallis 1958, 1959, 1964; Oechel and Collins 1973; Hicklenton and Oechel 1976) have shown that they have strong affinities with lichens, whilst others, e.g. *Funaria hygrometrica* (Hoffman 1966) are ephemeral colonists of disturbed habitats. However, in addition to these stress-tolerators and ruderals there are many bryophytes, e.g. *Brachythecium rutabulum*, which are associated with mesic conditions and exhibit marked seasonal changes in biomass (Al-Mufti *et al.* 1977; Furness and Grime 1982a, b). This group includes many of the mosses and liverworts which, in temperate regions, colonise bare soil and plant litter beneath productive herbaceous vegetation. In such habitats, the bryophyte component usually shows well-defined peaks of growth in the moist, cool conditions of early spring and autumn (see Figure 94, page 273). The most likely explanation for the decline in biomass observed during the summer would appear to be desiccation, although under certain conditions effects of shade and high temperatures also may be implicated. The results of laboratory studies (see Figure 95, page 274) show that, in accordance with their phenology, bryophytes of this type have lower temperature optima for growth than the vascular plants with which they are associated in the field.

STRESS-TOLERANT COMPETITORS

A large number of plant species are typically associated with vegetation types which exhibit moderate productivity and experience very low intensities of disturbance. As we might expect, the characteristics of these plants fall between those of the competitors and the stress-tolerators. They may be conveniently classified into herbaceous and woody plants.

Herbaceous Plants

In derelict grassland, heath, and marshland in temperate regions there are many grasses, sedges, and rushes which fall into the category of the stress-tolerant competitors. In common with the 'competitors' (e.g. *Urtica dioica*, *Epilobium hirsutum*) these plants are robust perennials which have the capacity for lateral vegetative spread by means of rhizomes or expanding tussocks. However, points of difference from the competitive herbs are the lower maximum potential relative growth-rates (Grime and Hunt 1975), the longer life-span of the leaves, and the shoot phenology which in many species shows an exact compromise between that of the competitor and the stress-tolerator, i.e. they are evergreens with a marked decline in shoot biomass during the winter and a pronounced peak in the summer (Figure 45, page 134). European grasses which exemplify this strategy occur in a variety of habitats; they include *Brachypodium pinnatum*, *Bromopsis erecta*, *Festuca arundinacea*, *F. rubra* (semi-derelict grassland), *Molinia caerulea* (marshes), and *Ammophila arenaria* and *Elymus arenarius* (sand-dunes). In marshland there are numerous examples among both sedges, e.g. *Eriophorum vaginatum*, *Cladium mariscus*, *Scirpus sylvaticus*, and *Carex acutiformis*, and rushes, e.g. *Juncus effusus*, *J. acutus*, and *J. subnodulosus*. Among dicotyledonous herbs a most familiar example is provided by the woodland plant, *Mercurialis perennis*. This shade-tolerant species is strongly rhizomatous and forms extensive stands with a leaf canopy of moderate density in which individual leaves may remain alive from March to November. Another woodland species of rather similar ecology is the monocotyledon, *Convallaria majalis*.

Stress-tolerant competitors are strongly represented among the grasses and sedges exploiting the semi-arid conditions of the North American prairies and the Russian Steppes, where typical species include *Stipa viridula*, *S. spartea*, *Sporobolus heterolepis*, and *Carex pensylvanica* (Weaver and Albertson 1956). Another species which appears to provide a particularly good example is *Carex lacustris*, a North American wetland species. Intensive studies of this plant (Bernard and McDonald 1974; Bernard and Solsky 1977) have revealed a shoot phenology typical of the stress-tolerant competitor and a maximum lifespan for the leaves of approximately one year.

Woody Plants

As explained on page 132, there is reason to suspect that the majority of the trees and shrubs which occur in unproductive habitats or in the later stages of vegetation succession in more fertile terrain fall into the category of the stress-tolerant competitor. Because of the shortage of comparative data on the life-histories and physiology of trees and shrubs there is insufficient evidence with which to examine this hypothesis. It is apparent, however, that many woody species exhibit, in particular attributes, degrees of specialisation which lie between the two extremes recognised in Table 6 as characteristic of the competitor and the stress-tolerator. Examples here include the intermediate longevity of many deciduous forest trees in genera such as *Quercus*, *Carya*, *Castanea*, and *Fagus* and their fluctuating but fairly continuous seed output (Harper and White 1974). A feature of certain deciduous shrubs, e.g. *Euonymus europaeus* and *Vaccinium myrtillus*, which may indicate their intermediate status is their capacity to maintain some photosynthetic activity throughout the year by the possession of evergreen stems (Perry 1971). Until much more information is available, however, it is probably unwise to attempt to classify precisely with respect to strategy even the most common shrubs, woody climbing plants, and trees.

'C-S-R' STRATEGISTS

In certain environments it is possible to identify vegetation types in which plants conforming to very different strategies are growing together. In Chapter 9 the circumstances which allow this to occur will be analysed; here it will suffice to observe that co-existence of widely-different strategies is associated with habitats in which the respective effects of competition, stress, and disturbance are confined to particular times in the year and/or show local spatial variation in intensity within the habitat. However, not all of the habitats which experience major impacts of competition, stress, and disturbance are subject to pronounced seasonal or spatial variation in the equilibrium between the three phenomena. In many unfertilised pastures in temperate regions, for example, mineral nutrient-stress and moderate intensities of defoliation by grazing animals are more or less constant features of the habitat. In these circumstances the bulk of the vegetation is usually composed of species with characteristics intermediate between those of the competitor, the stress-tolerator, and the ruderal. Table 11 contains a list of examples of 'C-S-R strategists' of common occurrence in unproductive pastures and grazed marshes of the British Isles. The majority of the grasses are perennials of rather small stature and moderate maximum potential relative growth-rate (e.g. *Festuca ovina*, *Briza media*, *Helictotrichon* spp.). The sedges include a large number of small, shortly rhizomatous species (e.g. *Isolepis setacea*, *Luzula campestris*, *Carex panicea*). A variety of growth-forms and phenologies are apparent in the dicotyledons, but the majority possess rosettes of leaves (e.g.

Table 11. Some C-S-R strategists of the British flora classified into six sub-groups according to taxonomic and morphological criteria

(a) Small tussock grasses	
Festuca ovina	*Anthoxanthum odoratum*
Koeleria macrantha	*Sesleria albicans*
Helictotrichon pratense	*Briza media*
(b) Small deep-rooted forbs with rosettes	
Sanguisorba minor	*Centaurea nigra*
Pimpinella saxifraga	*Scabiosa columbaria*
Leontodon hispidus	*Silene nutans*
(c) Small stoloniferous species	
Fragaria vesca	*Pilosella officinarum*
Potentilla sterilis	*Veronica chamaedrys*
Ajuga reptans	*Glechoma hederacea*
(d) Forb with short rhizome or stock	
Stachys officinalis	*Potentilla erecta*
Prunella vulgaris	*Leucanthemum vulgare*
Succisa pratensis	*Serratula tinctoria*
(e) Legumes	
Lotus corniculatus	*Trifolium medium*
Anthyllis vulneraria	*Astragalus danicus*
Lathyrus linifolius	*Hippocrepis comosa*
(f) Small sedges and rushes	
Carex caryophyllea	*Juncus bulbosus*
Isolepis setaceus	*Luzula campestris*
Juncus squarrosus	*Carex nigra*

Succisa pratensis, *Lotus corniculatus*, *Sanguisorba minor*). As more information becomes available it seems inevitable that a range of types will be distinguished within the C-S-R strategists. As an illustration of the variety of substrategies, three types of common occurrence in unproductive grasslands of the British Isles will now be considered.

Small Tussock Grasses

Where pastures occur on shallow or sandy soils and are subject to mineral nutrient stress and possibly also to desiccation, a high proportion of the standing crop is composed of tussock grasses with narrow, folded, or rolled leaves. Certain of these species, e.g. *Anthoxanthum odoratum*, *Festuca ovina*, and *Koeleria macrantha*, retain green shoots throughout the year but also produce a well-defined peak in shoot weight which coincides with flowering in the late spring. Grasses of this type tend to be rather shallow in root penetration and it is not uncommon for the tussocks to suffer partial or total mortality among the tillers during the summer.

Small tussock grasses are a ubiquitous feature of unproductive pastures. In the North American prairies, for example, the ecological niche which in Britain is occupied by *Festuca ovina* is filled by the closely-similar species, *Bouteloua gracilis*, which survives on ridges and dry uplands where few other grasses can grow (Weaver and Albertson 1956).

Small, Deep-rooted Forbs

A second group which may be recognised is composed of forbs such as *Sanguisorba minor*, *Centaurea nigra*, and *Leontodon hispidus*, the phenology of which resembles that of the competitors in that there is a well-defined summer peak in shoot production. However, in certain other respects these plants differ considerably from the competitors. In each, the shoot biomass is relatively small and is composed of one or more rosettes of leaves, a morphology which severely restricts both the height of the leaf canopy and the capacity of the shoot for lateral spread above and below ground. An additional characteristic of these plants is the presence of a long tap-root system. As Walter (1973) pointed out, such roots usually penetrate deep fissures and allow the species to exploit reserves of moisture which are inaccessible to grasses and other shallow-rooted plants, many of which often tend to become severely desiccated during the summer.

An additional feature of many of these plants is the capacity for regeneration of the shoot from buds situated near the top of the tap-root. In relatively palatable species such as *Sanguisorba minor*, *Leontodon hispidus*, and *Hypochoeris radicata*, this characteristic appears to play an important role in allowing the persistence of these plants in pastures subject to close grazing by sheep, rabbits, and invertebrates.

Small, Creeping, or Stoloniferous Forbs

A recurrent morphology in certain pastures containing C-S-R strategists is that of the creeping plant. In some species (e.g. *Veronica chamaedrys*, *Teucrium* scorodonia) the creeping shoots bear leaves and occasional adventitious roots along their entire length, whilst in others (e.g. *Fragaria vesca*, *Pilosella officinarum*, *Rubus saxatilis*) the shoots are in the form of stolons which are capable of producing daughter rosettes, which although often situated at considerable distance from the parent plant remain connected at least during the first year of their existence. It is tempting to interpret these growth-forms mainly in relation to the process of vegetative propagation and this function has been confirmed in some circumstances by field experiments in which the fate of intact and severed clonal fragments has been compared (e.g. Pennings and Callaway 2000). However, without denying this function, another possibility should be considered. A characteristic of the ecology of many, if not all, of these species is the association with habitats such as rock outcrops, screes, and quarry heaps, in which a high proportion of the ground surface is covered by stone. In this type of habitat opportunities for rooting are extremely localised and dense leaf canopies tend to develop above the areas where soil is accessible. In such circumstances it is apparent that a selective advantage may accrue to species which can absorb water and mineral nutrients from one part of the environmental mosaic, and photosynthesise in another, i.e. plants with growth forms which cause leaves to be subtended over areas of bare rock situated at a considerable distance from the roots. It seems reasonable to suggest that such growth forms may allow exploitation of gaps in herbaceous canopies which are inaccessible to the majority of neighbouring herbs. In an excellent experimental test Friedman and Alpert (1991) were able to demonstrate the validity of this hypothesis by measuring reciprocal transport of photosynthate and nitrogen between patches exploited by connected ramets of *Fragaria chiloensis*.

In passing it is interesting to observe that the same explanation can be applied on a larger scale to the creeping growth forms associated with the pools and bare mud in marshes (*Ranunculus repens*, *Agrostis stolonifera*), woodland hollows filled with persistent tree litter (*Rubus fruticosus*, *Lamiasatrum galeobdolon*), and walls, cliffs, and tree trunks (*Hedera helix*, *Clematis vitalba*).

STRATEGY AND LIFE-FORM

From the evidence reviewed in this chapter it would appear that plant strategies in the established phase of the life-cycle may be classified by reference to a defined range of equilibria between stress, disturbance, and competition. This classification introduces more subtlety in the recognition of strategies and allows some observations to be made concerning the strategic range of particular life-forms and taxonomic groups. In Figure 44 an attempt has been made to describe the approximate strategic range of selected life-forms. The widest range of strategies is that attributed to perennial herbs and ferns. Annual herbs are predominantly ruderal whilst biennial herbs become prominent in the areas of the triangular model corresponding to the competitive-ruderals and the stress-tolerant ruderals. Trees and shrubs comprise competitors, stress-tolerant competitors, and stress-tolerators. Although most lichens are confined to the stress-tolerant corner of the model, bryophytes are more wide-ranging with the centre of the distribution in the stress-tolerant ruderals. Reference has been made already to evidence (Grime 1984b, 1985; Hodgson 1986c; Bond 1989) that the strategic range of some plant families and even some higher taxa is limited by powerful phylogenetic constraints. This factor is particularly evident in some ancient groups such as the gymnosperms, and ferns where the failure to evolve species with short life-histories appears to have had a profoundly limiting effect on the ability of such taxa to exploit intensively-disturbed landscapes. A similar but less pronounced constraint on the development of R- and C-strategists can be detected when families of angiosperms are compared; in the survey conducted by Hodgson (1986c) a strong correlation was detected between the evolutionary advancement (*sensu* Sporne 1982) of plant families and their capacity to exploit the heavily-disturbed landscape of the Sheffield region.

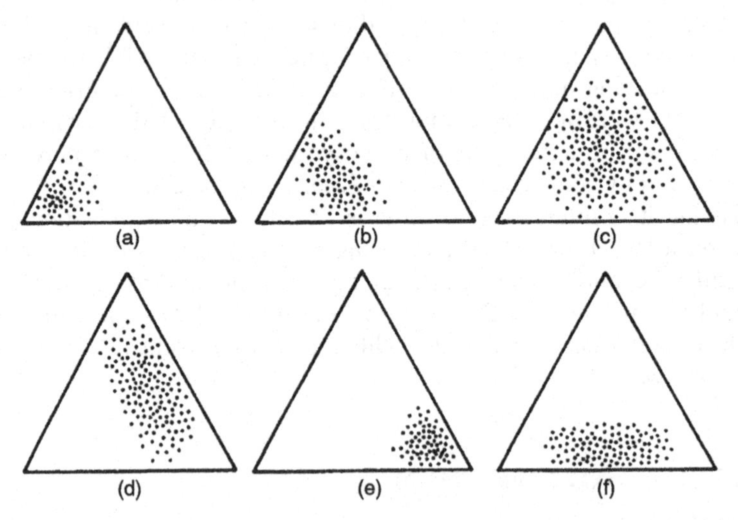

Figure 44. Diagrams describing the range of strategies encompassed by (a) annual herbs, (b) biennial herbs, (c) perennial herbs and ferns, (d) trees and shrubs, (e) lichens, and (f) bryophytes. For the distribution of strategies within the triangle, see Figure 40. (Reproduced from Grime 1977, *American Naturalist* **111**, by permission of the University of Chicago Press. © 1977. The University of Chicago Press.)

TRIANGULAR ORDINATION; 'HARD' AND 'SOFT' ATTRIBUTES

If the triangular model is an accurate general summary of the range of contingencies to which the established phase may be adapted then it should provide a basis upon which to classify both plants and vegetation types. Moreover, as explained in Chapter 6, useful predictions of the impacts of management can be made in circumstances where all the component species in a plant community or area of landscape such as a nature reserve can be classified with respect to C-S-R strategy. There is therefore a requirement for measurable traits that are diagnostic of C-S-R plant functional type. The results of the Integrated Screening Programme (ISP) described on page 103, suggest that precise determinations of strategy are possible when resources and time are available for 'hard' tests involving screening of traits such as seedling relative growth rate, leaf nutrient concentrations, palatability to generalist herbivores, and decomposition rates. Unfortunately, such effort is not practicable for large numbers of species and there is a requirement for 'soft' tests that are easier to apply in field and laboratory.

In the cool temperate zone where there is a sharply-defined growing season, the shoot phenologies of herbaceous species are a useful indicator of their CSR functional types. Figure 45 summarises the patterns detected in quantitative investigations such as that of Al-Mufti *et al.* (1977). It is reassuring to note from these studies that shoot phenology not only distinguishes the three primary strategies but also identifies the four intermediate (secondary) strategies (C-R, S-R, C-S, and C-S-R). However, it must be emphasised that these patterns apply only to herbaceous species and care would be necessary in any effort to extend the use of shoot phenology as an identifier of C-S-R strategy in regions other than the cool-temperate zone.

Efforts to find sets of easily-measured traits that are reliable indicators of strategy constitute a currently very active area of research (e.g. Westoby 1998; Gitay *et al.* 1999; Hodgson *et al.* 1999; Valerio DePatta 1999; Weiher *et al.* 1999). Encouragement in this quest can be drawn from comparison of the ordination of species by hard tests in the ISP (Figure 37) with the distribution of species in an extremely rudimentary ordination using field data and growth analysis results (Figure 46). The method involved a triangular ordination in which species were classified with respect to two criteria. The first of these was the potential maximum rate of dry matter production (R_{max}) measured under a standardised productive environment, whilst the second was a morphology index reflecting the maximum size attained by the plant under favourable conditions. This approach was based therefore upon the hypothesis that in herbaceous plants the primary strategies correspond to three permutations between R_{max} and morphology, i.e. rapidly-growing and large (competitors), rapidly-growing and small (ruderals), and slow-growing and small (stress-tolerators). Particularly in its use of provisional estimates of R_{max} based upon

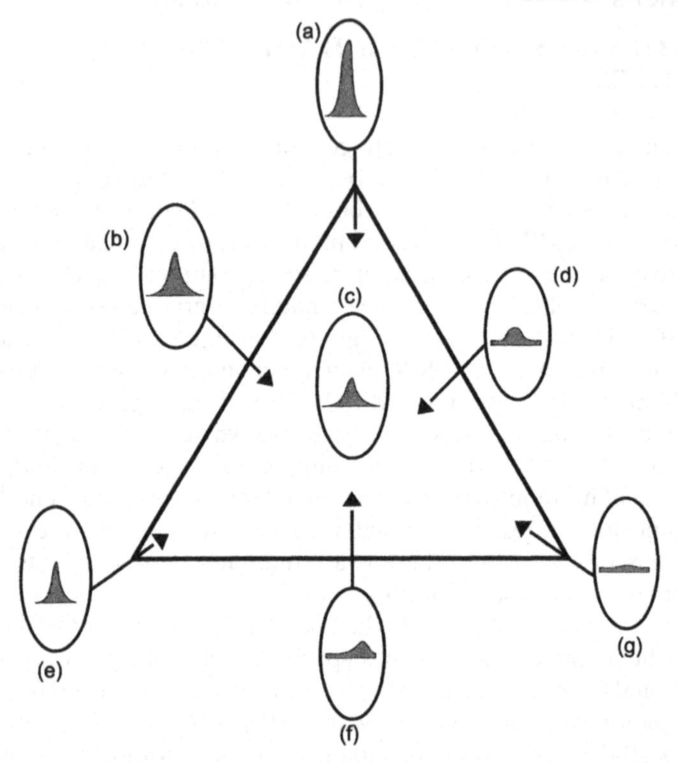

Figure 45. A scheme relating the pattern of seasonal change in shoot biomass to strategy type in herbaceous plants of cool temperate regions: (a) competitor, (b) competitive ruderal, (c) C-S-R strategist, (d) stress-tolerant competitor, (e) ruderal, (f) stress-tolerant ruderal, and (g) stress-tolerator

samples from single populations, this form of classification is extremely unsubtle. Despite these limitations, consistent patterns of distribution were obtained and, with few exceptions, species with strong ecological affinities were found to be located in close proximity to each other. The species which in Figure 46 are closest to the left-hand corner of the triangle include evergreens tolerant of desiccation (*Sedum acre*, *Thymus polytrichus*, *Helianthemum nummularium*), cold (*Dryas octopetala*, *Nardus stricta*), shade (*Viola riviniana*, *Sanicula europaea*, *Deschampsia flexuosa*), or frequent in mineral nutrient-deficient habitats (*Helictotrichon pratense*, *Sesleria caerulea*, *Danthonia decumbens*). As one might expect, annual plants of productive severely-disturbed habitats such as arable land (*Poa annua*, *Stellaria media*) are located in the 'ruderal' corner and perennial herbs of fertile, derelict environments such as *Urtica dioica*, *Epilobium hirsutum* and *Chamerion angustifolium* occur towards the apex of the triangle.

Many of the species included in Figure 46 have been cited in earlier sections of this chapter as examples of particular secondary strategies, and it is

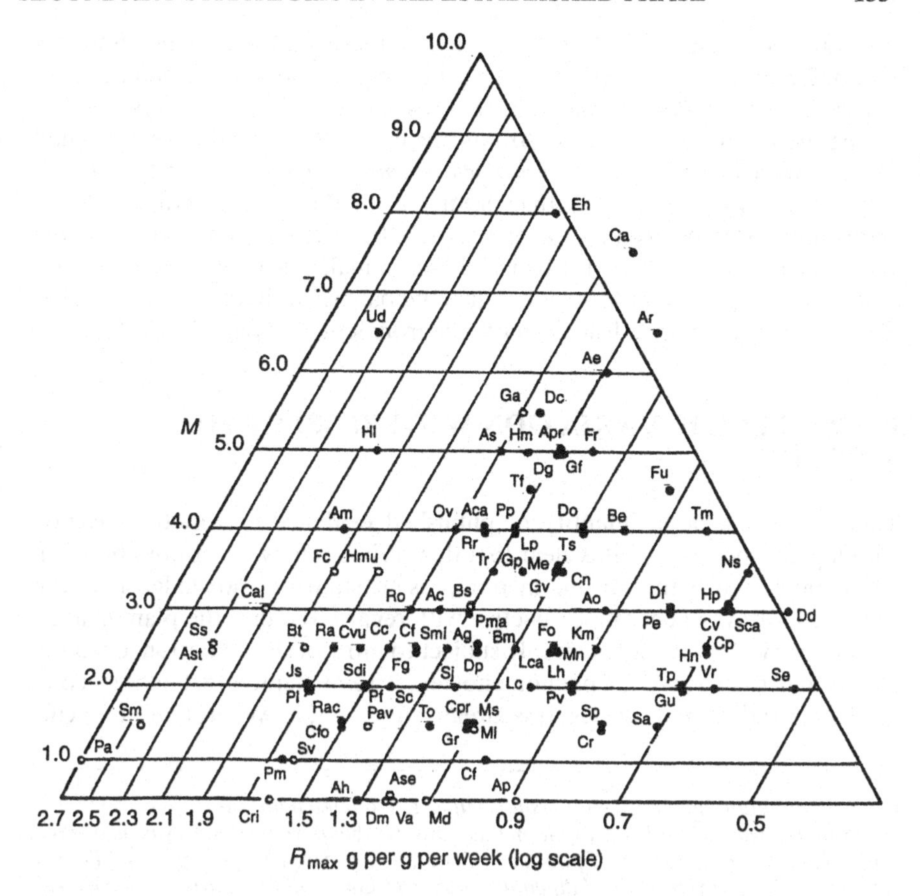

Figure 46. A triangular ordination of herbaceous species. Key: ○ annuals, ● perennials (including biennials). The morphology index *(M)* was calculated from the formula $M = (a + b + c)/2$ where *a* is the estimated maximum height of leaf canopy (1, <120 mm; 2, 120–250 mm; 3, 250–370 mm; 4, 370–500 mm; 5, 500–620 mm; 6, 620–760 mm; 7, 750–870 mm; 8, 870–1000 mm; 9, 1000–1120 mm; 10, >1120 mm); *b* is the lateral spread (0, small therophytes; 1, robust therophytes; 2, perennials with compact unbranched rhizome or forming small (<100 mm diameter) tussock; 3, perennials with rhizomatous system or tussock attaining diameter 100–250 mm; 4, perennials attaining diameter 260–1000 mm; 5, perennials attaining diameter > 1000 mm); *c* is the estimated maximum accumulation of persistent litter (0, none; 1, thin discontinuous cover; 2, thin, continuous cover; 3, up to 10 mm depth; 4, up to 50 mm depth; 5, >50 mm depth). The species are: Ac, *Agrostis canina*; Ae, *Arrhenatherum elatius*; Ag, *Alopecurus geniculatus*; Ah, *Arabis hirsuta*; Am, *Achillea millefolium*; Ao, *Anthoxanthum odoratum*; Ap, *Aira praecox*; Apr, *Alopecurus pratensis*; Ar, *Elytrigia repens*; As, *Agrostis stolonifera*; Ase, *Arenaria serpyllifolia*; Ac, *Agrostis capillaris*; Bm, *Briza media*; Bs, *Brachypodium sylvaticum*; Ast, *Anisantha sterilis*; Bt, *Bidens tripartita*; Ca, *Chamerion angustifolium*; Cal, *Chenopodium album*; Cc, *Cynosurus cristatus*; Cf, *Carex flacca*; Cfl, *Cardamine flexuosa*; Cfo, *Cerastium fontanum*; Cn, *Centaurea nigra*; Cp, *Carex panicea*; Cpr, *Cardamine pratensis*; Cro, *Campanula rotundifolia*; Cri, *Catapodium rigidum*;

(*continued over*)

interesting to observe that the majority of these plants also fall into the expected patterns of distribution. Typical examples here are the competitive-ruderals, *Chenopodium album* and *Holcus lanatus*, and the stress-tolerant competitors, *Festuca rubra* and *Bromopsis erecta*. The location of the small winter annuals, *Aira praecox*, *Veronica arvensis*, *Arenaria serpyllifolia*, and *Draba muralis* coincides with that predicted for the stress-tolerant ruderals and is quite distinct from that of the C-S-R strategists such as *Festuca ovina* and *Koeleria macrantha*, with which these annuals occur on limestone outcrops. This suggests that in such habitats the patches of bare soil occupied by the annuals constitute a distinct micro-environment (see page 268).

INTRASPECIFIC VARIATION WITH RESPECT TO STRATEGY

There is now a large quantity of published evidence of genetic variation within plant species, and it is clear that this variation may occur both between populations and within them. In our present state of knowledge it is not possible to draw a general perspective with regard to either the plant characteristics, which are most commonly subject to intraspecific variation, or to the circumstances in which the phenomenon exercises a major effect on the ecological amplitude of the species. It is already apparent, however, that in specific

---

Cv, *Clinopodium vulgare*; Cvu, *Cirsium vulgare*; Dc, *Deschampsia cespitosa*; Df, *Deschampsia flexuosa*; Dg, *Dactylis glomerata*; Dm, *Draba muralis*; Do, *Dryas octopetala*; Dp, *Digitalis purpurea*; Eh, *Epilobium hirsutum*; Fg, *Festuca gigantea*; Fo, *Festuca ovina*; Fr, *Festuca rubra*; Fu, *Filipendula ulmaria*; Ga, *Galium aparine*; Gf, *Glyceria fluitans*; Gp, *Galium palustre*; Gr, *Geranium robertianum*; Gu, *Geum urbanum*; Gv, *Galium verum*; Hn, *Helianthemum nummularium*; Hl, *Holcus lanatus*; Hm, *Holcus mollis*; Hmu, *Hordeum murinum*; Hp, *Helictotrichon pratense*; Js, *Juncus squarrosus*; Km, *Koeleria macrantha*; Lc, *Lotus corniculatus*; Lca, *Luzula campestris*; Lh, *Leontodon hispidus*; Lp, *Lolium perenne*; Me, *Milium effusum*; Ml, *Medicago lupulina*; Md, *Matricaria discoidea*; Mn, *Melica nutans*; Ms, *Myosotis sylvatica*; Ns, *Nardus stricta*; Ov, *Origanum vulgare*; Pa, *Poa annua*; Pav, *Polygonum aviculare*; Fc, *Fallopia convolvulus*; Pe, *Potentilla erecta*; Pl, *Plantago lanceolata*; Pm, *Plantago major*; Pp, *Poa pratensis*; Pma, *Persicaria maculosa*; Ps, *Sanguisorba minor*; Pt, *Poa trivialis*; Pv, *Prunella vulgaris*; Ra, *Rumex acetosa*; Rac, *Rumex acetosella*; Ro, *Rumex obtusifolius*; Rr, *Ranunculus repens*; Sa, *Sedum acre*; Sac, *Sesleria caerulea*; Sc, *Scabiosa columbaria*; Dd, *Danthonia decumbens*; Sdi, *Silene dioica*; Sj, *Senecio jacobaea*; Sm, *Stellaria media*; Sp, *Succisa pratensis*; Ss, *Senecio squalidus*; Sv, *Senecio vulgaris*; Tp, *Thymus polytrichus*; Tf, *Tussilago farfara*; Tm, *Trifolium medium*; To, *Taraxacum officinalis*; Tr, *Trifolium repens*; Ts, *Teucrium scorodonia*; Ud, *Urtica dioica*; Va, *Veronica arvensis*; Vr, *Viola riviniana*; Be, *Bromopsis erecta*. Estimates of R_{max} are based on measurements during the period 2–5 weeks after germination in a standardised productive controlled environment conducted on seedlings from seeds collected from a single population in northern England. (Reproduced from Grime 1974 by permission of Macmillan (Journals) Ltd.)

instances (e.g. Böcher 1949) genetic variation is sufficient to enlarge substantially the strategic and ecological range of the species. From the investigations of Law *et al.* (1977) and Law (1979), for example, it is evident that within the common grass, *Poa annua*, there are populations which differ considerably in life-history. In severely-disturbed habitats *P. annua* occurs as ephemeral plants which are typical representatives of the ruderal strategy. In marked contrast populations of the same species in productive pastures contain a high proportion of biennial or possibly even perennial plants which may be described more accurately as competitive-ruderals.

A second example of intraspecific variation with respect to strategy is available from several studies conducted on the perennial grass, *Agrostis capillaris*. From investigations such as those of Jowett (1964) it is apparent that whilst pasture populations are usually composed of potentially fast-growing plants of moderately high competitive ability, the species is represented on infertile mine-waste by stress-tolerant individuals of smaller stature and slower potential growth-rate.

Further evidence of the development of stress-tolerant populations in species more usually associated with mesic environments has been obtained from the studies of Böcher and Larsen (1958) and Böcher (1961), from which it is apparent that, in tundra conditions, species such as *Holcus lanatus* and *Dactylis glomerata* may have considerably extended life-histories.

3 Regenerative Strategies

INTRODUCTION

There is enormous variety in the mechanisms whereby plants regenerate, and there can be little doubt that this fact accounts for many of the differences in ecology observed between species. It follows therefore that in order to understand fully the ecology of a plant species or population it is necessary to characterise the strategies adopted in both the regenerative (immature) and the established (mature) phases of the life-cycle. Often the analysis is complicated even further by the fact that plants may possess more than one method of regeneration. The objective of this chapter is to recognise the main types of regenerative strategies in plants. First, however, it is necessary to examine the distinction between vegetative reproduction and regeneration by seed.

REGENERATION BY VEGETATIVE OFFSPRING AND BY SEXUALLY-DERIVED SEEDS OR SPORES

In many plants, population expansion is capable of occurring, at least in the short term, without the involvement of sexual methods of regeneration. The majority of perennial herbs, ferns, and bryophytes together with some trees and shrubs are able to produce new individuals either by proliferation and subsequent fragmentation of the plant or through the formation of vegetative propagules such as bulbils, tubers, or gemmae. This type of regeneration is associated particularly with local consolidation of populations and has the effect of producing populations of genetically-uniform plants. It would be a mistake, however, to assume that all cohorts of sexually-derived offspring are genetically-diverse. Inbreeding is extremely common in ephemeral plants and it has been reported recently that ascospores released from individual ascomata in the pioneer lichens, *Graphis scripta* and *Ochrolechia parella*, are genetically uniform and are clearly the result of self-fertilisation (Murtagh *et al.* 2000).

The importance of sexual reproduction derives only in part from the basis which the frequently resulting genotypic variety provides for the responses of populations and species to natural selection. In flowering plants it is also characterised by the peculiar properties of seeds. Relative to the majority of vegetative propagules, seeds and spores are numerous, independent, and stress-tolerant, and these characteristics confer, respectively, the potential for rapid multiplication, dispersal, and dormancy.

Seedlings of the majority of plants are small in size and many are incapable of survival in the conditions experienced by established plants of the same genotype. The dependence of many species upon local and unusually favourable sites for seedling establishment has been known for a considerable period of time and a comprehensive review of this phenomenon has been assembled (Grubb 1977). It is evident that many plants possess mechanisms of seed production, dispersal, and dormancy which, together with the germination requirements, facilitate seedling establishment in circumstances which are quite distinct from those experienced during the later stages of the life-cycle and are local and/or intermittent in occurrence. Moreover, in long-lived or vegetatively-reproducing species, only low rates of regeneration from seed may be adequate for the maintenance of populations and in these circumstances populations may remain viable despite the fact that opportunities for seedling establishment occur relatively infrequently.

It is clear, therefore, that in terms of their capacity to influence the size, dispersion, survival, and genetic adaptability of plant populations, some broad distinctions can be drawn between the processes of vegetative reproduction and regeneration from seed. However, it is not possible to base a functional classification of regenerative strategies upon a simple dichotomy between vegetative reproduction and regeneration by seed. When the full range of plants is examined it is clear that a great deal of convergent evolution in structure and physiology has occurred between the two forms of regeneration, with the result that it is appropriate to include examples from both categories in the description of certain regenerative strategies.

TYPES OF REGENERATIVE STRATEGY

Information concerning mechanisms of regeneration by native plants is extremely fragmentary and it would be unwise to attempt a comprehensive and detailed account of this subject. However, certain basic forms of regeneration are now recognisable, and these may be provisionally classified into five types.

Vegetative Expansion (V)

Under this heading it is convenient to assemble many of the regenerative mechanisms which involve the expansion and subsequent fragmentation of the vegetative plant through the formation of persistent rhizomes, stolons, or suckers. The most consistent feature of vegetative expansion is the low risk of mortality to the offspring (Ginzo and Lovell 1973; Allessio and Tieszen 1975; Slade and Hutchings 1987). This is achieved through prolonged attachment to the parent plant and mobilisation of resources from parent to offspring in quantities adequate to sustain the offspring during establishment.

Vegetative expansion is associated particularly with the maintenance and local consolidation of populations of perennial herbs and shrubs. Because of the lower risks of mortality this regenerative strategy is often successful in vegetation types in which establishment from seed is precluded by the presence of a dense cover of vegetation and litter. In temperate regions vegetative expansion is very common among tall herbs, e.g. *Epilobium hirsutum*, *Chamerion angustifolium*, and *Urtica dioica*, and also allows certain woody plants, e.g. *Populus tremula*, *Prunus spinosa*, *Hippophae rhamnoides*, and *Symphoricarpos albus*, to form extensive thickets in relatively undisturbed environments.

Regeneration by vegetative expansion, although less conspicuous, is also of frequent occurrence in unproductive habitats. As Billings and Mooney (1968) and Callaghan (1976) have recorded, it is particularly common in arctic and alpine environments and northern heathlands, where species such as *Gaultheria shallon*, *Salix herbacea*, *Dryas octopetala*, Vaccinium myrtillus, *V. vitis-idaea*, and *Carex bigelowii* may develop clonal patches of a considerable size. Vegetative sprouting is also a common form of regeneration in circumstances where shrubs are subject to intermittent damage by fire (Hanes 1971; Hansen 1976). In the coastal and desert chaparral of California, for example, vegetative expansion is evident in dominant species such as *Adenostoma fasciculatum* and *Ceanothus leucodermis.* Another vegetation type of low productivity in which vegetative expansion plays a most important role is tropical rainforest. From a number of sources there is evidence (see page 291) that where local disturbance causes relatively small openings in the forest canopy, vegetative sprouts from neighbouring established trees usually provide the major source of invading plants.

In many of the herbs and woody species which exhibit vegetative expansion the breakdown of the connections between parent and offspring is delayed for many years, and it is often exceedingly difficult to draw a distinction between these plants and the large number of trees and shrubs in which vegetative reproduction does not normally occur.

In vegetative expansion, a low risk of mortality to the offspring is achieved through attachment to the parent and a necessary condition for success is the survival of the parent during the period of establishment. In rhizomatous grasses and forbs in productive habitats establishment may be completed within one growing season, whereas in stressed environments, and particularly in woody plants, the process may extend over many years. It is hardly surprising, therefore, to find that vegetative expansion is most apparent in relatively undisturbed habitats. In contrast, the remaining strategies to be described are all adapted to exploit disturbance and differ from each other according to the nature and frequency of this disturbance and the environmental context in which it occurs.

Seasonal Regeneration in Vegetation Gaps (S)

In a wide range of habitats, herbaceous vegetation is subjected to seasonally-predictable damage by phenomena such as temporary drought, flooding, trampling, and grazing. Under these conditions, the most common regenerative strategy is that in which areas of bare ground or sparse vegetation cover are created every year and are recolonised annually during a particularly favourable season. This form of regeneration can lead to stable co-existence between monocarpic and polycarpic life-histories (Crawley and May 1987) and reaches its highest frequency in the temperate zone where two main types can be distinguished with respect to the season in which recolonisation occurs. In the first, establishment takes place in the autumn whereas in the second it is delayed until spring.

Autumn Regeneration

In regions in which rainfall is mainly restricted to a cool season, herbaceous vegetation is subjected to highly predictable damage by drought; this effect is most pronounced in grasses, the majority of which are shallow-rooted. In many grasslands there is additional seasonal disturbance resulting from trampling and defoliation by wild or domestic animals.

In habitats of this type detached viable seeds of the majority of the grasses are present in the habitat only during the dry season, and sampling of the surface soil at intervals throughout the year (Figure 47) reveals a strong annual fluctuation in seed content. The numbers of germinable seeds reach a minimum during the wet season, at the beginning of which synchronous germination of the entire seed population results in the appearance of large numbers of seedlings of grasses in the areas of bare ground developed during the preceding dry season.

Grasses of this type possess relatively large seeds which are characterised by lack of innate dormancy and by the ability to germinate in light and in darkness over a wide range of temperatures (Table 12). All these characteristics are consistent with the conclusion that in these species moisture supply is the overriding determinant of the timing of germination. It is interesting to note that many of the commonest grasses of pastures and meadows in Europe (e.g. *Arrhenatherum elatius*, *Bromopsis erecta*, *Anisantha sterilis*, *Cynosurus cristatus*, *Dactylis glomerata*, *Festuca pratensis*, *Festuca rubra*, *Hordeum murinum*, *Lolium multiflorum*, *Lolium perenne*) conform to this type. The failure of these species to develop persistent reserves of buried seeds in the soil is well-documented (Brenchley and Warington 1930; Chippindale and Milton 1934; Milton 1939; Champness and Morris 1948; Thompson and Grime 1979) and is of profound importance to pasture management in Europe since it appears to play a crucial role in the process whereby sown species such as *Lolium perenne* and *L. multiflorum* tend to be replaced by native grasses.

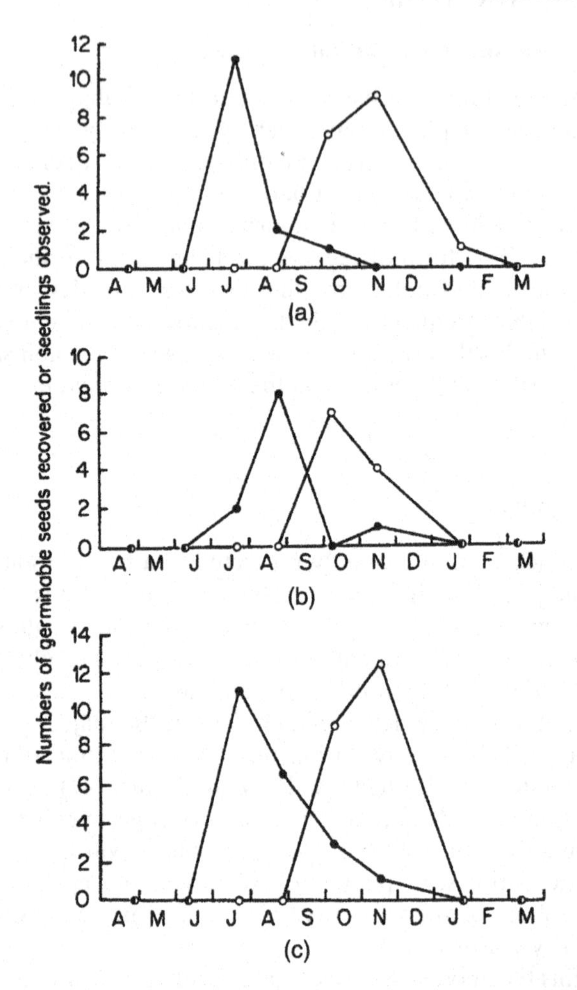

Figure 47. Seasonal variation in the numbers of detached viable seeds (●-●) and newly-germinated seedlings (○-○) of three tussock grasses in dry calcareous grassland in northern England, (a) *Briza media*, (b) *Helictotrichon pratense*, (c) *Koeleria macrantha.* Seed number refers to seeds germinating from a standard volume of surface (0–3 cm) soil removed to a controlled environment. Seedling number refers to freshly-germinated seedlings (i.e. seedlings were removed after each count) in field plots of total area 0.2 m^2. (Reproduced by permission of Thompson 1977.)

Although the most familiar type of autumn regeneration in seasonally droughted habitats involves the seeds of grasses it is interesting to note that there are similar mechanisms which depend upon the production of vegetative propagules. An example of this phenomenon is provided by *Allium vineale*, a species in which regeneration is effected by bulbils which remain dormant until the onset of moist conditions in the autumn. In addition, there

Table 12. Estimation of mean seed weight and germination response to constant temperatures in the light in British populations of 10 grasses which are predominantly autumn germinators and do not develop buried seed banks. (Reproduced from Grime *et al.* 1981 by permission of *Journal of Ecology*.)

Species	Mean seed weight (mg)	Temperature range for 50% germination	
		Lower limit	Upper limit
Arrhenatherum elatius	2.4	<5	29
Anisantha sterilis	8.4	<5	37
Bromopsis erecta	4.2	10	34
Cynosurus cristatus	0.7	<5	33
Dactylis glomerata	0.5	<5	31
Festuca ovina	0.4	6	33
Festuca rubra	0.8	6	35
Hordeum murinum	6.6	<5	33
Lolium perenne	1.8	<5	34
Phleum pratense	0.4	11	36

are less specialised mechanisms of vegetative regeneration whereby established plants which have been damaged by drought and/or predation during the summer may be rehabilitated in the autumn and winter. Examples here include the grass, *Poa trivialis*, and various bryophytes (e.g. *Brachythecium rutabulum*, *Eurhynchium praelongum*) in which new individuals frequently originate from surviving meristems.

Spring Regeneration

In northern or more continental regions within the temperate zone, plant growth in the late autumn and winter is often severely restricted by low temperatures and frost damage, and for this reason gaps in the vegetation which originated during the summer are likely to remain open until the following spring. Moreover, in many habitats, additional patches of bare soil arise during the winter and early spring as a result of frost-heaving, solifluction, and water erosion following snow-thaw. It is not surprising, therefore, to find that in plants of more northern latitudes there is a tendency for regeneration to be delayed until the spring. An excellent example of this phenomenon is provided by the comparative study of the germination characteristics of seeds from a moist, tall grassland community in Japan by Washitani and Masuda (1990), who have shown that seeds are shed late in the year and remain ungerminated until spring; in some species this delay arises from a requirement for chilling to break dormancy but in others it is a consequence of a high temperature requirement for germination.

In north-temperate and cold climates many trees, shrubs, and herbs are incapable of germination until the imbibed seed has experienced temperatures in the approximate range of 2–10°C. The length of the required chilling period varies from several days (e.g. *Impatiens glandulifera*) to months (e.g. *I. parviflora*).

Table 13 contains a list of the species which were found to have a chilling requirement when a screening of germination characteristics was carried out on the common plants present in a local flora in northern England (Grime *et al.* 1981). In a small number of plants (e.g. *Polygonum aviculare*, *Galium palustre*) the chilling requirement appears to be associated with characteristics which cause many of the seeds to become incorporated into a buried seed bank (see page 152). However, for the majority of these species there is evidence from laboratory and/or field investigations that germination of the entire populations of seeds occurs soon after the requirement for chilling has been satisfied. Seeds of these plants differ in several respects from those characteristic of buried seed banks (Table 14). The majority are quite large, none is inhibited from germination by darkness, and once the chilling requirement is fulfilled a high proportion is capable of germinating rapidly at low (<10°C) temperatures. The main consequence of this physiology is that the seeds are prevented from germinating during the summer and autumn and exhibit synchronous germination at a time in the winter or early spring determined by the length of the chilling requirement and the temperatures prevailing in the habitat.

Trees, shrubs, and herbs of deciduous woodlands, scrub, and hedgerows form a high proportion of the species listed in Table 13, and it seems likely that the main advantage of the chilling requirement in such plants is related to the fact that germination is caused to occur during the unshaded phase in the early spring at which time conditions are most favourable for seedling establishment. This phenomenon is particularly obvious in British woodlands where observations prior to the expansion of the tree canopy in the early spring reveal large, even-aged populations of germinating seedlings of woody species (e.g. *Acer pseudoplatanus*, *Fraxinus excelsior*) and herbs (e.g. *Hyacinthoides non-scripta*, *Lamiastrum galeobdolon*). Evidence of the importance of early germination in woodland plants is available from studies such as those of Wardle (1959) and Gardner (1977) which have included reports of exceedingly high rates of seedling mortality during the summer maximum in shading by the canopy of trees and herbs.

A similar principle may be applied to explain the presence of a chilling requirement in many of the commonest forbs of productive herbaceous vegetation types such as meadows, roadsides, and river banks. In such habitats the density of living plant material and litter is at a minimum in the early spring and germination at this time allows seedling growth to precede that of the established perennials. In Europe, numerous examples of this phenomenon occur in annuals (e.g. *Impatiens glandulifera*, *Galeopsis tetrahit*, *Galium*

Table 13. Some flowering plants for which there is evidence in British populations of a chilling requirement for germination. The values refer to the mean seed weight in mg of the populations investigated. (Reproduced from Grime *et al.* 1981 by permission of *Journal of Ecology*.)

(a) Woodland plants			
Acer pseudoplatanus	33.47	*Allium ursinum*	4.78
Clematis vitalba	1.27	*Bromopsis ramosa*	7.37
Corylus avellana	63.68	*Campanula latifolia*	0.06
Frangula alnus	17.05	*Conopodium majus*	2.26
Fraxinus excelsior	46.19	*Festuca gigantea*	3.12
Ligustrum vulgare	14.19	*Hyacinthoides non-scriptus*	6.17
Lonicera periclymenum	5.21	*Impatiens parviflora*	6.91
Prunus spinosa	118.13	*Lamiastrum galeobdolon*	1.98
*Quercus petraea**	744.70	*Mercurialis perennis*	2.17
Rosa pimpinellifolia	15.84	*Oxalis acetosella*	0.99
Sambucus nigra	1.90	*Sanicula europaea*	2.97
Sorbus aucuparia	2.58	*Viola riviniana*	1.01
(b) Plants of grassland, heathland, and other types of herbaceous vegetation in unshaded habitats			
Aegopodium podagraria	2.73	*Myrrhis odorata*	35.01
Agrimonia eupatoria	23.78	*Odontites verna*	0.15
Alchemilla filicaulis	0.46	*Pimpinella major*	2.12
Anthriscus sylvestris	5.18	*Pimpinella saxifraga*	1.19
Chaerophyllum temulum	1.44	*Rhinanthus minor* agg.	2.84
Empetrum nigrum	0.75	*Smyrnium olusatrum*	20.46
Euphrasia officinalis	0.13	*Torilis japonica*	1.98
Heracleum sphondylium	5.52	*Viola hirta*	2.79
(c) Marsh plants			
Angelica sylvestris	1.15	*Pinguicula vulgaris*	0.02
Caltha palustris	0.99	*Saponaria officinalis*	1.46
Eleocharis palustris	0.96	*Silaum silaus*	1.09
Galium palustre	0.91	*Solanum dulcamara*	1.49
Impatiens glandulifera	7.32		
(d) Arable weeds			
Aethusa cynapium	1.07	*Papaver rhoeas*	0.09
Anagallis arvensis	0.40	*Polygonum aviculare*	1.45
Atriplex prostrata	0.86	*Persicaria hydropiper*	1.24
Atriplex patula	1.33	*Viola arvensis*	0.40
Chenopodium album	0.77		
Fallopia convolvulus	1.28		
Papaver dubium	0.12		

*Chilling requirement specific to plumule extension

aparine), biennials (e.g. *Anthriscus sylvestris*, *Heracleum sphondylium*, *Torilis japonica*), and perennials (e.g. *Myrrhis odorata*, *Pimpinella major*, *Silaum silaus*). In all these species the early spring germination is allied to a comparatively large seed and it seems likely that both contribute to the impetus to

Table 14. Some flowering plants which, in Britain, have seeds which are partially or totally inhibited by darkness and tend to accumulate seed-banks in the soil. Values are mean seed weights. (Reproduced by permission of Thompson 1977.)

(a) Herbaceous plants of grassland, heathland and woodland			
Agrostis capillaris	0.06	*Danthonia decumbens*	0.87
Agrostis vinealis	0.05	*Origanum vulgare*	0.10
Anthoxanthum odoratum	0.45	*Poa trivialis*	0.09
Arabidopsis thaliana	0.02	*Potentilla erecta*	0.58
Arenaria serpyllifolia	0.06	*Sagina procumbens*	0.02
Calluna vulgaris	0.03	*Saxifraga tridactylites*	0.01
Campanula rotundifolia	0.07	*Silene nutans*	0.27
Digitalis purpurea	0.07	*Thymus polytrichus*	0.11
Holcus lanatus	0.32	*Urtica dioica*	0.20
Milium effusum	1.20		
(b) Marshland plants			
Agrostis canina ssp. canina	0.05	*Juncus effusus*	0.01
Cirsium palustre	2.00	*Juncus inflexus*	0.03
Deshampsia cespitosa	0.31	*Ranunculus sceleratus*	0.08
Epilobium hirsutum	0.05	*Rumex sanguineus*	1.13
Juncus articulatus	0.02		
(c) Arable weeds			
Matricaria discoidea	0.08	*Poa annua*	0.26
Polygonum aviculare	1.45	*Stellaria media*	0.35

seedling growth which enables these plants to regenerate in close proximity to established perennials.

All of the examples considered so far have involved regeneration by seed and most of the species concerned have been annuals or biennials. However, we may also include in this category a large number of perennial herbs in which spring regeneration is a vegetative process. Although in certain of these plants propagation is by means of specialised structures such as the tubers of *Ranunculus ficaria*, in the majority of species new individuals originate each spring from fragments of rhizomes or stolons which were produced during the previous growing season and have remained dormant during the winter. The latter form of regeneration occurs in both grasses (e.g. *Agrostis stolonifera*, *Glycera fluitans*, *Poa pratensis*) and forbs (e.g. *Tussilago farfara*, *Cirsium arvense*, *Trifolium repens*) and differs from vegetative expansion in that (1) establishment of the offspring usually coincides with (or is preceded by) the death or senescence of the parent and (2) there is a higher rate of mortality among the offspring.

General Characteristics of Seasonal Regeneration

A feature common to all the plants which exhibit seasonal regeneration is the capacity to invade vegetation gaps at a specific period in the year, and so far in this account attention has been concentrated on the mechanisms which control the timing of regeneration. However, there is also a need to take account of the interaction between seed size and gap size. In a classic investigation of seedling regeneration in monocarpic perennials (Gross and Werner 1982; Gross 1984) and in a comparative study of regeneration in various Umbelliferae (Thompson and Baster 1992) it is clearly demonstrated that whilst all species benefit from gaps the dependence upon them is considerably greater in small-seeded species.

An essential characteristic of vegetation gaps is their localisation in space and it is interesting to observe that most of the plants which belong to this category possess mechanisms of regeneration which appear to increase the chances that a proportion of the offspring will be dispersed into gaps. In species in which seasonal regeneration is effected by seed, the flowering shoot is usually a large structure which is either compound and spreading (e.g. the majority of Umbelliferae) or tall and flexuous (e.g. *Anisantha sterilis*, *Bromopsis erecta*). The most likely adaptive value of such inflorescences is that they ensure that, despite their large size, the seeds of these plants become dispersed within the habitat to an extent which ensures that a proportion fall onto bare soil or into areas where the density of established vegetation and litter is insufficient to prevent establishment. In certain herbaceous species such as *Impatiens glandulifera* and *Geranium robertianum* the same effect is achieved by means of explosive dehiscence of the seeds.

An essentially similar mechanism of gap-exploitation can be recognised in many of the species (e.g. *Tussilago farfara*, *Trifolium repens*, and *Ranunculus repens*) in which seasonal regeneration is primarily a vegetative process. In these plants, however, the dispersal of propagules is achieved by the proliferation and fragmentation of rhizomes and stolons.

It seems reasonable to conclude that the strategy of seasonal regeneration represents a rather unsophisticated mechanism of gap exploitation in which, after propagules have been dispersed within the habitat and have germinated simultaneously (or recommenced growth, in the case of fragments of rhizomes or stolons), survival is limited to those individuals which, by chance, occur in gaps. It is evident that this method of regeneration depends upon the creation each year of a high density of suitable gaps. As we shall see from later sections of this chapter, in vegetation in which gaps occur more rarely seasonal regeneration gives way to more discriminating types of regenerative strategy.

Regeneration Involving a Persistent Bank of Seeds or Spores (B_s)

When flowering plants are compared with respect to the fate of their seeds, two contrasting groups may be recognised. In one, most if not all of the seeds germinate soon after release even if they have become buried in the soil (Cook 1980; Pons 1989a, 1991a), whilst in the other group many become incorporated into a bank of dormant seeds which is detectable in the habitat at all times during the year and may represent an accumulation of many years. These two groups are, of course, extremes and between them there are species and populations in which the seed bank, although present throughout the year, shows pronounced seasonal variation in size. Nevertheless, as indicated in Figure 48, it is convenient to draw an arbitrary distinction between 'transient' and 'persistent' seed banks. A transient seed bank may be defined as one in which none of the seed output remains in the habitat in a viable condition for more than one year. In the persistent seed bank, some of the component seeds are at least one year old.

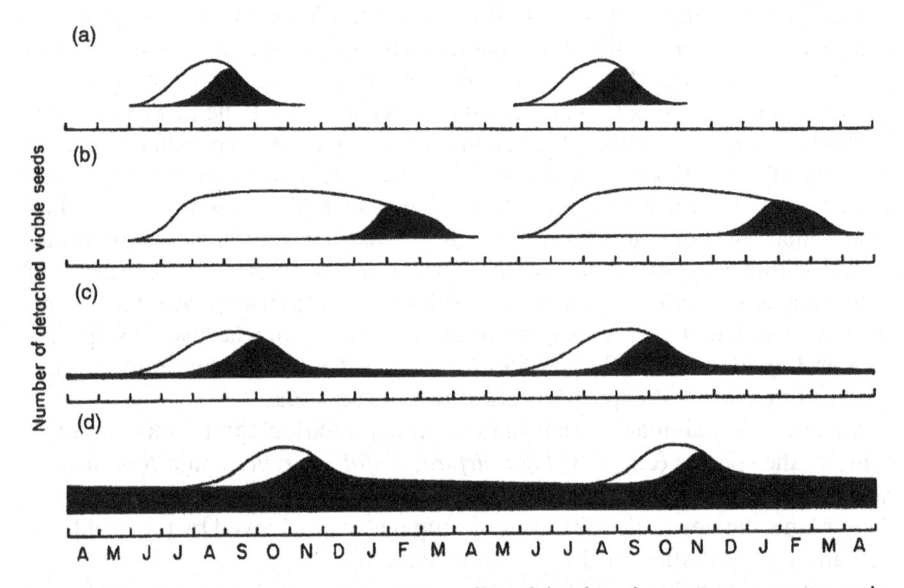

Figure 48. Scheme describing four types of seed banks of common occurrence in temperate regions. ■, seeds capable of germinating immediately after removal to suitable laboratory conditions; □, seeds viable but not capable of immediate germination. (a) Annual and perennial grasses of dry or disturbed habitats (e.g. *Hordeum murinum*, *Lolium perenne*, *Catapodium rigidum*); (b) annual and perennial herbs, shrubs, and trees colonising vegetation gaps in early spring (e.g. *Impatiens glandulifera*, *Anthriscus sylvestris*, *Acer pseudoplatanus*); (c) winter annuals mainly germinating in the autumn but maintaining a small seed bank (e.g. *Arenaria serpyllifolia*, *Saxifraga tridactylites*, *Erophila verna*); (d) annual and perennial herbs and shrubs with large, persistent seed banks (e.g. *Stellaria media*, *Origanum vulgare*, *Calluna vulgaris*). (Reproduced from Thompson and Grime 1979 by permisson of *Journal of Ecology*.)

Another difference which may be recognised between transient and persistent seed banks relates to the location of the seeds within the habitat. Those workers who have split up the soil profile into horizons of different depths (Chippindale and Milton 1934) or who have excluded surface soil (Wesson and Wareing 1969a, b) before estimating the germinable seed content have demonstrated conclusively that the majority of persistent banks consist of buried seeds. Although in this account it is convenient to restrict attention to persistent seed banks it is important to recognise that analogous phenomena occur through the accumulation of spores in the soil. There are well-documented examples of persistent spore banks in both bryophytes and ferns (Furness and Hall 1981; During and ter Horst 1983). In order to examine the role of persistent seed banks in regeneration it will be helpful to identify the range of habitats with which they are associated.

The Distribution of Persistent Seed Banks in Relation to Habitat

Where an attempt is to be made to characterise the seed banks of the various species present in a habitat it is necessary to sample the soil and litter at intervals throughout the year and to carry out germination procedures, or laboratory tests, to estimate seasonal variation in the numbers of viable seeds. The extremely laborious techniques which are required for accurate estimations of the numbers of seeds in soil samples have been described by Kropac (1966) and Major and Pyott (1966) and these are impracticable for large-scale studies, especially those involving small-seeded species. However, where the purpose is merely to detect the presence of a persistent seed bank in the soil it is not necessary to adopt such exhaustive methods. Data obtained by simplified procedures of seasonal sampling and germination tests (e.g. Ter Heerdt *et al.* 1996) provide ample confirmation of the presence of viable seeds throughout the year.

The best-documented examples of species which accumulate large reservoirs of seeds in the soil are the ruderals of arable fields (Brenchley 1918; Brenchley and Warington 1930; Barton 1961; Roberts and Stokes 1966). For these plants there is abundant evidence of the capacity of buried seeds to survive for long periods and to germinate in large numbers where the habitat is disturbed. The advantage of the persistent seed bank is obvious in ruderal species of the intermittently-open habitats of farmland, since long-term survival of their populations frequently depends upon the ability to remain in a dormant condition through periods in which the habitat is occupied by a closed cover of perennial species. However, persistent seed banks are not confined to the ruderals of arable land; they occur in a wide range of ecological groups.

Large seed banks are particularly characteristic of shrubs (e.g. *Calluna vulgaris)* and perennial herbs (e.g. *Hypericum perforatum*, *Origanum vulgare*) occurring in habitats subjected to intermittent damage by fire (Thompson and Band 1997). In a very different context, enormous banks of buried seeds are produced by some marshland species (e.g. *Juncus effusus*, *Juncus articulatus*, *Epilobium hirsutum*). Persistent seed banks are uncommon in meadow species (Akinola *et al.* 1998) but they are characteristic of certain pasture plants (e.g. *Agrostis capillaris*, *Anthoxanthum odoratum*, *Holcus lanatus*, *Plantago lanceolata*) and they are especially common in species colonising seasonally water-logged grasslands (e.g. *Deschampsia cespitosa*, *Poa trivialis*, *Ranunculus repens*, *Potentilla erecta*). Although some of the largest and more persistent seed banks are to be found in wetland habitats (Darwin 1859; Milton 1939; van der Valk and Davis 1976; Thompson and Grime 1979) the phenomenon is not restricted to such conditions. In vegetation subjected to summer drought, persistent seed hanks are accumulated by both perennial herbs (e.g. *Silene nutans*, *Thymus polytrichus*, *Campanula rotundifolia*) and certain winter annuals (e.g. *Arabidopsis thaliana*, *Arenaria serpyllifolia*, *Saxifraga tridactylites*) (King 1976; Thompson and Grime 1979).

Regeneration from banks of persistent buried seeds in shrubs and trees is restricted to species which are associated with frequently-disturbed vegetation. In second-growth tropical forest seed banks of woody species are a common feature (Gómez-Pompa 1967; Webb *et al.*, 1972) whilst in temperate deciduous forest in North America this form of regeneration has been described in the pin cherry (*Prunus pensylvanica*). From his investigations with this species Marks (1974) concluded that 'sufficient numbers of viable pin cherry seeds reside in the soils of second-growth forests in central New Hampshire to account for the dense stands frequently observed after cutting or burning'.

In temperate and wet tropical regions seeds that persist in the soil spend most of the time in an imbibed state and in this condition they have greater longevity than those stored in the laboratory. This is because with the passage of time seeds experience forms of molecular damage that are continuously repaired during periods in which the seeds are imbibed (Villiers 1974; Priestley 1986; Vàzquez-Yanes and Orozco-Sergovia 1990). A high proportion of the persistent seeds in arid regions have a very different mechanism in which dormancy is imposed by a hard seed coat. Through mechanisms that are not understood such hard seeds show remarkable capacities for persistence in dry storage in the laboratory (Youngman 1951) and in the soil (Shen-Miller *et al.* 1995).

A different example of seed bank regeneration in woody species is provided by the closed-cone pines of California (*Pinus attenuata*, *P. contorta*, *P. muricata*, *P. radiata*, and *P. remorata*) and the Appalachian Mountains (*P. pungens*). From the results of several investigations (Munz 1959; Zobel 1969; Vogl 1973) it is clear that the seed bank is specifically adapted for regeneration after periodic destruction by fire. In these relatively short-lived trees which occur in even-aged stands dating back to the most recent fire, the large cones remain firmly attached to the parent and accumulate throughout the life of each tree. Release of seeds is dependent upon the opening of the cones by heat generated during forest fires. A broadly similar 'aerial' seed bank has been recognised (Gutterman 1993) in some desert annuals where ripe seeds are retained within fruits on the dead plant and released only after substantial rain.

It is apparent that persistent seed banks are associated with an extremely wide range of species and habitats. It is interesting to note, however, that large populations of buried seeds do not appear to accumulate in arctic regions or in areas occupied by mature tropical or temperate forest (Livingstone and Allessio 1968; Thompson *et al.* 1998). In these habitats, seed banks tend to be replaced by banks of persistent seedlings (see page 163). Interesting exceptions to this general rule are to be found in a small number of shade-tolerant plants of European woodlands which have been found to accumulate persistent seed banks (Piroznikow 1983; Metcalfe and Grubb 1995; Jenkowska-Blaszezuk and Grubb 1997). It would appear that these species exploit gaps in the litter arising from activities of mammals such as wild boar.

What can we conclude from this brief survey and on theoretical grounds about the circumstances conducive to the evolution of persistent seed banks? Both the theoretical models (Venable and Lawlor 1980; Venable and Brown 1988; McPeek and Kalisz 1998) and the field evidence strongly point to selection mechanisms dominated by a scenario in which seed production and opportunities for seedling establishment are intermittent but spatially predictable.

The Mechanism of a Persistent Bank of Buried Seeds

Although the functional significance of a bank of buried seeds varies in detail according to species and ecological situation, the same basic problems are encountered in any attempt to discover how the bank arises and is involved in seedling establishment. Firstly, it is necessary to describe the mechanism by which seeds become buried. Secondly, the factors must be identified which prevent the germination of seeds before burial and during their period of survival in the soil. Thirdly, since many seeds appear to germinate in places and at times propitious for seedling establishment, it is of considerable interest to examine the mechanisms which initiate the germination of buried seeds.

One approach to these problems has been to examine the morphology and germination physiology of seeds from species which develop persistent seed banks and to compare the results with those obtained for species with transient seed banks. From such comparisons it is possible to recognise certain characteristics which are consistently associated with the tendency to form a persistent seed bank. This suggests that for many plant species it may be possible to predict the presence of a seed bank from the laboratory characteristics of the seeds.

Seed burial. In Table 14, a number of common British plants known to accumulate buried seed banks are listed, together with measurements of their average seed weight. A striking feature of this data is the high incidence of species with extremely small seeds. This applies particularly to species such as *Juncus effusus*, *Calluna vulgaris*, *Agrostis capillaris*, and *Origanum vulgare*, which are known to accumulate huge populations of buried seeds in their respective habitats (Thompson 1987, 1992; Bekker *et al.* 1998a, b). Small seeds are more likely than larger ones to be washed into small fissures in the soil surface and to be buried by the activities of the soil fauna (McRill and Sagar 1973; McRill 1974; Peart 1984; Thompson *et al.* 1993) and evidence in support of this hypothesis has been obtained by Mortimer (1974), who observed the fate of marked seeds introduced into a grassland habitat. The same relationship of seed size and shape to seed persistence in the soil was not observed in a study conducted in Australia by Leishman and Westoby (1998) although a survey conducted in New Zealand by Moles *et al.* (2000) did find size and shape to be correlated with persistence—but these relationships were not as well-defined as in European studies.

The correlation between small size and seed persistence does not extend to weeds of cultivated land (Thompson 1998; Thompson *et al.* 1998) where the burial and unearthing of seeds occurs mainly through ploughing: many arable weeds such as *Avena fatua*, *Elytrigia repens*, and *Fallopia convolvulus* have banks composed of comparatively large seeds.

The probability that a seed will become buried is likely to be increased by any dormancy mechanisms which delay germination in the period immediately after seedfall. Following the terminology of Harper (1957), we may distinguish two mechanisms of seed dormancy (innate and enforced), both of which appear to facilitate seed burial. Examples of innate dormancy include the need for an extended period of incubation in warm moist conditions for maturation of the embryo (e.g. *Ranunculus repens*, *R. acris*, and *Potentilla erecta*) initiation of germination by light (e.g. *Elytrigia repens*, *Cirsium arvense*, *Poa pratensis*), chilling requirements (e.g. *Polygonum aviculare*, *Galium palustre*), and the presence of an impermeable testa (e.g. *Helianthemum nummularium* and the majority of leguminous herbs). Enforced dormancy occurs in circumstances where seeds are prevented from germination by the conditions prevailing in the habitat during the period following seedfall. In some British species (e.g. *Calluna vulgaris*, *Rumex sanguineus*, *Urtica dioica*) the release of the seeds from the parent plant is so delayed, and the temperature requirements for germination are so high, that it is inevitable that dormancy will be enforced by low winter temperatures.

A quite different mechanism of enforced dormancy may operate in habitats in which imbibed seeds on the soil surface experience light which has been filtered by a dense leaf canopy. Laboratory experiments conducted by Taylorson and Borthwick (1969), King (1975), Grime and Jarvis (1975), and Gorski (1975) suggest that modification of the spectral composition of the incident radiation may be sufficient to enforce dormancy in certain species, and from data such as those in Table 15 there is some evidence that germination of the seeds of ruderal herbs is more strongly inhibited by a dense leaf canopy than that of closed-turf species.

An additional complication which must be taken into account arises from the observation that plant species differ considerably in the light-requirement of freshly-dispersed seeds. Whilst some species produce populations of seeds

Table 15. The influence of a leaf canopy upon the percentage germination of seeds of two groups of herbaceous plants of contrasted ecology. (Reproduced from Fenner 1978 by permission of *Journal of Ecology*.)

		Full sunlight		Shaded by *Festuca rubra*	
		Filter paper in closed petri dish	Bare soil in plant pot	Short turf in plant pot	Tall turf in plant pot
Species of open habitats	*Inula conyza*	94	70	72	16
	Reseda luteola	100	52	57	30
	Sonchus oleraceus	98	27	36	12
	Spergula arvensis	96	44	35	10
	Verbascum thapsus	76	45	45	27
Closed-turf species	*Galium verum*	70	56	57	39
	Hypochoeris radicata	98	80	62	29
	Leontodon hispidus	100	81	79	59
	Plantago media	88	36	65	38
	Rumex acetosa	100	75	75	42
	Scabiosa columbaria	92	60	66	40

that without exception are capable of immediate germination in darkness, others release some or all of their seeds in a state in which exposure of the imbibed seeds to unfiltered sunlight is necessary before germination can occur. In a comparative investigation (Cresswell and Grime 1981) it was shown (Figure 49) that seed light requirements are dependent upon the persistence of chlorophyll in the tissues investing the seed in the later stages of its development on the mother plant. This is because the state of the photoreversible phytochromes that control germinability of a mature seed depend upon the red/far red ratio of the light experienced at the time in seed ripening when phytochrome changes are arrested by drying out of the seed. By premature drying of seeds in species that lose their chlorophyll from the inflorescence before seed maturation it is possible to induce a light requirement in seeds which are usually capable of germination in darkness. The ecological implications of this mechanism are (1) that the fate of seeds that become buried in the soil or beneath litter may be strongly affected by the interplay between light quality and seed moisture content both before and after release from the parent plant and (2) that there is often considerable scope for variation in seed germinability and persistence within a cohort of seeds produced by individuals of the same species.

Dormancy in buried seeds. In many of the species which develop persistent seed banks it is necessary to draw a distinction between the mechanisms which inhibit germination prior to burial and those which become important once the seed is incorporated into the soil. In early studies some authorities

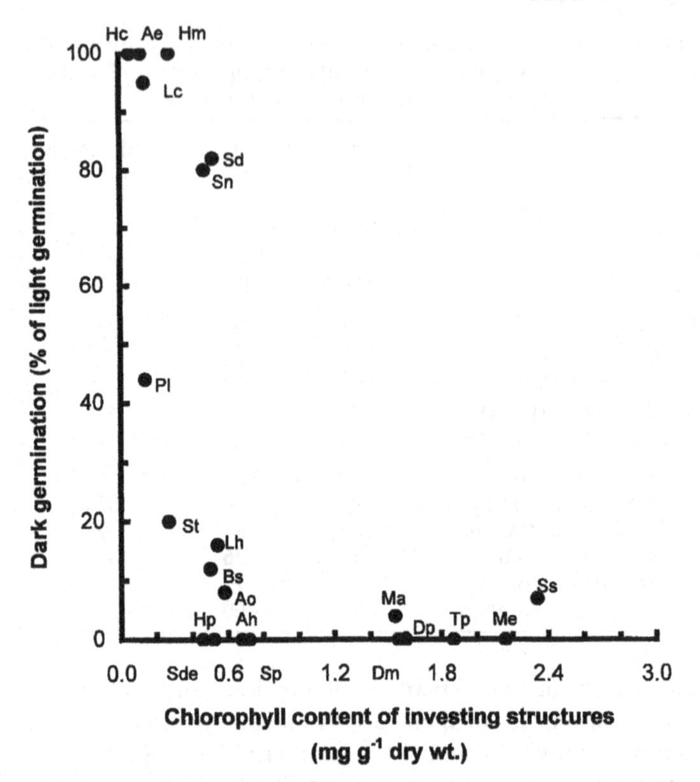

Figure 49. Relationship between chlorophyll concentration of the investing structures and dark germination of mature seeds in various species of flowering plants. Values for chlorophyll refer to the concentration at the mid-point in the range of seed moisture contents. The species are: Ae, *Arrhenatherum elatius*; Ah, *Arabis hirsuta*; Ao, *Anthoxanthum odoratum*; Bs, *Brachypodium sylvaticum*; Dm, *Draba muralis*; Dp, *Digitalis purpurea*; Hc, *Helianthemum nummularium*; Hm, *Hordeum murinum*; Hp, *Hypericum perforatum*; Lc, *Lotus corniculatus*; Lh, *Leontodon hispidus*; Ma, *Myosotis arvensis*; Me, *Milium effusum*; Pl, *Plantago lanceolata*; Sd. *Silene dioica*; Sde, *Danthonia decumbens*; Sn, *Silene nutans*; Sp, *Succisa pratensis*; Ss, *Senecio squalidus*; St, *Serratula tinctoria*; Tp, *Tragopogon pratensis*. (Reproduced from Cresswell and Grime 1981 by permission of *Nature*.)

such as Bibbey (1948) and Harper (1957) contended that the composition of the soil atmosphere was a major factor restricting the germination of buried seeds. However, as the results of large-scale comparative experiments on the germination requirements of native species have become available (Baskin and Baskin 1998; Table 14), it has become clear that a most consistent feature of the species forming persistent seed banks is the inhibition of germination by darkness. It seems reasonable to suspect, therefore, that in the majority of buried seeds dormancy is enforced primarily by lack of light. It is interesting to note, moreover, that in some persistent seeds (e.g. *Plantago lanceolata*) the requirement for light is not apparent immediately after the seeds are shed from the plant, but is induced later by a period of burial in the soil (Wesson and Wareing 1969a; Pons 1991b; Milberg and Anderson 1997).

Defences of persistent seeds. Seed predators such as beetles and ants are active at the soil surface but once small seeds are buried the main threat to their survival is likely to be microbial pathogens. In one of the few investigations of the relationship between persistence and chemical defence, Hendry *et al.* (1994) compared the levels of ortho-dihydroxyphenols in seeds of 81 species and found that concentrations were significantly higher in species with persistent seeds. However, not all of the persistent seeds investigated in this study were rich in ortho-dihydroxyphenols and we must suspect that other chemical defences may be important. There is a requirement for more research on this subject.

Initiation of germination in buried seeds. It is well known that cultivation of agricultural soils usually results in the appearance of large numbers of seedlings of annual weeds, and it seems likely that this is at least in part due to the unearthing of buried seeds and the triggering of germination by exposure to light. The same explanation may be applied to the frequent observation that gaps in pastures caused by local damage to the turf are often rapidly colonised by seedlings. However, scrutiny of the seedlings which appear in response to disturbance of the soil (Thompson 1977) has revealed that in some situations a high proportion of them originate from seeds which are situated beneath the soil surface below the depth to which light can penetrate. From investigations with naturally-buried seeds (Figure 50) and seeds placed in incubators simulating the temperature regimes experienced in the soil (Thompson and Grime 1983; Figure 51), there is evidence that the germination of buried seeds *in situ* is often brought about by a response to the increased diurnal fluctuations in temperature which may result from the removal of the insulating effect of foliage, litter, or humus layers from the soil (Thompson *et al.* 1977).

Figure 52 provides an illustration of the alteration of soil temperature regime which may result from the creation of gaps in an herbaceous canopy. From such data, it is apparent that the effect of a gap upon the diurnal fluctuations of temperature experienced by a buried seed is strongly affected by the depth of burial and the size of the gap. We may suspect therefore that the capacity to respond to particular amplitudes of temperature fluctuation in darkness acts as a depth-sensing mechanism and may also cause the germination of buried seeds to be restricted to gaps of a particular range in size.

Circumstantial evidence of the importance of sensitivity to fluctuating temperatures as a mechanism whereby gaps in vegetation may be detected by buried seeds is available from experiments (Figure 51) in which the capacity to respond to fluctuations in temperature in darkness has been found to occur

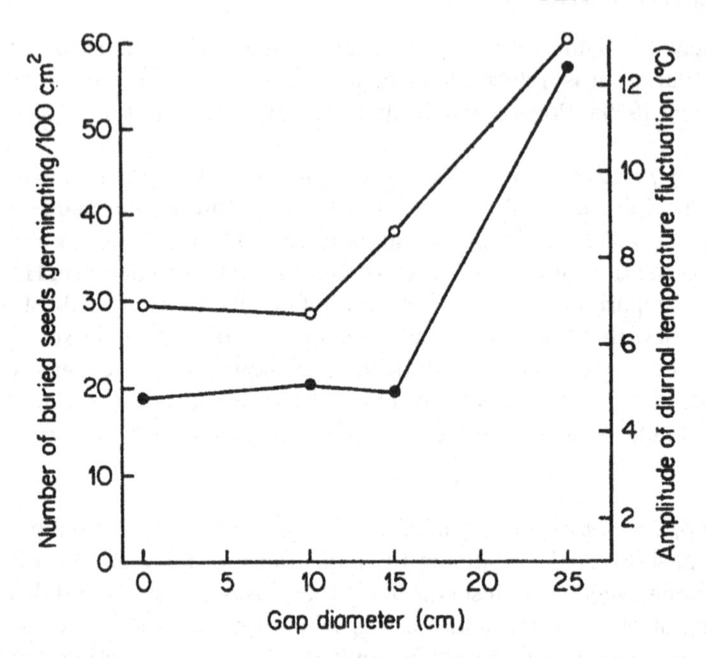

Figure 50. Comparison of soil temperature fluctuations (○) and germination (●) of naturally-buried seeds of *Holcus lanatus* in canopy gaps of different diameters in a sown pasture of low productivity in northern England. Temperature fluctuations are based upon the mean amplitudes of diurnal fluctuation measured in the centre of circular gaps at a soil depth of 10 mm over the period 11–23 May 1977. Germination counts were carried out during the period 6 April–30 May (Reproduced by permission of Thompson 1977.)

in grasses such as *Poa annua*, *P. trivialis*, *P. pratensis*, *Holcus lanatus*, and *Deschampsia cespitosa*, all of which are prominent in the buried seed populations of pastures (Chippindale and Milton 1934) and frequently exploit gaps in turf arising from excessive trampling and poaching by domestic animals.

Whilst the ability to respond to fluctuating temperatures in darkness is of widespread occurrence among species forming persistent seed banks (Thompson and Grime 1979) this potential is consistently absent in species with extremely small seeds, all of which appear to have an obligatory light requirement. The significance of this phenomenon may be related to the fact that seedlings originating from small seeds are likely to be severely limited in their capacity to penetrate to the soil surface.

The mechanisms whereby persistent seeds may detect conditions propitious for seedling establishment vary considerably according to the ecological situation. Some desert annuals germinate following the leaching of germination inhibitors (Went 1949) whereas many legumes with hard-coat dormancy become imbibed only following exposure to repeated cycles of temperature and moisture fluctuation (Moreno-Casasola *et al.* 1994). Seeds may also be caused

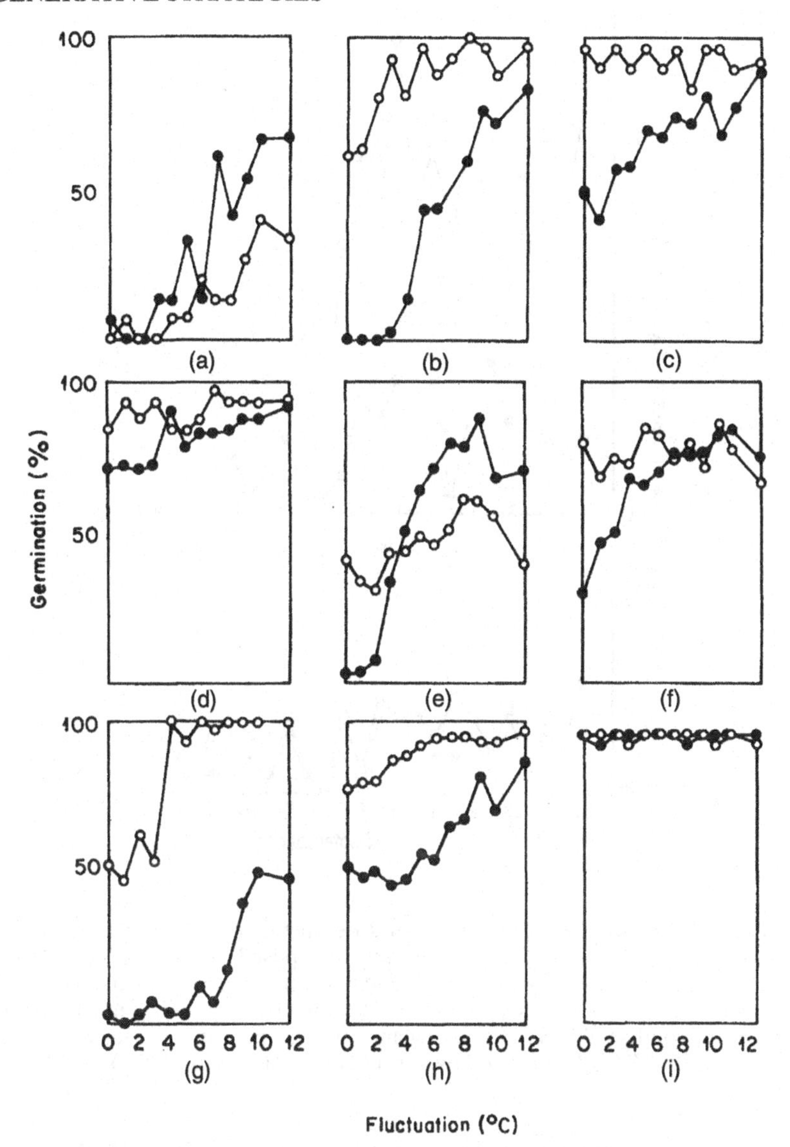

Figure 51. Germination responses to various amplitudes of diurnal fluctuations of temperature in light (○) and in continuous darkness (●) in nine herbaceous plants of common occurrence in pastures and arable land. Fluctuating temperatures were applied in the form of a depression below a base temperature of 22°C extending for six in every 24 h. Diurnal fluctuations were maintained throughout each experiment. Light intensity during the light period was 1846 μW cm^{-2}. Prior to testing, seeds were stored dry at 5°C. (a) *Elytrigia repens*, (b) *Deschampsia cespitosa*, (c) *Holcus lanatus*, (d) *Poa annua*, (e) *Poa pratensis*, (f) *Poa trivialis*, (g) *Rumex obtusifolius*, (h) *Stellaria media*, (i) *Lolium perenne*. (Reproduced from Thompson, Grime, and Mason 1977 by permission of Macmillan (Journals) Ltd.)

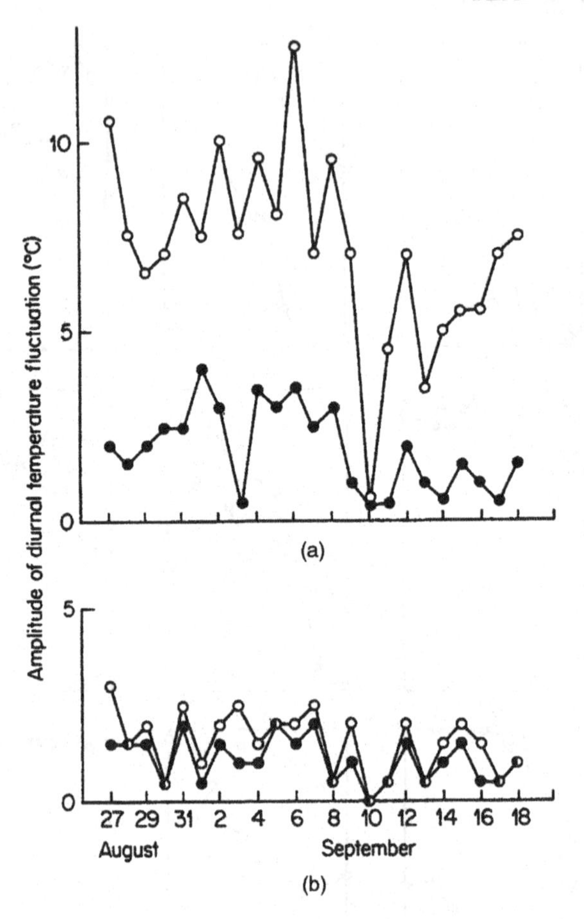

Figure 52. The amplitude of diurnal temperature fluctuations in soil temperature recorded beneath a dense leaf canopy mainly composed of the grasses, *Dactylis glomerata* and *Arrhenatherum elatius* (●-●), and beneath artificially created gaps of 200 mm diameter (○-○), (a) at a soil depth of 10 mm, (b) at a soil depth of 50 mm. Measurements conducted in Sheffield, England, over the period 27 August–18 September 1977. (Reproduced by permission of Thompson 1977.)

to germinate by an increase in soil nitrate concentration (Pons 1989b), and in habitats subject to fire there have been reports of seed germination in response to compounds released from burning plant material (Thanos and Rundel 1995; Keeley and Fotheringham 1998).

A feature of many of the herbaceous species with persistent seed banks is the tendency to exhibit polymorphism with respect to germination requirements. Seeds produced by the same population of plants (New 1958; Grouzis *et al.* 1976) or even one individual (Cavers and Harper 1966; Gutterman 1980) may show considerable differences in response to such factors as light and temperature and there may be an extremely wide variation with respect to the amplitude of diurnal fluctuation in temperature required to initiate germination (Figures 50 and 51). The significance of this 'polymorphism' is particularly obvious in the ruderal species. Because these plants occupy disturbed environments, the risks of seedling mortality are high and synchronous germination of the seed bank could cause the population to be vulnerable to extinction. It would appear therefore that in this context polymorphism not only provides the possibility of regeneration in a range of different spatial and temporal niches but also ensures the persistence of a reservoir of germinable seeds in the soil.

Dispersal of persistent seeds. So far in this account the significance of persistent seed banks had been reviewed strictly in relation to the dispersal through time of opportunities for regeneration. However, as explained by Poschlod and Bonn (1998), in intensively-managed landscapes such as those of Central Europe traditional methods of grassland and arable farming involved movements of seeds in a dormant state either as contaminants of crop seed, as hayseed, in manure, in or attached to farm animals, or through irrigation ditches. It is interesting to note that although dispersal through the agency of these traditional practices must now be in sharp decline, new avenues for the dispersal of a rather different set of species with persistent seeds are developing through the increased movement of soil-borne seeds on motor vehicles and through long-distance transport of vegetables and potted plants (Hodkinson and Thompson 1997).

Regeneration Involving Numerous Wind-dispersed Seeds or Spores (W)

In landscapes subject to erratic and large-scale disturbance (e.g. soil erosion, fire, tree felling, mining, quarrying, and road construction) it is not unusual for the flora to contain a high proportion of species which produce numerous wind-dispersed seeds. In temperate latitudes herbaceous plants of this type are particularly common in the *Compositae* (e.g. *Cirsium*, *Taraxacum*, *Crepis*, *Tussilago*, *Senecio*, *Petasites*), while shrubs and trees of the genus *Salix* also fall into this category. Under favourable circumstances the fecundity of some of these plants may be such that each year extensive areas are 'saturated' by wind-dispersed seeds and each patch of bare ground is rapidly colonised by populations of seedlings. Such rapid colonisation has been described by Bakker (1960) for newly-drained polders in Holland, by Nakashizuka *et al.* (1993) for avalanche debris in Japan, by del Moral and Wood (1993) for the aftermath of the Mount St Helens eruption, and by Stöcklin and Bäumler (1996) for glacial forelands in Switzerland. Regeneration by means of small wind-dispersed seeds is also well-documented for the North American prairies where it is associated particularly with the initial phase of colonisation of earth mounds produced by burrowing animals such as prairie dogs (Koford 1958) and badgers (Platt 1975). It seems reasonable to conclude that primarily this form of regeneration is adapted to exploit forms of disturbance which are spatially unpredictable.

Regeneration involving the copious production of wind-dispersed seeds also occurs among large, comparatively fast-growing trees, which in temperate regions are to be found in genera such as *Fraxinus*, *Acer*, *Ulmus*, *Ailanthus*, *Liriodendron*, and *Populus*. In aboriginal tropical forests this regenerative strategy is restricted to the so-called 'emergents' or 'nomads' (Keay 1957; van Steenis 1958; Knight 1975), tall, fast-growing trees which appear to be adapted to colonise large gaps created in the forest canopy by windfalls (see page 292).

In addition to inferences based upon direct observations of colonising ability, theories of wind-dispersal also originate from mathematical models and measurements in the laboratory (Askew *et al.* 1997). Relatively simple models have been devised that rely upon wind-speed, the height of release, and the terminal velocity of the dispersule (Sharpe and Fields 1982; Andersen 1991; Green and Johnson 1996). In field situations, however, there are additional factors that modify the effectiveness of wind-dispersal mechanisms; in particular, neighbouring plants not only reduce wind-speed but may directly intercept dispersing propagules (Green and Johnson 1996).

A further source of evidence concerning wind-dispersal is direct observation of the distances achieved by individual propagules (e.g. McEvoy and Cox 1987; Vegelin *et al.* 1997). However, the results of such investigations show that a high proportion of most seed populations are dispersed only very short distances and indicate that many critical long-distance colonisations depend on rare events that elude direct or experimental observation (Portnoy and Willson 1993).

A detailed classification of the mechanisms of wind-dispersal is beyond the scope of this book but is available from the works of Ridley (1930) and van der Pijl (1972). However, for ecological purposes some broad functional classes may be recognised. Table 16 contains a list of some common plants of the British flora which regenerate by means of numerous wind-dispersed seeds. These can be conveniently subdivided into three classes which differ with respect to the mechanism of dispersal. The first category includes a number of ferns and orchids in which the seeds (or spores in the pteridophytes) are exceedingly small and are capable of long-distance transport in air currents. In the ferns, it would seem likely that this form of dispersal resembles that of many bryophytes and lichens, in that it facilitates colonisation of habitats inaccessible and/or inhospitable to the majority of flowering plants (crevices in cliffs, rocks, tree trunks, and walls). Such habitats present a small target for a colonising species but once located they constitute 'safe-sites' (Sagar and Harper 1961) amenable even to establishment by dispersules with such very small reserves. However, some of the orchids, with 'dust' seeds, have a different type of ecology and are components of grassland and other types of herbaceous vegetation. Here, as pointed out by Harper (1977), the sacrifice of seed reserve is consistent with the peculiar process of seedling establishment which in these orchids involves a dependent subterranean phase with a mycorrhizal associate. Natural selection for high fecundity and dispersal efficiency in such plants may be interpreted, therefore, as a mechanism increasing the chance of effective contact in suitable environments between orchid seed and fungal symbiont.

Table 16. Some plants which regenerate by means of numerous wind-dispersed seeds or spores and/or common, widespread, or expanding distribution in the British Isles. The values (mg) are the mean weights of dispersules in populations sampled in northern England. All species in category (a) have mean seed or spore weights less than 0.01 mg

(a) Species with 'dust' seeds or spores			
Asplenium ruta-muraria		*Dryopteris filix-mas*	
Asplenium trichomanes		*Equisetum arvense*	
Athyrium filix-femina		*Lycopodium clavatum*	
Dactylorhiza fuchsii		*Orchis mascula*	
Dryopteris dilatata		*Pteridium aquilinum*	
(b) Species with plume or pappus			
Chamerion angustifolium	0.05	*Leontodon hispidus*	0.85
Cirsium vulgare	2.64	*Mycelis muralis*	0.34
Crepis capillaris	0.21	*Petasites hybridus*	0.26
Epilobium hirsutum	0.05	*Phragmites australis*	0.12
Epilobium montanum	0.13	*Salix cinerea*	0.09
Epilobium brunnescens	0.02	*Senecio squalidus*	0.21
Eriophorum angustifolium	0.44	*Taraxacum officinale*	0.64
Pilosella officinarum	0.15	*Tussilago farfara*	0.26
Hypochoeris radicata	0.96	*Typha latifolia*	0.03
Leontodon autumnalis	0.70		
(c) Species with winged seeds			
Acer pseudoplatanus	34.47	*Linaria vulgaris*	0.14
Alnus glutinosa	1.30	*Pinus sylvestris*	5.59
Betula pubescens	0.12	*Rhinanthus minor*	2.84
Fraxinus excelsior	46.19	*Ulnus glabra*	9.93

The second class of species in Table 16 consists of herbaceous plants in which dispersal of a larger seed is facilitated by structures such as a plume or pappus which increases the buoyancy of the dispersule. An interesting example of this phenomenon is provided by the composites, *Tussilago farfara* and *Petasites hybridus*, in which extremely rapid germination occurs over a wide range of temperatures in both light and darkness, the seeds being very short-lived. The seeds of these two plants are released during the early spring and their germination physiology appears to facilitate the rapid colonisation of bare soil created during the winter by flooding of river terraces and exposure of glacial moraines.

Many of the other species cited in the second category in Table 16 resemble *T. farfara* and *P. hybridus* in that the seeds germinate rapidly over a wide temperature range, decline in viability also rather rapidly, and do not appear to be incorporated into persistent seed banks. This does not mean that germination in all of these species invariably occurs immediately after contact of the seed with the soil surface. It is interesting to observe that, in Table 15, the wind-dispersed composite, *Hypochoeris radicata*, is among the species found by Fenner (1978) to be inhibited from germination in light filtered by a dense leaf canopy. This suggests that in at least some of the species in this category there may be mechanisms which by temporary delay of germination optimise the timing of germination by responding to seasonal changes in the canopy of established plants.

A more elaborate post-dispersal sequence may be suspected in a number of wind-dispersed marsh-plants (e.g. *Epilobium hirsutum*, *Typha latifolia*) and ruderals (e.g. *Sonchus asper*, *Crepis capillaris*, *Conyza canadensis*, Senecio viscosus) in which germination is inhibited by darkness (Grime *et al.* 1981). This finding suggests that, in such species, effective wind-dispersal may be allied to the capacity of the seed to remain dormant in the soil. Studies of the composition of natural seed banks provide confirmation that, particularly in the case of marshland habitats, some wind-dispersed species such as *Typha glauca* (van der Valk and Davis 1976, 1978) and *Epilobium hirsutum* (Thompson 1977) may form a persistent and major component of buried floras. As suggested by Grime (1979) further research is required in order to determine how commonly this potentially powerful combination of attributes (for dispersal through space and time) occurs in nature.

When comparisons are drawn across large data-bases (e.g. Rees 1993) it is apparent that there is a negative relationship between capacity for spatial dispersal and seed persistence. However, it may be unwise to assume that this pattern arises from morphological or physiological constraints. In fact, as described later in this chapter, some of the most widespread species in modern landscapes appear effectively to combine attributes conducive to both dispersal and persistence. If there are no fundamental 'design constraints' operating we must consider the possibility that the relationship detected by Rees (1993) does not reflect inevitable tradeoffs but instead involves phylogenetic inertia and different balances during the recent evolutionary history between selection forces promoting dispersal in space and immediate germination and those associated with seed persistence.

The third class of wind-dispersed species in Table 16 is made up of plants in which seed buoyancy depends upon planar extensions of the fruit wall into membranes or wings. Perhaps the most interesting members of this group are trees such as *Acer pseudoplatanus* and *Fraxinus excelsior* in which the rather large winged seeds have only a modest potential for wind-dispersal and depend heavily upon the fact that many of them are released from a considerable height during windy conditions. It is interesting to consider why natural selection in these species has failed to produce smaller seeds such as those of

Betula, capable of dispersal over greater distances. The explanation seems to be that dispersule size in these trees represents a compromise between the capacity for short distance dispersal into clearings or beyond expanding forest margins and the benefits of an embryo and seed reserve large enough to facilitate establishment in herbaceous vegetation (see page 192).

Regeneration Involving a Bank of Persistent Seedlings (B_{sd})

In the majority of forest trees there is a comparatively short interval between seed-drop and germination, and in tropical species germination is often extremely rapid (van der Pijl 1972). Among temperate forest trees it is not uncommon for either germination itself (e.g. *Fagus sylvatica*) or plumule extension (e.g. *Quercus petraea*) to be temporarily delayed, but there is no evidence that such species accumulate persistent seed banks. However, for some considerable time it has been recognised (Sernander 1936) that in mature temperate and tropical forests a very common regenerative strategy is that in which even-aged populations of tree seedlings (Figure 53) and saplings

Figure 53. A dense population of persistent seedlings of *Nectandra ambigens* on the floor of primary tropical rainforest at Los Tuxtlas, Mexico

persist for long periods in an extremely stunted or etiolated condition and a small percentage survive to maturity by expanding into canopy gaps when these arise through the senescence and death of established trees. In Table 17, a list is presented of trees for which regeneration by such 'advance reproduction' (Marks 1974) has been documented.

Table 17. A list of tree species regenerating by means of persistent seedlings

Species	Source of information	
	Country	Observers
Tsuga canadensis (Hemlock)	North America	Marshall (1927)
Picea abies (Norway spruce)	Sweden	Sernander (1936)
Quercus alba (White oak)	North America	Oosting and Kramer (1946)
Ilex aquifolium (Holly)	England	Peterken and Lloyd (1967)
Acer saccharum (Sugar maple)	North America	Keever (1973)
Fagus grandifolia (Beech)	North America	Marks (1974)

In the population dynamics of these trees, the reservoir of seedlings and saplings functions in a way which is in some respects analogous to that of a seed bank. The similarity extends even to the critical role of disturbance of the established vegetation by senescence, pathogenic fungi, and windfalls in releasing individuals from the bank. In many forest trees, seeds are not produced each year and the capacity of the seedlings to survive for long periods under unfavourable circumstances ensures that the potential for regeneration is maintained. A factor which may contribute to the persistence of the seedlings of many trees in temperate and tropical forest is the size of the seed (Salisbury 1942), which in many species is exceptionally large in relation to the rate of seedling growth and provides an effective buffer during initial establishment in heavily-shaded and/or nutrient-deficient environments (Grime and Jeffrey 1965).

Woody species with persistent seedlings tend to be strongly mycorrhizal and situations arise in which there are fungal connections (Figure 54) between seedlings in dense shade on the forest floor and the root systems of mature trees of the same or different species. There is experimental evidence (Simard *et al.* 1997) that seedlings of tree species can benefit from the transfer of carbon from the network of ectomycorrhizal fungi under natural field conditions.

In forests, reproduction by means of slow-growing seedlings is not confined to large-seeded trees. In his pioneer studies of plant demography, Tamm (1956) recorded the presence of a bank of extremely persistent slow-growing seedlings in a population of the shade-tolerant evergreen herb, *Sanicula europaea*, and from several laboratory investigations (Bordeau and Laverick 1958; Hutchinson 1967; Loach 1970; Mahmoud and Grime 1974) there is

Figure 54. Microcosm containing small seedlings of *Pinus sylvestris* connected through the mycelium of the ectotrophic mycorrhizal fungus, *Suillus bovinus*, to a two-year old sapling of the same species. (Reproduced by permission of DJ Read.)

abundant evidence of the role of persistent, slow-growing seedlings in the regeneration of shade plants. The results of a number of field observations and experiments suggest that the same regenerative strategy is also characteristic of arctic–alpine species (Billings and Mooney 1968) and plants of nutrient-deficient soils (Kruckeberg 1954; Beadle 1954; Rorison 1960; Rabotnov 1969; Grime and Curtis 1976).

THE ROLE OF ANIMALS IN REGENERATION

In the description of the mechanism of a persistent seed bank (page 152) reference was made to the involvement of animals both in the burial of seeds in the soil and in the initiation of germination in buried seeds. Animals also exert two other major impacts upon regeneration. One is the result of the predation of seeds and seedlings by animals (see page 174) whilst the second concerns the dispersal of seeds or vegetative propagules.

The mechanisms whereby animals may affect the dispersal of seeds are many and various. From work such as that of Ridley (1930) and van der Pijl (1972) considerable progress has been made in the task of describing and classifying seeds and fruits which are dispersed by animals. In a large number of plants cases have been reported where transport of the seed by particular animals occurs and a broad difference can be recognised between dispersal mechanisms (e.g. burrs, hooked fruits, glutinous seeds) which involve attachment to the exterior of the animal and those in which the dispersule is attractive and all or part of it is eaten by vertebrate or invertebrate animals. Within this second category a further distinction can be drawn between certain plants (e.g. *Chenopodium rubrum*, *Vaccinium myrtillus*, *Persicaria maculosa*, Urtica dioica) in which some or all of the seeds are capable of surviving the passage through the gut of various animals and others, including many large-seeded forest trees (e.g. *Quercus* spp., *Castanea* spp., *Carya* spp., *Fagus* spp.), in which the seeds are readily digested and successful dispersal and establishment depends upon the small proportion of seeds which are 'lost in transit' or escape predation in 'unclaimed caches' (Reynolds 1958; Van der Wall and Balda 1977; Handel 1978; Van der Wall 1990).

In the context of a general description of the main types of regenerative strategies a detailed analysis of the role of animals in seed dispersal need not be attempted. However, it is necessary to point out that adaptation allowing dispersal by animals is frequently associated with three of the five regenerative strategies described in this chapter. From detailed reviews (Howe and Smallwood 1982; Janzen 1983; Jordano 1992; Murray *et al.* 1994; Traveset 1998) and the examples cited in Table 18, it is clear that dispersal by animals can occur as a prelude to seasonal regeneration and may precede incorporation into a buried seed bank or a bank of slow-growing seedlings.

Despite several determined attempts to obtain a more precise understanding of the influence of birds or mammals as dispersal agents of particular plant species (e.g. Herrara 1998; Murray 1988; Jordano 1982; Masaki *et al.* 1998;

Table 18. Examples of the role of animals as dispersal agents in three types of regenerative strategies

	Plant species	Dispersal agent	Authority
(a) Regenerating by seasonal colonisation of vegetation gaps	*Hordeum* spp.	Various agents	Stebbins (1971)
(b) Regenerating from a persistent seed bank	*Pedicularis sylvatica*	Ants	Berg (1954)
	Persicaria lapathifolia *Persicaria maculosa*	Cottontail rabbit (*Sylvilagus floridanus*)	Stainforth and Cavers (1977)
	Prunus pensylvanica	Various birds	Olmsted and Curtis (1947)
(c) Regenerating from a bank of persistent seedlings	*Casearia corymbosa*	Masked Tityra (*Tityra semifasciata*)	Howe (1977)
	Pinus edulis	Clark's Nutcracker (*Nucifraga columbria*)	Van del Wall and Balda (1977)
	Ilex aquifolium	Various birds	Peterkin and Lloyd (1967)

Brewer and Rejmánek 1999; Wenny 2000) there has been remarkably little progress beyond the collection of circumstantial evidence towards a coherent evolutionary and ecological perspective. However, this is hardly surprising because, particularly in the case of long-lived plants, a vital contribution to fitness by a disperser may depend upon extremely rare events involving only one or two seeds among the millions produced by a particular individual. This is a research field requiring the attention of dedicated and long-lived scientists!

MULTIPLE STRATEGIES OF REGENERATION

A complication of major significance arises from the fact that several regenerative strategies may be exhibited by the same population or genotype. This phenomenon has important consequences for both the ecological and evolutionary potential of the plant. Before examining the ecological consequences of multiple regeneration it is necessary to explain why this phenomenon occurs. First it is helpful to draw a fundamental distinction between regenerative strategies and those recognised in the C-S-R classification for the established phase of plant life-histories. As explained in Chapter 1 the distinctions between C-, S-, and R-strategists involve inescapable tradeoffs between sets of traits implicated in fundamental aspects of plant physiology and development. Some tradeoffs are also apparent in the regenerative phase. For example, although there is some variation within seed crops, and even between seeds on the same inflorescence (Datta *et al.* 1970), most species have a seed size that falls within a narrow range and it is unknown for the same species to produce both very large and small seeds. As Hodgson and Mackey (1986) point out, variation in seed size appears to be severely constrained by phylogenetically-inflexible reproductive characters such as carpel structure, placentation, presence or absence of endosperm, and pattern of embryogenesis. However, several other traits associated earlier in the chapter with different regenerative strategies are mutually compatible and are frequently expressed in the same genome. Many species are simultaneously capable of producing vegetative and sexually-derived offspring and because both wind-dispersed and persistent seeds tend to be small in size there appears to be no fundamental barrier to their co-existence in the same genotype. Why then do not more species evolve multiple forms of regeneration? Perhaps the answer lies in the strongly contrasted circumstances of regeneration in different types of habitat and the evolutionary and ecological inertia that characterises some taxa (Hodgson and Mackey 1986). In Figure 55 the five regenerative strategies recognised in this chapter are mapped broadly with respect to their association with spatial and temporal intensities of disturbance. A challenge for the future is to explain why some species and higher taxa are confined to one cell of this matrix whilst others occupy all four.

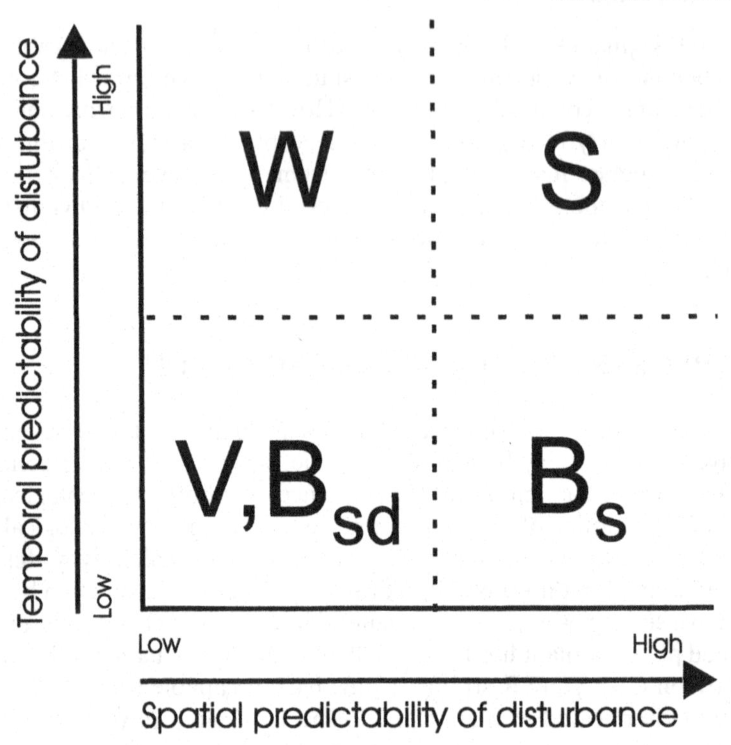

Figure 55. Theoretical relationship between the predictability of disturbance events in space and/or time and strategies of regeneration from seed. *W*, widely-dispersed seeds; B_{sd}, persistent juveniles; *S*, seasonal regeneration; B_s, persistent seed bank

In order to illustrate some of the combinations of regeneration strategies which may occur, a number of examples referring to populations examined in Britain have been assembled in Table 19. Six of these plants will now be considered in a series which is one of increasing complexity with respect to mechanisms of regeneration.

Example 1. *Impatiens glandulifera.* References have been made already (pages 118 and 144) to this summer annual which regenerates exclusively by means of large seeds which germinate synchronously in the spring. The result of this restriction to a single regenerative strategy is that there is no capacity for vegetative expansion nor for the development of a buried seed bank. The ecological consequence is the confinement of populations to streamsides and to other habitats in which there is a major impact of seasonally-predictable disturbance.

Example 2. *Digitalis purpurea.* In this biennial or short-lived perennial regeneration is by means of small seeds which are produced in very large

Table 19. The regenerative strategies exhibited by various flowering plants of common occurrence in the British Isles

Species	Regenerative strategies				
	Vegetative expansion	Seasonal regeneration	Persistent seeds	Numerous wind-dispersed seeds	Persistent seedlings
Impatiens glandulifera		*			
Dactylorhiza fuchsii				*	
Quercus petraea					*
Salix cinerea	*			*	
Ilex aquifolium	*				*
Tussilago farfara		*		*	
Digitalis purpurea		*	*		
Bromopsis erecta	*	*			
Poa annua (ruderal population)		*	*		
Poa annua (pasture population)	*	*	*		
Elytrigia repens	*	*	*		
Epilobium hirsutum	*	*	*	*	

numbers at the end of the vegetative phase. Where seeds fall onto bare soil germination and seedling establishment can occur directly, and in habitats where there is extensive and continuous disturbance, such as trackways in commercial forests and plantations, very high densities of established plants develop. However, the seeds of *D. purpurea* are inhibited from germination by darkness and are capable of extended periods of dormancy after burial in the soil. It is not surprising therefore to find that in many woods, copses, and hedgerows small populations often persist in vegetation in which disturbance is infrequent and more localised.

Example 3. *Poa annua* (ruderal populations). In arable land, gardens, paths, lawns, and intensively-grazed pastures, this small grass is a major vegetation component. Reference to studies of the life-cycle and reproductive biology in disturbed habitats (Wells 1974; Thompson 1977; Law *et al.* 1977) suggests that in this type of habitat *P. annua* is represented by populations in which regeneration is mainly by seed and, as in the case of *Digitalis purpurea*, incorporates two regenerative strategies, i.e. there is a large burst of seedling germination and establishment from freshly-shed seed but regeneration also involves subsequent recruitments from a persistent seed bank in the soil.

Example 4. *Poa annua* (pasture populations). In grassland populations of *P. annua* it is not unusual for a high proportion of individuals to be robust plants which are capable of expanding laterally into neighbouring gaps by means of tillers which root at the nodes. Experiments in which seed progeny of *P.*

annua from pasture populations have been grown under standardised conditions (Law *et al.* 1977) have revealed that flowering and seed production is considerably delayed in comparison with ruderal populations and there is a corresponding increase in longevity and vegetative vigour. It would appear, therefore, that under certain pasture conditions a third dimension has been added to the regenerative capacity of *P. annua* through the development of vegetative reproduction by rooting tillers.

Example 5. *Elytrigia repens.* Couch-grass (*E. repens*) is a coarse, strongly rhizomatous grass which is a persistent weed in arable land but is also of widespread occurrence in less disturbed habitats such as pastures, roadsides, and derelict land. Three forms of regeneration may be recognised in this plant. The first consists of vegetative expansion by means of the extensive rhizomatous system which allows the plant to form large patches in areas of productive herbaceous vegetation dominated by perennial species. The two remaining regenerative strategies of *E. repens* are particularly important in disturbed habitats. One involves regeneration from a persistent seed bank whilst the other depends upon vegetative propagation from pieces of rhizome fragmented and dispersed during cultivation of the soil.

Example 6. *Epilobium hirsutum.* Regeneration in this tall rhizomatous wetland herb is characterised by an extraordinary array of strategies. In relatively stable habitats, *E. hirsutum* is capable of rapid clonal spread by extensive rhizomes. The species also regenerates by means of numerous small wind-dispersed seeds, some of which germinate immediately whilst others are incorporated into a persistent seed bank (Thompson 1977). In disturbed habitats *E. hirsutum* may also regenerate from rhizome fragments.

These few examples and those cited in Table 19 illustrate only a small proportion of the combinations of regenerative strategies which exist in natural populations. Certain combinations are characteristic of plants from particular habitats and here perhaps the best example is the consistent association in stress-tolerant herbs, shrubs, and trees between vegetative expansion and regeneration involving a bank of persistent seedlings.

We may predict that possession of several strategies will enlarge the range of circumstances in which regeneration can occur and this, in turn, may be expected to widen the ecological range of the plant and to confer a greater degree of persistence under fluctuating environmental conditions. The ecological range of a plant is also, of course, a function of characteristics manifested during the established phase of the life-history. Moreover, the abundance and ecological amplitude will be affected also by the degree of genetic variability within and between populations. Nevertheless, we may expect that in an environmentally heterogeneous landscape there will be a general relationship between the frequency of occurrence of a species or ecotype and the number of regenerative strategies which it employs. Evidence consistent with this hypothesis has been obtained from calculations based upon surveys of the grassland flora of the Sheffield region (Figure 56). From these data it is apparent that multiple regeneration is associated with a marked expansion in abundance and ecological amplitude.

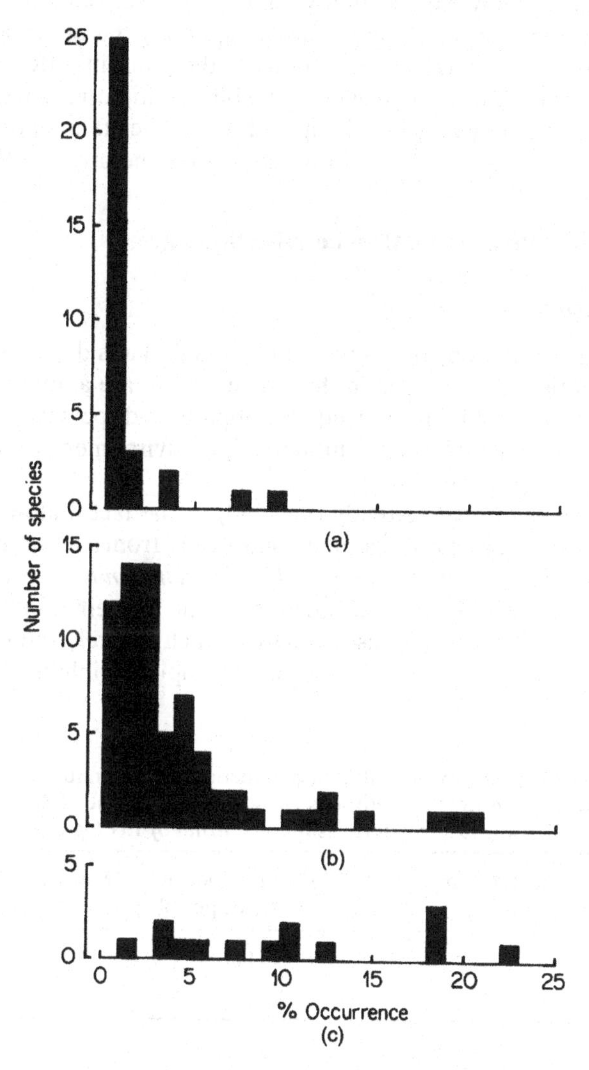

Figure 56. Histograms illustrating the frequency of occurrence of grassland plants in the Sheffield area: (a) species with a single regenerative strategy, (b) species with two regenerative strategies, (c) species with three or more regenerative strategies. Frequency is based upon the percentage occurrence of the species in 2748 m^2 quadrats distributed within the 32 major habitats occurring in an area of 2400 km^2

REGENERATION FAILURE

From earlier sections of this chapter it is apparent that each type of regenerative strategy is attuned to a limited range of habitat conditions. We may predict, therefore, that the failure of a plant to regenerate outside this range will be a major determinant of its ecological amplitude and geographical distribution. In addition we may expect that, within the habitat range, there will be marked effects of physical and biotic factors upon regenerative capacity and these will modify the abundance and status of the plant in different types of vegetation. Moreover, in a plant which exhibits multiple regeneration, profound changes in ecology may be anticipated where the effect of environment is to alter the relative effectiveness of its various regenerative mechanisms.

Failure in Plants with a Single Regenerative Strategy

Annual and Biennial Herbs

The survival of populations of annual and biennial plants depends upon successful regeneration by seed, and in this process there are a number of stages (flower production, pollination, seed development, dispersal, germination, seedling establishment) where an unfavourable environment may intervene to prevent effective regeneration.

Reference has been made already (page 50) to the fact that annual plants are relatively scarce in arctic and alpine regions, and from experiments such as those of Davison (1977) with the annual, *Hordeum murinum*, it is evident that major limiting effects in these environments are the low temperatures and the short duration of the growing season, factors which severely reduce the level of seed production (Table 20) and which may completely inhibit regeneration in particularly unfavourable years.

Table 20. Comparison of flower and fruit production in the annual grass, *Hordeum murinum*, growing at contrasted altitudes in northern England. (Reproduced from Davison 1977 by permission of *Journal of Ecology*.)

Site	Altitude (m)	Mean number of inflorescences per plant	Mean number of fruits per plant
Newcastle	91	26.0	728
Moorhouse	558	9.8	235

Even in geographical regions where the climate allows high rates of seed production, annual and biennial plants are unable to colonise productive, relatively undisturbed herbaceous vegetation, and here the breakdown of the regenerative process usually occurs at the stage of seedling establishment. It is clear from earlier parts of this chapter that seedling establishment in many, if not all, plants is adversely affected by the presence of a dense cover of established plants and litter, and in a large number of annual and biennial species successful regeneration by seed occurs only by the exploitation of gaps. One consequence of this dependence is that, in grasslands, annual plants become most important in areas of low rainfall where the effect of desiccation is to restrict the development of a continuous cover of foliage and litter. At northern latitudes, grassland annuals are often restricted in occurrence to areas of shallow soil on south-facing slopes (Ratcliffe 1961; Perring 1959; Grime and Lloyd 1973). A classic study of the dependence of an annual species upon vegetation gaps created by drought is that of Carter and Prince (1981), and Prince *et al.* (1985) were able to explain the metapopulation dynamics of the wind-dispersed annual, *Lactuca serriola*, in terms of year-to-year variation in rainfall at the northerly limit of its range.

In Britain, unfertilised pastures on chalk or limestone contain a diverse assemblage of annual and biennial plants, many of which declined in abundance during the last century (Perring and Walters 1962; Perring 1968). The disappearance of these plants from many localities has been associated with fertiliser applications and/or reduced grazing by sheep and rabbits, changes which have produced a more dense and uniform cover of foliage and litter and which, as pointed out by Grubb (1976), have obliterated the gaps upon which many of the annual and biennial species depend for seedling establishment.

Trees and Shrubs

There are many trees and shrubs in which seedling establishment is the only effective method of regeneration and we may expect therefore that the ecology of woody species will be restricted by effects of environment upon seed production and seedling mortality. As in the case of the annuals and biennials, discussed under the last heading, seed production in many trees and shrubs is severely reduced at northern latitudes (Pigott and Huntley 1981). The causes of this phenomenon vary with species and location but they include frost damage to flowers and the failure of fruits to reach maturity.

Within the regions in which seed production is relatively unrestricted by climate, a factor limiting the regeneration of many shrubs and trees is predation by animals and fungi. Seeds and fruits of trees are consumed by a wide range of animals and they are the primary food source of many birds, rodents, and primates. It is to be expected, therefore, that in addition to their incidental role as agents of dispersal, animals will often function as the most important factor controlling the proportion of the seed output which survives and germinates (Smith 1970; Batzli and Pitelka 1970; Gardner 1977). Predation also operates during the seedling phase and the results of experiments (Vaartaja 1952; Vaartaja and Cran 1956; Grime and Jeffrey 1965) and observations under natural conditions (Wardle 1959; Ross *et al.* 1970; Janzen 1970) suggest that fungal attack and selective feeding by small mammals and invertebrates provide an inconspicuous but potent influence on seedling survival.

Effects of predation by animals are likely to be particularly important in mature forest or scrub where regeneration often depends upon the maintenance of a bank of slow-growing seedlings and saplings. In many parts of the world, including Australia (Dixon 1892; Crisp and Lange 1976) and North America (Roughton 1972), regeneration of woody species is restricted by domestic animals such as sheep and goats. This phenomenon is also well known in Britain where regeneration in many of the surviving fragments of unfenced northern oakwoods is prevented by the destructive effect of sheep grazing (Pigott 1983; Figure 57).

Figure 57. Two saplings of *Betula pubescens* established inside the boundary fence of a sheep exclosure at Padley Wood, North Derbyshire, England

Failure in Plants with More than One Regenerative Strategy

In perennial species with several mechanisms of regeneration it is not uncommon to find populations in which the effect of the environment is to reduce the number of 'effective' regenerative strategies to one. This phenomenon is frequently observed in populations which lie near to the limits of the geographical distribution of a plant species. One of the earlier documented examples of this phenomenon is that of Mooney and Billings (1961), who showed that, the further north their situation, populations of the herbaceous perennial, *Oxyria digyna*, show decreasing amounts of seed production and increased dependence upon vegetative reproduction. In Britain there are several perennial herbs, such as *Cirsium acaulon* and *Filipendula vulgaris*, in which the most northerly populations rely almost exclusively upon vegetative expansion and only rarely regenerate by seed. After careful comparison of populations at contrasted latitudes over several years Pigott (1968) concluded that the very low frequency of regeneration from seed by *C. acaulon* at its most northerly stations is due to reductions in the number of mature seeds produced and more especially to infection by the fungus, *Botrytris cinerea*, which causes the seeds to rot whilst they are still attached to the parent plant.

It is not uncommon, as a consequence of partial regenerative failure, for populations of the same species in neighbouring habitats to depend upon different regenerative strategies. As we have seen in the case of *Poa annua* (page 98), genetic differences between populations may be involved in this phenomenon. However, in many instances, the explanation lies in the fact that local differences in the physical environment or in vegetation management cause changes in the relative effectiveness of alternative forms of regeneration. For example, in many British woodland herbs and shrubs, e.g. *Deschampsia flexuosa*, *Lonicera periclymenum* and *Rubus fruticosus* spp., regeneration in shade is almost exclusively restricted to vegetative expansion, although in clearings nearby it is not unusual to find high rates of seed production. Similar contrasts abound in the British landscape where closely adjacent sites are often subjected to very different forms of management. In roadsides and disturbed habitats in suburban areas common grasses such as *Poa trivialis*, *Lolium perenne*, and *Holcus lanatus* frequently regenerate from seed, whereas in neighbouring lawns and pastures the same species may be almost exclusively dependent upon vegetative forms of regeneration.

A particularly interesting form of regeneration failure is commonly observed in dioecious plants where seed production may be inhibited by geographical isolation of the sexes. In Britain, two well-known examples of species illustrating this phenomenon are *Petasites hybridus* and *Mercurialis perennis.* These herbs are capable of forming extensive stands by vegetative expansion and, in both species, it is not uncommon for local populations to be composed exclusively of either male or female plants.

REGENERATIVE STRATEGIES IN FUNGI AND IN ANIMALS

There is now an extensive literature concerning the possible adaptive significance of differences in the timing and form of reproductive effort in animal life-histories. Much of this work is theoretical and controversial and, as Stearns (1976) has pointed out, often considerable effort has been devoted to 'the development of ideas for which no one has established connections with the real world'. In this situation, no useful purpose would be served by attempting a detailed comparison of regenerative strategies in plants and animals. It is possible to observe, however, that several of the regenerative strategies recognised in this chapter appear to be closely similar to mechanisms occurring in heterotrophic organisms.

One of the most obvious parallels between plants and animals is that which may be drawn between 'fugitive' herbs and trees and many of the sedentary marine animals (e.g. barnacles, oysters). In both groups migration and establishment in 'safe-sites' is achieved through the production of numerous small, widely-dispersed offspring, most of which suffer early mortality.

Close similarities in regenerative mechanism are also apparent between autotrophic and heterotrophic organisms in environments subject to frequent and temporally unpredictable disturbance. Under such conditions, the role of the seed bank in the survival and re-establishment of plant populations is closely similar to that of the various resting-stages produced by many bacteria, fungi, molluscs, insects, and fish. Here, one of the most remarkable parallels is that between the 'polymorphic' germination requirements of buried seeds (see page 158) and the complex polymorphism which has been described (Wourms 1972) in the hatching responses of the eggs of annual fish.

In both plants (large-seeded trees and shrubs) and animals (larger mammals and birds) there are regenerative strategies in which risk of juvenile mortality appears to be reduced through the production of a small number of large offspring (Salisbury 1942; Lack 1947, 1948; Cody 1966). It is interesting to note, however, that in functional rather than genetic terms, some of the closest parallels with the large dependent offspring of mammals and birds appear to exist in the asexually-derived progeny of plants regenerating by vegetative expansion.

The complementary roles of sexual and asexual regeneration observed in the life-cycles of many colonising plants (page 138) recur in certain heterotrophic organisms. They are particularly well known in rust-fungi and in aphids, in both of which colonising episodes involving the replication of favoured genotypes alternate with sexual forms of regeneration.

II Plant Strategies and Vegetation Processes

4 Dominance

INTRODUCTION

For both plants and animals there is abundant evidence of the advantage of large size in many of the interactions which occur either between species or between individuals of the same species (e.g. Black 1958; Bronson 1964; Miller 1968; Grant 1970; Grime 1973a; Aguilera and Lauenroth 1993). In natural vegetation, examples of the impact of plant size are widespread. A consistent feature of vegetation succession is the incursion of plants of greater stature and at each stage of succession the major components of the plant biomass are usually the species with the largest life-forms. It is necessary, therefore, to acknowledge the existence of another dimension—that of dominance—in the mechanism controlling the composition of plant communities and the differentiation of vegetation types. In attempting to analyse the nature of dominance it is helpful to recognise two components:

1. The mechanism whereby the dominant plant achieves a size larger than that of its associates; this mechanism will vary according to which strategy is favoured by the habitat conditions.
2. The deleterious effects which large plants may exert upon the fitness of smaller neighbours; these consist principally of the forms of stress which arise from shading, depletion of mineral nutrients and water in the soil, the deposition of leaf litter, and the release of phytotoxic compounds.

It is suggested, first, that dominance depends upon a positive feed-back between 1 and 2 and, second, that because of the variable nature of 1, different types of dominance may be recognised. A broad distinction can be drawn between 'competitive', 'stress-tolerant', and 'ruderal' dominants. In 'competitive dominants' (usually perennial herbs, shrubs, or trees of the early stages of succession in productive habitats) large stature depends upon a high rate of uptake of resources from the environment, whilst in 'stress-tolerant dominants' (usually perennial herbs, shrubs, and trees of unproductive habitats or of the late successional stages of productive habitats) large stature is related to the ability to sustain slow rates of growth under limiting conditions over comparatively long periods. Large stature in 'ruderal dominants' (usually annual herbs) depends upon such characteristics as large seed reserves and rapid rates of germination, growth, and seed production.

The impact of a dominant plant may be exerted upon neighbours at various stages of their life-cycles. Mineral nutrient depletion and shading, together with the physical and chemical effects of litter, not only are capable of preventing seedling establishment but may also reduce the seed output of established plants and bring about premature mortalities among their populations. However, there can be little doubt that the most profound and widespread effects of dominance operate upon the seedling stage. As we shall see later in this chapter, this phenomenon is well illustrated by trees and shrubs, many of which as seedlings are subject to dominance by established perennial herbs but are themselves capable of dominance (often over the self-same herbs) at a later stage of their lifespan.

DOMINANCE IN HERBACEOUS VEGETATION

Dominance by Competitive-ruderals

In severely disturbed productive habitats the vegetation usually consists of a heterogeneous assemblage of ephemeral plants, none of which is capable of functioning as a dominant in the short intervals between successive effects of disturbance. However, where disturbance occurs rather less frequently and conditions favour the persistence of competitive-ruderals (see page 117), dominance may be observed. A competitive-ruderal which habitually attains the status of a dominant is *Impatiens glandulifera* (Figure 41), a large summer annual which in Europe colonises extensive areas where the margins of water courses have been disturbed by erosion, flooding, and silt deposition. During the summer *I. glandulifera* produces dense colonies and it is not uncommon for other species, including perennial herbs, to be submerged beneath a continuous tall leafy canopy composed of a large number of plants. It is clear therefore that the tendency of *I. glandulifera* to suppress the growth of other species colonising its habitat is due to the concerted effect of neighbouring plants which, as isolated individuals, would be incapable of exerting a significant degree of dominance.

It is interesting to note that the objective of many forms of arable farming, especially cereal cultivation, is to achieve weed control by creating conditions in which the crop plant attains the status of a dominant. As in the example of *I. glandulifera*, dominance by a cereal crop depends primarily upon the synchronous germination of a high density of large seeds followed by the rapid development of a dense vegetation cover composed of a large number of plants of comparable age and maturity. It is not unreasonable therefore to describe the events preceding ruderal dominance as a race between seedlings, the outcome of which is measured in terms of relative seed output and is mainly determined firstly by the frequency, size, and germination characteristics of contending seed populations and secondly by the growth-rates and morphologies of the seedlings and established plants.

In the discussion on page 85, reference has been made to the overriding importance of events early in the establishment of ruderal populations in determining the fate of component species. This phenomenon is highly relevant to arable farming where it is desirable to maximise the cost-effectiveness of weeding by selecting the optimal times for mechanical or chemical control measures. Agriculturists such as Nieto, Brondo, and Gonzalez (1968), Dawson (1970), and Roberts, Bond, and Hewson (1976) have, in fact, recognised a 'critical period', usually extending for a few weeks after crop emergence, during which depletion of resources by the presence of weeds may exercise a major effect upon yield. Provided that weeds are eliminated during the critical period, it has been shown that those appearing subsequently have an insignificant influence upon crop yield. This we may presume to be due to two factors. The first is the tendency of the crop to dominate late-germinating weeds, whilst the second is the 'ruderal' response (i.e. condensation of the life-history) exhibited by weeds subjected to the more limiting conditions experienced during this more advanced stage of vegetation development.

Dominance by Competitors

In derelict productive grassland and wasteland where perennial herbs are allowed to grow without major disturbance, there is a well-documented tendency (Watt 1955; Smith *et al.* 1971; Grime 1973a) for certain of the larger plants to expand and to suppress the growth of smaller neighbours. Despite the presence, under such conditions, of a large reservoir of resources, the effect of the activity of the plants during the growing season is to produce expanding zones of depletion, the most conspicuous of which are for light (expanding upwards from the soil surface) and for water and mineral nutrients (expanding downwards from the soil surface). In this type of vegetation high rates of mortality during the growing season and low rates of reproduction are characteristic of those plants which are outstripped by their neighbours and become 'trapped' in the depleted zones. If the growth of the larger species remains unchecked a process of exclusion occurs and this may eventually result in the vegetation approaching a state of monoculture. It seems reasonable to conclude that the dramatic reduction in species densities in meadows and pastures observed over the last 50 years in Europe (Thurston 1969) is to a large extent the result of an increasing intensity of competitive dominance brought about by stimulating the yield of the more robust and productive species and genotypes through the application of high rates of mineral fertilisers.

The ability of large, fast-growing perennial herbs to suppress the growth of smaller neighbours is particularly evident in vegetation dominated by tall herbs. Dominance by these plants is invariably associated with fertile or moderately fertile soil conditions and with circumstances in which the vegetation is subject to little disturbance. Studies of the seasonal changes in biomass in tall herb communities (Al-Mufti *et al.* 1977; Figure 58) reveal that where

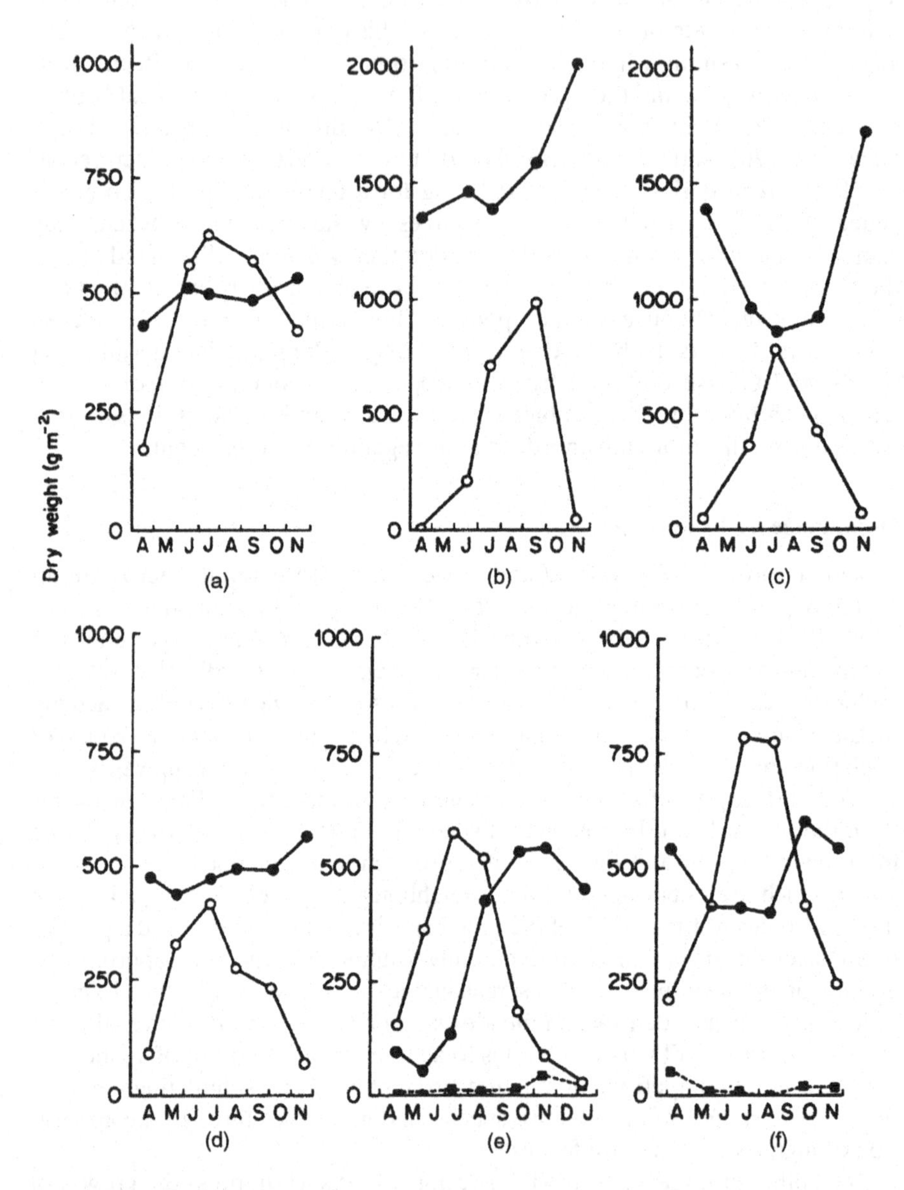

Figure 58. Seasonal change in the total shoot biomass (○), herbaceous litter (●), and tree litter (■) in six tall-herb communities in productive, relatively undisturbed conditions in northern England. The dominant species at the sites were (a) *Urtica dioica*, (b) *Pteridium aquilinum*, (c) *Chamerion angustifolium*, (d) *Filipendula ulmaria*, (e) *Petasites hybridus*, (I) *Urtica dioica* (Reproduced from Al-Mufti *et al.* 1977 by permission of Blackwell Scientific Publications Ltd.)

species such as *Urtica dioica*, *Pteridium aquilinum*, *Chamerion angustifolium*, and *Petasites hybridus* are dominant there is a high peak (>400 g m^{-2}) in standing crop during the summer and at this time most of the living shoot material is accounted for by the dominant species.

As explained in Chapter 1, these plants have a number of morphological and phenological characteristics in common which allow them to maximise the capture of resources during the summer. However, the dominant impact of such species cannot be explained simply in terms of the ability to monopolise the environment during the season of high potential productivity and, in particular, it is necessary to explain why species with shoot phenologies complementary to the dominants (i.e. vernal or evergreen plants) tend to be suppressed or excluded. From data such as those illustrated in Figure 58, there is strong evidence suggesting that a major component of the exclusion mechanism is the presence throughout the year of a high density of herbaceous litter. The impact of litter is particularly well exemplified by the strongly rhizomatous fern, *Pteridium aquilinum.* Although it is frost-sensitive and has a very short-lived canopy, this species forms single-species stands over very large areas of derelict land. Invariably, sites dominated by *P. aquilinum* are characterised by the presence of a very dense accumulation (1000–2000 g m^{-2}) of litter which shows comparatively little variation with season and it seems reasonable to conclude that this litter, either by shading or by physical impedance of germination, establishment, and growth, restricts the frequency of smaller or slower-growing species. Confirmation of the importance of a constant cover of herbaceous litter in excluding other species may be obtained by reference to another competitive dominant, *Petasites hybridus*, which, like *P. aquilinum*, produces copious quantities of litter. In comparison with that of *P. aquilinum*, however, the litter of *P. hybridus* is highly palatable to decomposing organisms and, as shown in Figure 58, declines rapidly to a minimum in the spring, at which time subsidiary species such as *Poa trivialis* and *Galium aparine* often make a temporary but major contribution to the standing crop.

Another observation which suggests the importance of litter in mechanisms of competitive dominance concerns the morphology and vernation of plants such as *Pteridium aquilinum*, *Urtica dioica*, and *Chamerion angustifolium.* Without exception the shoots of these plants are robust structures capable of penetrating a thick layer of litter and expanding the leaf laminae above it. This feature also occurs in a number of tall grasses, e.g. *Arrhenatherum elatius* and *Alopecurus pratensis*, which are capable of functioning as competitive dominants in productive meadows, derelict grasslands, road verges, and wasteland.

Although the effect of herbaceous litter is apparent at northern and temperate latitudes (Gross and Werner 1982; Facelli and Pickett 1991; Reader 1993) it is not confined to these regions. A strong impact of litter is described by Egunjobi (1974a) in an area of unburned savannah grassland in Western Nigeria dominated by the tall perennial grasses, *Andropogon gayanus* and

Imperata cylindrica. Measurements at this site showed that in early spring, at the onset of the rainy growing period, the litter density was high (1000 g m^{-2}) and growth of the vegetation during April and May was retarded, probably as a result of the choking effect of the litter.

In many types of herbaceous vegetation competitive dominance is a component of the mechanism controlling species density (number of species per unit area) and in Chapter 9 the practical significance of this fact will be considered.

Dominance by Stress-tolerant Competitors

Effects of dominance upon species density in herbaceous vegetation are not confined to highly productive environments. Even where grassland or heathland is established on shallow infertile soils, a gradual trend towards monoculture is frequently observed provided conditions allow the plant biomass to remain relatively undisturbed. Here, a well-known example is the widespread fall in species density which has been associated with reductions in grazing intensity by rabbits and sheep in unfertilised calcareous grasslands of the British Isles (Watt 1957; Thomas 1960; Duffey *et al.* 1974). In this example, the disappearance of the smaller low-growing herbs is correlated with the expansion of grasses such as *Brachypodium pinnatum*, *Festuca rubra*, and *Dactylis glomerata.* As in the case of the dominants associated with more fertile habitats (e.g. *Urtica dioica*, *Epilobium hirsutum*, *Chamerion angustifolium*) these species develop a dense leaf canopy over the period June–August and produce a high density of persistent litter.

Dominance by stress-tolerant competitors is particularly common in unfertilised grasslands in semi-arid continental climates. In phenological studies in North Dakota, USA, for example, Redman (1975) has described vegetation in which grasses and sedges such as *Stipa viridula*, *S. comata*, and *Carex pensylvanica* exercise local dominance through moderately-rapid rates of dry matter production and litter accumulation

An Index of Dominance

In the preceding sections dominance by various types of herbaceous plants has been associated with particular plant characteristics, e.g. a high leaf canopy and litter accumulation. We may predict, therefore, that in natural vegetation there will be a correlation between the appearance of these characteristics and the tendency for species density to fall as subordinate species are eliminated by dominance. One method of testing this hypothesis relies upon classification of herbaceous plants by means of a dominance index (Grime 1973a). The one illustrated in Table 21 is based upon four plant attributes consisting of the observed maxima for the species in height of leaf canopy, lateral spread, relative rate of dry matter production (under standardised productive laboratory conditions), and extent of litter accumulation. Each species has been scored with respect to the four attributes and the sum of the scores has been used to derive a dominance index with a scale from 0–10.

Table 21. Examples illustrating the derivation of the dominance index

Species	Attributes				Dominance index (Total/2)
	(a)	(b)	(c)	(d)	
Chamerion angustifolium	5	5	5	2	8.5
Arrhenatherum elatius	5	4	4	3	8.0
Brachypodium pinnatum	3	4	3	5	7.5
Ranunculus repens	3	5	3	1	6.0
Helictotrichon pratense	3	2	3	2	5.0
Taraxacum officinale	3	1	4	1	4.5
Festuca ovina	2	1	3	2	4.0
Campanula rotundifolia	2	2	3	0	3.5
Arenaria serpyllifolia	1	0	4	0	2.5

KEY TO SCORING SYSTEM. (a) Maximum plant height: 1, <260 mm; 2, 260–500 mm; 3, 510–750 mm; 4, 760–1000 mm; 5, >1000 mm. (b) Morphology: 0, small therophytes; 1, robust therophytes; 2, perennials with compact unbranched rhizome or forming small (<100 mm diameter) tussock; 3, perennials with rhizomatous systems of tussock, attaining diameter 100–250 mm; 4, perennials attaining diameter 260–1000 mm; 5, perennials attaining diameter >1000 mm. (c) Relative growth rate (g/g/wk): 1, <0.31; 2, 0.31–0.65; 3, 0.66–1.00; 4, 1.01–1.35; 5, >1.35. (d) Maximum accumulation of persistent (i.e. from one growing season to the next) litter produced by the species: 0, none; 1, thin, discontinuous cover; 2, thin, continuous cover; 3, up to 10 mm depth; 4, up to 50 mm depth; 5, >50 mm depth.

Although it ignores genetic variation within the species with respect to the potential for dominance and is rather subjective, especially with regard to the estimation of the potential for litter accumulation, the dominance index appears to be informative. In Figure 59, it has been used to seek evidence of effects of dominance upon species density in several types of vegetation sampled from the same geographical area in the north of England. For each square metre of herbaceous vegetation examined, a mean value has been calculated from the dominance indices of the component species and the contribution of each species to the mean has been weighted according to the frequency of the species in the vegetation sample. In Figure 59a, the mean dominance index for each sample has been plotted against the species density (number of species per square metre). The results show that where species of high dominance index are abundant (that is, where the mean DI exceeds 6.0) species densities are relatively low (less than 20 species per m^{-2}). From the data presented in Figures 59a and b, it would appear that in the area of study there is a low incidence of exclusion of species by dominance in the sparse vegetation characteristic of sheep pastures and rock outcrops over limestone. In contrast, there is strong evidence that dominance is a causal factor in the maintenance of the rather low species densities encountered in the samples from road verges (Figure 59f). The three remaining vegetation types examined in Figure 59 are enclosed pastures, derelict limestone grassland, and meadows. In each, species richness varies widely but shows a marked decline with increasing dominance index.

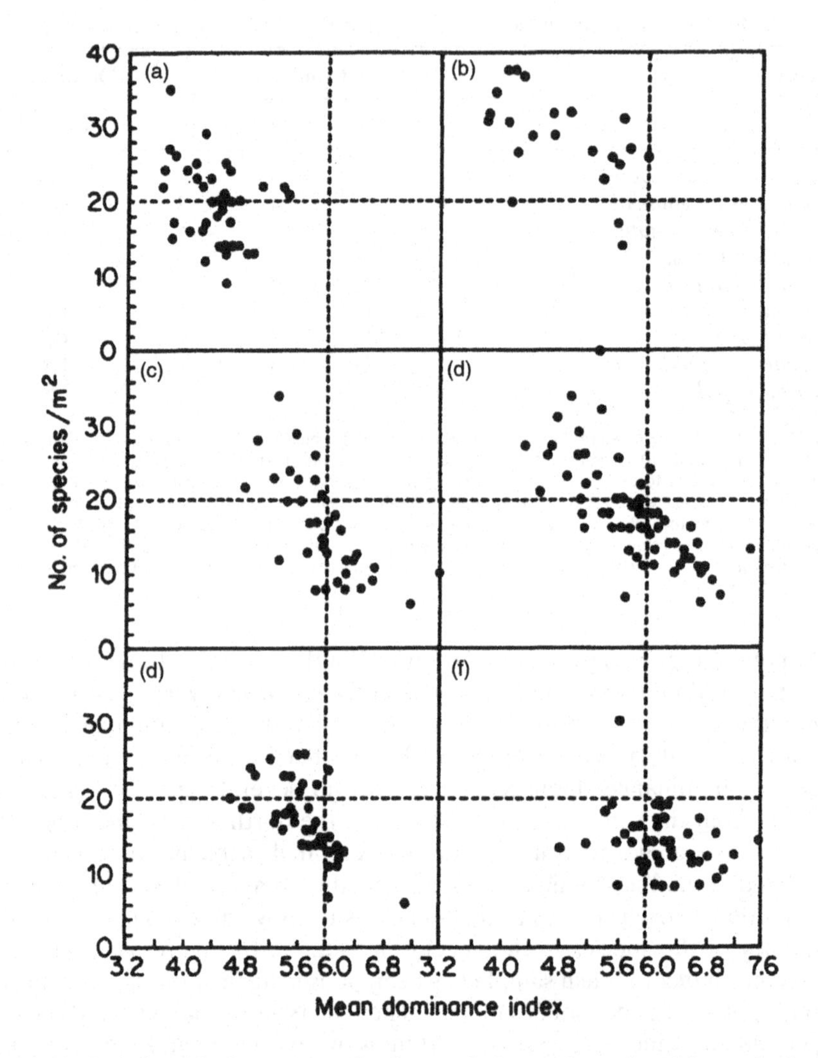

Figure 59. Relationship between a dominance index and species richness in herbaceous vegetation in six habitats sampled widely from the same geographical area in northern England. (a) Limestone outcrops with discontinuous soil cover; (b) unenclosed limestone pastures; (c) enclosed pastures; (d) derelict limestone grassland; (e) meadows; (f) road verges. The contribution of each species to the mean dominance index is weighted in proportion to its frequency in the m² sample of vegetation. (Reproduced from Grime 1973a by permission of Macmillan (Journals) Ltd.)

These results appear to confirm that the development of a marked degree of dominance in herbaceous vegetation is associated with the appearance of a high frequency of particular plant characteristics. For the practical purposes of vegetation management it seems possible, therefore, that criteria such as those used in the dominance index may be used to assess or anticipate the impact of dominance either at individual sites or in different types of vegetation.

DOMINANT EFFECTS OF HERBACEOUS PERENNIALS UPON TREE SEEDLINGS

With the exception of environments subjected to continuous or repeated disturbance there is a tendency for herbaceous vegetation to be colonised by the seedlings of shrubs and trees. The most obvious advantage of seedlings of woody species establishing in vegetation dominated by perennial herbs is the cumulative nature of height growth which, in successive seasons, elevates the leaf canopy of the tree seedling above that of the tallest herbs. However, despite their rapid growth in height, tree seedlings and saplings may become severely stunted in the presence of herbaceous species such as *Pteridium aquilinum* and *Chamerion angustifolium* which, because of their rapid growth-rates, capacity for lateral spread above and below ground, and copious litter production, may subject tree seedlings to major effects of dominance. The ability of perennial herbs to retard the growth of tree seedlings and saplings is well known to foresters and it is now common practice, as a preliminary to tree planting, to check the growth of adjacent herbs by treatment with herbicides.

Apart from their practical and economic importance, the strong effects which productive herbaceous vegetation may exert upon the growth of tree seedlings are of considerable theoretical interest, particularly in relation to the process of vegetation succession (see page 243). Three characteristics may be identified, which may account for the vulnerability of many tree seedlings. These are: low relative growth rate, delay in consolidation and lateral spread of the leaf canopy, and failure at least during the initial phase of establishment to accumulate a deep and persistent layer of leaf litter. Of these characteristics, the most consistent is the low relative growth rate which usually falls well below that of the seedlings of perennial herbs (Jarvis and Jarvis 1964; Grime and Hunt 1975; Ampofo *et al.* 1976). The comparatively slow growth of tree seedlings has been attributed to the expenditure of photosynthate on woody tissue, a process concomitant with a slow rate of expansion of leaf area (Jarvis and Jarvis 1964).

As explained on page 244, a characteristic which appears to increase the chance of successful tree seedling establishment in vegetation composed of perennial herbs is the possession of a large seed.

DOMINANT EFFECTS OF TREES UPON HERBACEOUS PLANTS

Some of the most dramatic effects of trees upon herbaceous vegetation occur in commercial forests and plantations. In mature, even-aged stands, both coniferous and deciduous species are capable of developing an extremely dense canopy and it is not unusual in these circumstances for the regeneration of herbaceous and woody species to be suppressed and for the herb layer to be extremely sparse. A particularly good system in which to detect the impact of trees upon herbaceous vegetation occurs in coppiced woodlands where the tree canopy is removed at regular intervals varying approximately between 10 and 20 years, and there are associated cycles of regeneration and suppression in the ground flora. Sampling at various stages in the coppice cycle (Figure 60) reveals a close correlation between the biomass of coppice shoots and the level of production in the herb layer.

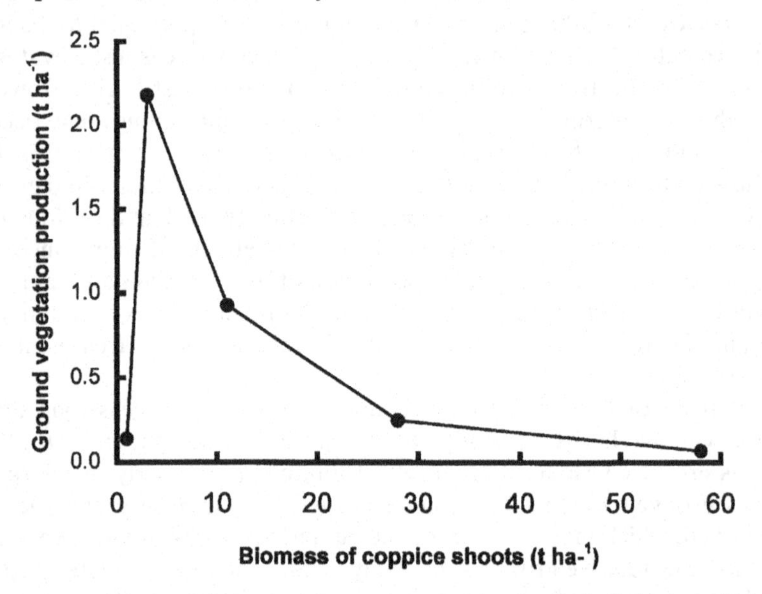

Figure 60. The relationship between annual production in the herb layer and biomass of the coppice shoots at different stages of the coppice cycle in a sweet chestnut *(Castanea sativa)* stand in south-eastern England. (Reproduced from Ford and Newbould 1977 by permission of Blackwell Scientific Publications Ltd.)

In some temperate woodlands (e.g. Figure 43, page 123) there is a very pronounced seasonal fluctuation in tree litter, the density of which remains relatively high (approximately 200 g m^{-2}) during the winter, only to fall extremely sharply in the early spring to a minimum of about 10 g m^{-2}. It is not uncommon, however, for the density of tree litter in temperate woodlands to attain seasonal maxima considerably greater than this and, in both deciduous and coniferous woods, there are local situations in which the density of tree litter exceeds 500 g m^{-2} throughout the year. From both field studies and experiments there is evidence that high densities of tree litter exert a profound effect upon the structure and species composition of woodland herb layers.

In an investigation in northern deciduous woodland (Sydes and Grime 1981a, b), studies were been made of the response of various woodland herbs to differences in the quantity and type of litter deposited upon them by deciduous trees. Experiments such as that shown in Figure 61 reveal that there are marked differences between herbaceous plants in their capacity to tolerate the physical impact of tree litter. In the five species investigated, the order of tolerance was *Lamiastrum galeobdolon*, *Hyacinthoides non-scripta*, *Viola riviniana*, *Holcus mollis*, *Poa trivialis*, and this sequence matches closely the correlations recorded in natural woodland between the abundance of the species and the quantity of tree litter (Figure 62). From the results of this investigation and from field observations, it is clear that the ability to tolerate persistent tree litter is related to shoot morphology. Emergence is achieved in *L. galeobdolon* by creeping stems from which a small number of erect shoots are supported above the litter. In *Allium ursinum* and *H. non-scripta* the litter is penetrated by robust enciform leaves whilst in *V. riviniana* emergence depends upon the combined effect of an erect shoot-base and elongated petioles. The susceptibility of species such as *P. trivialis* and *H. mollis* to persistent tree litter appears to be due to the fact that both the shoots and the

Figure 61. The effect of plastic 'litter' on the development of an assemblage of woodland herbs created by transplanting a standardised mixture of herbaceous species into garden plots. In the control plot (left) the grasses *Holcus mollis* and *Poa trivialis* are prominent but in the plot with litter (right) only the geophyte *Hyacinthoides non-scripta* and the erect shoots of the rhizomatous *Lamiastrum galeobdolon* are apparent

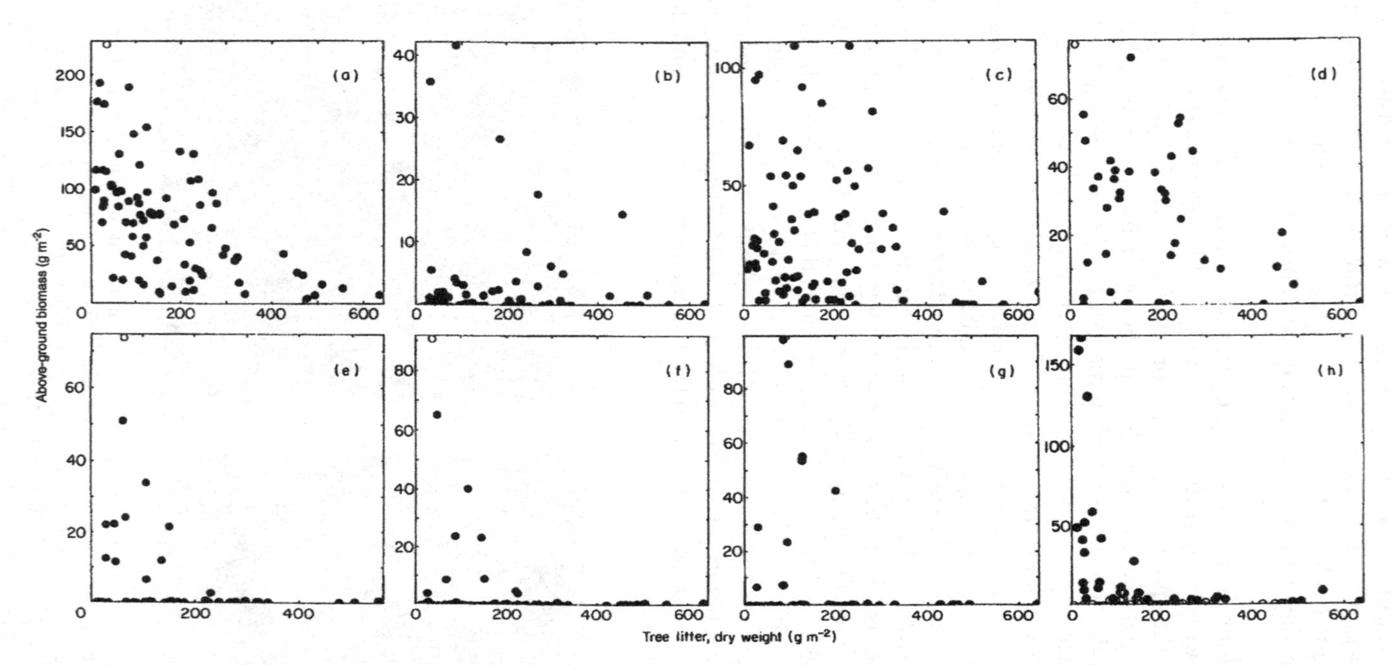

Figure 62. Scatter diagrams of above-ground biomass of individual herbaceous species and dry weight of tree leaf litter in 250 x 250 mm samples from oak woodlands in northern England. (a), Total ground flora; (b), *Lamiastrum galeobdolon;* (c), *Hyacinthoides non-scriptus;* (d), *Anemone nemorosa;* (e), *Holcus mollis;* (f), *Milium effusum;* (g), *Poa trivialis;* (h), *Mnium hornum* (bryophyte). (Reproduced from Sydes and Grime 1981a by permission of *Journal of Ecology.*)

individual leaves are insufficiently robust to resist physical impedance and to escape from the shading effect of dense litter.

It seems likely that physical effects of tree litter upon forest herbs will be commoner in cold and temperate regions. In a number of studies (Jenny *et al.* 1949; Olsen 1963; Hopkins 1966; Egunjobi 1974b) very rapid rates of litter decomposition in tropical forest have been measured and in certain sites litter has been found to disappear completely within periods varying from 2–7 months (Nye 1961; Madge 1965; Bernhard 1970).

So far discussion of the effects of trees upon herbaceous plants has been restricted to those arising from depletion of resources or the physical impact of litter. However, in a wide range of studies which have been reviewed by Whittaker and Feeny (1971), Muller and Chou (1972), and Rice (1974), evidence has been obtained of circumstances in which compounds released from living foliage or litter have been found to be capable of inhibiting seed germination and/or plant growth. Research workers differ widely in the extent to which they have attributed a direct ecological significance to these phenomena. Major difficulties arise when an attempt is made to determine whether such effects arise from natural selection for allelopathic ability or are merely a consequence of quite different adaptations such as the production of compounds which provide a defence against predation or microbial attack.

An example of the difficulties in the interpretation of quasi-allelopathic effects can be seen in the investigations of Sydes (1981) in which litter from a range of common British trees and shrubs was compared with respect to the rate of decay and the ability of leachates to inhibit seedling growth. From these data it was apparent that, in short-term experiments with seedlings, phytotoxic effects can be detected in leachates from litter of species such as *Fraxinus excelsior*, *Sambucus nigra*, and *Sorbus aucuparia.* However, release of toxins from these species appears to be a transient effect associated with the rapid decay and high initial efflux of solutes from freshly-deposited litter. It is clear that the species (e.g. *Fagus sylvatica*, *Quercus robur*) which form the main residue of persistent litter in deciduous woodlands are relatively non-toxic. Moreover, even *in Acer pseudoplatanus*, which combines moderate toxicity with an intermediate degree of persistence, it would appear that the main impact of the litter is due to physical rather than chemical causes. This conclusion is prompted by experimental evidence which suggests that freshly collected litter of *A. pseudoplatanus* was less pronounced, in its effect on herbaceous plants, than litter which had experienced leaching and decomposition in a woodland floor. Further evidence suggesting the overriding importance of the physical impact is the fact that the deleterious effects of persistent types of tree litter in woodland herbs are closely matched by those produced by an equivalent area of artificial litter (Figure 61) composed of opaque plastic 'leaves' (Sydes and Grime (1981b).

DOMINANT EFFECTS OF TREES UPON OTHER TREES AND SHRUBS

Most of the evidence and theories concerning the interactions between woody plants are based upon field observations in woodlands and plantations (Watt 1919; Wilde and White 1939; Ashton and Macauley 1972; Vaartaja 1952; Vaartaja and Cran 1956; Wardle 1959). These studies have provided insights into the mechanisms whereby the presence of established trees affects the regeneration of trees and shrubs and they have focussed attention on two critical phenomena. The first, known to foresters as the seed-bed condition, refers to the physical nature of the substratum in which the seed has germinated. The second concerns the role of fungi in seedling mortality.

In view of the profound effects upon woodland herbs described under the last heading it is not surprising to find that tree litter exerts a major influence upon the establishment of the seedlings of trees and shrubs. In temperate woodlands, seedling establishment in relatively small-seeded genera, e.g. *Betula*, *Pinus*, *Salix*, and *Tsuga*, is adversely affected by a continuous cover of tree litter and there is a strong tendency for seedling survival in such species to be restricted to local areas where mineral soil remains exposed or to emergent sites such as rotting logs or tree stumps (Keever 1973). A marked contrast to this phenomenon occurs in the case of larger-seeded species such as *Quercus*, *Castanea*, *Carya*, and *Fagus*, many of which are capable of emergence from beneath litter and appear to experience lower rates of seed predation in such circumstances.

From studies such as those of Vaartaja (1952) and Wardle (1959) it is evident that fungal pathogens account for a high proportion of the fatalities which occur among seedlings germinating in temperate woodlands and there is evidence that small seedlings may be predisposed to fungal attack by the low light intensities and high humidities which arise in deep litter and beneath dense leaf canopies of tree and herb layers. 'Damping off' is especially prevalent among the etiolated seedlings of small-seeded, potentially fast-growing trees such as *Betula populifolia* and *Betula lenta* (Grime and Jeffrey 1965). The capacities to emerge through shaded strata, such as those created by tree litter or herbaceous canopies, and to resist fungal attack are considerably greater in the large-seeded species which frequently occur as the dominants of mature forest (Figures 63 and 64).

In addition to effects of shading and fungal attack a variety of other possible mechanisms have been considered (e.g. Janzen 1970; Fox 1977) whereby trees might exercise a controlling influence on the regeneration of woody species. The phenomena suggested here include alterations in the availability of soil moisture and mineral nutrients, release of organic toxins from leaf canopies, and encouragement of insect populations which destroy seeds and seedlings. Only when the results of further research are available will it be possible to judge the validity of these hypotheses.

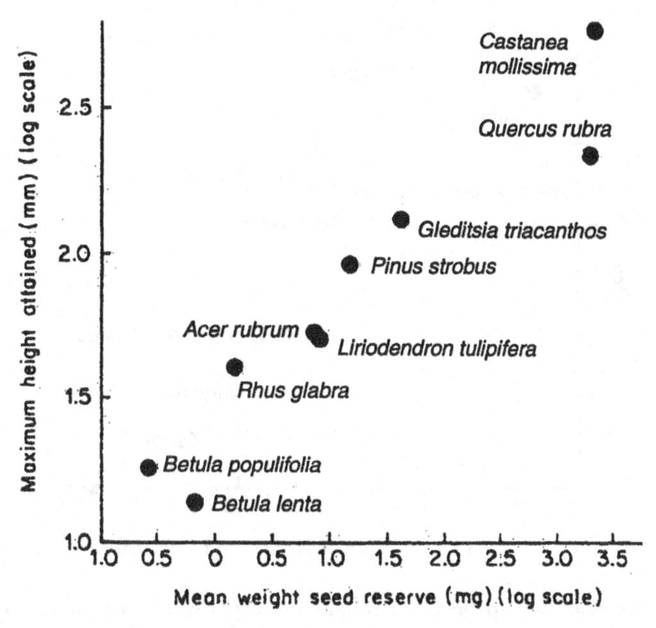

Figure 63. The relationship between weight of seed reserve and maximum height attained after 12 weeks in dense shade by seedlings of nine North American tree species. (Reproduced from Grime and Jeffrey 1965 by permission of Blackwell Scientific Publications Ltd.)

Seedling mortalities represent only one stage of the process by which the establishment of shrubs and trees may be affected by the presence of mature trees. As explained in Chapter 3, regeneration in many species depends upon exploitation of gaps in the canopy arising from windfalls and senescence. During the colonisation of gaps there may be intense competition between saplings, some of which may have originated from banks of persistent seedlings, others from wind-dispersed seeds. In some forests, interactions between the species may result in two or more phases in the recolonisation of the gap. In oak-birch woodlands of northern Britain, for example, gaps in the canopy are usually rapidly colonised by the wind-dispersed species, *Betula pubescens*, the saplings of which grow relatively rapidly and enjoy a temporary advantage over the slower-growing saplings of oak (*Quercus petraea*). In the long term, however, oaks usually replace the birches, and shading appears to be involved in this process. The ability of the oaks to suppress saplings of neighbouring birch trees appears to be related not merely to the fact that it is a taller, longer-lived species. It would seem also that the greater lateral spread and density of the leaf canopy causes the species to cast a deeper, more extensive shade than that of birch, which has a narrow and rather diffuse canopy.

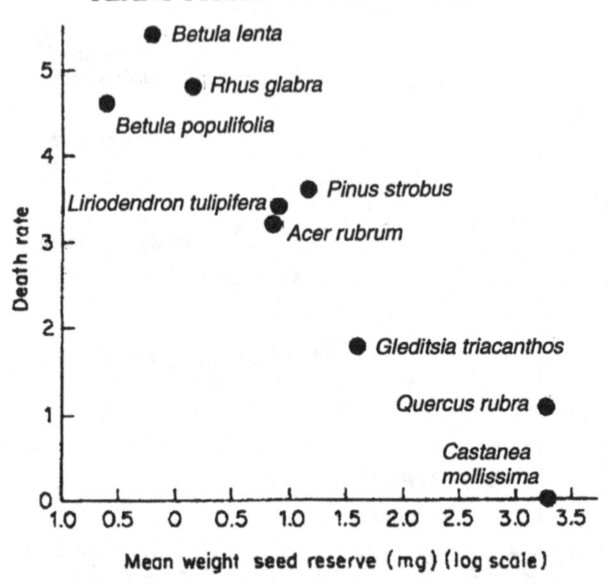

Figure 64. The relationship between weight of seed reserve and death-rate (mean number of fatalities per container in 12 weeks) in dense shade in seedlings of nine North American tree species. (Reproduced from Grime and Jeffrey 1965 by permission of Blackwell Scientific Publications Ltd.)

A similar pattern of species replacement occurs during the re-vegetation of gaps in northern hardwood forests of the USA. In the sequence illustrated in Figure 65, for example, the primary colonist, pin cherry (*Prunus pensylvanica*), is a short-lived species which is excluded in the later stages of re-colonisation by the arrival of larger, long-lived forest trees.

Shading of pioneer trees and shrubs by forest dominants is also a conspicuous feature of the later stages in the process of woodland development on abandoned fields in the Eastern United States. In this habitat primary colonists such as red cedar (*Juniperus virginiana*) and grey birch (*Betula populifolia*) display extreme sensitivity to shade (Lutz 1928). As the forest cover closes, shaded sectors of the canopies show rapid die-back and this is followed by high rates of mortality among established trees.

SUCCESS AND FAILURE IN DOMINANT PLANTS

At the beginning of this chapter dominance was defined by reference to the stresses which are produced by large plants and the effects of these stresses upon smaller neighbours. It is clear, however, that dominance depends not only upon the generation of stresses but also upon the capacity to avoid or resist their effects. Here, a familiar example is the ability of the shoots of

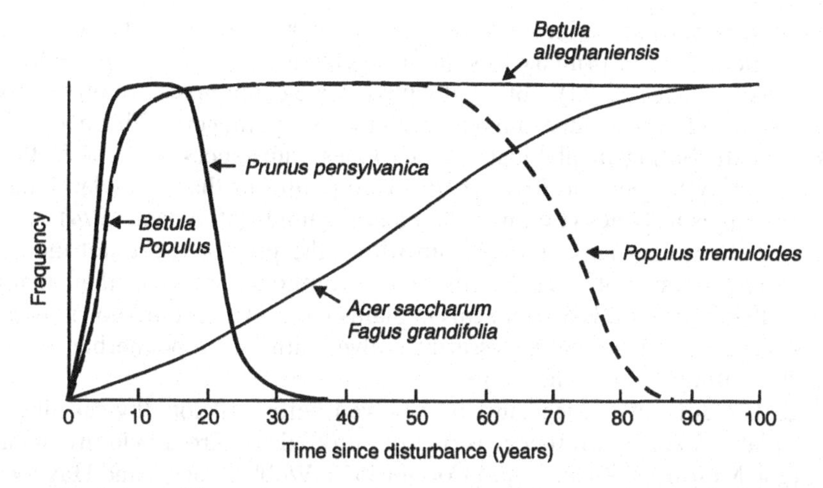

Figure 65. Scheme illustrating a common sequence of vegetation recolonisation following local disturbance in northern deciduous woodland in North America. (Reproduced from Marks 1974 by permission of the Ecological Society of America. © 1974 The Ecological Society of America.)

herbaceous dominants such as *Pteridium aquilinum* and *Chamerion angustifolium* to penetrate through the litter which they have themselves accumulated.

In the 'competitive' dominants of productive environments, stresses originating from local depletion of resources are avoided by means of growth responses which permit a redistribution of the absorptive surfaces of the plant and exploitation of new reserves within the habitat (see page 23). However, as the dominant plant enlarges in size, and resources (especially mineral nutrients) are sequestered in the biomass, we may predict that the competitive dominants will often decline in vigour and will tend to be replaced by 'stress-tolerant' dominants which are better adapted to withstand such 'autogenic' stress. Phases of dominance, decline, and replacement are, in fact, frequently observed in nature (Watt 1947) and are an integral part of the processes of succession and cyclical change considered in Chapter 8.

DOMINANCE AND ALLELOPATHY

Both living and dead plants have constituents which either directly or by microbial transformation in the soil are capable of exerting an inhibitory effect on plant growth. Many of these compounds, e.g. lignin, are essential to the structure or metabolism of the plant and their phytotoxic properties appear to be incidental.

In addition there are 'secondary' chemicals in plants, some of which when released into the environment inhibit germination or plant growth. Since many of these substances have no known function within a plant it is tempting to conclude that their primary function is allelopathic. However, as we have seen already (pages 71–73), potentially phytotoxic compounds are often produced as a defence against animal predators or pathogens. The need for caution in attributing an allelopathic role to these substances has been further emphasised by Siegler and Price (1976), who point out that secondary compounds such as terpenes (Loomis 1967) and alkaloids (Robinson 1974) often exist in a state of dynamic equilibrium within the plant and are not merely static end-products of metabolism. These authors conclude that many potentially-toxic secondary chemicals in plants have an as yet unknown metabolic function and have not necessarily evolved either to repel herbivores or to inhibit pathogens or competitors.

A further difficulty which must be resolved before allelopathy can find a secure place in ecological theory is that of autotoxicity. From field investigations (e.g. Milton 1943; Curtis and Cottam 1950; Webb, Tracey, and Haydock 1967; Hanes 1971; Newman and Rovira 1975) it is apparent that many phytotoxic compounds are effective against the plants which produce them. This does not mean, however, that we should assume that these plants derive no advantage from the production and release of toxins. Quite clearly, contingencies may arise in nature which differ substantially from those occurring in laboratory experiments or at local field sites.

One set of conditions in which a plant may escape the harmful effect of its toxins is that in which there is spatial separation between the root system and the zone of contaminated soil. An example of this phenomenon occurs in the chaparral of North America when toxins originating from the foliage and leaf litter of shrubs such as *Adenostoma fasciculatum* tend to be confined to the surface layers of the soil profile. However, there is evidence (McPherson and Muller 1969; Hanes 1971) that, in mature chaparral, the level of toxin accumulation reaches a point where autotoxicity occurs and regeneration of *A. fasciculatum* is inhibited. In the normal cycle of events in chaparral, the toxins are destroyed by fire and after an initial phase of colonisation by annual plants and grasses there is a phase of renewed toxin accumulation and dominance by shrubs. It is interesting to note that cycles of colonisation, dominance, and reduced productivity have been recognised in acidic heathlands and grasslands of the British Isles (Watt 1947, 1955; Grime 1963; Grubb, Green, and Merrifield, 1969), although, in these vegetation types, it remains to be shown what part, if any, is played by phytotoxins.

CO-EXISTENCE WITH DOMINANT PLANTS

It is unusual for the process of dominance to result in stands of vegetation composed of a single species. Even in circumstances where most of the biomass belongs to one species it is not uncommon for others to remain as a relatively minor but constant component of the vegetation. Studies of these plants suggest that there is a variety of mechanisms whereby subordinate species may co-exist with the dominants. Some, by virtue of their morphology or phenology, escape the main impact of the stresses generated by the dominant (e.g. vernal herbs beneath deciduous trees) whilst others (e.g. evergreen herbs in woodland) are adapted to tolerate these stresses. Co-existence in certain other plants is achieved by exploitation of areas within the habitat where the environment is locally unfavourable to the dominant. In addition there are subordinate species which owe their presence to micro-habitats created by the dominant (e.g. epiphytic angiosperms, terns, mosses, and lichens). A more detailed examination of these phenomena is attempted in Chapter 9.

ANALOGOUS PHENOMENA INVOLVING ANIMALS

The search for examples of dominance in animal communities can be complicated by the difficulties inherent in studies of interactions between mobile organisms and has sometimes been frustrated by the tacit assumption by some evolutionary biologists that competitive interactions generate niche differentiation and species co-existence rather than exclusive occupation of habitat by one or few species. However, as Paine (1984) points out, such monopoly is common in sessile animals and plants in the marine environment:

> Single sessile species on some marine rocky shores are capable of forming resource monopolies. . . . The outcome of these interspecific interactions seems deterministic, perhaps entirely so in the case of mussels. . . . A comparable determinism characterises the spatial domination by the brown algae *Hedophylla* and *Alaria*.

Dominance has been the subject of many investigations by animal ecologists, and the results of these studies suggest that different mechanisms of dominance, resembling in certain respects those observed in plants, can be recognised. 'Ruderal' dominance is apparent in populations of certain birds, rodents, and insects in which relatively small animals acting in concert exercise a dominant, if temporary, effect upon the fate of other animals exploiting the habitat. Familiar examples here include insect eruptions such as those of locusts (*Anacridium*, *Locusta*, *Schistocerca* spp.) and gypsy-moth (*Lymantria dispar*) and the huge flocks of small grain-eating birds which congregate in cereal-growing regions. A specific instance of 'ruderal' dominance in birds is that described by Orians and Collier (1963), who observed that tricoloured blackbirds (*Agelaius tricolor*) through force of numbers are able to exclude a larger species, the redwinged blackbird (*A. phoeniceus*).

'Competitive' dominance appears to be analogous in certain respects to the process of social dominance recognised by animal ecologists such as Allee

(1938), Elton and Miller (1954), Park (1954), Miller (1967), Morse (1971), and Wilson (1971). Social dominance has been described by Morse (1974) as 'the priority of access to resources that results from successful attacks, fights, chases, or supplanting actions, present or past'. The appropriateness of this definition will not be lost on anyone who has kept a bird table or who has observed the feeding behaviour of species such as the starling (*Sturnus vulgaris*) or herring gull (*Larus argentatus*). As in the case of competitive dominance in plants, social dominance in animals is achieved at the cost of a high rate of reinvestment of captured resources and it is most characteristic of animals exploiting productive habitats.

Parallels between animals and stress-tolerant dominants are rather more tenuous. Nevertheless, it is interesting to note that, in mature ecosystems, dominant effects of large mammalian herbivores upon smaller animals appear to be exerted not through direct social interactions but through resource depletion and habitat destruction arising from the presence of a high density of large animals. Janzen (1976) has suggested that in countries such as Kenya and Uganda the scarcity of reptiles may be related to the dominant effect of large mammals. He postulates that 'through intensive grazing, browsing and trampling, especially near water courses during severe dry seasons, large herbivores should greatly reduce the cover available for reptiles, the small vertebrate prey of snakes, and the insects available to reptiles'. Janzen also points out that where there is a large biomass of herbivores there is likely to be a high density of carnivores which may exert an 'incidental depressant impact on the reptile biomass'.

5 Assembling of Communities

INTRODUCTION

In preceding chapters of this book it has been argued that conflicting selection pressures and severe internal constraints have limited the paths of functional specialisation available to plants with the result that characteristic types of ecologies recur and are associated with particular combinations of established and regenerative strategies. One of the ultimate objectives of ecologists, addressed in the remainder of this book, is to understand how and to what extent the sets of traits identified with the primary plant strategies control community and ecosystem properties. First, however, we need to consider how plant functional types aggregate into local assemblages, some of which have such consistent structure and species composition that for more than a century ecologists have been inspired to construct systems of vegetation classification for the purposes of mapping, conservation, and management (Schouw 1832; Raunkiaer 1913; Clements 1916; Braun-Blanquet 1932; Tansley 1939; Curtis 1959; Ellenberg 1963: Rodwell 1991). Underlying much of the research on plant community assembly is an old conundrum: how similar must species be to exploit the same community and how different to permit their co-existence?

With the exception of the vegetation in recently and heavily disturbed habitats, or in conditions of extremely low productivity, plant communities contain dominant species and, as we have seen in Chapter 4, these differ not only in the mechanisms whereby dominance is attained, but also in their positive and negative effects on neighbouring species. We should not be surprised, therefore, that many investigations of plant community structure and dynamics focus on dominants. However, in order to understand plant communities fully it is necessary also to examine the ecology of the plant populations that are present as minor contributors to the biomass. Are these species merely exploiting opportunities to co-exist with dominants or have they a crucial involvement in community assembly and function?

The investigation of the processes involved in the assembly of plant communities must not be restricted to studies of those species that occur as major or minor components of communities. A complete understanding of the mechanisms of community assembly will also take account of the various filters that operate in the past and present to exclude other species, some of which may be extremely abundant in the surrounding landscape.

There are three main approaches whereby ecologists have sought to identify the processes involved in the assembly of plant communities. Insights have been obtained by direct observation in the field, by experimental manipulations of natural communities, and by synthesis of communities under controlled conditions.

DIRECT OBSERVATIONS IN THE FIELD

In seeking to interpret data collected from direct observation of plant communities a very useful first step is to adopt the procedure of Whittaker (1965) by ranking the plant species within a community in order of decreasing abundance to construct a dominance diversity curve (Figure 66). Whittaker suspected that often a functional significance could be attached to the position

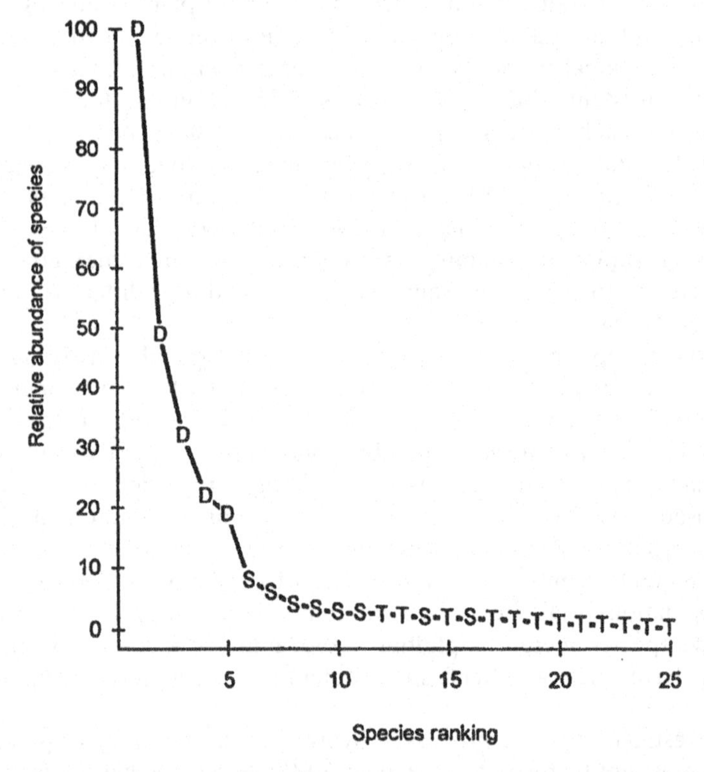

Figure 66. An idealised dominance-diversity curve (*sensu* Whittaker 1965) for a small sample of herbaceous vegetation. The distinction between dominants and subordinates is based upon relative abundance and has been set at an arbitrary value (10%). Transients are distinguished from subordinates by their failure to regenerate and persist in the vegetation under scrutiny. D, dominant; S, subordinate; T, transient

of a species in the hierarchy depicted by a dominance diversity curve. Recently this idea has been extended by the suggestion (Grime 1998) that there are three types of contributions—dominants, subordinates, and transients—made by species to plant communities and that these occupy characteristic sections (see Figure 66) of the dominance diversity curve.

Dominants and Subordinates

Early efforts to describe and interpret herbaceous vegetation (e.g. Clements 1916; Tansley 1939; Ramenskii 1938) involved listing of the plants present in selected stands of vegetation and estimates of the abundance of each species. Even before the widespread adoption of experimental approaches there was a keen awareness of the potential of certain species to occupy a high proportion of the plant biomass and to control the abundance and fitness of other minor contributors. A classic example is the rapid expansion of tall coarse grasses and the coincident suppression of small herbs noted by Tansley and Adamson (1925) in their study of the consequences of excluding rabbits from chalk grassland (Figure 67).

A more quantitative approach followed as plant ecologists adopted stricter sampling methods and measured the relative abundance of plant species by point analysis (Figure 68) or by harvesting, sorting, and weighing vegetation samples. This allowed more formalised attempts to examine the functional relationships between dominant and subordinate members of plant communities (Whittaker 1965; McNaughton 1978a). Subsequently, it was recognised that the potential to dominate vegetation could be associated with specific plant traits such as tall canopies, extensive lateral spread, and the capacity to project shoots forcefully through litter and herbaceous cover (Al-Mufti *et al.* 1977; Sydes and Grime 1981a, b; Grubb *et al.* 1982; Campbell *et al.* 1992). A further step in defining the functional differences between potentially dominant and subordinate members of plant communities became possible through the application of foraging theory to plants. From a study of the development of the roots and shoots of isolated plants exposed to standardised patchy environments (Campbell *et al.* 1991a), it was found possible to predict the dominance hierarchy that developed when eight herbaceous species were grown together in an experimental assemblage (Figure 14, page 28). In this investigation it was found that dominance was associated with relatively imprecise foraging both above and below ground. Such coarse-grained foraging allowed exploitation of a large volume of habitat but was associated with an imprecise concentration in resource-rich sectors. In marked contrast, subordinate species were characterised by precise but local placements of roots and shoots in resource-rich patches, a foraging pattern permitting temporary co-existence with dominants but apparently committing them to a minor status in the community

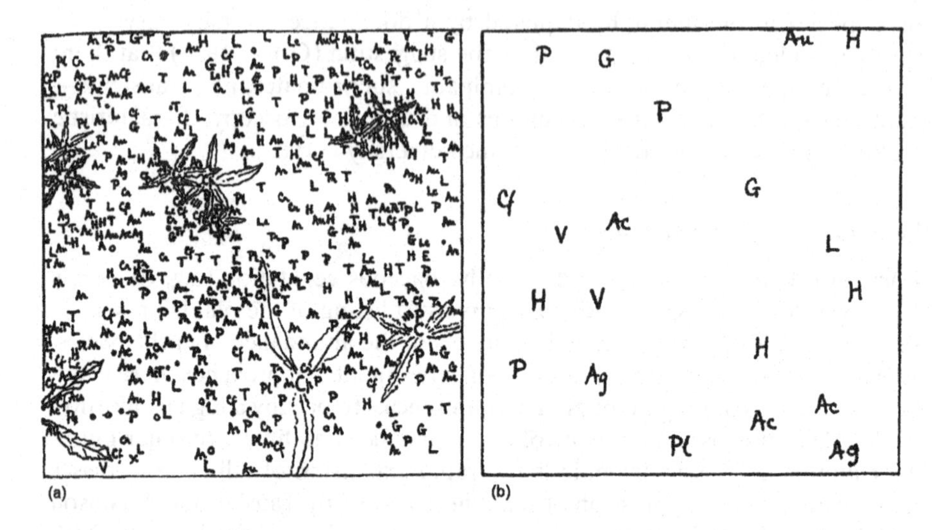

Figure 67. Sketch maps showing the response to rabbit exclusion of species-rich chalk grassland of the Sussex Downs, England. The authors described the vegetation as follows: (a) Quadrat 2.5 dm square (= 6.25 square dm) outside the Downley 'B' enclosure in August 1914. Herbage 4–5 cm high. *Festuca ovina* (not represented on the chart) was dominant, filling all spaces between the other plants. The flat rosettes of *Cirsium acaule* and *C. palustre*, whose leaves are diagrammatically represented, were conspicuous, and occupied a considerable area. The total number of rooted shoots of flowering plants, other than *Festuca ovina*, is approximately 431. (b) Quadrat of same size inside the Downley 'B' enclosure, Sept. 1914. Herbage about 18–20 cm high (mosses not recorded). *F. ovina* (not represented on the chart) was dominant, filling all spaces on the surface of the turf between the other plants, though there were bare spaces on the soil surface, concealed by the mass of vegetation. The total number of other flowering plants was only 20 (in some other quadrants of the same size as many as 40 occurred), but the individual plants were much bigger, as is roughly indicated by the size of the letters used as symbols. The effect of the complete withdrawal of grazing and nibbling was primarily to allow the growth of the dominant grass to the suppression of the great majority of the other flowering plants. (Reproduced from Tansley and Adamson 1925 by permission of *Journal of Ecology*.)

and leaving them vulnerable to extinction in circumstances where the dominants continued to enjoy unrestricted growth.

These findings raise important questions concerning the ecology and evolution of the very large numbers of plant species that consistently occupy subordinate positions within the hierarchies of herbaceous communities. Are these plants merely 'also rans' in the struggle for existence or are there fitness benefits associated with playing a minor part in plant communities? What is the precise role of subordinates in the long-term functioning and

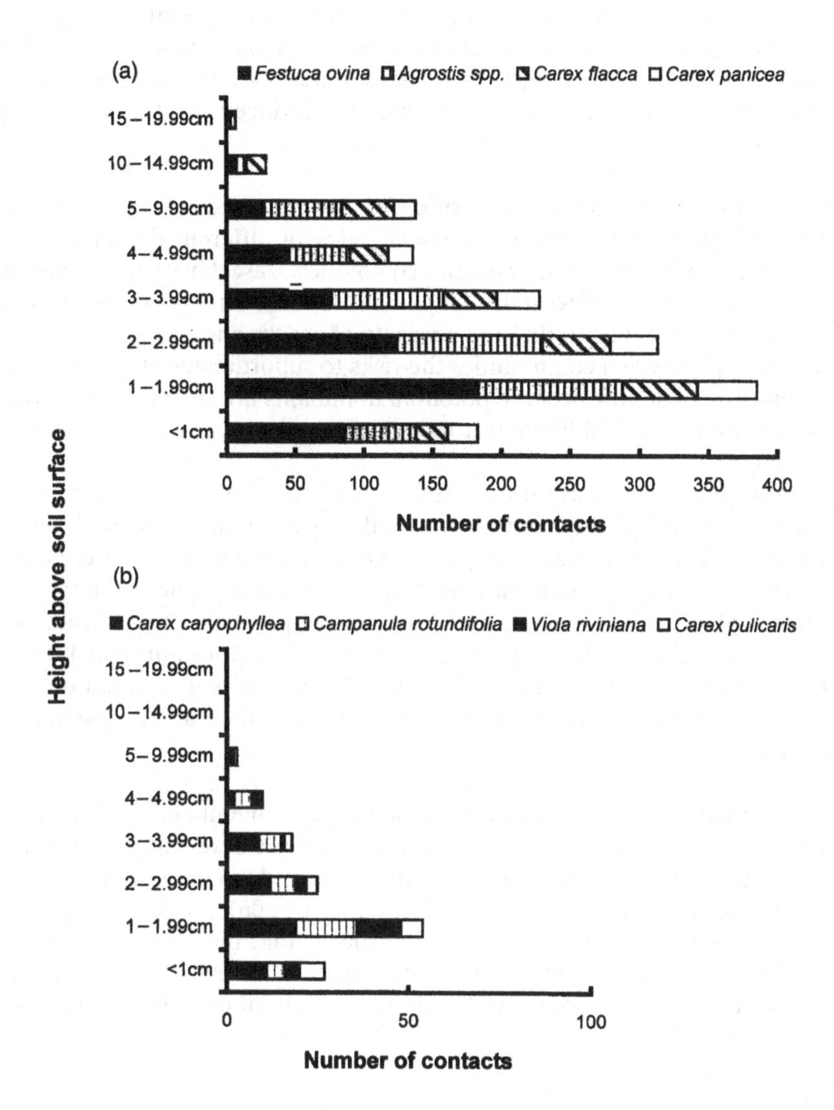

Figure 68. Vertical distribution of the leaf canopy of (a) four selected dominant and (b) four selected subordinate component species in an ancient limestone pasture at Buxton, North Derbyshire, England. Canopy distribution in June was estimated by measuring contacts with 375 randomly distributed, vertical pins. (Reproduced by permission of SH Hillier.)

dynamics of communities? These are not trivial questions in view of the fact that it is the subordinates rather than dominants that account numerically for the existence of high species-richness in plant communities (see Chapter 9, page 263). Detailed attempts have been made to address these issues (Grime 1987, 1998); here argument will be reduced to the following propositions:

1 Many subordinates achieve widespread occurrence because they are capable of exploiting similar niches associated with different dominant species. For example, *Poa trivialis*, the commonest vascular plant species of lowland Britain, occupies a diversity of shaded microhabitats beneath tall grasses, broad-leaved herbs, and a variety of shrubs and trees.
2 In many species-rich communities the risks to subordinates of competitive exclusion remain low because potential dominants are restricted in vigour by environmental and biotic factors (e.g. mineral nutrient stress, drought, grazing, fire).
3 Because they do not experience the costs in support materials and construction time incurred during the building of robust root and shoot architectures, subordinates enjoy a temporary advantage when communities are subject to episodes of biomass removal. Where damage is catastrophic as in tree felling and coppicing or ploughing and burning of grasslands (Skutch 1929; Marks 1974; Platt 1975; Bormann and Likens 1979; Pickett and White 1985; Pons 1989a), this can lead to a temporary but massive expansion of subordinates until such time as the dominants recover.

We may conclude, therefore, that many subordinate members of plant communities achieve high levels of fitness, particularly where natural or man-inspired interventions restrict the vigour of potential dominants either continuously or intermittently. Later in this chapter (page 206), it will be suggested that, in the long term, this rather opportunistic ecology displayed by subordinates may be sufficient to allow them an indirect but crucial involvement in the recruitment of dominants and the determination of community and ecosystem properties.

Transients

Many published records of the species composition of vegetation are incomplete. This can arise from cursory sampling and recording or from a deliberate policy of discounting minor constituents. Where the objective is to recognise recurring plant communities, data analysis often involves procedures excluding species of low frequency or inconsistent occurrence in vegetation samples. The result of these various decisions can be a divergence between published data and field reality and it has been suggested (Grime

1998) that important information relevant to the long-term dynamics of ecosystems may be lost. In order to examine this hypothesis it is necessary to refer to surveys in which a strenuous attempt has been made to include all the plant species present in the vegetation.

During the period 1965–75 a series of vegetation surveys was conducted to produce an inventory of the herbaceous plant communities of the Sheffield region in north central England (Grime *et al.* 1988). In each of approximately 10 000 metre square quadrats a complete list of species was made and the frequency of each species was recorded. When the resulting data were examined it became apparent that, in addition to dominants and subordinates, the majority of the samples contained species that were represented only as scattered seedlings and small immature individuals. Most of these species were present as dominants or subordinates in neighbouring vegetation associated with different environmental conditions, successional states, or management regimes. In Table 22, examples of this phenomenon are provided for three contrasted sets of vegetation samples; in each case it is evident that the plant communities examined were harbouring juveniles of species that occur as persistent, reproductive populations elsewhere in the landscape in vegetation of a different type. It seems reasonable to conclude therefore that in addition to the dominant and subordinate members of a community a third contributor can be identified in the form of transients originating either from the seed rain from the surrounding landscape or from seed banks occurring as a legacy of previous vegetation types occupying the site. Few ecologists have commented on the significance of transients but a notable exception is the scarcely-cited paper by Janzen (1986) in which transients are perceptively and memorably described as 'lost plants'.

Table 22. Ecological classification of the species recorded in three distinctive and highly contrasted habitats sampled widely in an area of 2400km^2 around Sheffield, England. All species encountered within a particular habitat were classified in terms of their primary habitat (columns 3–6). Details of the sampling, recording, and habitat classification procedures are provided in Grime *et al.* (1988)

Sampled habitat	Number of m^2 samples	Woodland species	Grassland species	Arable species	Others	Total
Woodland on limestone	51	65	23	1	5	94
Meadows	40	7	64	8	0	79
Cereal arable	55	5	38	69	2	114

The Role of Subordinates and Transients in Communities

In Chapter 4 and earlier in this chapter, dominants have been defined in relation to their capacity to monopolise resource capture, to occupy a high proportion of the above- and below-ground environment, and to exercise controlling effects on the abundances and niche-dimensions available to subordinate species. If immediate control of the functional properties of communities rests with dominant plants and if species-richness depends upon the numbers of subordinates and transients (Figure 66), it is pertinent to ask 'Does a decline in species richness matter?' To address this question it is necessary to consider the significance of losses of subordinates and transients in the long-term dynamics of plant communities and ecosystems.

Filter Effects of Subordinates?

One mechanism whereby losses in subordinates could affect the assembling of communities is through alteration of the filter controlling the recruitment, identity, and relative abundance of dominants. In order to review the opportunities for subordinates to control the admission of dominants into communities it is necessary to consider the long-term dynamics of vegetation and the regenerative phases in the life-cycles of dominants. Studies of vegetation succession conducted in the first half of the 20th century (e.g. Watt 1925, 1947) established that continued dominance by particular species is frequently determined by the success of seedling or vegetative re-establishment following disturbance events. Often the early course of events following a disturbance is a temporary expansion in the cover and vigour of subordinates. This is most obvious in forest clearings where a dense low cover of shrubs, herbs, and bryophytes characterises the environment of regenerating trees (Watt 1925; Skutch 1929; Marks 1974; Bormann and Likens 1979) but similar phenomena have been described for grasslands and heathlands (Oosting 1942; Keever 1950; Hillier 1990). Establishment following disturbances involves complex interactions of seedlings and vegetative shoots with substratum conditions, and contributions to the ground cover by subordinate plants may be expected to have both positive and negative effects (Cavers and Harper 1967; Ross and Harper 1972; Grubb 1977; Connell and Slatyer 1977; Nobel and Slatyer 1979; Pickett and White 1985; Bazzaz 1986; Crawley 1987; Burke and Grime 1996). Benefits to establishment have been described in circumstances where seedlings survive in the shelter afforded by low-growing shrubs, herbs, and bryophytes (Lawrence and Hulbert 1960; Ward 1990; Hillier 1990). Detrimental effects of shrub, herbaceous, and bryophyte cover on the establishment of grassland and forest dominants have been observed (Wardle 1959; Niering and Goodwin 1962; Webb *et al.* 1972; Pons 1989a) and it is widely-recognised (Fenner 1992) that many small-seeded herbs, trees, and shrubs are incapable of establishment in a closed cover of vegetation. We may deduce, therefore, that there is a potential for subordinate members of a plant community to act as a filter selecting between different potential dominants during the early phases of recolonisation following a disturbance event. Selection could operate on the basis of variation in the seed reserves of dominants and on the capacity of their seedlings to penetrate a low canopy (Grime and Jeffrey 1965; Westoby *et al.* 1992). The filter might also discriminate between dominants that rely upon rapid emergence and those which regenerate by persistent juveniles (Marshall 1927; Chippindale and Milton 1934; Marks 1974). Subordinates could also control regenerating dominants through more indirect mechanisms, such as provision of sites in which seed predation is reduced (Thompson 1987; van Tooren 1988) or through more complex phenomena such as the maintenance of critical pests, pathogens, herbivores, or mutualists (Gilbert 1977; Huston and Gilbert 1996).

The most definitive demonstration of the potential of subordinates to control the recruitment of dominants is that available from the excellent investigations by George and Bazzaz (1999a, b) on the impacts of the fern understorey on the growth and survival of canopy-tree seedlings in forests in north-eastern USA. From earlier comparative studies of the sensitivity of deciduous tree seedlings to herbaceous shade (Grime and Jeffrey 1965) it was known that hardwood species differed considerably in their reactions to the presence of understorey vegetation but George and Bazzaz have now shown by direct observation of seedling recruitment and by experimental manipulations of the ground vegetation that composition of the seedling bank in the forests is subject to a filter effect exercised by the understorey. This elegant investigation provides strong circumstantial evidence of a critical role for subordinates in community assembly. There is now a requirement for carefully-designed, long-term experiments that document the precise role of subordinates throughout the process of ecosystem reassembly in a range of vegetation types and following various disturbance events.

Founder Effects of Transients?

If, as suggested earlier, the sources of the transients are seed-banks in the soil and the seed-rain from the surrounding landscape it would appear that they are an index of the pool of potential colonising species at each site. On this basis a diversity of transients occupying the tail of a dominance-diversity curve signifies a high probability that, in the event of habitat disturbance or changes in management, there will be a rapid ingress of different plant functional types, some of which are likely to be capable of exploiting the new conditions. An obvious example is the benefit to woodland development where an abandoned grassland already contains a diverse assortment of tree seedlings.

Current losses in biodiversity in Europe and in many other parts of the world are taking place in a complex landscape mosaic continuously disturbed by natural events and by urbanisation, arable cultivation, forestry, and various forms of grassland management. The continued survival of many communities therefore depends in part upon relocation of populations and reassembly of vegetation types. The extent to which communities are rapidly reconstituted is likely to be related to the reservoir of colonisers, many of which should be detectable prior to disturbance as transient constituents of the existing vegetation. As Egler (1954) recognised, we may suspect that the speed and completeness with which community reassembly occurs will depend upon early colonisation by appropriate dominants and subordinates; late arrivals will be delayed in their establishment and some may be excluded completely (Keever 1950; Niering and Goodwin 1962; Holt 1972; Platt 1975). It is not difficult to envisage how circumstances could then arise whereby efficiently dispersed plant species with 'poor fit' to habitat and management conditions could assume dominance with damaging consequences for community and ecosystem function. There is an urgent need to discover to what extent failure in the processes of plant dispersal and ecosystem reassembly can be predicted from the decline through time in the density and species-richness of transients in plant communities.

MANIPULATIONS OF NATURAL COMMUNITIES

Many early hypotheses about the factors controlling the structure and composition of plant communities originated from observations following natural events (windstorms, droughts, floods, late frosts) or unwitting ecological experiments conducted by man (e.g. forest clearance, eutrophication, exclusion of grazers). These useful insights have prompted some ecologists to pursue this 'interventionist' line of enquiry by conducting deliberate, controlled field manipulations in an attempt to understand how communities are assembled and maintained. Some of these experiments such as the Park Grass fertiliser treatments at Rothamsted (Brenchley and Warington 1958) involve gross manipulations over long periods of time, whereas others rely upon more subtle operations such as sowing of seeds into existing communities, insertion of seedlings or vegetative transplants, or removal of particular species or functional components (Silander and Antonovics 1982; Goldberg and Werner 1983; Wilson and Keddy 1986; Goldberg 1987; Reader and Best 1989; Goldberg and Barton 1992; Zobel *et al.* 1994; Hulme 1994, 1996a, b; Herben *et al.* 1997).

Reviews of the results obtained from numerous experiments removing or adding species to communities (e.g. Fowler 1981, 1986; Goldberg and Barton 1992; Shipley and Keddy 1994; Goldberg 1997) have not provided a clarification of the processes involved in assembling and structuring plant communities. Why have manipulative experiments not provided a more powerful tool? It seems that the weaknesses are both philosophical and technical and are mainly related to the fact that many of the experiments fall short in terms of both realism and scope for mechanistic analysis; often the manipulations compromise the reality of the natural system without allowing the insights available (see page 210) from the use of other approaches such as the construction of synthetic communities in controlled environments. The limitations of manipulative experiments are particularly evident in those studies where attempts have been made to understand the role of competition in determining the structure of plant communities by selective removal of particular species; there appear to be at least three reasons why such investigations have not been more revealing:

1 Removal experiments are usually restricted in application to species mixtures. This excludes from consideration many common situations (e.g. Al-Mufti *et al.* 1977; Figure 6, page 15) where extensive areas are occupied by monospecific stands. It seems likely that this limitation may have resulted in undersampling of vegetation types in which competitive interactions have already played a major formative role and, by resisting invasions, continue to do so.
2 Within plant communities, component species may differ considerably in seasonal patterns of resource capture and growth. Such phenological differences strongly condition the abilities of the species to take advantage of the removal of neighbours and make the timing of such manipulations critical. The possibility must be considered that the different responses of various neighbours to species removal (e.g. Silander and Antonovics 1982) may be determined by phenological sequence rather than release from differential competitive suppression.
3 Interpretation of removal experiments usually involves comparison of a neighbouring species' subsequent growth against that achieved over the same period in intact vegetation. If a reliable inference is to be achieved with regard to the role of competition in the community this must rest on the assumption that the effect of the experimental intervention has been to remove the competitive effects of a particular vegetation constituent whilst preserving other aspects of resource supply to the remaining species at the levels prevailing in the intact system. Whilst the sun and clouds can be relied upon to maintain comparable inputs of solar energy and moisture to the manipulated system no such assumption can be made with respect to nutrient release from a soil system in which roots have been partially removed by hand-pulling or killed by herbicide treatment. This is a particularly important consideration on infertile soils where considerable quantities of mineral nutrients and carbohydrate may be sequestered in underground storage organs, tap-roots, extensive fine roots, and mycorrhizal networks, some of which provide connections between neighbouring species (Chiariello, Hickman, and Mooney 1982; Whittingham and Read 1982; Grime *et al.* 1987b). It seems likely that species removal will often cause a major disruption to infertile soils, not least through the effect of released carbohydrates on microbial populations. The risks of artefactual enrichment or depletions are such that it may be unwise to regard a positive response to neighbour removal as unequivocal evidence of competitive suppression.

These difficulties and others reviewed by Bender, Case, and Gilpin (1984) and Keddy (1989) suggest the need for caution in using removal experiments to analyse the role of competition in the assembly and maintenance of plant communities. A specific example of the difficulties in interpreting removal experiments arises in the case of the forest root-trenching experiments already discussed on page 69. There is little reason to doubt that the resulting stimulus to the ground flora recorded in most of these investigations (Coomes and Grubb 2000) confirms that soil resources were limiting productivity. However, it may be unwise to assume that the response is a measure of current competition for nutrients between trees and herbs in the intact system. The mineral nutrients released by trenching may represent an accumulation over many years and give little indication of the current and past roles of competition in the mechanism whereby trees have assumed dominance and currently maintain it.

SYNTHESIS OF COMMUNITIES: I MODEL ASSEMBLAGES

Particularly in the case of ancient, species-rich vegetation the assembly of a plant community may depend upon a complex chain of events involving competitive, complementary, and facilitating interactions between plants and numerous interactions with herbivores, pathogens, soil fauna, and micro-organisms.

Some of these processes may be revealed by detailed field study but the majority remain inaccessible to direct observation because they are either lost in the earlier history of the community or else embroiled in complex species interactions and resource transfers, most of which occur below the ground surface. Faced with this complexity, some ecologists (e.g. Harper 1982; Woolhouse 1982) have taken the view that in order to understand how plant communities assemble and are maintained our only course is to undertake a laborious reductionist analysis of the whole ecosystem. Indeed, the first steps in this direction are recognisable in attempts to construct complete food webs (Bengtesson *et al.* 1995) and to trace the many pathways of carbon and mineral nutrients within ecosystems using isotopes.

An alternative approach is to assemble simplified model plant communities and ecosystems experimentally under natural field conditions or to synthesise them under the more strictly controlled conditions of laboratory microcosms. Several authors including May and Seger (1986) Heal and Grime (1991), Byers and Odum (1993), and Lawton (1995) have concluded that community and ecosystem synthesis often provides insights that are not currently available from direct observations or manipulations of natural communities and ecosystems. The opportunities for recognition of the processes affecting community assembly that arise from this approach are summarised below.

Recruitment of Plant Communities from the Species Pool

The plant species which recur together in familiar assemblages in particular ecological situations in various parts of the world are small subsets of the pool of species available at a regional or local scale. Hypotheses can be developed concerning the nature of the sieves that operate to prevent the majority of candidates in the species pool from gaining admittance to a community (Keddy 1992b). Often the mechanism of exclusion is suggested by consistent differences in the traits exhibited by successful and unsuccessful species. Such theories can be tested by experiments (Grime *et al.* 1987b; Campbell and Grime 1992; Stockey and Hunt 1994; Weiher and Keddy 1995) in which plant communities are allowed to assemble on bare soil from a standardised inoculum of seeds, seedlings, or vegetative material containing species predicted to be admitted or excluded from the community. An example of this approach is illustrated in Figure 69 which describes the predictable sieve effects of soil fertility and the presence of generalist invertebrate herbivores on recruitment from a seed pool of 48 species containing a wide array of plant functional types. The results show, as we might expect, that the introduction of high soil fertility leads to the rapid exclusion of the majority of small or slow-growing species. Filter effects on species recruitment are also apparent in Figure 69 as a consequence of the presence of the herbivores; at low soil fertility there is evidence of the selective predation of fast-growing species such as *Poa annua* and *Lolium perenne* and at high soil fertility the palatable grass *P. trivialis* is eliminated.

The possibility of an additional filter controlling admission and exclusion from plant communities has been recognised through the discovery that within local species pools there are differences in the extent to which species are colonised by mycorrhizal fungi (Harley and Harley 1987). It is well established that, in the case of vesicular-arbuscular mycorrhizas, individuals of different neighbouring plant species may be connected through a common mycelial network from which benefits in mineral nutrition are derived (Janos 1980; Francis and Read 1984; Grime *et al.* 1987). There is also evidence (Francis *et al.* 1995) that some species suffer negative effects of mycorrhizal infection. In theory, at least, therefore a filter mechanism can be proposed in which depending upon soil nutrient status plant species tend to aggregate into VA-compatible or VA-incompatible communities.

Control of Relative Abundance within Communities

The dominance hierarchies within plant communities arise from the expression of genetic traits in the component species and the modifying effects of the local environment, its plants and animals, upon this expression. In conditions of high soil fertility and low levels of damage to the vegetation we might expect that the relative abundance of the species in a community will be most

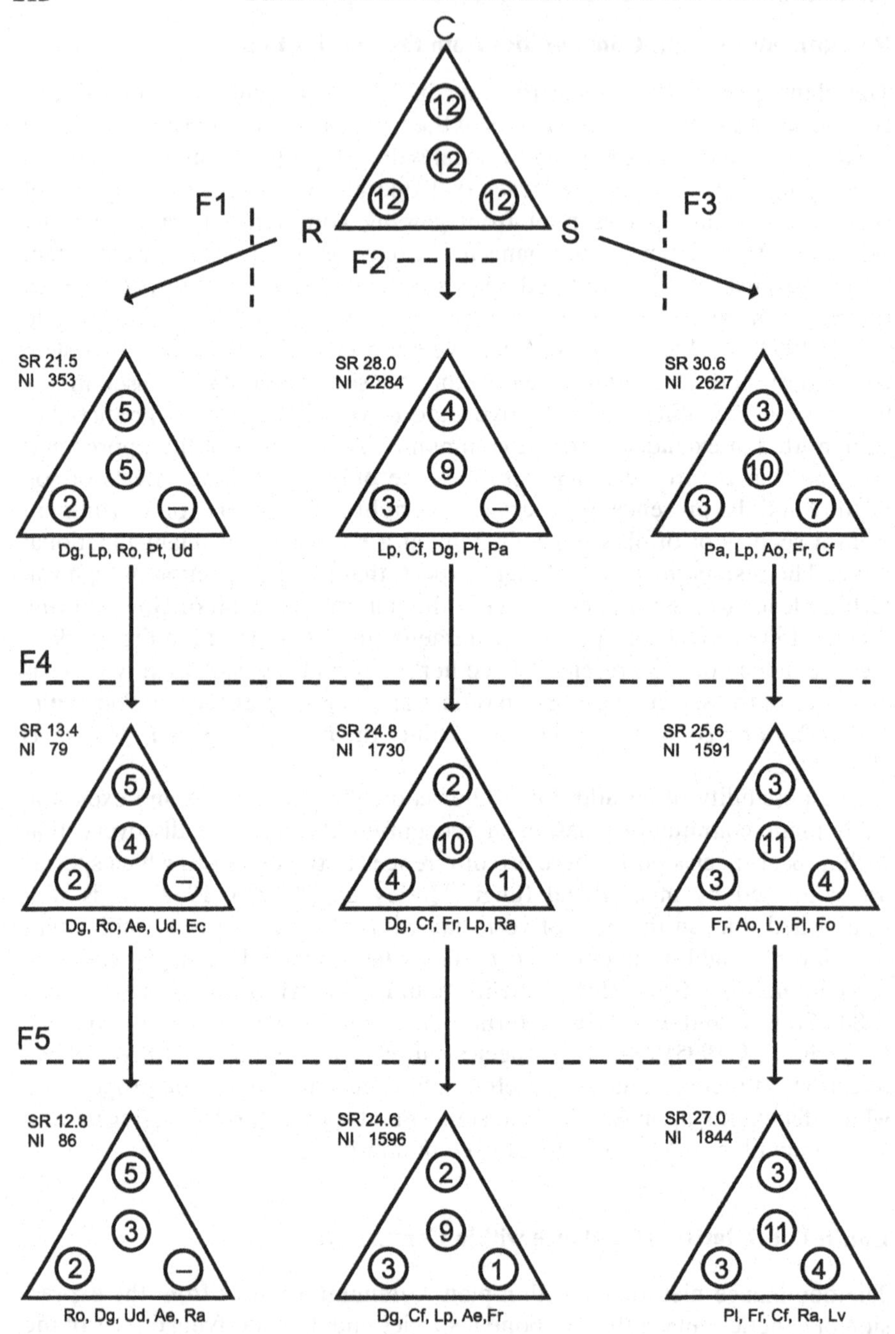

Figure 69. Summary of the effects of five filters (F1–F5) on the assembly of plant communities from an initial pool of 48 herbaceous species comprising a balanced spectrum of plant functional types (12C, 12S, 12R, and 12 CSR) sown into 72 1.0 m ×

easily predicted because the size, morphology, and foraging characteristics of shoots and roots (see page 23) will be fully expressed. Confirmation of our ability to predict hierarchies in multispecies mixtures under these simplified conditions is available from a number of laboratory experiments (Mahmoud and Grime 1976; Campbell *et al.* 1991a; Figure 14; Fraser and Grime 1998b). However, these represent a circumstance where both the prediction and its test are relatively easy to apply. The greater challenge it to use community synthesis to investigate the impacts upon plant community composition and relative abundance of more subtle ecosystem variables and processes such as extreme climatic events, soil heterogeneity, selective grazing, the protective effects of carnivores and parasitoids, and the influence of pathogens, endophytes, mycorrhizas, and free-living soil invertebrates and micro-organisms. Several studies have been reported using this approach demonstrating effects of snails (Fraser and Grime 1999), earthworms (Thompson *et al.* 1993), beetles (Buckland and Grime 2000), and vesicular-arbuscular mycorrhizas (Grime *et al.* 1987b) and endophytes (Clay and Holah 1999) on plant community structure and composition. More investigations of this type are likely to follow as progress is made in developing microcosm techniques that allow precise control of the physical and chemical conditions and the species composition and trophic complexity of synthesised communities and ecosystems.

SYNTHESIS OF COMMUNITIES: II THE MATRIX APPROACH

As explained under the last heading some of the most fundamental questions concerning the structure of plant communities can be addressed by assembling communities and ecosystems in controlled conditions. Often the most direct and informative way by which to explore the role of particular factors is simply to examine the consequences upon assembling communities of gross manipulations of species input or environmental conditions. Many of these

---

0.5 m × 1.5 m insect-proof outdoor microcosms and allowed to develop for two years. F1 = high fertility, F2 = moderate fertility, F3 = low fertility, F4 = generalist herbivores (slugs and grass aphids) present, F5 = predators (carnivorous beetles) of the herbivores present. The encircled values refer to the number of species of each functional type achieving more than 1% of the total above-ground biomass at the end of the experiment. SR = mean species richness and NI = mean number of individual plants surviving in each treatment after two years. Below each triangle the dominant plant species are listed in order of descending biomass. The species are: Ae, *Arrhenatherum elatius*; Ao, *Anthoxanthum odoratum*; Cf, *Cerastium fontanum*; Dg *Dactylis glomerata*; Ec, *Epilobium ciliatum*; Fo, *Festuca ovina*; Fr, *Festuca rubra*; Lp, *Lolium perenne*; Lv, *Leucanthemum vulgare*; Pa, *Poa annua*; Pl, *Plantago lanceolata*; Pt, *Poa trivialis*; Ra, *Rumex acetosa*; Ro, *Rumex obtusifolius*; Ud, *Urtica dioica*. (Reproduced from Buckland and Grime 2000 by permission of *Oikos*.)

experiments can be conducted on a small scale and yield clear results within a short period of time. However, very different circumstances prevail in many areas of applied research where the objective is to test our understanding of how current changes in the environment are likely to affect plant community composition. Particularly severe logistical problems arise where the need is to test predictions of the impact on ecosystems of gradual changes (slow forcing) brought about by increased atmospheric deposition of reactive forms of nitrogen or elevating CO_2 or progressive shifts in temperature, moisture supply, or UVB irradiance. In perennial vegetation, responses to experimental manipulation of these factors may be considerably delayed and reliable inferences may be possible only after long exposures. It would appear that we have little alternative but to conduct large-scale, long-term experiments. These entail high costs in consumables such as CO_2 and electricity and in technical support. The scale of this research falls outside that previously attempted by agricultural researchers and ecologists, and is likely to depend upon government support or perhaps direct or indirect financial inputs from those industries currently driving global change.

If such long-term manipulations of ecosystems are to be attempted it is vital that these experiments are cost-effective. Even in a small area such as the British Isles there are numerous ecosystems on which we might wish to conduct environmental change experiments. Since it is clearly impossible to finance work at more than a handful of sites it is imperative (a) that manipulative experiments are dispersed in a strategically efficient network with respect to climate, soils, and human usage, and (b) that the experimental designs which are employed maximise the scope for extrapolation of the results to other ecosystems.

Only a moment's reflection is required to recognise the difficulties involved in (b) above. We need multifactorial designs which permit study of the interacting effects of numerous environmental factors. Often it may be necessary to compare the effects of different levels and seasonal distributions of climatic variables. In addition it is likely to be essential to understand how soil factors and management will modify the impact of changing conditions. It is inevitable that the eventual solution to a research requirement of such scale and complexity will involve models of vegetation and ecosystem dynamics; the rudiments of such models are already available as cellular automata (Wolfram 1984; Green 1989; Colasanti and Hunt 1997). However, **models are no substitute for experimentation**; to be useful models will need repeated validation against field reality.

Clearly, therefore, the crucial problem in model development and testing is to maximise information gain from experiments which manipulate environmental factors in the field over long periods of time. One approach to this problem is to introduce gradients in soil or management factors within replicated plots which are themselves subject to manipulations (Campbell and Grime 1992; Burke and Grime 1996). Some results from an experiment using this approach to study the interaction of climate and management procedures in meadow plots initially sown with 40 species of vascular plants are presented in Figure 70. In each of the experimental plots there were orthogonally-opposed five-step gradients in fertiliser application and date of cutting (mowing at 50 mm). At the whole-plot level, manipulations of temperature were conducted using a novel technique (Hillier, Sutton, and Grime 1994) to lengthen or shorten the growing season. The results (Figures 70b and c) reveal some consistent differences between *Conopodium majus* and *Plantago lanceolata* in their sensitivity to the interacting effects of climate and management. A particularly distinctive pattern is apparent in *C. majus*; this species is

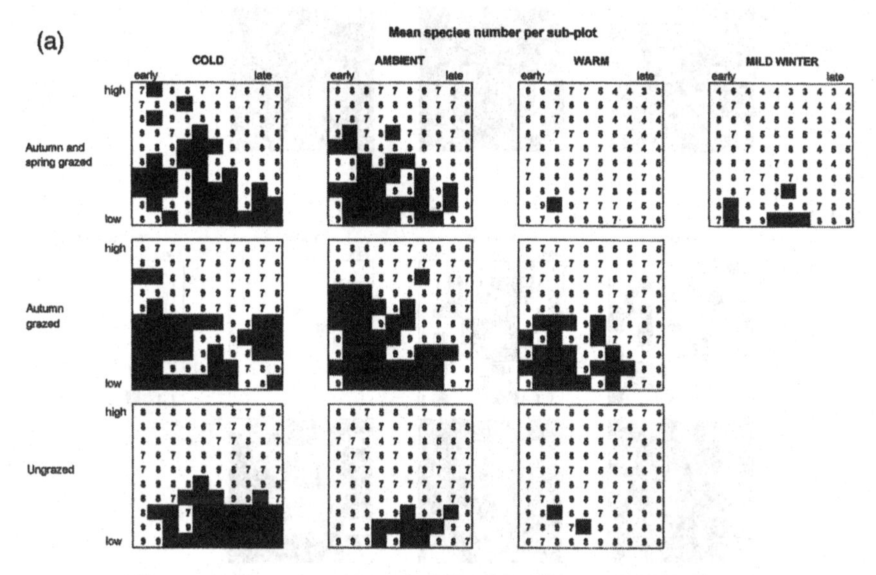

Figure 70. Effects of temperature and simulated grazing treatments (applied to whole plots) with fertiliser and cutting date treatments (applied as five-step gradients within plots) on (a) species richness and (b) (*overleaf*) the distribution of *Plantago lanceolata* and *Conopodium majus*. Key: Early—Late = 5-point cutting-date gradient (late May, mid-June, mid-July, early September, late September. Low—High = 5-point fertility gradient (no addition of 20:10:10, N:P:K fertiliser; 40 Kg per hectare; 80 Kg per hectare; 120 Kg per hectare; 160 Kg per hectare. COLD = shortened growing season (cooling 3°C below ambient temperatures, March–May and Sept–Nov.). AMBIENT = ambient temperatures. WARM = extended growing season (warming 3°C above ambient temperatures, March–May and Sept–Nov.). MILD WINTER = warming 3°C above ambient temperatures, Sept–May.
In (a) species richness after 6 years is presented for each location on the matrix with values of 10 or more indicated by shading. The patterns for individual species in (b) and (c) are indicated as follows:
■—species present in this location on the matrix in at least 2/3 replicates.
1—species present in this location on the matrix in 1/3 replicates.
□—species absent in this location on the matrix in all 3 replicates.
(Reproduced by permission of Hillier and Grime.)

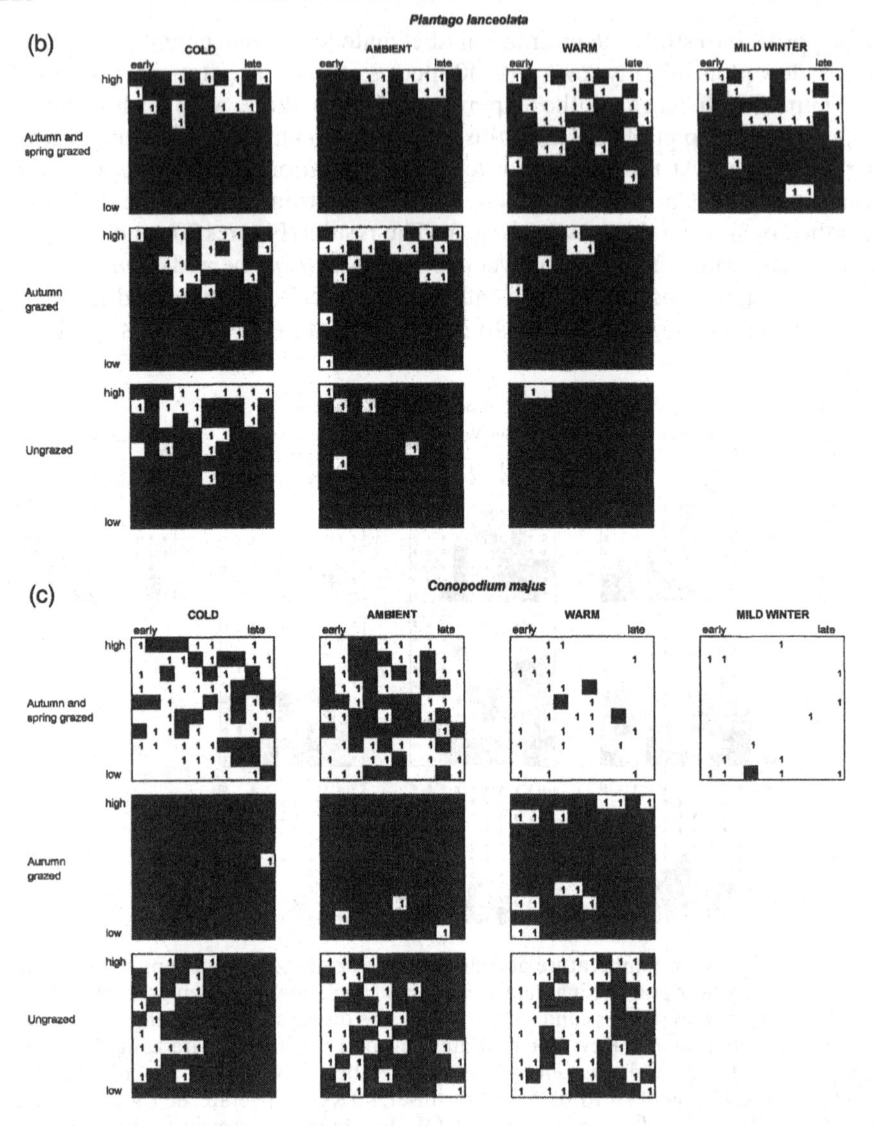

rapidly eliminated by spring grazing and by early cutting and this effect is increased by elevating spring and autumn temperatures or introduction of consistently mild winters.

In Figure 70a patterns of species-richness across the plots are depicted for the same set of treatments six years after commencement of the experiment. These data indicate that under the conditions of this experiment species richness is promoted by low fertiliser application, early mowing, simulated intensive grazing, and shortening of the growing season. A more detailed analysis of the results of the experiment strongly suggest that the effects on species richness are strongly conditioned by effects of the climate and management treatments on the vigour of potentially dominant grasses such as *Dactylis glomerata* and *Festuca rubra*.

For the purpose of illustrating the benefits of the matrix approach the exact details of these results need not concern us. The key finding is related to the fact that when gradients are introduced to the plots it is possible to construct a matrix in which the consequences for the plant community and for individual species of numerous interactions between climatic and non-climatic factors can be examined within a small area and with modest cost and effort in experimental maintenance and recording.

6 Rarification and Extinction

INTRODUCTION

It is widely recognised that large numbers of plant species are experiencing global, regional, or local decline in abundance and many face extinction (Ehrlich and Ehrlich 1981). The explanation for decline is often related to specific circumstances such as the application of selective herbicides in arable fields, the removal of attractive flowering plants from their natural refuges, or the destruction of habitats by extractive industries. In such cases, assessment of conservation priorities and development of protective measures can often proceed quite effectively on an *ad hoc* basis. However, for the purposes of long-term conservation and management of rare or declining plant species it is necessary to examine their vulnerability in the wider context of the large-scale changes now affecting vegetation on a world scale. The objective in this short chapter is not to comment in detail on the possible consequences of global change on the world's flora. Instead, attention will focus on the opportunities that now exist to recognise and interpret some major causes of plant rarification and extinction in terms of predictable, measurable, functional shifts within both regional or national floras and individual plant communities.

FUNCTIONAL SHIFTS AND RARIFICATION WITHIN FLORAS

In a ground-breaking investigation involving a complete inventory of the ecologies of both common and rare species in the vascular plant flora of an area of extremely varied landscape in north central England, Hodgson (1986a, b, c) drew the following broad conclusions:

1 The abundances of most plant species are broadly predictable from the extent of their habitats. Many common plants are ruderals or competitors exploiting extensive habitats in agricultural, industrial, or urban areas, whereas a high proportion of rare species are stress-tolerators occupying small parcels of land in more ancient parts of the landscape.
2 Although climate and soils have played a historical role in controlling the identity and abundance of plant species, current changes in land-use have now become the dominant controllers of rarity and abundance.

3 Some constraints on the ability of plants to tolerate modern and intensive forms of land-use can be traced to ancestral specialisations. Hodgson (1986c, 1989) found that plant families of greater antiquity such as the Papilionaceae and Rosaceae were becoming less abundant than families of more recent origin such as the Chenopodiaceae and Brassicaceae and he interpreted this as a consequence of (a) the greater representation of ruderal and competitive traits in families that emerged later in the angiosperm phylogeny, and (b) the compatibility of these traits with the disruptive land-use and eutrophication characteristic of intensively-exploited landscapes.

It should be emphasised that the region in which this study was conducted contains areas of high human population density and has experienced a long history of agricultural and industrial development. It would be unwise to expect the detected functional shift away from stress-tolerance and towards ruderal and competitive dominance to be observed in all regional floras. However, the detailed nature of Hodgson's investigation has made it an excellent model for subsequent comparative studies of functional shifts in local or national floras (Hodgson 1991; Thompson 1994; Wisheu and Keddy 1994; Bunce *et al.* 1999; Thompson and Jones 1999).

A particularly interesting comparison is contained in the study of Thompson (1994), who examined records of the changing abundances of vascular plants in national data-bases resulting from monitoring studies conducted in six countries in Western-Europe. In this study Thompson compared the functional characteristics of species that are currently in decline with those which are expanding in abundance. Only in two heavily-populated countries, The Netherlands and England, was a consistent functional difference detected between declining and expanding species. This difference, expressed in Figure 71 as a simple numerical index, classifies the two groups of species with respect to their proximity to the S corner of the CSR classification. From these data we may tentatively conclude that although all floras contain species that are currently changing in abundance it is only in conditions of intensive land-use, corresponding roughly to human population densities above 100 persons per km^{-2}, that a predictable functional shift is usually detectable.

Human population density promises to be an extremely useful, readily-available, surrogate for the collective impact of a wide range of phenomena driving the functional shift in the floras of intensively-exploited landscapes. Confirmation of its predictive value is illustrated in Figure 72 which shows the general relationship between species extinctions and human population density detected by Thompson and Jones (1999) across local areas of the British Isles. It is clearly useful not only to be able to predict in such broad terms the threshold at which functional shifts occur and many species become vulnerable but also to monitor the functional shift as an index of the effectiveness of remedial actions.

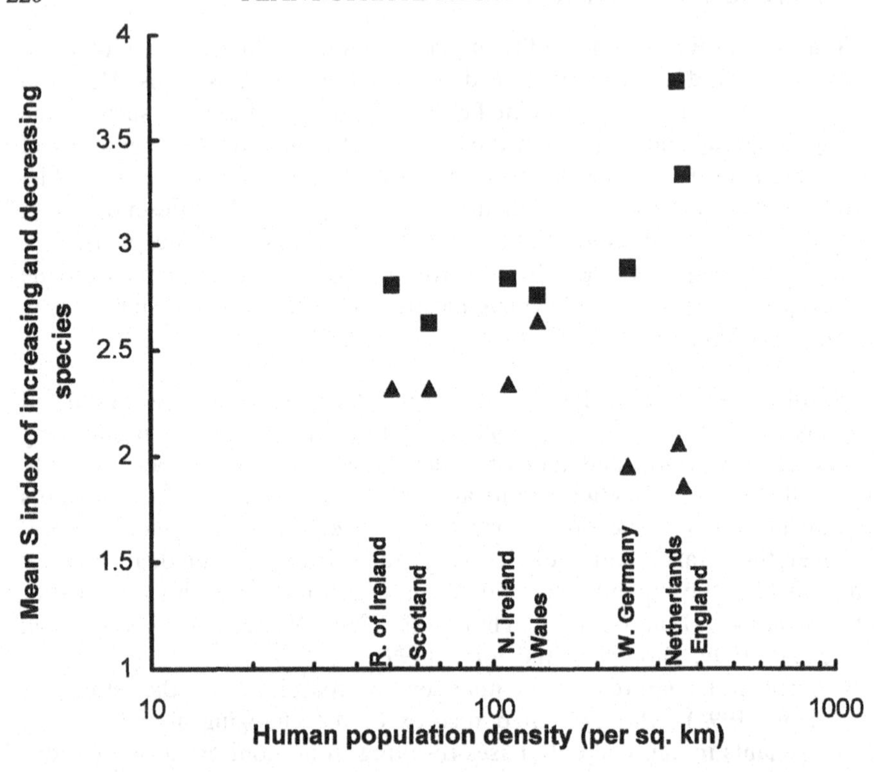

Figure 71. Relationship between mean S index of increasing (▲) and decreasing (■) species and human population density in seven European countries where the S index reflects the proximity of the strategy of the species to the S corner of the CSR model calculated following the procedures of Hodgson *et al.* (1998). S class of the two groups was not significantly different in Scotland, N. Ireland, or Wales, but the two groups do differ significantly in the R. of Ireland ($p = 0.049$), England, W. Germany, and The Netherlands (all <0.001). (Reproduced from Thompson 1994 by permission of Springer-Verlag.)

FUNCTIONAL SHIFTS AND RARIFICATION WITHIN PLANT COMMUNITIES

Monitoring of change in the species composition of plant communities has become an established procedure for the detection of long-term effects of climate change, air and groundwater pollution, and impacts of vegetation management. Botanical recording is also used to test the effectiveness of measures designed to protect nature reserves or to restore biodiversity in agricultural landscapes. Although most ecologists recognise the value of monitoring there is a diversity of opinion with regard to techniques of interpretation of the resulting data. Some workers focus upon 'indicator species' suspected to be unusually sensitive to particular environmental factors. A

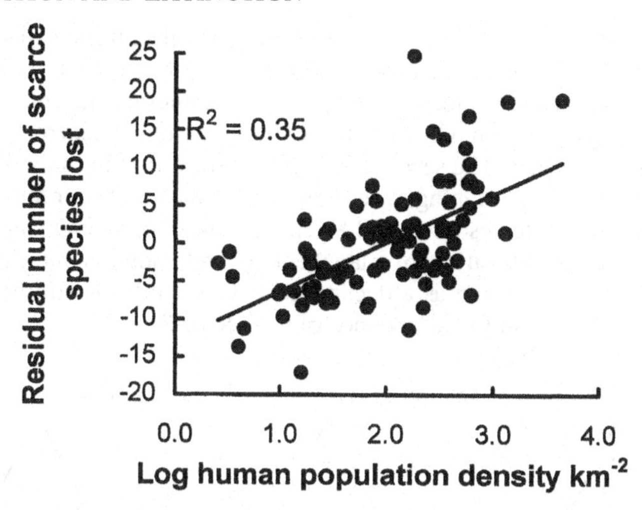

Figure 72. The relationship between residual scarce plant loss, after the effects of the original number of scarce plants are corrected for, and human population density in Britain. Each point represents a single vice-county. N. Ireland, the Isle of Man, and those vice-counties for which scarce plant data were missing or inadequate were omitted from the analysis. (Reproduced from Thompson and Jones 1999 by permission of *Conservation Biology*.)

specific example of this approach is the system developed by Ellenberg (1974) in which a majority of the common vascular plants species of Central Europe were allocated numerical scores reflecting different aspects of their edaphic and climatic tolerance such as nitrogen supply, soil pH, drought, and shade. Ellenberg Numbers rely upon field observations relating the distribution and abundance of each species in natural vegetation to variable characteristics (measured or subjectively assessed) of the environment. Despite their essentially correlative basis, Ellenberg Numbers can provide many useful indications of environmental change and recently efforts have been made to test their value (Thompson *et al.* 1993b) and to adapt them for use at the western extremities of Europe (Hill *et al.* 1999).

An approach complementary to that of Ellenberg is to interpret floristic changes by reference to shifts in the representation of individual traits or sets of traits in the plant communities. Here a specific example is the system known as FIBS (*F*unctional *I*nterpretation of *B*otanical *S*urveys) which has been used to analyse changes over time in both ecological and archaeological records (Hodgson 1991; Charles *et al.* 1997; Wilson 1998). FIBS uses a variety of plant attributes for interpretation and prediction including several from the regenerative phase such as seed size and persistence. However, the most basic and informative component of the FIBS analysis consists of a test for functional shifts in the plant community in the relative abundance of primary and secondary strategies of the established phase. As illustrated in Figure 73, the objective is to discriminate between types of functional shift indicative of three different agencies forcing change in plant communities. The first is eutrophication arising from groundwater or aerial inputs of mineral nutrients and is detectable from an increase in C, C-R, and R strategies and concurrent decline in the remaining strategies. The second, increased disturbance by activities such as trampling, grazing, or ploughing, is recognisable from the greater abundance of ruderals or allied strategies (C-R and S-R). The third agent of change, dereliction, arises from a relaxation of management and biomass removal and is expected to be recognisable as a decline in R, C-R, and S-R strategies and a simultaneous increase in the abundance of C, C-S, or S.

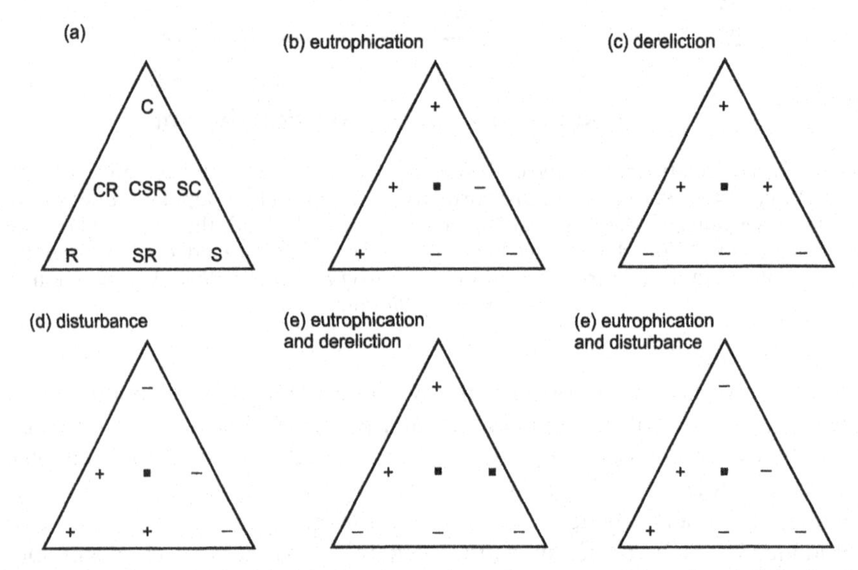

Figure 73. Theoretical basis for the diagnosis of the causes of functional shifts in the species composition of vegetation used in the FIBS system. In (b) to (e) diagnosis depends upon a characteristic and simultaneous expansion (+), decline (–) or stasis (■) in the primary strategies identified. (Reproduced from Hodgson 1991 by permission of *Pirineos*.)

Successful applications of the approach include the detection of climatically-driven functional shifts in the long-term monitoring study of grassland vegetation at Bibury (Grime *et al.* 1994; Dunnet *et al.* 1998) and recognition of contrasted trajectories and causes of vegetation change between the upland and lowland grasslands of the Sheffield region (Grime 1999). A thorough test of the usefulness of this approach is taking place through the application of FIBS (Wilson 1998) to interpretation of the Countryside Survey (Bunce *et al.* 1999) in which, during the 21st century, changes in the botanical composition of the vegetation will be monitored at intervals of 10 years in permanent plots distributed in all major habitats and landscape types of the British Isles. Figure 74 illustrates the very large differences in functional composition that distinguish familiar types of vegetation in Britain; there is an opportunity here not only to record future losses of species from communities but also, using the FIBS approach, seek to diagnose the causes of floristic change in different regions and parts of the British landscape.

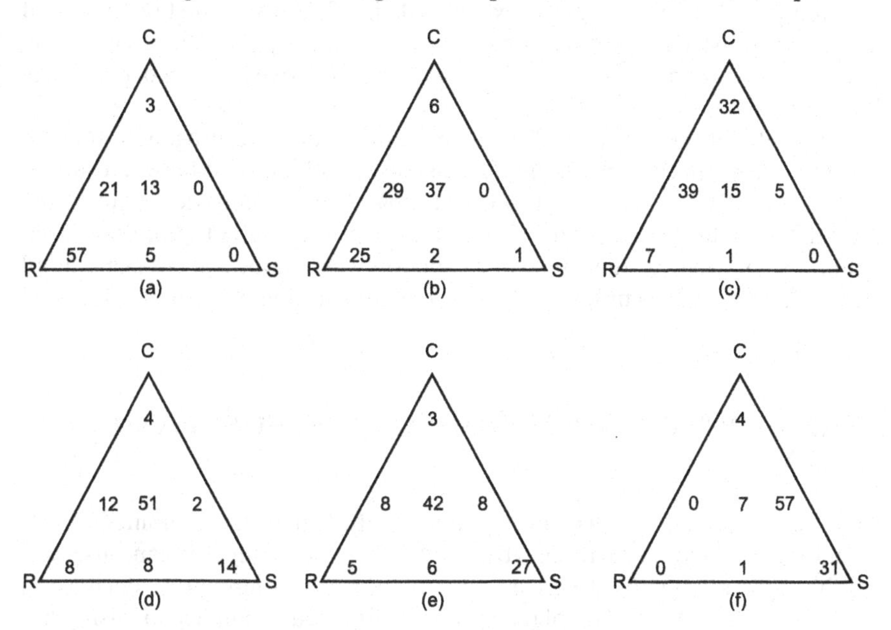

Figure 74. The relative abundance of plant functional types in various vegetation types sampled widely in the British flora in the Countryside Survey. (a) Cereal crops with mixed weeds; (b) rye-grass and clover grassland; (c) wet streamsides and tall herbs in wetland; (d) neutral grassland; (e) species-rich acidic grassland; (f) dry podzolic grassland. (Redrawn from Bunce *et al.* 1999.)

ENDANGERED SPECIES: AN EARLY WARNING SYSTEM

The conventional method of monitoring local populations of rare plant species is to visit each of their sites at regular intervals and, where possible, to record population size and frequency of flowering. One of the limitations of this approach is that individuals of many of the most endangered species are extremely long-lived and may survive long after conditions in their habitats have become unsuitable for regeneration and persistence of their populations.

An alternative method, using the approach summarised in Figure 73, has been designed by Wilson (1998) to alert nature reserve managers and statutory authorities to circumstances where although populations of rare species remain in existence, conditions have changed and they have become liable to extinction. The method relies upon monitoring of the whole communities occupied by the rare plants. The objective is to detect functional shifts in the commoner species that indicate processes of change (eutrophication, disturbance, or dereliction) that threaten the survival of the associated rarities. In order to test this approach records of the composition of plant communities containing particular rare species were made in 1997 at sites first documented in the 1970s. At each site a formal procedure was applied to measure the extent of any functional shifts that had occurred over the two preceding decades.

The results of the Wilson (1998) investigation were encouraging in that for a range of rare species the method consistently distinguished between surviving and recently extinct populations in terms of the magnitude of functional shifts detected in the associated vegetation. It is interesting to note that even within a single species the cause of extinction was not always the same at all sites, indicating the complexity of the processes causing extinctions at a local scale.

ANALOGOUS PHENOMENA INVOLVING FUNGI AND ANIMALS

The broad conclusion drawn in this chapter is that under the combined effects of regional and global eutrophication and disruptive changes in land-use predictable functional shifts (increases in ruderals and competitors, declines in stress-tolerators) are taking place in floras subjected to impacts of rising human population densities. It would be surprising if these massive changes were not paralleled by similar shifts in the functional spectrum of microorganisms and animals, some arising from direct effects of man on the environment and others as a consequence of the floristic change. In fact, Myers (1996) has forecast an apocalyptic future scenario in which the faunas of heavily-exploited landscapes will be dominated by animals that 'rapidly exploit newly vacant niches by making widespread use of food resources; they are generally short-lived, with brief gaps between generations; they feature high rates of population increase; and they are adaptable to a wide range of environments'. Myers cites the expansions in species such as starlings (*Sturnus vulgaris*), herring gull (*Larus argentatus*), and rabbit (*Oryctolagus caniculus*) as examples of this phenomenon and points out that there have been coincident declines in the populations of many longer-lived species with more specialised habitat requirements.

7 Colonisation and Invasion

INTRODUCTION

In Chapter 3 a wide range of phenomena have been described whereby plant populations are sustained in the same place or move about the landscape through processes of regeneration. However, a general principle that emerges from Chapters 1, 3, 4, and 5 is that, with the notable exception of the continuously-disturbed habitats exploited by ruderal (ephemeral) plants, the establishment of offspring is a hazardous process which frequently ends in complete failure. In view of this evidence of the great difficulties that usually attend regeneration it is a matter of considerable interest when a particular plant species displays a propensity for expansion or redistribution within its existing geographical and ecological range (colonisation) or, more spectacularly, appears to 'break out' beyond its former boundaries (invasion).

Many authorities have concluded that the world is currently experiencing an increasing frequency of plant (and animal) invasions. Some invading plants have achieved the status of notorious aliens as a consequence of their large size and the speed with which they have recently extended their geographical ranges and usurped native species from their natural habitats. Well-known examples in Europe include the shrub, *Rhododendron ponticum*, the clonal herb, *Fallopia japonica*, and the remarkable umbellifer, *Heracleum mantegazzianum*. Many theories have been put forward to explain the current high incidence of invasions and there has been considerable speculation about its causation. Global changes in climate and land-use have been implicated and searches have been conducted to determine whether successful invaders have distinctive attributes. Unfortunately, to date, basic questions concerning invasions have remained unanswered and it has not been possible to predict invasions with any reliability. A difficulty here may be related to the tendency of some ecologists (e.g. Elton 1958; Dukes and Mooney 1999) to regard invasions as rather exotic events. As explained by Davis *et al.* (2000b) this has had the undesirable effect of dissociating the study of alien invaders from many relevant areas of ecological theory and research. The main purpose of this chapter is to introduce a general theory of invaders and invasibility (Davis *et al.* 2000a) that explains these phenomena in conventional terms; that is in the context of existing knowledge of vegetation and resource dynamics, regeneration, and colonisation. To begin the analysis it is helpful to ask two fundamental questions:

1. Is there any consistent difference between the characteristics of colonising and invading plant species and the circumstances encouraging their expansions?
2. What determines the invasibility of a plant community?

COLONISERS AND INVADERS: A COMPARISON

Successful colonisers depend upon dispersal and it is therefore understandable that efforts to recognise their defining characteristics have often placed strong emphasis on the mobility of individual seeds and spores. It is perhaps surprising to find that this line of enquiry has not identified a consistent link between propagule type and colonising ability. In a review of the attributes of plant species colonising spoil habitats in the Sheffield region (Grime 1986) the production of numerous, widely-dispersed propagules was found to be conspicuously associated with colonising success in only a small, specialised element within the spoil flora. This consisted of rarer species (e.g. *Chaenorhinum minus* and *Senecio viscosus*) exploiting local habitats such as cinder tips and railway ballast of small total contribution to the landscape. When the 20 most successful colonisers of spoil are examined (Table 23) it is immediately evident that most of these are exceedingly common British plants. This is confirmed in Table 23 by presenting for each species an index of spoil colonisation (CI) where:

$$\text{CI} = \frac{\text{percentage occurrence on spoil}}{\text{percentage occurrence in all habitats}}$$

Most values exceed 1.0, indicating some potential to expand differentially on spoil, but since few of the indices are greater than 3.0 it is also apparent that the commonest species colonising spoil habitats are plants of high abundance in the region as a whole. When a detailed examination was made of these 20 species they were found to be widely divergent with respect to flowering season, breeding system, capacity for wind dispersal, seed dormancy, seed persistence, and germination requirements. Many of the most effective spoil colonists were found to have seeds that had no well-defined mechanism of dispersal and several had a strong tendency to accumulate large numbers of dormant seeds in the soil. This suggests the most important determinant of colonising success was seed output and that above a certain threshold access to spoil sites was largely independent of the mobility of individual seeds and occurred through transfer with soil, in the spoil itself, on human feet, or in the tread of vehicle tyres (Hodkinson and Thompson 1997).

Caution must be applied in seeking to generalise from this study to colonisers of other habitats and in making comparisons with invaders. However, strong

Table 23. The 20 most common vascular plant species of spoil habitats in the Sheffield region of England. (Reproduced from Grime 1986 by permission of Royal Society, London.)

		F	CI	A	P	G	C	W
1	*Holcus lanatus*	36.4	2.0	•	+	+	•	•
2	*Chamerion angustifolium*	32.8	2.6	•	+	•	•	•
3	*Agrostis stolonifera*	32.0	1.8	•	+	+	•	•
4	*Dactylis glomerata*	29.4	1.6	•	+	+	•	•
5	*Festuca rubra*	28.5	1.5	•	+	+	•	•
6	*Agrostis capillaris*	26.0	1.4	•	+	+	•	•
7	*Poa annua*	24.9	2.0	(+)	(+)	+	•	•
8	*Poa pratensis*	24.8	1.7	•	+	+	•	•
9	*Arrhenatherum elatius*	24.1	1.7	•	+	+	•	•
10	*Poa trivialis*	23.5	1.1	•	+	+	•	•
11	*Tussilago farfara*	23.5	4.1	•	+	•	+	+
12	*Taraxacum officinale*	22.0	1.8	•	+	•	+	+
13	*Deschampsia flexuosa*	18.3	0.9	•	+	+	•	•
14	*Cerastium fontanum*	18.2	2.0	•	+	•	•	•
15	*Plantago lanceolata*	17.3	1.8	•	+	•	•	•
16	*Cirsium arvense*	16.1	2.4	•	+	•	+	+
17	*Elytrigia repens*	14.4	1.6	•	+	+	•	•
18	*Trifolium repens*	14.2	1.5	•	+	•	•	•
19	*Senecio vulgaris*	14.2	3.6	+	•	•	+	+
20	*Senecio squalidus*	13.7	4.2	+	•	•	+	+

F, percentage of spoil samples containing the species; CI, index of colonisation (see text); A, annual; P, perennial; G, grass; C, composite; W, wind-dispersed. The brackets for *Poa annua* denote the fact that annual and perennial phenotypes of this species occur in the Sheffield region.

parallels can be observed between the conclusions of this study and those drawn from a large number of autecological investigations (e.g. Salisbury 1932, 1953, 1964; Kent 1956, 1960; Davey 1961; Conolly 1977; Simpson 1984) documenting the spread of particular alien species. The extent to which, as in the case of colonising species, studies of invaders have not identified a critical role for dispersal mechanisms but have instead recognised a critical threshold in propagule output is amply illustrated by two quoatations from Salisbury (1953):

> Perhaps the most interesting feature of the changes in our weed population has been the diverse manner in which these various species have spread.
>
> . . . all exhibited a slow rate of spread followed by a rapid one which suggest it is a phenomenon analogous to the 'infection pressure' of epidemic disease. In each instance, the rapid spread followed upon local increase due to the artificial provision of suitable habitats.'

At this point we can draw two conclusions:

1. Both colonisers and invaders comprise a diverse assortment of plant functional types (Figure 75) with widely different methods of regeneration and dispersal.

Figure 75. Alien invaders of the British flora. The main picture shows the massive shoots of the giant hogweed (*Heracleum mantegazzianum*) on the banks of the River Teviot, Roxburghshire, Scotland. Inset is the diminutive *Epilobium brunnescens*, from a population invading damp brickwork at Whirlow Park, South Yorkshire, England

2 A prerequisite for colonisation and invasion is the production of large numbers of propagules.

COMMUNITY INVASIBILITY

Strong clues about the circumstances conducive to invasion are available from the case histories of plant species such as *Senecio squalidus*, *Buddleja davidii*, *Impatiens glandulifera*, and *Heracleum mantegazzianum* that have spread rapidly over large geographical areas during the 20th century. All of these species have exploited continuous and extensive linear features such as riverbanks, roadsides, and railways. These corridors have provided access to invaders through their continuity but it may be more important to recognise that rapid spread along such conduits also depends upon the high levels of disturbance (flooding, herbicide treatments, burning, earth-moving operations) which create areas of bare soil open to colonisation.

These accounts of rapid invasion and spread by alien species are in sharp contrast to the highly intermittent nature of successful establishments by aliens in relatively undisturbed habitats. However, evidence is accumulating that such invasions are occurring with increasing frequency (Vitousek *et al.* 1996; Lonsdale 1999; Stohlgren *et al.* 1999) and the challenge for ecologists is to identify the causes.

Recently Davis *et al.* (2000) have proposed a new theory explaining the intermittent nature of invasibility that appears to characterise the majority of plant communities. This theory is based upon insights from experiments and from long-term monitoring studies of vegetation and identifies fluctuation in resource availability as the key factor controlling invasibility. The theory can be stated simply as follows: **A plant community becomes more susceptible to invasion when there is an increase in the amount of unused resources**.

This theory rests on the assumption that an invading species must have access to available resources, e.g. light, nutrients, and water, and that a species will enjoy greater success in invading a community if it does not encounter intense competition for these resources from resident species. This assumption is based on the theory that competition intensity should be inversely correlated with the amount of unused resources (Davis *et al.* 1998) and means that circumstances conducive to invasion may be expected to arise periodically in a wide range of habitats and vegetation types. We may suspect that it is this diversity in the range of resource-release mechanisms that partly explains the delay in formulating a general theory of invasibility, the failure to discover many consistent ecological correlates of invasibility (Mack 1996; Williamson and Fitter 1996; Rejmánek 1996; Lonsdale 1999), and the difficulty in predicting invasions (Williamson 1999). The elusive nature of the invasion process appears to arise from the fact that it depends upon conditions of resource enrichment or release that have a variety of causes but occur only intermittently and, to result in invasion, must coincide with an availability of invading propagules.

An increase in resource availability can occur in two basic ways: use of resources by the existing vegetation can decline, or resource supply can increase at a rate faster than the resident vegetation can sequester it. Resource use could decline due to a number of factors. A disturbance could damage or destroy some of the resident vegetation, reducing light, water, and nutrient uptake. Heavy herbivory due to grazing or a pest outbreak, or a widespread disease among the resident vegetation, would also reduce resource use. Increases in gross resource supply could arise in a particularly wet year (increased water supply), as a consequence of eutrophication (increased nutrients), or following removal of a tree canopy (increased light for the understorey vegetation). Whether resource use goes down for a time, or gross supply goes up, there are more resources available to invaders, and this is when the community is particularly vulnerable to invasion (Figure 76).

One important corollary of this theory is that a community's susceptibility to invasion is not a constant attribute, but a condition that can fluctuate over time. Many of the questions that have guided thinking about invasibility of plant communities have been misleading because they tend to characterise

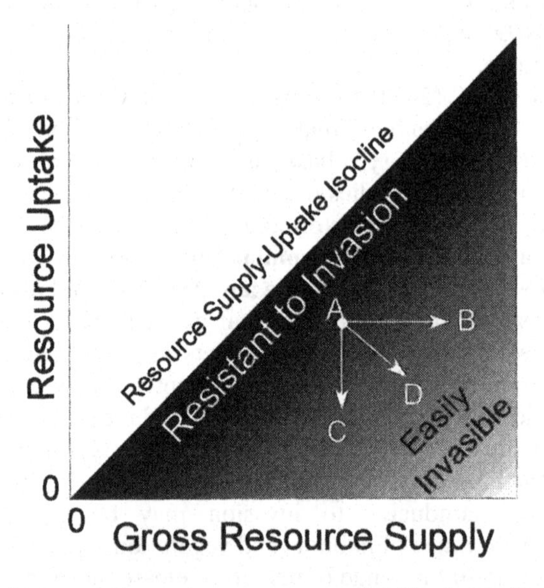

Figure 76. The theory of fluctuating resource availability holds that a community's susceptibility to invasion increases as resource availability (the difference between gross resource supply and resource uptake) increases. Resource availability can increase due to a pulse in resource supply (A–B), a decline in resource uptake (A–C), or both (A–D). In the plot shown, resource availability, and hence invasibility, increases as the trajectory moves further right and/or below the supply/uptake isocline (where resource uptake = gross resource supply)

invasibility as an invariable attribute of communities (e.g. Case 1990). Rather than being an inherent property, it seems likely that invasibility of many communities changes from year to year and even within a given year, as the amount of unused resources fluctuates. As pointed out by Johnstone (1986) this means that successful species invasions are likely to occur as episodic events.

The model of invasibility summarised in Figure 76 is consistent with a large body of field observations and experimental evidence. The important role played by disturbance in facilitating invasions has long been recognised (Elton 1958; Crawley 1987; Lodge 1993) and data exist supporting this view (Hobbs and Atkins 1988; Burke and Grime 1996). It has been proposed that disturbance facilitates invasions by eliminating or reducing the cover or vigour of competitors or by increasing resource levels (D'Antonio 1993; Hobbs 1989; Symstad 2000; White *et al.* 2000a). In such cases, the increase in invasibility following a disturbance can be explained by the theory of fluctuating resource availability. Whether the disturbance inputs new resources into the community (e.g. during a flood), or there is a decline in resource uptake by the resident vegetation due to mortality or debilitation of the resident species, resource availability will increase, and thus, according to the theory, invasibility should increase. Disturbances need not be community-wide to increase invasibility. Frequent small scale disturbances, e.g. by burrowing animals, can create localised patches of unexploited resources, and thereby may facilitate invasions (Hobbs and Mooney 1985).

On a world-wide basis, phosphorus and nitrogen are often the limiting resources in natural vegetation and several recent studies have shown that soil nutrient levels may play an important role in determining a community's invasibility. A survey of North American grassland sites revealed that invasions by exotic species were often highest in nutrient-enriched sites (Stohlgren *et al.* 1999). Nitrogen addition in a California serpentine grassland increased the invasion success of several alien grass species (Huenneke *et al.* 1990). An experimental study of plant invasions into a limestone grassland in the UK showed that invasion was highest in sites that were nutrient-enriched, and particularly rapid when this enrichment was accompanied by disturbance (Burke and Grime 1996; Figure 77). Similar findings were found by Hobbs and Atkins (1988), who also found that disturbance combined with eutrophication increased a community's invasibility. The combination of disturbance and eutrophication involves both a reduction in resource uptake by resident vegetation and an increase in gross resource supply. As depicted in Figure 77, this combination will inevitably result in the largest increase in resource availability and is therefore consistent with these published reports of additive effects of disturbance and eutrophication.

Experimental manipulations of resource supply, such as by root-trenching to remove the impact of tree roots, has been observed to permit invasions by 'weedy' species such as *Galeopsis tetrahit* (Slavikova 1958). Other studies

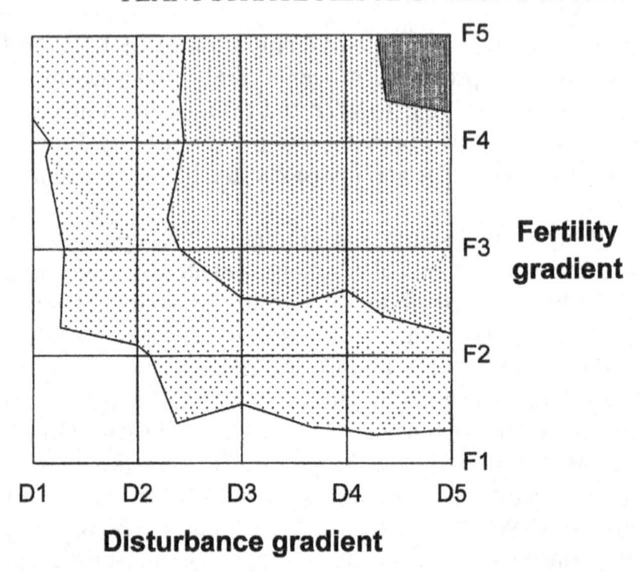

Figure 77. The pattern of invasion recorded after five years following the sowing of seeds of 54 species, not originally present at the site, into 18 replicate plots, each with an annually imposed fertility × disturbance matrix, established in unproductive limestone grassland at the Buxton Climate Change Impacts Laboratory, Derbyshire, England, in 1990. The mean cover of successful invaders is shown in arbitrary units: darker shading represents more invasion. Invasion of the sown species was clearly promoted by increased resource supply (fertiliser) and by reduction in resource uptake by the resident vegetation (disturbance). Note that since an identical inoculum was applied to all parts of all replicates, variation in propagule pressure did not contribute to the observed pattern of invasion

(Hobbs and Mooney 1991; Burgess *et al.* 1991; Harrington 1991; Li and Wilson 1998; Dukes and Mooney 1999; Davis *et al.* 1999) have shown that in dry regions, increase of water supply, whether by natural rainfall or by experimental additions, increases the invasibility of vegetation, either as a direct effect of water supply or through improved access to mineral nutrients. Invasions of the California serpentine grasslands by exotic grasses were reported to increase following wet years (Hobbs and Mooney 1991), and similar findings were recorded in the Sonoran Desert (Burgess *et al.* 1991). Experimental studies have shown that water supplements increase the availability of soil water and hence the invasibility of herbaceous communities by both woody and herbaceous vegetation (Davis *et al.* 1998; Harrington 1991; Li and Wilson 1998). Imposed drought conditions reduced the availability of soil water and hence decreased the invasibility of the same communities (Davis *et al.* 1998, 1999). Note however that drought severe enough to cause mortality and thus create gaps in previously closed vegetation will increase invasibility. This suggests that this capacity of a single environmental variable to have opposing effects according to its intensity has been a further source of confusion in attempts to devise general theories of invasibility.

It has been argued that global environmental changes may accelerate species invasions (Dukes and Mooney 1999). In most instances, these arguments can be subsumed under the theory of fluctuating resource availability. Some authors have predicted that the increase in atmospheric CO_2 will favour invasions by certain species by increasing soil water availability due to more efficient use of water by the resident plants (Dukes and Mooney 1999; Idso 1992; Johnson *et al.* 1993). This is an example of increased resource availability due to reduced uptake by the resident vegetation. Others have argued that invasions may be facilitated by increases in precipitation (Dukes and Mooney 1999), an example of increased resource availability due to increased resource supply. Still others have argued that the global nitrogen eutrophication resulting from anthropogenic activities is already facilitating invasions (Wedin and Tilman 1996), another example of enhanced supply increasing resource availability.

Some theories regarding community invasibility are based on equilibrium assumptions. For example, it has been argued that invasion-resistant communities are inhabited by resident species that have already monopolised and partitioned the key resources. Thompson (1991) argued that 'the invader must have advantageous properties not held by pre-existing species', implying that the only way to invade a community is to be fundamentally different from the resident species since the resident species have already filled up certain niches. Tilman (1997) argued similarly, concluding that invaders in his experimental plots were filling 'empty niches'. This view is contested by Crawley *et al.* (1999) and Davis *et al.* (2000a), who argue that many, if not most, plant communities are seldom in equilibrium with their resources due to periodic fluctuations in resource supply and uptake by the resident vegetation, caused by meteorological fluctuations as well as by site-specific events such as large- or small-scale disturbances, pest outbreaks, changes in grazing pressure, and anthropogenic eutrophication. If a community, or a component of a community, experiences a pulse of resource supply, or a decline in resource uptake, an invading species may be able to exploit the unused resource even though its ecology is not fundamentally different from that of the resident species. The same non-equilibrium conditions that permitted invasion in the first place may also permit ongoing persistence in the new community. That is, fluctuating availability of resources in space and/or time will lead to a fluctuation in the intensity of competition (Davis *et al.* 1998), which may prevent competitive exclusion from occurring. This reasoning is consistent with that of Huston and De Angelis (1994), who concluded that if water, light, and nutrients were not limiting due to spatial and temporal variation in these resources, exotic species might be able to invade and co-exist in species-rich communities. A similar emphasis upon resource availability rather than species richness and niche-filling as controllers of invasibility is expressed as follows:

> sites high in herbaceous foliar cover and soil fertility and hot spots of plant diversity (and biodiversity), are invasible in many landscapes and the pattern may be more closely related to the degree to which resources are available in native communities, independent of species richness.
>
> Stohlgren *et al.* (1999)

Invasions by exotic species are often studied and discussed as if they were a distinct ecological phenomenon (Elton 1958; Dukes and Mooney 1999). However, it seems more likely that the basic processes that admit exotic plant species are essentially the same as those that facilitate colonisations by native species or allow occasional regeneration of long-lived species at the same site. It has been suggested (Grime *et al.* 1994) that conditions favouring invasions by exotic species may be detectable as temporary expansions in the relative abundance within communities of subordinate resident species that rely upon seedling establishment to sustain their populations. An example of this phenomenon is presented in Figure 78 which shows that over the duration of the Bibury investigation (Dunnett *et al.* 1998) there have been two occasions in which the vegetation has allowed temporary expansions of the biennial *Anthriscus sylvestris*. The second outbreak followed the exceptional drought of 1976 but the cause of the first expansion has not been identified. Further examples include the encroachment of native woody plants into adjacent old fields or grassland environments, events that have been shown to be episodic (Myster 1993) and to be associated with increases in available resources (Harrington 1991; Myster 1993; Davis *et al.* 1999). Succession in Dutch heathlands has been accelerated experimentally by the addition of mineral nutrients (Berendse *et al.* 1994), while temporary expansions of range and abundance of monocarpic species often follow the debilitation of perennials by drought (Prince *et al.* 1985; Dunnett *et al.* 1998). An experimental study of the invasion by native plants into a limestone grassland also found that invasion was enhanced by resource availability (Burke and Grime 1996). If exotic invaders exhibited unique functional attributes compared to native colonisers, one might also conclude that establishment by the two groups involved different processes. However, a detailed comparison of the attributes of exotic invaders and native colonisers concluded that the two groups were functionally indistinguishable (Thompson *et al.* 1995). Hobbs and Atkins (1988) found that in a nutrient-poor environment, both native and exotic species diversity increased when nutrients were added and concluded that the success of both groups was nutrient-limited. Control of the establishment of exotic and native plant species by the same ecological processes may explain why some natural communities with high native species diversity also appear to be very susceptible to invasion by exotic species (Stohlgren *et al.* 1999).

In various ways, aspects of the theory of fluctuating resource availability have been outlined by others. Huenneke *et al.* (1990) argued that 'invasibility may be directly influenced by nutrient availability'. Stohlgren *et al.* (1999) concluded more generally that invasibility 'may be closely related to the

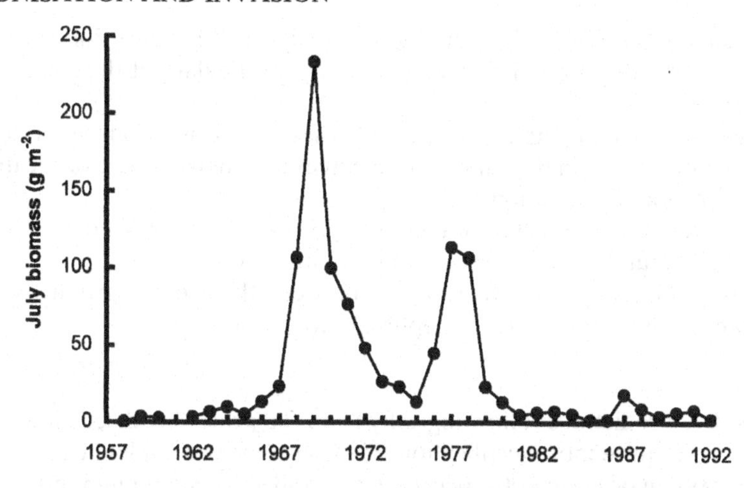

Figure 78. Figure 58 Fluctuations in the shoot biomass of the biennial umbellifer, *Anthriscus sylvestris*, in the Bibury plots. (Reproduced from Grime *et al.* 1994 by permission of CAB International.)

degree that resources are available'. Thompson *et al.* (1995) emphasised the importance of a variable environment, proposing that 'the alternation of episodes of gap creation by drought and fire with wet periods more suitable for seedling establishment may render such habitats particularly susceptible to invasion'. Huston and DeAngelis (1994) emphasised the importance of a variable environment in permitting exotic species to invade species-rich communities. In reviewing tree invasion into old fields, Myster (1993) emphasised the importance of 'windows of opportunity', which periodically 'open' (e.g. during a wet year) and 'close' (during a drought year), and Davis *et al.* (1999) emphasised the role of 'wet and dry spells' in tree invasions of herbaceous environments. Thus, the theory of fluctuating resource availability integrates various published reports implicating resource availability and fluctuating environmental conditions in invasibility.

The theory leads to testable predictions:

1 Communities subject to pronounced fluctuations in resource supply, either by enrichments from external sources or by release from the resident organisms, will be susceptible to invasion.
2 Invasions will be most likely in the period immediately following an abrupt increase in the rate of supply or decline in the rate of use of a limiting resource.
3 Invasions will be facilitated by mechanical, chemical, or climatic disturbances or by disease and pest attacks that increase resource supply either by direct leakage from damaged tissues or by reducing the rate of capture by the resident vegetation.

4 The susceptibility of a community to invasion will increase following the introduction of grazers into the community, particularly if the community is nutrient-rich.
5 Invasions are more likely to succeed where there is a long interval between an increase in the supply of resources and their capture or recapture by the resident vegetation.
6 There will be no consistent relationship between the species diversity of a plant community and its susceptibility to invasion.
7 There will be no general relationship between the average productivity of a plant community and its susceptibility to invasion.

The first five predictions follow logically from the preceding discussion, while the last two require more justification. First, since near complete exploitation of resources can occur in both species-rich and species-poor communities (e.g. in stable environments with consistent resource supply rates) and since very incomplete resource exploitation can also occur in both communities (e.g. in environments with widely fluctuating resource supply rates), there is no reason to expect any consistent pattern between the species richness or diversity of a community and its susceptibility to invasion. As demonstrated by Crawley *et al.* (1999) resistance to invasion is likely to be determined by the characteristics of the resident dominants rather than by the number of species already present in the vegetation. Second, resource availability is subject to periodic increase in both productive and unproductive habitats, with consequent effects on invasibility in both types of habitat. Both Rejmánek (1989) and D'Antonio (1993) have pointed out that while persistently harsh environments are likely to inhibit invasion, even here there may be more benign interludes. In a similar way the intense competition for resources characteristic of productive communities may render them proof against invasion for most of the time but they may become susceptible to invasion if there are periodic disturbances.

It is unlikely that any single theory will be universally successful in explaining why plant invasions are specific to certain times and places. The theory of fluctuating resource availability focuses on the essential fact that the opportunities for invading species to capture photosynthate, water, and nutrients at rates allowing their success are often severely limited in space and time. This dependence of invasion upon rare events is likely to have some more specific implications for particular colonising species. As Salisbury (1953) recognised, invasion success is often crucially limited by the number and dispersal range of invading propagules. In addition, the theory is primarily intended to apply to situations where a community is being invaded by several or many different plant species. If a community is being heavily invaded by just a single species, the explanation may lie more in the specific attributes of the invading species than in the characteristics of the community.

Species invasions are likely to be one of the main ecological consequences of global changes in climate and land-use. To respond effectively ecologists must now begin the essential task of transforming the study of invasions from a diffuse anecdotal subject into predictive ecological theory. A unifying emphasis on fluctuating resource availability may provide some of the predictive power needed by land managers and conservationists.

8 Succession

INTRODUCTION

From the work of numerous ecologists, most notably that of Clements (1916) and Watt (1947), it is now generally accepted that most types of vegetation are subject to temporal changes both in species composition and in the relative importance of constituent life-forms. These changes may be classified broadly into two kinds: successional and cyclical. During successional change there is a progressive alteration in the structure and species composition of the vegetation, whereas in cyclical change similar vegetation types recur in the same place at various intervals of time. In the case of succession, a further distinction may be drawn between the changes which occur during the colonisation of a new and skeletal habitat usually initially lacking in soil and vegetation (primary succession) and those which characterise the much more common circumstance in which succession is a feature of the process of recolonisation of a disturbed habitat (secondary succession).

More recent investigations and reviews (e.g. Egler 1954; Watt 1960; Whittaker 1975; Connell and Slatyer 1977; Fox 1977; Osbornova *et al.* 1989; Olff 1992; Van Andel *et al.* 1993; Turner *et al.* 1997; Hughes *et al.* 1999; Donnegan and Rebertus 1999) have progressed beyond the descriptive phase and have achieved considerable progress in identifying the mechanisms which cause various types of vegetation change. The objective in this chapter will be to augment these studies by considering the involvement of plant strategies in successional and cyclical vegetation changes. Although, initially, attention will be mainly confined to strategies of the established phase, the last section will be concerned with the role of regenerative strategies.

Some important clues to the role of strategies in successional change may be found in the relationships between the primary strategies and life-forms summarised in Figure 44, page 132. From this figure it is apparent that whereas the ruderal strategy comprises a fairly homogeneous group of ephemeral plants with many similarities in life-history and ecology, the competitors consist of a wide range of plant forms including perennial herbs, shrubs, and trees. However, the most remarkable conjunctions in plant size, form, and ecology occur within the stress-tolerators where we find such apparently diverse organisms as lichens and certain forest trees. There is no mystery here; as illustrated on page 66 the mineral nutritional constraint that characterises vegetation dominated by stress-tolerators is common to circumstances of absolute deficiency and low biomass and to ecosystems where the plant biomass is considerable but the large stock of mineral nutrients is tightly recycled and scarcely available.

In order to explain how this morphological variety among the ranks of the competitors and stress-tolerators is a key to certain aspects of the mechanism of vegetation succession it is necessary to examine the sequence of strategies and life-forms which appear in primary and secondary successions in productive and unproductive environments.

PRIMARY SUCCESSION

An early example of a detailed description of primary succession is that of Evans (1953), who examined the sequence of events associated with vegetation development and soil formation on the large gritstone boulders in oak woods on the Pennine escarpments of northern England (Figure 79). Initial colonisation is by small lichens and bryophytes which are slowly replaced by foliose lichens and mosses that are capable of forming large cushions. As organic matter accumulates and moisture and nutrients are retained the slow-growing evergreen tussock grass, *Deschampsia flexuosa*, appears and the abundance of bryophytes and lichens declines. Eventually a soil profile becomes defined consisting of weathered gritstone capped by a layer of organic matter and the early stages of woodland development occur through colonisation by bilberry (*Vaccinium myrtillus*) and seedlings of oak (*Quercus petraea*), rowan (*Sorbus aucuparia*), and birch (*Betula pubescens*).

The succession at Padley Wood can be represented by a simple model (Figures 80a, b) which consists of an equilateral triangle in which variation in the relative importance of competition, mineral nutrient stress, and disturbance as determinants of the vegetation is indicated by three sets of contours. At their respective corners of the triangle, competitors, stress-tolerators, and ruderals become the exclusive constituents of the vegetation. The relative biomass of the vegetation at each stage is indicated by the size of the circles in Figure 80b, and the role of different life-forms can be predicted by reference to their strategic ranges in Figure 84, page 252. It is interesting to note that although this example of primary succession involves a slow increase in the plant biomass and in the total stock of mineral nutrients held in the vegetation, litter, and soil, the whole sequence occurs under conditions of severe nutrient limitation, all the plant species involved are stress-tolerators, and the successional process clearly conforms to the model of facilitation proposed by Connell and Slatyer (1977). The entire sequence from bare rock to woodland can be represented as a simple short upward trajectory in CSR space (Figure 80b). Variants on this pattern are described by Evans (1953) and are correlated with the angle of slope of the rock surface colonised. Where the slope is moderate succession may proceed for many years only for the process to return to the starting point as the soil and vegetation is dislodged. On a

Figure 79. Primary succession on gritstone boulders at Padley Wood, North Derbyshire, England. (a) The lichen phase; (b) the moss phase; (c) *Vaccinium myrtillus*, and *Deschampsia flexuosa*; (d) tree litter accumulation and colonisation by a sapling of *Sorbus aucuparia*

vertical surface, however, gravity intervenes much earlier and the succession does not progress beyond the stage dominated by lichens and bryophytes.

At Glacier Bay, Alaska, Crocker and Major (1955) were able to reconstruct a slightly more complicated sequence of events whereby freshly-exposed, nutrient-depleted moraines are colonised. Again cryptograms, mainly mosses, are the primary colonists but these are then followed by nitrogen-fixing species of *Dryas* and *Alder* (*Alder crispus*). In the period 50–100 years after initial colonisation there is a dramatic increase in the biomass of vegetation, and following the rapid accumulation of nitrogen the ecosystem is colonised by larger trees. When the primary succession at Glacier Bay is depicted in CSR terms (Figure 81) and compared with boulder colonisation in Pennine

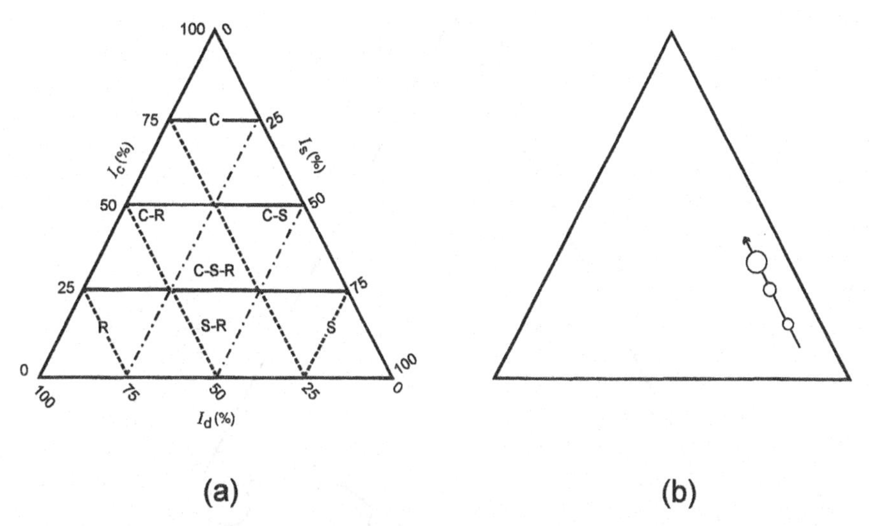

Figure 80. (a) The various equilibria between competition, mineral nutrient stress, and disturbance as determinants of vegetation. (b) The path of primary succession on gritstone boulders in Padley Wood, North Derbyshire, England, depicted in terms of the shifting equilibrium between competition, mineral nutrient stress, and disturbance. At each stage the changing strategies of the dominant plants are indicated by the path of the arrowed line, while the size of the biomass is reflected by the size of the circles; the involvement of different life-forms can be inferred from their distributions in Figure 84 on page 252

oakwoods (Figure 80) several differences are apparent. Despite the low nitrogen status, a rooting medium containing moisture and some mineral nutrients is present at the outset and the accumulation of plant biomass is quite rapid once nitrogen-fixation gathers momentum. However, the later stages of the succession involve a functional shift in which the relatively fast-growing *Alnus crispus*, which has a rapid turnover of tissues and releases nitrogen into the soil, is replaced by the more stress-tolerant Sitka spruce (*Picea abies*) and Hemlock species (*Tsuga heterophylla* and *T. mertensiana*) which achieve dominance by sequestration and tighter cycling of mineral nutrients.

SECONDARY SUCCESSION IN PRODUCTIVE ENVIRONMENTS

Already in the two simple cases of primary succession that have been described central importance has been granted to mineral nutrients and the influential role of dominant plants in affecting their quantity and availability. A similar emphasis is necessary in examining secondary succession but this does not usually apply in its earliest stages. This is because the effect of the disturbance events that initiate secondary succession (tree felling, wind-

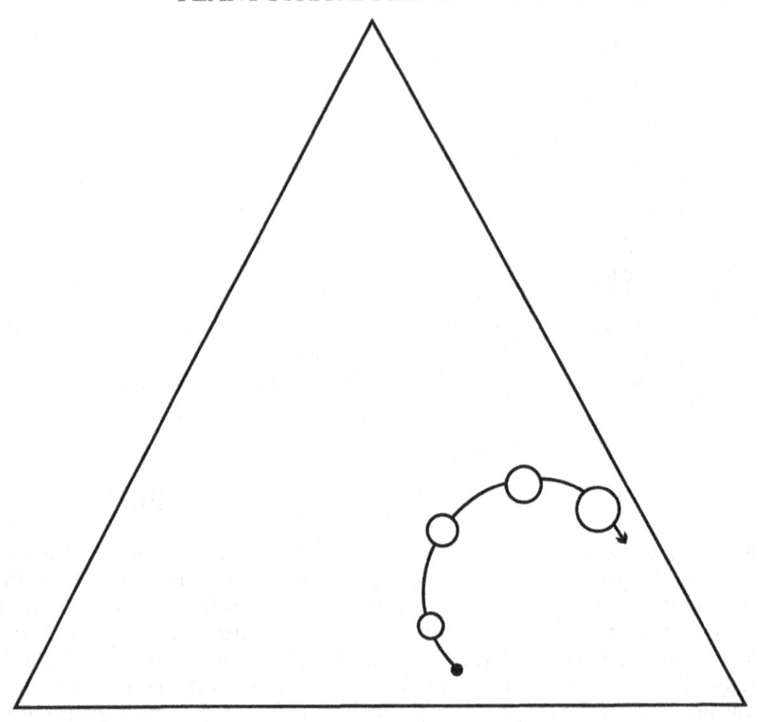

Figure 81. The path of primary succession on glacial moraines at Glacier Bay, Alaska (Crocker and Major 1955) depicted in CSR strategic terms. The size of the biomass at each stage is reflected by the size of the circles and the involvement of different life-forms can be inferred from their distributions in Figure 84 on page 252

storms, fires, floods, etc.) is to release mineral nutrients from the biomass, litter, and soil.

In localities where the soils and climate are conducive to high productivity, clearance of woodland is followed, in the absence of further major disturbance, by a characteristic process of recolonisation in which annuals, perennial herbs, shrubs, and trees are successively represented in the vegetation (Bazzaz 1979; Carson and Barrett 1988; Carpenter *et al.* 1990). Although some plants are more effectively dispersed than others it is unlikely that this sequence is determined primarily by the differences in rates of colonisation. As argued by Egler (1954) and Holt (1972) the observed patterns are more consistent with the hypothesis that most species enter the successional process early but grow at different rates. This interpretation, which is compatible with the inhibition and tolerance models of vegetation dynamics proposed by Connell and Slatyer (1977), involves the proposition that late successional species pass through a long juvenile phase in which they are usually subjected to suppression by faster-developing species.

Succession in productive habitats is a complex phenomenon involving progressive modification of soil and micro-climate by the changing vegetation, changes in the rate at which resources are captured, cycled, and released and, as discussed later (pages 247–248), impacts of herbivores. There can be little doubt also that competitive-dominance (page 181) plays a major part in the process by which the annuals are displaced by perennial herbs. Mechanisms of dominance during the later stages of succession are more varied and complex and require careful analysis.

It is tempting to explain the sequence perennial herbs → shrubs → trees simply in terms of the longer period required for the development of the elevated leaf canopies of woody species. Reference has been made already to the comparatively slow growth-rates of shrubs and trees, an inevitable consequence of the expenditure of photosynthate upon supporting structures (trunks, branches, etc.) at the expense of leaf area. However, the possibility may be considered that the process of succession is further attenuated by the effects of perennial herbs upon seedlings and saplings of woody species. As explained on page 187 the establishment of tree seedlings may be strongly inhibited by the presence of competitive herbs which, because of their capacity for lateral spread above and below ground, are better equipped, in the short-term, to dominate the vegetation. As an extension of this hypothesis, we may suppose that the tendency of shrubs to dominate the intermediate phase of vegetation succession in productive habitats is related to the branching form of the shoot which, in contrast to that of most trees, allows, at an early stage of growth, lateral spread of the leaf canopy and effective competition with herbs for light.

The woody species which appear early in the course of vegetation succession in productive habitats resemble in several respects the competitive herbs which they usurp. Their competitive characteristics include rapid (in comparison with other trees) rates of dry matter production, continuous stem extension and leaf production during the growing season, and rapid phenotypic adjustments in leaf area and shoot morphology in response to shade. These features are particularly conspicuous and have been documented (see page 23) among the deciduous trees and shrubs such as *Rhus glabra*, *Ailanthus altissima*, *Betula populifolia*, *Populus tremuloides*, and *Liriodendron tulipifera* which occur at the early phase in reforestation of disturbed woodland, abandoned pastures, and arable fields on the eastern seaboard of North America. Plants which appear to play an equivalent role in Europe include *Sambucus nigra* and species of *Salix*, *Populus*, and *Betula*.

An additional characteristic of competitive trees and shrubs is their intolerance of deep shade: as the tree canopy closes, seedling establishment in these species diminishes and many of the smaller saplings become severely etiolated and die. By this stage in vegetation succession it is usually evident that changes have also occurred in the composition of the herb layer. Although the competitive herbs are still represented, most of them are extremely reduced in vigour and tend to be replaced by evergreen, shade-tolerant herbs.

As the larger competitive trees reach maturity most of the shrubs succumb to shade or reach the end of their life-span and become senescent. A new element then becomes prominent, consisting of trees which are, to varying extents, shade-tolerant. Seedlings of many such trees may have been present in relatively small numbers since an early stage of the succession, but because of their slow growth-rates they have been outstripped by the competitive trees and shrubs and have remained inconspicuous. The shade-tolerant trees differ from their predecessors in characteristics of both the established and regenerative phases of their life-cycles; these differences have been reviewed at some length in earlier chapters of this book. Many possess large seeds which, by providing an initial source of energy and structural materials, buffer the seedlings against the stresses experienced during establishment beneath a closed canopy of herbs, shrubs, or trees.

Although the shade-tolerant trees of productive habitats are slow-growing, they include certain species which have a relatively long life-span, attain a large size, and eventually, in the later stages of succession, become the dominant component of the forest. Over a large area of Central Europe, beech (*Fagus sylvatica*) plays this role, in North America sugar maple (*Acer saccharum*) and hemlock (*Tsuga canadensis*) occupy a similar niche, whilst in New Zealand evergreens belonging to the genus *Nothofagus* are often the dominant trees in the final phase of succession in initially productive habitats. Some insight into the slow-dynamics of successional replacements late in the process is provided by the speculation that:

> Hemlock may be the ultimate competitive dominant in most sites but may require well over a millennium without major disturbance to displace sugar maple.
>
> Woods (2000)

A problem of great theoretical interest and practical significance concerns the mechanism whereby these climax trees attain their dominant status. Some ecologists, such as Walter (1973) and Begon *et al.* (1996), have attributed success to their competitive ability. However, characteristics of many of these trees, such as their infrequent fruiting (McClure 1966; Burgess 1972; Harper and White 1974), suggest that they should be classified as stress-tolerant competitors (page 128) rather than competitors. Consideration of the circumstances prevailing in the later stages of succession (Hellmers 1964; Whittaker 1966; Odum 1969) leads one to suspect that retention and recapture of mineral nutrients is likely to be a major selective mechanism in climax forest. A further consequence of the development of mature forest is the imposition of limitations in carbon and energy during the establishment phase as a result of shading by a dense, continuous leaf canopy (Horn 1971) and the declining ratio of photosynthetic production to respiratory burden (Lindeman 1942; Odum 1971). However, these constraints do not apply to the canopy dominants; it is therefore much more likely that stress-tolerant traits begin to appear because a high proportion of the mineral nutrients present in the habitat are sequestered in the plant biomass (Jordan and Kline 1972; Odum 1971; Coomes and Grubb 2000) and are conserved by very efficient and closed (*sensu* Odum 1971) systems of mineral nutrient cycling.

SECONDARY SUCCESSION IN UNPRODUCTIVE ENVIRONMENTS

Where the soil fertility is low, the role of ruderal plants and competitors in secondary succession is much contracted, and stress-tolerant herbs, shrubs, and trees become relatively important at an earlier stage. The early transition to slow-growing perennials appears to be related to the onset of mineral nutrient limitation. This interpretation is supported by experiments such as that of McLendon and Redente (1991), who showed that the addition of nitrogenous fertiliser to a sagebrush steppe community allowed annual plants to persist as vegetation dominants beyond the initial stage of secondary succession. The growth form and identity of the climax species varies according to the nature and intensity of the stresses occurring in the habitat. In semi-arid regions or on rather shallow nutrient-deficient soils, the climax vegetation is often composed of sclerophyllous shrubs (Figure 29, page 66) and small, relatively slow-growing trees, typical examples of which are the North American species, *Pinus aristata* (Currey 1965; Ferguson 1968) and *Quercus gambellii* (Cottam *et al.* 1959). It is not unusual for the canopy provided by such woody species to be sufficiently discontinuous to allow the presence of an understorey of stress-tolerant herbs (Siccama, Bormann, and Likens 1970; Arno and Habeck 1972; van Steenis 1972; Westman 1975).

Where productivity is even lower such as on the nutrient-deficient sandplain at Cedar Creek, Minnesota, secondary succession is exceedingly slow (Inouye *et al.* 1987) and tends to move directly from a sparse cover of annuals to vegetation comprised of slow-growing herbaceous perennials and scattered individuals of woody plants. The occlusion of C-strategists from the successional sequence at Cedar Creek is clearly a consequence of the low productivity of the site; when nitrogenous fertilisers were applied to field plots (Tilman 1987a) fast-growing clonal species such as *Elytrigia repens* were able to assume dominance at the expense of smaller slower-growing perennials.

Under more severe stress, such as that occurring in arctic and alpine habitats, ruderal and competitive species may be totally excluded and here both primary and secondary successions may simply involve colonisation by lichens, certain bryophytes, small herbs, and dwarf shrubs (Muller 1940; Shreve 1942; Viereck 1966).

MODELS OF SECONDARY SUCCESSION

From the preceding descriptions of secondary successions it is evident that a major factor determining the role of CSR strategies in vegetation succession is the potential productivity of the habitat. This effect may be summarised in the form of a simple model (Figure 82) which is basically the same as that used to depict the trajectories of particular primary successions (Figure 80, page 241). The curves S_1, S_2, and S_3 in Figure 82 describe, respectively, the paths of succession in conditions of high, moderate, and low potential productivity, and, as before, the circles superimposed on the curves represent the relative size of the plant biomass at each stage of the succession, and by reference to the distributions in Figure 80b–f, the curves may be used to identify the sequence of dominant life-forms likely to characterise each succession.

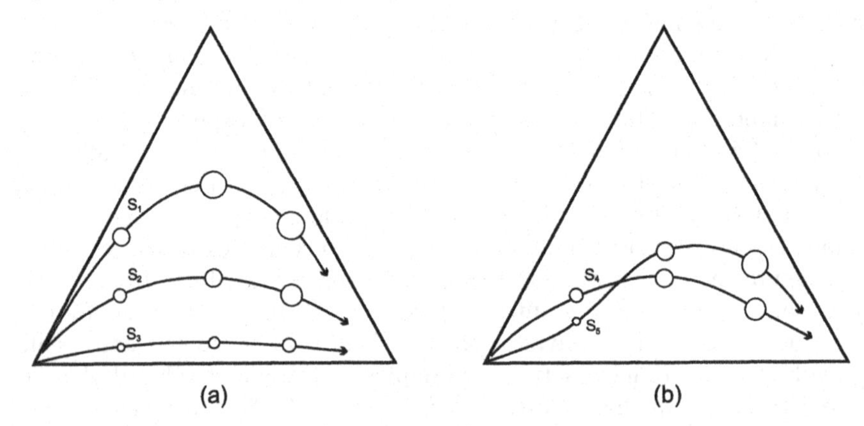

Figure 82. Diagrams representing the path of vegetation succession (a) under conditions of high (S_1), moderate (S_2), and low (S_3) potential productivity, and (b) under conditions of increasing (S_4) and decreasing (S_5) potential productivity. The size of the plant biomass at each stage of succession is indicated by the circles

In the most productive habitat, the course of succession (S_1) is characterised by a middle phase of intense competition in which first competitive herbs then competitive, shade-intolerant shrubs and trees dominate the vegetation. This is followed by a terminal phase in which stress-tolerance becomes progressively more important as mineral nutrient stress coincides with the development of a large plant biomass dominated by large, long-lived forest trees.

Where succession occurs in less productive habitats (S_2) the appearance of highly competitive species is prevented by the earlier onset of resource depletion and the stress-tolerant phase is associated with dominance by smaller slow-growing trees and shrubs in various vegetation types of lower biomass than those occurring in S_1. In an unproductive habitat such as Cedar Creek, the plant biomass remains low and the path of succession (S_3) moves directly from the ruderal to the stress-tolerant phase.

The curves of S_1–S_3 refer to succession under conditions of fixed potential productivity, a situation which is unlikely to occur in nature. In most habitats the process is associated with a progressive gain or loss in potential productivity. Increase in productivity, for example, may arise from nitrogen fixation, experimental additions of mineral fertiliser (Redente *et al.* 1992), or deposition of reactive nitrogen from the atmosphere. Conversely, losses of mineral nutrients may result from cropping or leaching of the soil by rainfall. In Figure 82b curves have been drawn to describe hypothetical pathways of succession under conditions of increasing (S_4) and decreasing (S_5) potential productivity.

THE ROLE OF HERBIVORES IN SUCCESSION

So far in this chapter it has been convenient to analyse the role of CSR strategists in successional processes by placing major emphasis on resource dynamics and highlighting the critical influence of the total stock of mineral nutrients in the ecosystem and the 'tightness' with which elements such as nitrogen and phosphorus are recycled. However, from the review of C-, S-, and R- attributes in Chapter 1 and from direct observations in the field (e.g. McNaughton 1978b; Bakker 1989; Jeffries 1999) it is apparent that there are two major opportunities for herbivores to intervene and strongly influence successional processes. These two contingencies are identified as points A and B in Figure 83 which reproduces the successional trajectory already discussed on pages 241–245 and describes the role of primary strategies in the re-colonisation of a productive habitat. Point A corresponds to the displacement of ruderals by competitors and point B indicates the stage in succession where stress-tolerant traits begin to be promoted.

At point A in Figure 83, the main impact of herbivores in succession is likely to be intensity-dependent and may not be substantially different from many other destructive forces acting on the vegetation. The attentions of vertebrate grazers (e.g. herds of mammals or burrowing species) or invertebrate infestations are often sufficient to damage both above- and below-ground vegetation severely and reverse succession (Kerbes *et al.* 1990; Hik *et al.* 1992). However, where, as in the case of a carefully-managed pasture, the influence of grazers is to encourage the development of a dense 'rapidly-repaired' perennial cover the effect of the herbivores is to prevent incursions by annuals and to drive succession forward.

Alternative impacts of herbivores on the direction of succession can also be predicted for point B in Figure 83. A reversal of succession, perhaps to a state of bare ground, can be envisaged for circumstances where migratory pest outbreaks (Rainey *et al.* 1990), heavy grazing by cattle (Olff 1992), or large herbivores such as elephants cause local devastation. However, one suspects that the much more common effect is more subtle and involves an acceleration of succession along the path described in Figure 83 and caused by

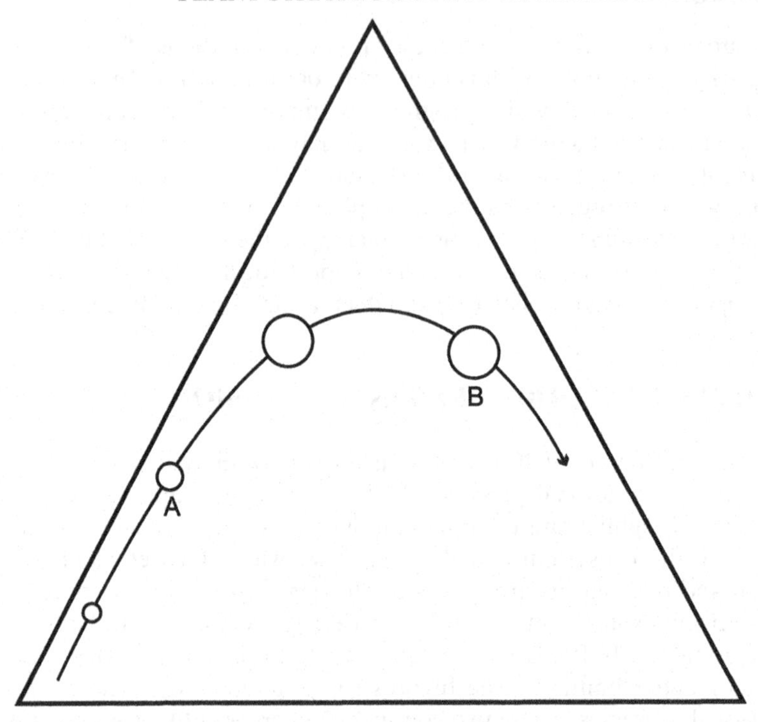

Figure 83. Identification of two critical points (A and B) at which herbivores may exert important effects on the course of secondary succession during the recolonisation of a productive habitat (see text)

selective herbivory by generalist invertebrates (Brown and Gange 1992; Carson and Root 1999), geese (Olff 1992), or small mammals (Moen *et al.* 1993). This effect can be predicted from field and laboratory comparisons of the palatability of early and late successional plant species (Grime *et al.* 1968, 1996; Reader and Southwood 1981) but strong support is also available from experiments in which pesticides have been used to eliminate above- and below-ground insect herbivores during the early stages of succession (Brown and Gange 1992; Brown *et al.* 1998; Carson and Root 1999, 2000) or sealed microcosms have been used to follow the course of succession in the presence and absence of generalist herbivores (Fraser and Grime 1999; Buckland and Grime 2000).

CLIMAX AND PROCLIMAX

In certain habitats the process of vegetation succession terminates in vegetation of a stable and predictable type—the so-called climax vegetation. In the most extreme type of climax vegetation, one species remains indefinitely as the dominant, and senescent individuals are replaced by members of the same species. A good example here is provided by the Central European forests dominated by beech (*Fagus sylvatica*). In many environments, however, there is no such well-defined end-point to succession and two or more species may co-exist in dynamic mixtures such as those described in Chapter 9.

Throughout the world, there are numerous examples of habitats in which succession has been arrested by severe forms of disturbance such as burning, grazing, mowing, or ploughing. In a given geographical region each form of disturbance when applied repeatedly in an orderly fashion results in a characteristic vegetation type usually described as a proclimax. Where the intensity of damage falls short of complete destruction of the established vegetation (e.g. grazing, mowing, burning) the result, in a potentially productive habitat, is to arrest succession at the stage where the dominant species are competitors. Under the most severe forms of constant disturbance, e.g. arable fields (page 83), succession does not progress beyond the ruderal phase. It has long been recognised (e.g. Tansley 1939) that the effect of orderly disturbance is not merely to arrest succession but to change the character of the vegetation. This arises because certain forms of disturbance when constantly applied tend to bring into prominence species which play a minor role or may not even be represented in the 'normal' (i.e. uninterrupted) course of succession. Examples of plants which are encouraged by constant application of particular types of disturbance are provided by the many genera of turf grasses (e.g. *Agrostis*, *Festuca*, *Lolium*, *Poa*) in which the established plant is promoted by regular grazing or mowing. Further examples include the poisonous or distasteful herbs (e.g. *Senecio jacobaea*, *Ranunculus acris*) which are favoured by persistent overgrazing and the shrub, *Calluna vulgaris*, which flourishes in heathlands managed by rotational burning.

It would be misleading, however, to interpret the ecology of proclimax species simply in terms of the adaptation of the established phase of the life-cycle to particular forms of regular disturbance. As we shall see under the next heading, some of the most consistent and important differences between successional and proclimax plants relate to their regenerative strategies.

REGENERATIVE STRATEGIES AND VEGETATION DYNAMICS

Already in Chapter 3 (pages 138–175), comments have been made concerning the ecological significance of particular regenerative strategies in various types of life-history. Here, therefore, only a brief reiteration of certain points is required in order to relate this information to the processes of vegetation change considered in this chapter.

Regenerative Strategies in Secondary Successions

Where succession is initiated by localised disturbance of mature forest, the initial phase of recolonisation usually involves those herbs, shrubs, and trees which produce an abundance of small, wind-dispersed seeds. By the time that succession has reached the stage where there is a dense vegetation cover (often as early as the second year following disturbance) conditions no longer favour the establishment of small, relatively fast-growing seedlings, and survival of the ephemeral element among the primary colonists depends critically upon its ability for renewed dispersal to freshly-disturbed sites.

A longer occupation is possible, however, among the perennial herbs, shrubs, and trees, in some of which regeneration by clonal expansion allows consolidation of the more vigorous of the genotypes surviving from the original population of colonising seedlings (Prach and Pysek 1994). Simultaneously or subsequently, these same plants release large quantities of seeds which facilitate establishment in new disturbed sites. The significance of this 'escape mechanism' in the primary colonists becomes apparent as succession proceeds to a stage where resource limitation exercises the dominant selective force with respect to strategies of both the regenerative and the established phase. At this late stage of succession, the most successful regenerative strategies appear to be vegetative expansion and that involving a bank of persistent seedlings, recruitment from which occurs mainly in response to mortalities among the established populations.

This description of the role of regenerative strategies in secondary succession is idealised and is based very largely upon patterns observed in the temperate forest environment. Quite clearly, more complex sequences will occur in circumstances where succession is interrupted more frequently; here one result may be to allow the persistence of species regenerating by wind-dispersed seeds. However, where disturbance is both frequent and extensive, e.g. in forests with a long history of damage by fire, grazing animals, or windfalls, it is not unusual to find represented in the flora species in which regeneration involves a persistent seed bank (see pages 149–151).

Regenerative Strategies in Proclimax Vegetation

The survival of proclimax vegetation depends upon continuous and, in some cases, severe disturbance and it is not surprising therefore to find that the two most common regenerative strategies (seasonal regeneration in gaps and regeneration from a persistent seed bank) in plants of arable land, pastures meadows, and heathlands both confer the ability for *in situ* expansion or re-establishment of plant populations.

In circumstances where disturbance is a constant but seasonal effect of climate or is a seasonally-predictable biotic effect, the predominant regenerative strategy is usually that involving the production of a crop of seedlings or vegetative offspring at the time of year most propitious for re-establishment in the denuded areas. Where disturbance is less predictable in time there is a strong tendency for regeneration from a persistent seed bank to become the most important strategy. The advantage of this form of regeneration is especially apparent where the vegetation is subject to intermittent disturbance (e.g. heather moorlands managed by rotational traditional burning). In such an environment the seed bank allows persistence of the population not only during periods of frequent disturbance but also over intervals in which the vegetation is relatively undisturbed and opportunities both for seedling establishment and even perhaps survival of the established plant are restricted.

Although the two most familiar strategies encountered in proclimax vegetation involve persistent seed banks or seasonal regeneration, other types are not excluded. Where the intensity of disturbance is relatively low, regeneration by vegetative expansion is often exceedingly common among proclimax herbs and shrubs (page 140) and the large populations of composites such as *Taraxacum officinale*, *Hypochoeris radicata*, and *Leontodon hispidus*, in temperate grasslands, provide conspicuous evidence that the production of numerous wind-dispersed seeds has remained a viable alternative strategy in some forms of proclimax vegetation

LIFE FORMS, STRATEGIES, AND PRODUCTIVITY: A SYNTHESIS

The general conclusion which may be drawn from this chapter is that processes of change in the structure and composition of vegetation can be interpreted as a function of the strategies of the component plant populations. Moreover, the relationships between established and regenerative strategies, life-forms, and habitat productivity provide a succinct basis upon which to explain the various sequences of plants observed in secondary successions. It would appear also that knowledge of the regenerative strategies of plants is particularly relevant to an understanding of the difference between successional and proclimax vegetation.

In Figure 84 the various arguments and functional classifications presented earlier in this book are condensed into a series of diagrams to summarise the involvement of plant strategies in contrasted successional phenomena. Figures 84a–e indicate the CSR strategic range of particular taxa or life-forms, Figures 84f–j describe the succession paths and development of the plant biomass in particular situations, and in Figures 84k–o the role played by various regenerative strategies is added.

A major contribution is predicted for W (the production of numerous widely-dispersed seeds or spores) during the early stages of both secondary (Figure 84k and l) and primary (Figure 84m) succession. This prediction,

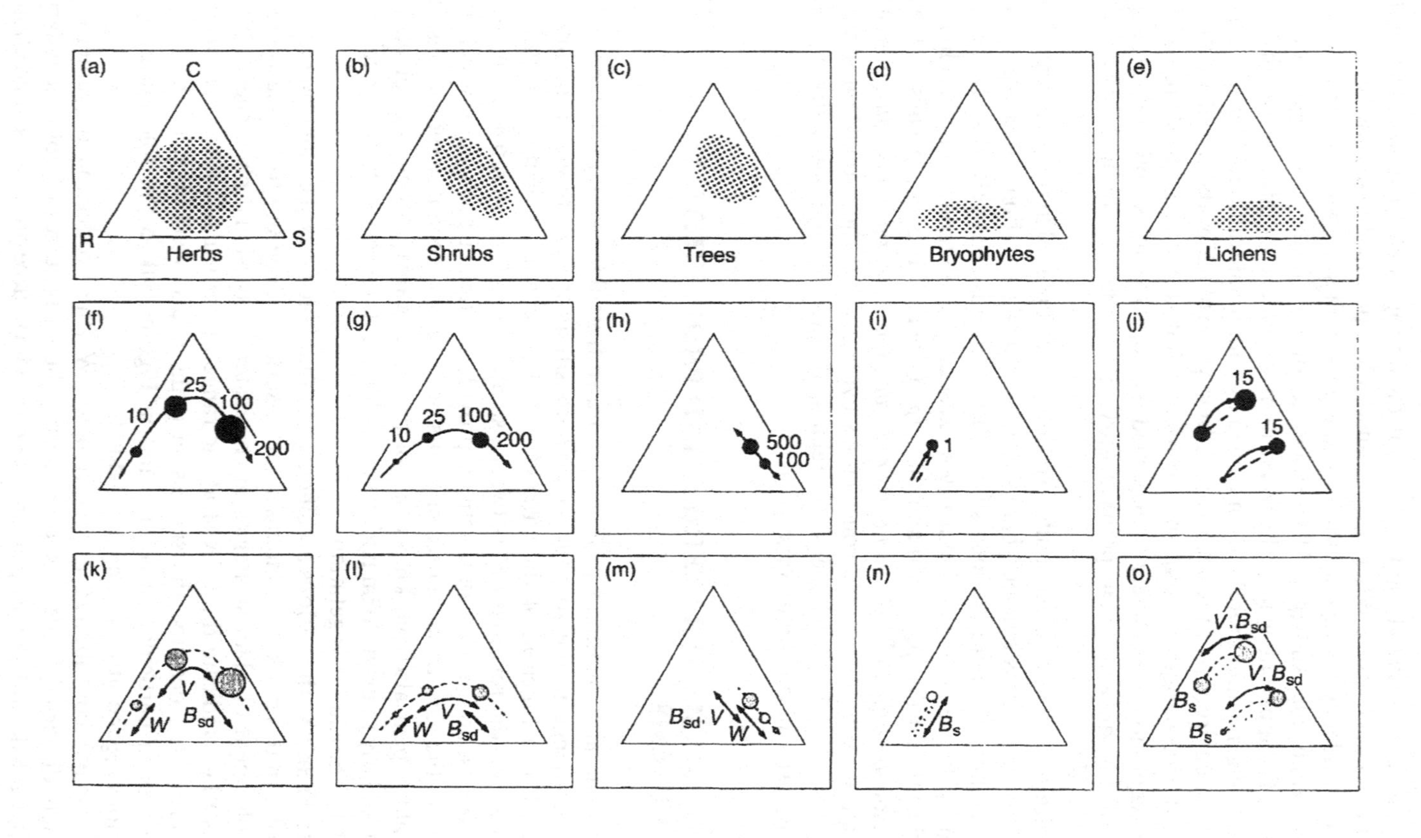

(a)
C
R
S
Herbs
(b)
Shrubs
(c)
Trees
(d)
Bryophytes
(e)
Lichens
(f)
10
25
100
200
(g)
10
25
100
200
(h)
500
100
(i)
1
(j)
15
15
(k)
V
W
B_{sd}
(l)
V
W
B_{sd}
(m)
B_{sd}, V
W
(n)
B_s
(o)
V, B_{sd}
V, B_{sd}
B_s
B_s

supported by a wealth of empirical evidence (Ridley 1930; van der Pijl 1972), rests on the assumption that a considerable selective advantage among potential early colonists will be enjoyed by those species which achieve 'saturating' densities of propagules across large areas of landscape. The small buoyant propagules associated with W contain low quantities of mineral nutrients and organic substrates and tend to produce offspring of low competitive ability and persistence in closed vegetation. As successions such as those described in Figures 84k–m proceed, therefore, it is expected that regeneration involving W will not provide an effective mechanism of persistence. Ephemerals dependent upon W will die out and be replaced by perennial herbs and shrubs which rely upon vegetative expansion (*V*). Although the single most important contribution of seedling regeneration is likely to be that of establishing founder populations during the initial 'open' stages of succession (Egler 1954) we can forecast that B_{sd} (regeneration involving a bank of persistent seedlings) will

---

Figure 84. Models using the CSR theoretical framework described by Grime (1974, 1979) to depict the strategic range of five life-forms (a–e), to represent various successional phenomena (f–j) and to indicate the role of four regenerative strategies in these same phenomena (k–o). The strategies of the dominant plants at particular points in time are indicated by the position of arrowed lines within the triangular model. The passage of time in years during succession is represented by numbers on each line and shoot biomass at particular points is reflected in the size of the circles. (f) Secondary succession in a forest clearing on a moderately fertile soil in a temperate climate. Biomass development is initially rapid and there is a fairly swift replacement of species as the community experiences successive phases of competitive dominance by rapidly growing herbs, shrubs, and trees. At a later stage, however, the course of succession begins to deflect downwards towards the stress-tolerant comer of the triangular model. This process begins even during stages where the plant biomass is expanding appreciably and reflects a change from vegetation exhibiting high rates of resource capture and loss, to one in which resources, particularly mineral nutrients, are efficiently retained or recaptured in the biomass. (g) Secondary succession for a site of lower soil fertility. Here the course of events is essentially the same as that described in (f) except that the successional parabola is shallower and the plant biomass smaller as a consequence of the earlier onset of mineral nutrient limitation of the initial phase of competitive dominance. (h) Primary succession in a skeletal habitat such as a rock outcrop. In this case the initial colonists, probably lichens and bryophytes, are stress-tolerators of low biomass and they occupy the site for a considerable period, giving way eventually to small slow-growing herbs and shrubs. This sequence coincides with the process of soil formation and provides an example of the facilitation model of vegetation succession (Connell and Slatyer 1977). (i and j) Vegetation responses to major perturbations. The range of models is extended to include loops representing the simple truncation of succession by annual harvesting and fertiliser input in an arable field (i) and the cycles of vegetation change associated with coppicing *Fraxinus excelsior* woodland (j, top) or rotational burning of *Calluna vulgaris* moorland (j, below). (k–o) The role of four regenerative strategies in the successional pathways represented in (f–j). See text for discussion. Functional types in the regenerative phase: *W*, numerous, small, widely-dispersed propagules; *V*, vegetative spread; B_s persistent seed bank; B_{sd}, persistent seedlings. (Reproduced from Grime 1987.)

play an inconspicuous but important role throughout primary succession (Figure 84m) and in the advanced stages of secondary succession (Figure 84k and l). This conclusion is based upon the field observation (e.g. Inghe and Tamm 1985) that many of the herbaceous and woody species which occupy mature ecosystems of low annual productivity (whether of low or high standing crop) produce seeds intermittently but maintain populations of seedlings; some of these may be capable of slowly graduating to reproductive maturity should disturbance events provide interludes of temporary release from the dominant effects of immediate neighbours.

A marked contrast with the uninterrupted successional sequences of Figures 84k–m is evident when attention is turned to the mechanisms of regeneration in the cyclical vegetation changes imposed by management (Figures 84n and o). In the simplest case, that of the arable field (Figure 84n), the dominant regenerative strategy predicted is that associated with a bank of persistent seeds. Here it is not difficult to recognise the selective advantage which populations of ephemeral weeds derive from the presence of high densities of persistent seeds in soils subjected to frequent and spatially predictable disturbance. In the coppice and heathland cycles persistent seed banks appear to play an essentially similar role but this is limited to phases of seedling recruitment initiated by occasional disturbance and followed by longer intervals in which biomass accumulates and other regenerative strategies (V, B_{sd}) may be expected to come into play.

In order to recognise principles of wide generality (in particular, the contrasting roles of W, B_s, V, and B_{sd} in vegetation dynamics), the examples selected have been relatively simple in character. It should be emphasised, however, that where vegetation is the result of a complex and continuing history of fluctuating conditions of environment and management (e.g. ancient hay pastures and meadows of the British Isles) it is not unusual to encounter plant communities in which there appears to be a stable coexistence between populations exhibiting each of the regenerative strategies identified in Chapter 3.

CHANGES IN VEGETATION STABILITY AND SPECIES RICHNESS DURING SECONDARY SUCCESSIONS

There appears to be general agreement between early and more recent authorities (e.g. Clements 1916; Tansley 1939; Margalef 1968; Whittaker 1975) that the progress of secondary succession is usually marked by a progressive decline in the rate of floristic change, and from certain studies such as that of Hanes (1971) it may be inferred that relatively slow rates of vegetation change characterise secondary succession in unproductive habitats.

Especially in highly productive environments, a rapid turnover of populations and species occurs during the phase of initial colonisation and vegetation development. This is followed by slower rates of species replacement once the ruderals have given way to competitors, stress-tolerant competitors, or stress-tolerators, the majority of which are comparatively long-lived. Other mechanisms which tend to limit the rate of vegetation change late in succession are the intermittent seed production, the increasing scarcity of conditions suitable for seedling establishment, and the slow dispersal rates of many late succession species.

It has been pointed out by Margalef (1963a, b, 1968) and Horn (1974) that some ambiguities may arise when the word 'stability' is used to characterise successional stages. In order to avoid confusion it is necessary to distinguish between *inertia*, i.e. resistance of the undisturbed vegetation to change (increasing during succession) and *resilience*, i.e. ability to recover rapidly from disturbance (decreasing during succession). The resilience of late successional vegetation is diminished not only by the slow rates of recolonisation and growth of its component plants but also by the fact that destruction of all or part of the biomass of a mature ecosystem may lead to irreversible changes in mineral nutrient status and soil structure such as those frequently observed after the felling of tropical rainforest. A more detailed discussion of the role of plant strategies as controllers of resistance and resilience is included in Chapter 10.

During secondary succession there are various developments, each of which is capable of influencing species richness. These include increasing modification of the soil and micro-climate by the vegetation, increasing interaction between plants, declining frequency of seedling establishment, and increasing stratification of the vegetation. Certain of these changes tend to increase species richness whilst others have the opposite effect. It is clear also that patterns of change in the species richness of one layer of the vegetation may be out of phase with changes in another. Both Margalef (1963b) and Odum (1971) have recognised that, during the intermediate stages of forest succession, species richness in the shrub and tree canopy may continue to increase at a time when effects of dominance by the woody plants are causing a progressive reduction in species richness in the herb layer.

A number of ecologists, including Bertalanaffy (1950), MacArthur (1955), Odum and Pinkerton (1955), Bray (1958), and Margalef (1963b, 1968), have suggested that the progress of succession is marked by increasing inertia, reduced dominance, and increasing species richness. Whilst there seems to be no reason to doubt the general validity of this hypothesis it is necessary to bear in mind that in modern landscapes two new factors are operating which drastically limit rates of succession and diversification.

The first of these limiting factors arises from the trend in most parts of the world towards dissected and depauperate floras. This change arises not only from increasing urbanisation but also from intensive methods of agriculture and forestry. One effect of such changes may be to restrict the distribution of late successional plants to islands, emigration from which is limited by their characteristically low reproductive output and dispersal efficiencies. In northern England this phenomenon may be observed in extreme form in derelict industrial landscapes such as those in the southern parts of Lancashire and Yorkshire (Grime 1999). Here refuges containing late-successional trees, shrubs, herbs, and cryptograms are extremely localised and in the intervening areas of wasteland succession may remain suspended indefinitely at the visually monotonous stage of competitive-dominance by a relatively small number of mobile early successional herbs and shrubs. In this type of situation it seems inevitable that landscape reclamation and nature conservation must involve procedures whereby succession is accelerated and diversity created through deliberate introductions of under-dispersed plants (Grime 1972; Bradshaw 1977).

A second development which may affect the course of vegetation succession in intensively-exploited landscapes is that of eutrophication. As we have seen earlier in this chapter, there is reason to suspect that many of the plants which occur in mature ecosystems of high floristic diversity are adapted to conditions of resource limitation. Where the effect of agriculture is to release mineral nutrients such as phosphorus and nitrogen into natural habitats the result may be the prevention of resource depletion and the arrest of succession at the stage of competitive-dominance. This may not be an exclusively modern phenomenon. Where vegetation on alluvial terraces is subject to mineral nutrient inputs from flood water it seems likely that conditions may arise which favour continuous occupation of the habitat by relatively fast-growing trees, shrubs, and herbs.

9 Co-existence

INTRODUCTION

As progress is made in understanding the primary mechanism of vegetation, the possibility arises that some generalisations can be attempted with regard to the processes which control the number of plant species per unit area of vegetation. Species richness is, of course, affected by many factors, and these vary in importance according to vegetation type and sample size. Where the area of the sample is large, high species richness may arise from the fact that the sample includes a mosaic of quite different environments and vegetation types corresponding to topographical variation, gradients in soil and climate, or patterns of vegetation management. However, as the size of the sample is reduced, major discontinuities in environment become less important and, in very small samples, the determinants of species richness are those which allow plants of different biology to establish and survive in close proximity. It is these more subtle mechanisms which provide the subject of this chapter.

CO-EXISTENCE IN HERBACEOUS VEGETATION

In the analysis of dominance (Chapter 4) reference was made to the tendency of larger herbaceous plants to suppress the growth and regeneration of smaller neighbours and it was concluded that, especially in fertile environments, a trend towards monoculture is frequently observed provided the biomass remains relatively undisturbed. We may suspect therefore that one of the preconditions for the co-existence of species in herbaceous vegetation is that factors are present which limit the expression of dominance. This limitation may operate through stress or disturbance or by a combination of the two, and its effect is usually to debilitate the potential dominants and to allow plants of smaller stature to regenerate and to co-exist with them.

The influence which stress and disturbance may exert (together and in isolation) upon the degree of dominance is illustrated in Figure 85, which describes the result of an experiment in which stress (represented by nitrogen deficiency) and disturbance (represented by mowing) were applied to a turf derived from a standardised seed mixture of common grasses. The data confirm that the effect of both nitrogen stress and frequent mowing was to suppress *Arrhenatherum elatius*, the dominant under fertile, undisturbed conditions. It is interesting to note, moreover, that the most equitable distribution of shoot dry matter between the component species occurred in the treatment involving both nitrogen stress *and* a high frequency of defoliation. From both field observations and experiments there is a wealth of additional evidence supporting the hypothesis that co-existence between a large number of herbaceous species (i.e. high species richness) is associated with the debilitation of potential dominants by effects of environment or management. The addition of mineral fertilisers to nutrient-deficient vegetation has been found to cause an expansion in species of large stature, with a coincident reduction in species richness (Thurston 1969; Willis 1963; Smith *et al.* 1971; Jeffrey 1971; Elberse *et al.* 1983; Berendse *et al.* 1992). Conversely, treatment of productive vegetation with the growth retardant, maleic hydrazide, tends to suppress potential dominants and to increase species richness (Yemm and Willis 1962); a similar effect has been achieved by grazing and herbage removal (Singh 1968; Singh and Misra 1969) and by treatment of annual communities with short-term doses of gamma radiation (Daniel and Platt 1968).

Measurement of species richness in a wide range of natural and semi-natural vegetation types in northern England (Grime 1973c) has provided circumstantial evidence of the importance of factors such as grazing, mowing, burning, and trampling in the maintenance of high species richness in certain habitats. From Figure 86 it is clear that types of perennial herbaceous vegetation subject to disturbance (e.g. pastures and meadows) in general display levels of species richness higher than those of unmanaged habitats.

The impact of management is most apparent when the numbers of species in derelict sheep pastures of the Derbyshire dales are compared with those from neighbouring dales in which grazing has been maintained to the present day. From comparisons such as that illustrated in Figure 87 there is evidence that the beneficial effect of continued grazing extends over the full spectrum of grasslands associated with the soil pH range 4.0–8.0. A similar effect of disturbance upon dominance and species richness has been observed in North America in areas of tall-grass prairie exploited by prairie dogs (Koford 1958) and badgers (Platt 1975).

From several sources, therefore, we have evidence substantiating the hypothesis that co-existence is encouraged by effects of stress and disturbance upon vegetation. However, in order to obtain a more complete assessment of the role of these factors upon species density it is necessary to consider the rather different circumstances which arise when the intensities of stress and/or disturbance become severe. Further scrutiny of the measurements of species richness in Figure 86 reveals that even in the mild temperate climate of northern England, habitats may be identified (acidic pastures, coal mine heaps, conifer plantations, shaded paths) in which the intensities of stress and/or disturbance experienced by herbaceous vegetation are sufficient not only to eliminate potential dominants but to produce local environments which are inhospitable to all except a few specialised plants.

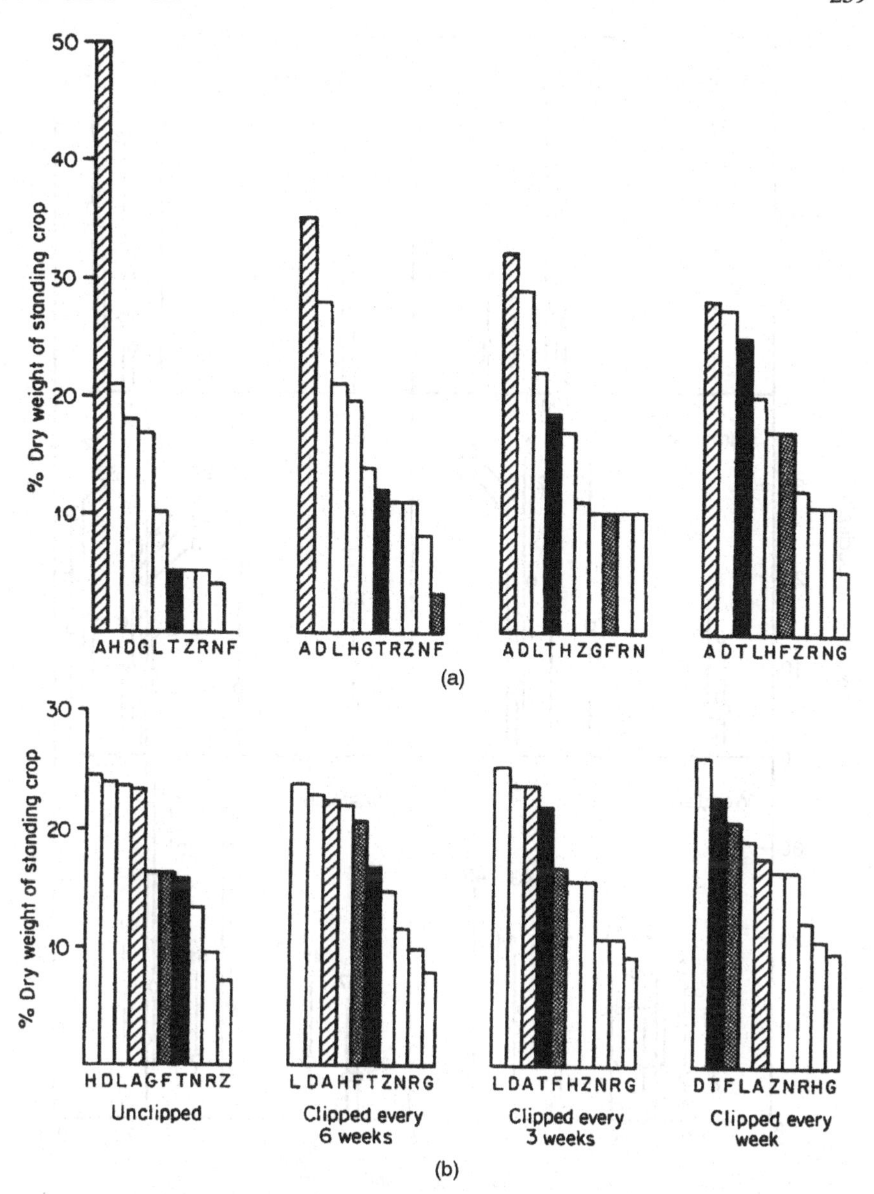

Figure 85. Histograms comparing the species composition of a turf synthesised from a standardised seed mixture and subjected to (a) high fertility (170 mg/1 N) and (b) low fertility (5 mg/1 N) and four intensities of defoliation. The experiment was conducted in 36-litre containers over a period of seven months and the results are expressed as percentages of standing crop. The species are: A, *Arrhenatherum elatius*, H, *Holcus lanatus*, D, *Dactylis glomerata*; G, *Elytrigia repens*; L, *Lolium perenne*; T, *Agrostis capillaris*; Z, *Bromopsis erecta*; R, *Festuca rubra*; N, *Anthoxanthum odoratum*; F, *Festuca ovina.* (Reproduced by permission of Mahmoud 1973.)

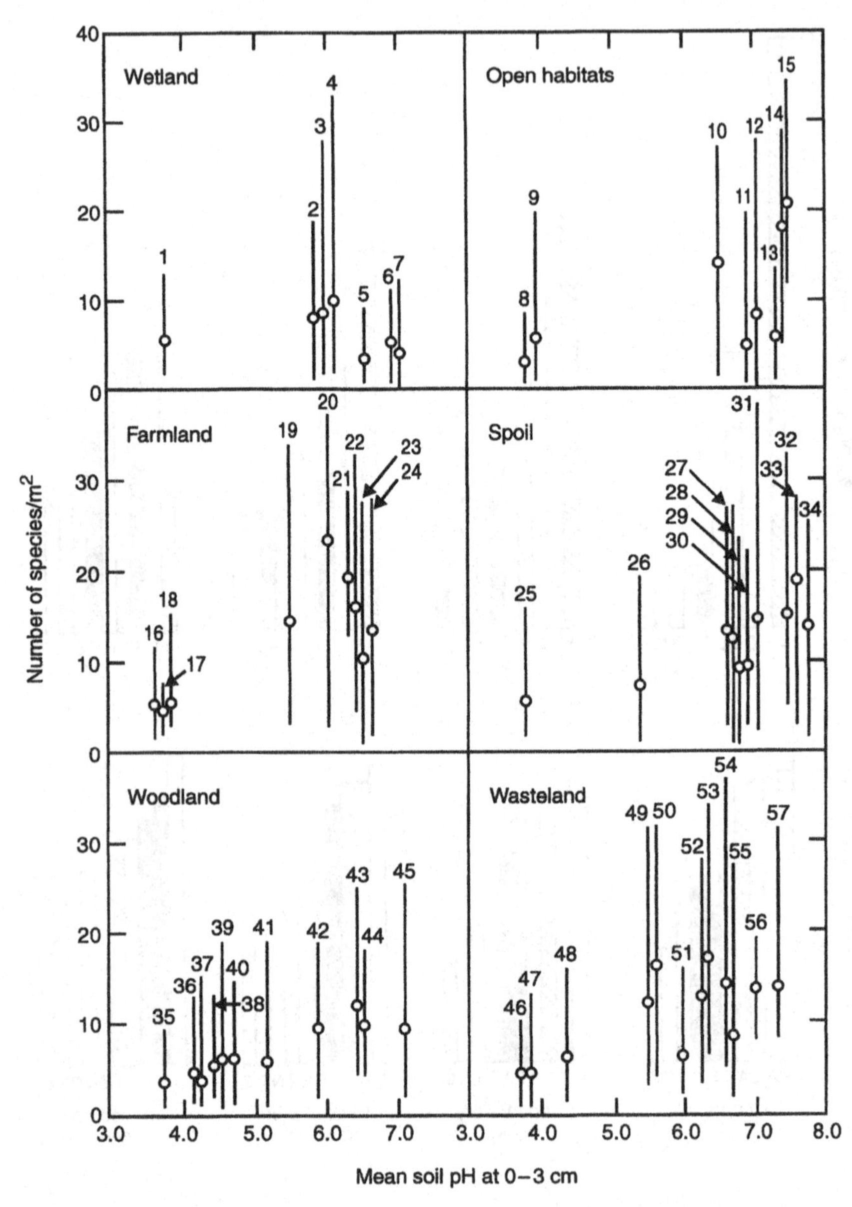

Figure 86. Mean and range in species richness in the one square metre samples of vegetation examined in the major habitats present in an area of 2748 km^2 around Sheffield, England. Bare ground was not sampled; hence the minimum value possible is one. Key to habitats: 1, unshaded mire on noncalcareous strata, soil pH <4.1; 2, shaded mire; 3, unshaded mire on calcareous strata; 4, unshaded mire on noncalcareous strata, soil pH >4.0; 5, lakes, ponds, canals, ditches, depth of water < 110 mm; 6, lakes, ponds, canals, ditches, depth of water > 100 mm; 7, rivers and streams; 8, noncalcareous cliffs; 9, rock outcrops on noncalcareous strata; 10, consolidated limestone scree; 11, walls;

It is clear, therefore, that if we are to describe the mechanism controlling species richness in herbaceous vegetation it will be necessary to recognise that the effects of stress and disturbance change according to their intensity. This non-linear relationship is the central feature of the simple model examined under the next heading.

A Model Describing the Control of Species Richness in Herbaceous Vegetation

In the 'humped-backed' model in Figure 88 it is suggested that the result of moderate intensities of either stress or disturbance (or both) is to increase species richness by reducing the vigour of potential dominants, thus allowing subsidiary species to co-exist with them. At the most extreme intensities of stress and/or disturbance, however, the model suggests that the number of co-existing species declines as conditions are created to which only a very small number of species are able to survive. This model is consistent with the observation of Odum (1963) that the greatest diversity occurs in the moderate or middle range of a physical gradient. Data conforming to the model have been obtained from transects along naturally-occurring gradients in stress and/or disturbance (Grime 1973a; Figure 89).

One of the most interesting and unusual examples of data conforming to this model is provided by the work of Woodwell and Rebuck (1967), who measured the changes in species richness in the ground flora along a gradient

---

12, limestone cliffs; 13, unconsolidated limestone scree; 14, rock outcrops on Magnesian Limestone; 15, rock outcrops on Carboniferous Limestone; 16, unenclosed pasture on Millstone Grit; 17, unenclosed pasture on limestone, soil pH <4.1; 18, unenclosed pasture on Coal Measures; 19, enclosed pasture; 20, unenclosed pasture on limestone, soil pH >4.0; 21, permanent meadows; 22, fallow arable; 23, arable, crop broad-leaved; 24, cereal arable; 25, coal mine heap, soil pH <4.1; 26, coal mine heap, soil pH >4.0; 27, lead mine waste, discontinuous vegetation cover; 28, cinders; 29, manure and sewage residues; 30, lead mine waste, continuous vegetation cover; 31, heaps of mineral soil (building sites, etc.); 32, spoil heaps in limestone quarries (Carboniferous Limestone); 33, spoil heaps in limestone quarries (Magnesian Limestone); 34, brick and mortar rubble; 35, plantations of broad-leaved trees, soil pH <4.1; 36, woodland on Bunter Sandstone; 37, coniferous plantations; 38, scrub on noncalcareous strata; 39, woodland on Millstone Grit; 40, woodland on Coal Measures; 41, plantation of broad-leaved trees, soil pH <4.1; 42, woodland on Magnesian Limestone; 43, scrub on limestone; 44, hedgerows; 45, woodland on Carboniferous Limestone; 46, derelict grassland and heath on Bunter Sandstone, soil pH <4.1; 47, derelict grassland and heath on Millstone Grit; 48, derelict grassland and heath on Coal Measures; 49, derelict grassland and heath on Bunter Sandstone, soil pH >4.0; 50, derelict grassland and heath on Carboniferous Limestone; 51, shaded paths; 52, river banks, etc.; 53, woodland on Magnesian Limestone; 54, unshaded paths with incomplete vegetation cover; 55, unshaded paths with complete vegetation cover; 56, road verges, mown frequently; 57, road verges, mown annually or unmown. (Reproduced from Grime 1973c by permission of *Journal of Environmental Management.*)

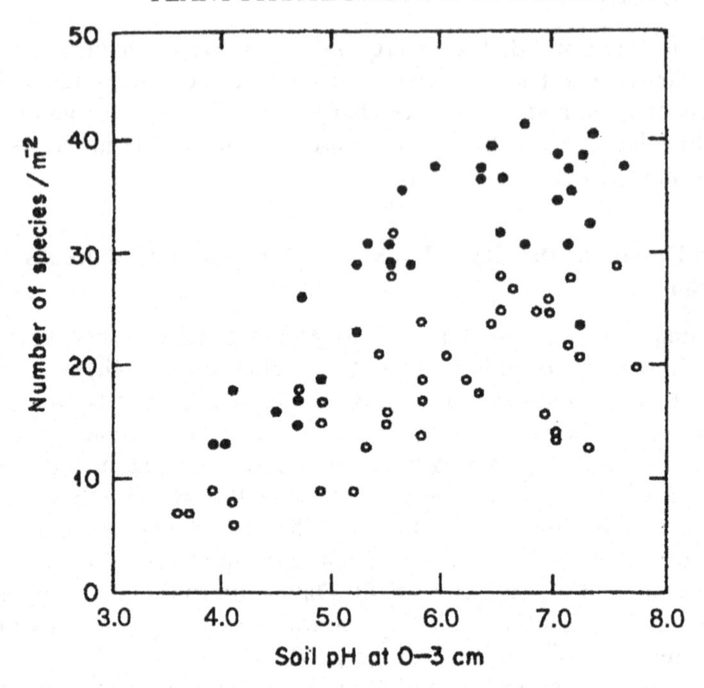

Figure 87. Comparison of species richness in unenclosed grassland in two Derbyshire dales in England differing in management (● Cressbrookdale; grazing by sheep and cattle. ○ Lathkilldale, ungrazed, burned sporadically). The samples refer to one square metre quadrats located at random. (Reproduced from Grime 1973c by permission of *Journal of Environmental Management.*)

of increasing exposure to gamma radiation. They conclude that 'the effects of irradiation do not appear to be simply a reduction in diversity but, at relatively low exposures where disturbance has been less, an increase in diversity as well'.

During the last 25 years a large number of syntheses (Huston 1979, 1994; Tilman 1988) and original investigations (e.g. Wheeler and Giller 1982; Bond 1983; Vermeer and Berendse 1983; During and Willems 1984; Moore and Keddy 1989; Wisheu and Keddy 1989; Wilson and Shay 1990; Shipley *et al.* 1991; Wheeler and Shaw 1991; Oomes 1992; Gough *et al.* 1994; Rapson *et al.* 1997; Janssens *et al.* 1998; Proulx and Mazumder 1998; Grace and Juntila 1999; Keddy and Fraser 1999; Gross *et al.* 2000; Gough *et al.* 2000) have yielded data consistent with the humped-back model, and a comprehensive review of research on the subject has been assembled by Grace (1999) recording some of the main contributions to the discussions that have been associated with testing and elaboration of the humped-back model.

It would appear therefore that the model provides a convenient summary of some of the phenomena which influence species richness, and this suggests that it may be used to predict or to explain the response of particular types of

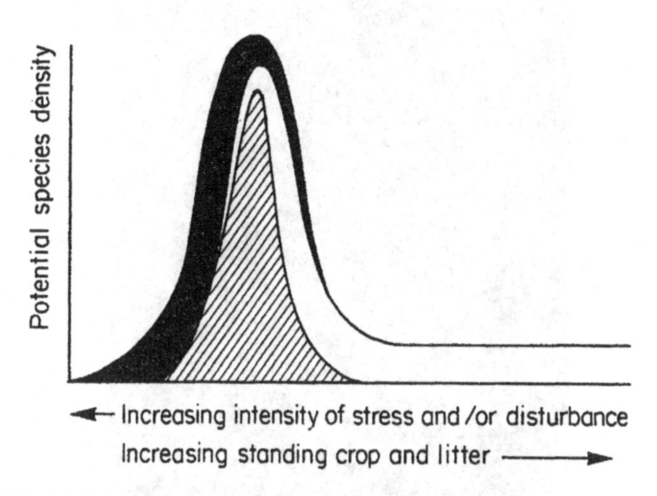

Figure 88. Model describing the impact of a gradient of increasing stress and/or disturbance upon the potential species richness in herbaceous vegetation. □ potential dominants; ■ species or ecotypes, highly adapted to the prevailing form(s) of stress or disturbance; shaded area species which are neither potential dominants, nor highly adapted to stress or disturbance. (Reproduced from Grime 1973a by permission of Macmillan (Journals) Ltd.)

herbaceous vegetation to changes in management (see page 297). However, before exploring this possibility it is necessary to consider the quantitative aspect of the mechanism described by the model and to review some additional factors which influence species richness.

Quantitative Definition of the Mechanism Controlling Species Richness in Herbaceous Vegetation

It has been argued (Al-Mufti *et al.* 1977) that certain of the phenomena which influence species richness in herbaceous vegetation (dominance, debilitation of dominants, incursion of subsidiary species, reduction of species richness by extremes of stress and/or disturbance) are characteristic of particular ranges in the total dry weight of shoot biomass and litter. This hypothesis was tested by examining the relationship which was obtained when species richness was plotted against the annual maximum in standing crop + litter (both measured at the time of the maximum in the former) in a range of contrasted vegetation types, including tall herb communities, woodland floor communities, and grasslands, of common occurrence in northern England. In order that the local reservoir of species (page 294) in the vicinity of the sites should not act as a major limiting factor upon species richness, each was selected from an area of landscape containing a diversity of long-established vegetation types.

Figure 89. An illustration of the humped-back model at the margin of a path at Littonslack, North Derbyshire, England. In the background of the photograph, low species richness coincides with dominance by coarse grasses and in the foreground all species are excluded by excessive trampling damage

The resulting graph (Figure 90) conforms to the model and contains a corridor of high species richness approximately in the range 350–750 g m^{-2}. From these results it would appear that above 750 g m^{-2} diversification tends to become restricted by dominance, whilst below 350 g m^{-2} species richness is limited by the small number of species capable of surviving the severity of environmental stress and/or disturbance experienced in such extreme habitats. Within the corridor, however, we may suppose that vegetation composition is determined by an equilibrium between the conflicting forces of natural selection associated with moderate intensities of competition, stress, and/or disturbance, with the result that habitats are created in which co-existence is possible between a variety of species and genotypes.

Before considering in more detail the mechanisms whereby species of different habitat requirements may co-exist within 'corridor environments' it is necessary to make certain qualifying statements concerning the relationship between species richness and the annual maximum in standing crop + litter. The precise form of the relationship in Figure 90 is clearly a function of

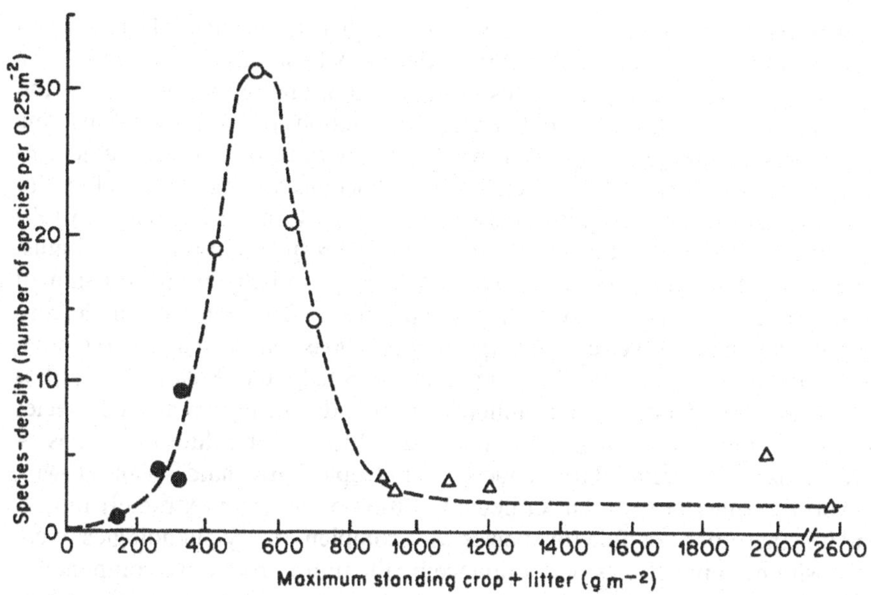

Figure 90. The relationship between maximum standing crop plus litter and species richness of herbs at 14 sites in northern England. ○ grasslands; ● woodlands; Δ tall herbs. (Reproduced from Al-Mufti *et al.* 1977 by permission of *Journal of Ecology*.)

sample size and will be affected also by factors peculiar to the sites and to the period of time at which this particular investigation was carried out. Moreover, for reasons which will be considered later (page 294), the maximum species richness attainable in corridor environments varies with latitude, regional location, and soil type.

Perhaps the most important qualification to be made with regard to the relationship in Figure 90 is that it is based upon measurements conducted on vegetation which has existed in its present state for some considerable time. In these circumstances we might expect, therefore, that the levels of species richness which have been attained correspond to an equilibrium between the habitat condition prevailing at each site and the reservoir of species in its vicinity. In herbaceous vegetation which is of recent origin or is subject to marked changes in management we would not expect such a stable equilibrium. In certain habitats (e.g. road verges, semi-derelict hill pastures, the margins of hay-meadows) the maximum in standing crop + litter often varies to such an extent that conditions favourable to high species richness in one year may be followed in the next by the development of a large and relatively undisturbed biomass. This effect is especially pronounced in the herb layers of woodlands where the density and persistence of deciduous tree litter at specific sites may vary considerably from year to year. Where vegetation is subject to such vicissitudes it is not to be expected that the maximum in biomass and litter measured in any one year will be predictably related to species richness. An illustration of the extent to which, in such circumstances, definition of the pattern may be lost is provided in Figure 91 which shows the relationship between species richness and estimations of the seasonal maxima in biomass + litter in samples drawn from a variety of habitats subject to erratic management. When species-rich herbaceous vegetation is subject to abandonment diversity usually declines quite rapidly and as explained by Olff and Ritchie (1998) and Hald and Vinther (2000) restoration procedures must take account of both the availability of species and the reintroduction of appropriate management. A further qualification which must be made concerning the quantitative relationships in Figure 90 is that they apply to temperate conditions. From the investigations of Singh and Misra (1969) it is apparent that in tropical environments the corridor of high potential species richness in herbaceous vegetation may extend to higher values of biomass + litter. This may be related to the fact that, in tropical grasslands, plant growth, flowering, senescence, and litter decomposition often occur extremely rapidly and it is therefore possible for species with complementary phenologies to coexist with minimal effects of dominance by the more productive components.

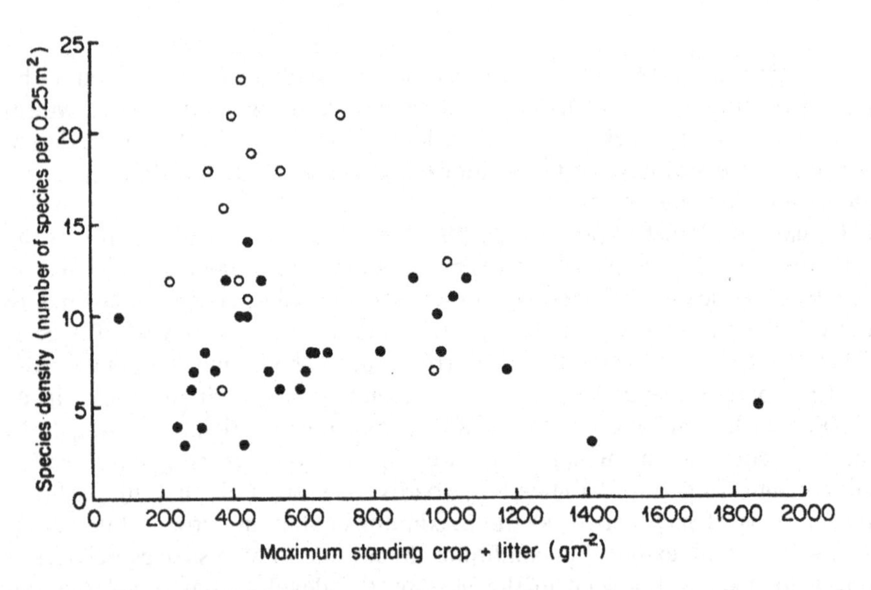

Figure 91. The relationship between estimates of the maximum standing crop plus litter and species richness of herbs in environments experiencing fluctuating patterns of vegetation management in northern England. ● road verges, subject to occasional mowing; ○ semi-derelict limestone pastures. (Reproduced by permission of Grime, Sydes, and Rodman)

Co-existence Related to Spatial Variation in Environment

Horizontal Variation

Even in apparently uniform terrain, most vegetation samples include a complex mosaic of micro-habitats. These arise from such factors as edaphic variation, interactions between micro-topography and climate, selective predation, local disturbance of the soil, and redistribution of nutrients by animals. To these must be added spatial differences in environment arising from the activity of the plants themselves. These include variation in nutrient availability (Snaydon 1962), water supply (Silvertown *et al.* 1999), degree of shading, accumulation of litter and organic toxins, and modification of the soil microflora. It is the task of the plant ecologist to identify the circumstances in which this small-scale spatial variation is sufficient to initiate or to maintain conditions favourable to the co-existence of different species.

Many of the events which play a critical role in maintaining high species richness in herbaceous vegetation occur in the regenerative phase and are related to complementary mechanisms of establishment in vegetation gaps. The importance of gaps for the survival of seedlings and some vegetative offspring has been emphasised in Chapter 3 and from various sources (e.g. Chippindale and Milton 1934; Champness and Morris 1948; Sarukhan 1974; Thompson and Grime 1979) there is conclusive evidence that co-existence frequently occurs between species with very different regenerative strategies. As suggested by Grubb (1977), it is not unreasonable to suspect that differences between neighbouring gaps in features such as size, soil texture, mineral nutrient status, desiccation, and degree of shading will be sufficient to encourage establishment by different species.

Although co-existence is frequently initiated by events occurring during the regenerative phase there are many instances where the pattern is reinforced by effects of spatial heterogeneity upon the vigour and survival of established plants. In the British Isles, some of the most conspicuous examples of this phenomenon occur in the herb layers of deciduous woodlands where differences in topography and depth of tree litter often give rise to micro-environments exploited by species of contrasted size and growth form (Scurfield 1953; Sydes and Grime 1981a). In North America also, circumstances have been described where co-existence between a variety of herbaceous plants may be related to the development of micro-habitats on the woodland floor. In one investigation described by Bratton (1976), for example, the range of micro-habitats and associated species included fallen logs (*Laportea canadensis*, *Sedum ternatum*, *Stellaria pubera*), the bases of large trees (*Osmorhiza longistylus*), the bases of small trees (*Dicentra cucullaria*, *D. canadensis*), and pockets of tree litter (*Phacelia fimbriata*).

In Britain, another vegetation type in which co-existence is related to spatial heterogeneity in the environment arises where sheep pastures are situated on shallow soils over fissured limestone (Figure 92a). In this habitat it is not uncommon to find four micro-habitats within areas of less than one square metre, each with a characteristic assemblage of herbaceous species. The first micro-habitat is that restricted to areas of extremely shallow soil and colonised by winter annuals such as *Arenaria serpyllifolia* and *Saxifraga tridactylites.* The second consists of patches where the soil is slightly deeper and the effects of summer desiccation are less severe, and here tussock grasses such as *Festuca ovina* and *Koeleria macrantha* are located. Narrow deep fissures in the limestone are usually exploited by species such as *Sanguisorba minor* and *Lotus corniculatus* which develop long, persistent tap roots (Sydes and Grime 1984). The fourth micro-habitat corresponds to the larger pockets of soil in which more productive species such as *F. rubra* and *Dactylis glomerata* are able to survive. Reference to Chapter 2, where the strategies exhibited by the four groups of plants just described are considered in more detail, reveals that in this example we are not concerned merely with an assemblage of basically

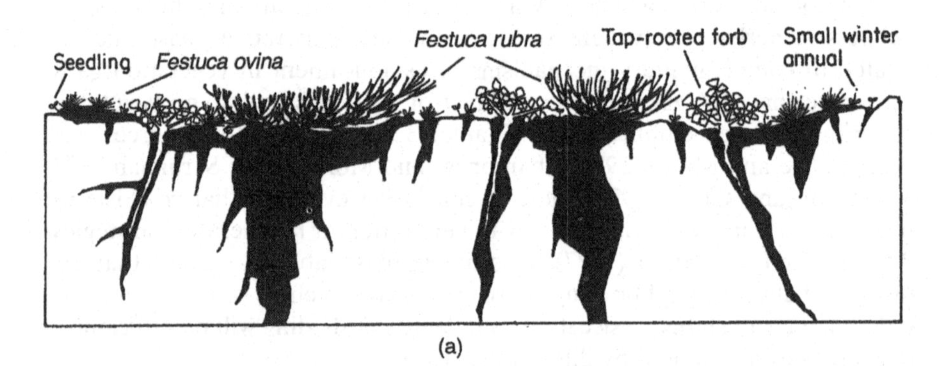

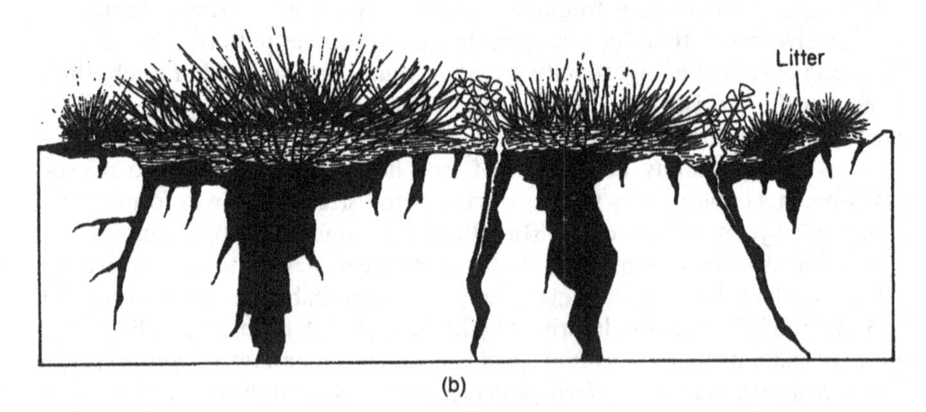

Figure 92. Diagrams illustrating the vegetation structure associated with limestone outcrops of varying soil depth in North Derbyshire, England: (a) an outcrop subject to low intensity of grazing by sheep or cattle; (b) the same outcrop after several years of dereliction

similar grassland plants. On the contrary, we find that within the eight plants which have been cited as examples there are profound differences in morphology, phenology, and life-history and there is a strategic range including stress-tolerant ruderals (the small winter annuals), stress-tolerant competitors (*F. rubra* and *D. glomerata*), and two different types of 'C-S-R' strategists (tussock grasses and tap-rooted plants).

It is unusual for high species richness to be maintained in herbaceous vegetation simply as a result of environmental heterogeneity. As we shall see under the next heading, spatial and temporal differences in environment frequently coincide. It is also of the utmost importance, particularly in designing management regimes for the purposes of nature conservation and amenity, to recognise that spatial variation in environment does not by itself provide a guarantee that high species richness will be maintained. This last point may be illustrated by reference to Figure 92b which depicts the kind of vegetation structure which has been observed to develop when limestone pastures over fissured limestone are allowed to become derelict. In the absence of grazing there is usually a marked increase in the vigour and lateral projection of the leaf canopy of the stress-tolerant competitors which are rooted in the larger pockets of soil. There is also a marked increase in accumulation of litter and, as the critical value of 750 g m^{-2} in maximum standing crop + litter (page 264) is approached, many of the areas of shallow soil exploited by the smaller plants become untenable. Reductions in species richness are also detectable where the change in biomass is less dramatic than that illustrated in Figure 92; in these circumstances we may suspect that the main impact of dereliction on species richness arises from the deposition of leaf litter over the areas of bare soil which, prior to the cessation of grazing, provided opportunities for seedling establishment.

Vertical Stratification

In productive, relatively undisturbed herbaceous vegetation, intense competition and the trend towards monoculture are associated with an extremely simple canopy structure in which the foliage of the dominant plants is concentrated in a dense, elevated layer (Figure 6) beneath which the frequency and vigour of smaller plants and seedlings are severely restricted by shading and deposition of litter (pages 180–187). However, where the productivity is lower and the canopy of the dominant plants is less complete it is not unusual for the shaded stratum to include additional layers of leaves mainly composed of shade-tolerant herbs and bryophytes.

In moderately productive herbaceous vegetation such as that in Figure 58, page 182 and Figure 93 the layers in the leaf canopy are reconstructed annually by extension of new herbaceous shoots from the ground surface and by expansion of relatively fast-growing bryophytes on to the most recent increment of plant litter (Rincon 1990). A more stable form of vertical stratification may

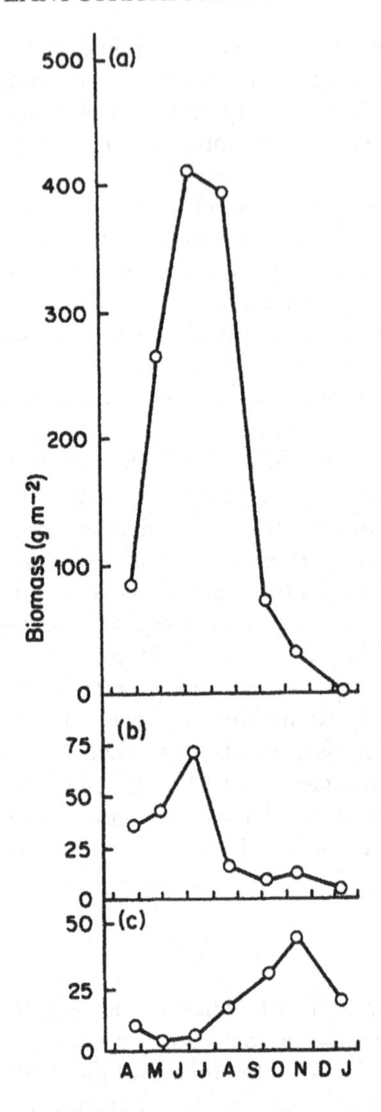

Figure 93. Seasonal changes in the shoot biomass of the main vegetation components in a stand of *Petasites hybridus* (Butterbur) situated on a lightly-shaded alluvial terrace in northern England: (a) *Petasites hybridus*, (b) *Poa trivialis*, (c) bryophytes. (Reproduced from Al-Mufti *et al.* 1977 by permission of Blackwell Scientific Publications Ltd.)

be observed in some of the less productive types of herbaceous vegetation; in species-rich limestone pastures it is usually apparent that, even within short turf, there is considerable layering in the predominantly evergreen canopy. In vegetation of this type there are often marked differences in the vertical distribution of the bryophytes, some of which (e.g. *Dicranum scoparium*) are in contact with the soil, whilst others (e.g. *Thuidium tamariscinum*) are attached to the standing litter of grasses and sub-shrubs or merely suspended in the herbaceous canopy (e.g. *Pseudoscleropodium purum*).

Co-existence Related to Temporal Variation in Environment

Seasonal Variation

Close scrutiny of small samples of vegetation often reveals situations in which species of contrasted ecology are growing in intimate association and in which it is difficult to explain their co-existence in terms of spatial heterogeneity in environment. In many of these cases, it is apparent that the species concerned have different seasonal patterns of shoot expansion and flowering and are adapted to different parts of the annual climatic cycle. Some of the best-documented examples of this phenomenon can be found in phenological studies conducted on warm temperate and semi-arid zone grasslands in which the growing season is characterised by a shift in relative abundance from grasses utilising the C_3 photosynthetic pathway to those reliant on the C_4 pathway. This shift usually coincides with the development of a leaf canopy capable of faster rates of photosynthesis at high temperatures (Cooper and Tainton 1968) and a lower rate of transpiration per gram of assimilated carbon (i.e. higher water-use efficiency). However, as explained in the comprehensive review by Risser (1985), there is enormous variety within both C_3 and C_4 species and the seasonal niches occupied by the two groups may be determined to some extent by differences in other traits which frequently coincide with the difference in photosynthetic mechanism. There may be differences in root penetration down the soil profile and there is also experimental evidence that some C_4 grasses are less palatable than C_3 species to insect herbivores (Caswell *et al.* 1973; Caswell and Reed 1976). A further set of differences may be related to the smaller genomes of most C_4 species (Levin and Funderburg 1979); as explained on pages 274–277, there is accumulating evidence that where C_3 and C_4 species co-exist in communities some aspects of niche differentiation may be attributable to the greater sensitivity of C_4 species to the limiting effect of low spring temperatures on the growth of species with small genomes.

Seasonal niches are not confined to communities in climates warm enough to support both C_3 and C_4 species. In the cool conditions of the British Isles a familiar example of the exploitation of temporal variation occurs in intensively-grazed lowland pastures where the sown grasses, *Lolium perenne* and *L. multiflorum*, exhibit peaks in biomass during the colder and moist conditions of spring and autumn, whilst White Clover (*Trifolium repens*), together with many of the naturally-invading indigenous species such as *Holcus lanatus* and *Elytrigia repens*, have complementary phenological patterns. Similar differences in phenology have been recorded in tall-herb communities where the simplest types of co-existence (Figure 93) are those in which the huge midsummer standing crop of dominants such as *Urtica dioica*, *Chamerion angustifolium*, or *Petasites hybridus* is associated with minor 'off-peak' contributions from bryophytes (Rincon and Grime 1989) or vernal herbs.

A more complex example is illustrated in Figure 94 which describes some of the phenological changes in a stand of Meadowsweet (*Filipendula ulmaria*). Here the well-defined summer peak in shoot biomass of the dominant herb (*F. ulmaria*) is combined with the broader distribution of *Mercurialis perennis*, a stress-tolerant competitor (see page 127) which is capable of surviving in the shaded stratum beneath *F. ulmaria.* The remaining contributors are the vernal herb, *Anemone nemorosa*, and the pleurocarpous moss, *Brachythecium rutabulum*, which grows on the herbaceous litter and shows a pronounced bimodal distribution with spring and autumnal maxima.

A rather different combination of phenologies is apparent in the study conducted by Tyler (1971) on a derelict sea-shore meadow where the co-dominants, *Juncus gerardii*, and *Agrostis stolonifera*, exhibit peaks of shoot biomass at quite different times in the spring and autumn, respectively. Similar differences in phenology have been reported in many other grassland types, and in the study conducted by Veresoglou and Fitter (1984) these were shown to coincide with differences in the timing of nutrient uptake from the soil.

In species-rich vegetation such as that which occurs in ancient calcareous grasslands of the British Isles a more varied array of shoot phenologies may be observed These range from the truncated vernal phenologies of geophytes, e.g. *Primula veris* and *Orchis mascula*, to certain of the evergreens such as *Carex flacca* and *Koeleria macrantha*, in which it is often extremely difficult to detect any seasonal changes in shoot biomass. In some calcareous pastures there is a tendency for certain grasses (e.g. *Anthoxanthum odoratum*, *Festuca ovina*) to peak in growth earlier than forbs, such as *Sanguisorba minor*, many of which possess long tap-roots and exploit reserves of moisture during periods in which the grasses are subjected to desiccation. A similar distinction in phenology between grasses and forbs has been observed in a wide range of investigations in temperate and semi-arid regions (Getz 1960; Golley 1960, 1965; Menhinick 1967; Barrett 1968; Precsenyi 1969; Shure 1971; Mellinger and McNaughton 1975).

A review of the environmental factors and the physiological mechanisms which control shoot phenology is beyond the scope of this book. It is clear, however, that differences in response to seasonal variation in daylength and temperature are particularly important as determinants of the timing of growth, flowering, and vegetative reproduction Where co-existence involves plants which grow in quite different seasons of the year, widely contrasted physiologies may be brought together in the same habitat. An example of this

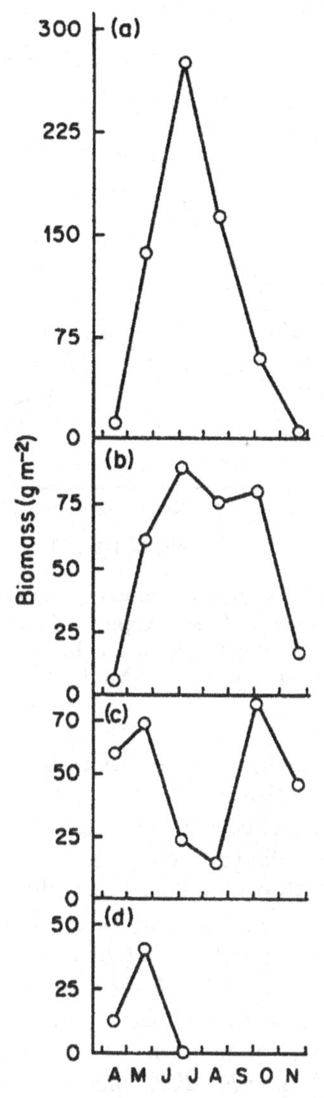

Figure 94. Seasonal changes in the shoot biomass of the main vegetation components in a stand of *Filipendula ulmaria* (Meadowsweet) situated on a damp, calcareous north-facing terrace in northern England: (a) *Filipendula ulmaria*, (b) *Mercurialis perennis*, (c) Bryophytes, (d) *Anemone nemorosa*. (Reproduced from Al-Mufti *et al.* 1977 by permission of Blackwell Scientific Publications Ltd.)

phenomenon is provided in Figure 95 from which it is apparent that the tall herb, *Urtica dioica*, and the moss, *Brachythecium rutabulum*, which are characteristically associated in the field, have markedly different temperature optima for growth (Furness and Grime 1982a).

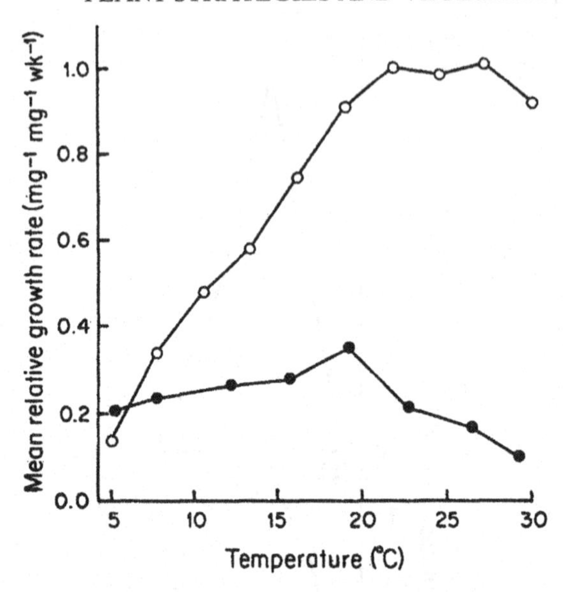

Figure 95. Comparison of the response to temperature in the relative growth-rates of the perennial herb, *Urtica dioica*, and the bryophyte, *Brachythecium rutabulum*. Measurements on both species were conducted at a light intensity of 25 W m^{-2}. (Reproduced by permission of Al-Mufti and Furness.)

In order to allow a more systematic approach to recognition of the circumstances where temporal niche differentiation allows species co-existence within plant communities it is necessary to find easily-measured traits that are reliable predictors of phenology. A candidate to fulfil this role is genome size—the total amount of DNA in the nucleus of the cell. Since the pioneering measurements and reviews of Van't Hof and Sparrow (1963), Britten and Davidson (1971), and Bennett and Smith (1976) it has been clear that in both animals and plants the quantity of DNA residing in the nucleus of eucaryotes varies considerably according to species. In the British flora, for example, estimates of DNA content range from 0.1 picograms (pg) per nucleus in *Cardamine amara* to 141.1 pg in *Fritillaria meleagris*. Even a cursory examination of the patterns of variation in plants and animals is sufficient to expose the fact that variation in nuclear DNA content is not related to organisational complexity nor to evolutionary antiquity. Failure to establish such correlations has come to be recognised as the 'C-value paradox' and has led to the discovery that much of the 'extra' DNA of organisms with large nuclear DNA contents consists of highly-repeated base sequences which are not transcribed into protein (Davidson and Britten 1973). This in turn has prompted various theories to explain the origin and function of the non-coding DNA. These may be classified very broadly into two groups. Some authors (Bennett 1971, 1972; Cavalier-Smith 1978; Olmo 1983, 1987; Bennett and Leitch 1997; Beaton and Cavalier-Smith 1999) have maintained that the total amount of DNA within the nucleus varies in association with other cell characteristics (e.g. cell size, length of the cell cycle) which exert important controls over development and function. Alternatively it has been suggested (Doolittle and Sapienza 1980; Orgel and Crick 1980; Hancock 1996) that the extra DNA is not directly a product of natural selection upon the phenotype but is merely a 'parasitic' constituent which has accumulated within the genomes of certain species.

An ecological dimension has become attached to the study of variation in DNA amount by the establishment of correlations between genome size and climate (Bennett 1976; Levin and Funderburg 1979; Bennett *et al.* 1982; Grime and Mowforth 1982; Grime 1983, 1989a; Wakamiya *et al.* 1993). From these investigations it has become evident that genome sizes tend to be low in tropical and consistently cold regions but vary widely in floras associated with strongly seasonal differences in temperature and rainfall. The largest genomes so far reported are in geophytes and grasses exploiting Mediterranean climates; most of these species grow in the cool moist conditions of winter and become inactive during the dry season.

Linkage between genome size and climate was further suggested by the establishment of a correlation between DNA amount and the timing of shoot growth in common British plants (Grime and Mowforth 1982). Measurements of the seasonal change in the shoot biomass of species established in natural herbaceous vegetation of the Sheffield region were used to relate the timing of shoot growth in the spring to the genome size of individual species (Figure 96). The results showed that from early spring to midsummer changes in the identity of the most actively growing species were associated with a progressive reduction in genome size. Grime and Mowforth (1982) suggested that the selection force determining this relationship between genome size and season of growth arose from a differential effect of low temperature on cell division and cell expansion. In cold conditions growth dependent upon cell division is inhibited by low temperatures (Kinsman *et al.* 1996) but shoot extension is still possible in species that grow by inflating large cells formed but not expanded in preceding warmer conditions. As temperatures rise during the spring the advantage of growth dominated by cell enlargement is likely to give way to that of growth involving high rates of cell division.

As the minimum duration of the mitotic and meiotic cycles is inescapably long in large cells (Van't Hof and Sparrow 1963; Commoner 1964; Bennett 1971) the advantage of growth by inflating large cells would not be expected to pertain at warm temperatures. Thus, small genomes are observed not only in species of continuously warm tropical or arid environments but also in plants in which growth is restricted to the summer period in temperate, continental, or polar regions.

In geophytes of very large genome size (such as *Fritillaria*, *Tulipa*, *Hyacinthoides*), growth in cool conditions occurs mainly by expansion of very large cells formed during warm relatively dry conditions of the preceding year

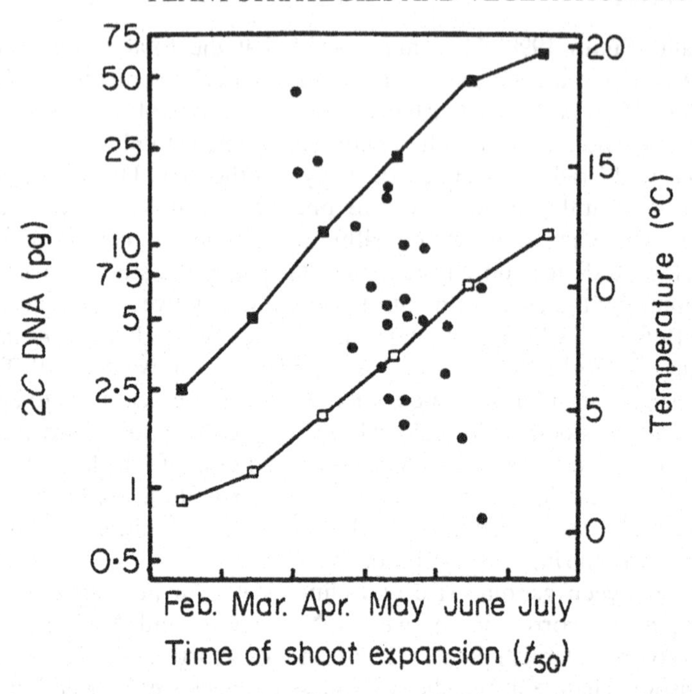

Figure 96. The relationship between DNA amount and the time of shoot expansion in 24 plant species commonly found in the Sheffield region. Temperature at Sheffield is expressed as the long-term averages for each month of daily minima (□) and maxima (■) in air temperature 1.5 m above the ground. (Redrawn from Grime and Mowforth 1982.)

(Hartsema 1961). A similar, if less extreme, temporal separation of cell division and expansion may occur in grasses such as *Briza media* and *Anthoxanathum odoratum*, which in Figure 96 follow the geophytes in the phenological sequence. Here spring growth involves the expansion of relatively large cells but it is less determinate and includes some cell division. In these species, spring development may be sustained by the capacity for growth by cell expansion during intermittent cold periods and in the cooler parts of the daily temperature cycle. An integral part of this mechanism may be the accumulation of unexpanded cells during periods in which cell expansion but not cell division is restricted by moisture stress.

In order to examine more closely the relationships between nuclear DNA content, growth, and spring climate Grime *et al.* (1985) conducted phenological measurements on a grassland community in North Derbyshire. Over the period March–July, measurements of leaf extension rate were made on each of the major constituent grasses, sedges, and forbs. The results established a consistent relationship between nuclear DNA content and phenology. In Figure 97, which refers to measurements during a period of cold weather in early

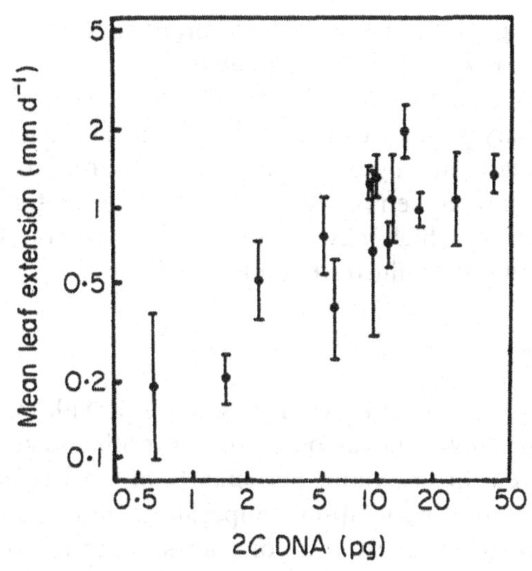

Figure 97. The relationship between DNA amount and the mean rate of leaf extension over the period 25 March to 5 April in 14 grassland species co-existing in the same turf. (Redrawn from Grime *et al.* 1985.)

spring, there is a clear association between genome size and leaf extension rate.

So far the relationship of seasonal variation in environment to the phenomenon of co-existence in herbaceous vegetation has been considered exclusively in relation to established plants. However, there are many circumstances in which the presence of a variety of different species within the same small area of vegetation can be explained satisfactorily only by reference to the regenerative strategies of the species concerned. In Chapter 3, it was concluded that most forms of vegetative regeneration and seedling establishment depend upon colonisation of vegetation gaps, and from the evidence which was reviewed it was apparent that, even between species which occupy similar habitats, there are major differences in the timing of gap exploitation.

In vegetation of high species richness such as that occurring in unfertilised calcareous grasslands of the British Isles, gaps arise throughout the year from a wide range of phenomena. These include solifluction and frost-heaving during the winter, and drought, rabbit-scraping, and the building of ant-hills during the summer. It is usually quite evident that the identities of the species which can most effectively colonise a gap change with the season. The colonists of bare ground created in limestone pastures during the summer include the small winter annuals (page 125) and many of the grasses such as *Festuca ovina*, *Helictotrichon pratense*, and *Koeleria macrantha* which produce large populations of seedlings in the early autumn. In the same habitat, colonisation of gaps arising during the winter is usually delayed until the appearance in the following spring of populations of quite different species, e.g. *Linum catharticum*, *Pimpinella saxifraga*, and *Viola riviniana*.

A wide range of mechanisms account for the interspecific differences in germination times observed in species-rich vegetation. These include differences in time of seed release, in length of after-ripening period, in chilling requirement, and in response to temperature. An example of the diversity of responses to temperature which may be observed in seed populations from one type of plant community is illustrated in Figure 98.

Short-term Variation

Although phenologies associated with season are of widespread importance, less conspicuous temporal niches operating on both shorter or longer time scales must be taken into account. Particularly in the British climate, daily, even hourly, fluctuations in radiation, temperature, and water potential of the atmosphere are likely to result in a constant shifting in the identity of the constituent species and genotypes for which conditions most nearly approximate to the optimum for photosynthesis and/or growth. Equally deserving of attention are the temporal niches which arise from short-term changes in vegetation structure resulting from grazing or mowing. Field observations and comparative studies (Milton 1940; McNaughton 1983) have shown, for example, that there are profound differences between species and varieties of grasses with respect to their response to defoliation; for example, there is evidence that the main reaction *of Lolium perenne* var. S23 to clipping consists of a rapid and almost vertical re-growth of the damaged leaves whereas, under the same treatment, *Agrostis capillaris* responds by producing a large number of small tillers and leaves, many of which are not projected into the clipped stratum but instead form a compact low sward which includes stolons capable of invading bare patches and infiltrating areas of short turf occupied by other species. From these results it is apparent that conditions allowing coexistence of species such as *L. perenne* var. S23 and *A. capillaris* are likely to arise where grazing is intermittent and fluctuations in the height of the turf alternately favour erect and prostrate growth-forms.

Long-term Variation

Floristic diversity in herbaceous vegetation may be dependent also upon year-to-year variation in climate and habitat conditions (Stampfli 1992). Cyclical fluctuations in the ratio of *Lolium perenne* to *Trifolium repens* associated with changes in nitrogen status are a familiar example known to grassland ecologists (e.g. Leith 1960) and similar effects correlated with fluctuations in rainfall have been reported (Watt 1960; Hopkins 1978; Buckland *et al.* 1997). An interesting study of the effect of year-to-year variation in habitat conditions upon the species composition of herbaceous vegetation is that of Muller and

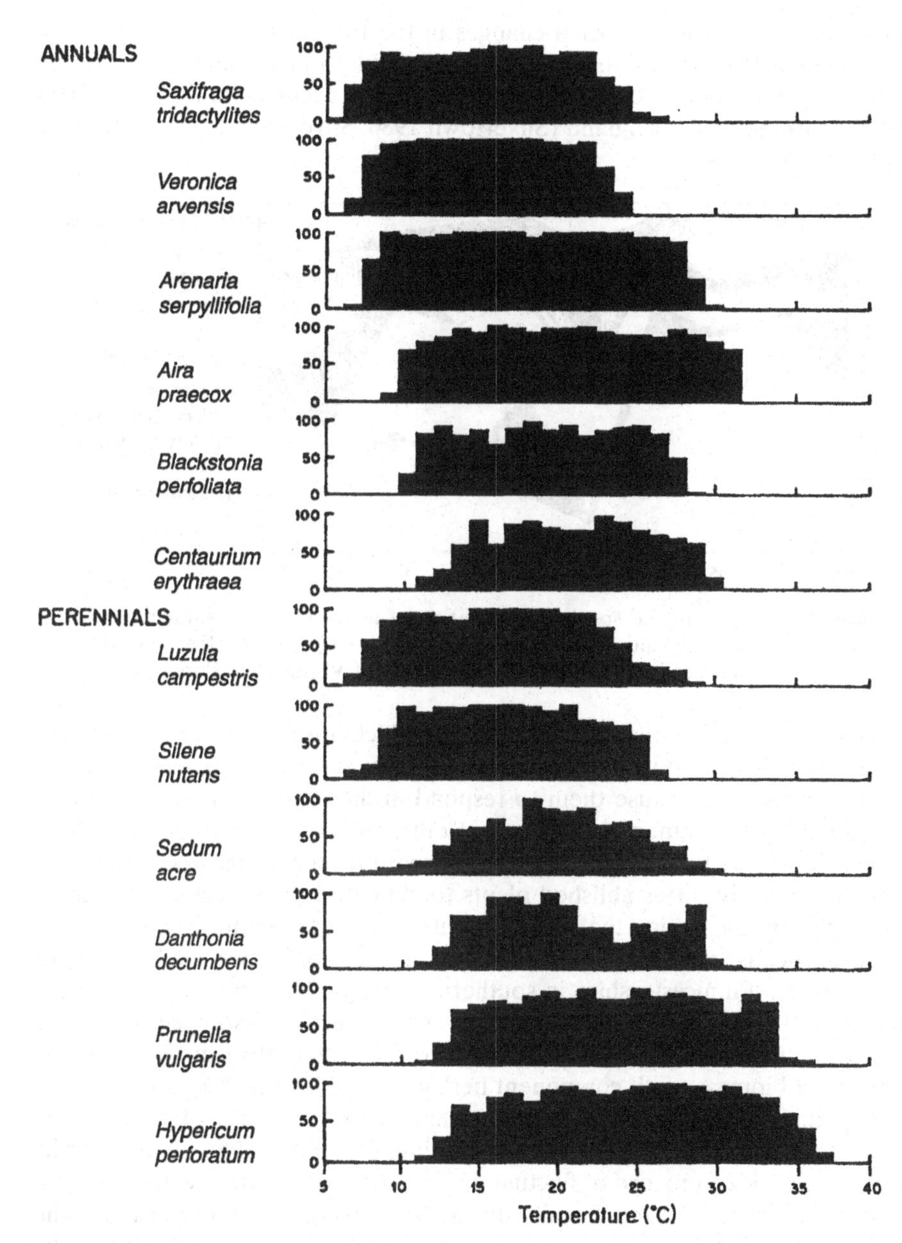

Figure 98. Comparison of the response to constant temperatures in the light in seeds collected from populations of herbaceous plants of common occurrence in calcareous grassland in northern England. The figures tabulated refer to the final percentage germination at 30 temperatures over the range 5–40°C (light intensity 38.0 W m^{-2}, daylength 15 h). (Reproduced from Thompson and Grime 1983 by permission of *Journal of Applied Ecology*.)

Foerster (1974), who recorded changes in the frequency of species in sown grasslands subject to varying intensities of flooding and silt deposition (Figure 99). Fluctuations in species composition have also occurred in the Park Grass plots at Rothamsted, England (Silvertown 1980; Silvertown *et al.* 1994; Dodd *et al.* 1995).

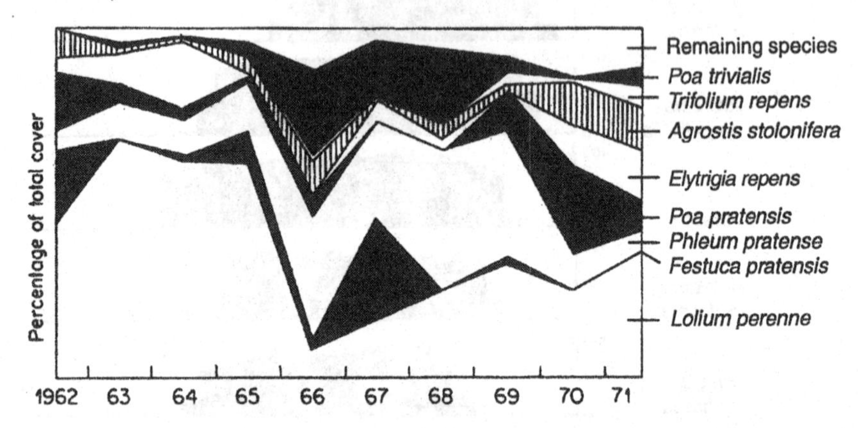

Figure 99. Changes in the species composition of a sown pasture subject to varying intensities of flooding and silt deposition in western Germany. (Reproduced from Muller and Foerster 1974 by permission of Verlag Paul Parey.)

In order to explain many of the long-term fluctuations in the composition of a plant community it is necessary to identify the differences between component species which cause them to respond in different ways to year-to-year variation in environmental conditions. Contrasts in response to the conditions prevailing in particular years may arise from differences between species in the susceptibility of established plants to climatic factors. An extraordinary opportunity to examine the role of fluctuating climatic conditions on species co-existence is available from the study conducted on mown roadside verges at Bibury in Gloucestershire in southern England. This unique investigation in which the monitored plots originated as the control (unsprayed) areas in a herbicide trial (Willis 1972, 1988) has involved quantitative measurement of the shoot biomass of all component herbaceous species by the same scientist (AJ Willis), who has collected data at the experimental site at the same time of year (the second week in July) continuously since 1958. The resulting dataset provides a record of fluctuations in productivity and species composition over a period of more than 40 years and has been used to examine the relationships between year-to-year variation in climate and the changes in the abundance of individual plant species. The results show that among the major co-existing perennial species at Bibury there are characteristisc responses to climate (Dunnett *et al.* 1998). These are revealed as different relationships to weather types (Lamb 1964) and as immediate or delayed responses (Willis *et al.* 1995) to fluctuations in the latitudinal position of the North Atlantic Gulf

Stream which exercises a moderating effect on the climate experienced in Western Europe.

The effects of the Gulf Stream on species composition at Bibury coincide with parallel variations in the zooplankton communities monitored in Lake Windermere, northern England and at marine stations in the North Sea (George and Taylor 1995). In view of the rapid turnover of plankton populations the sensitivity of the freshwater and marine records to changes in the Gulf Stream position and climate is not surprising but it is perhaps not obvious why the perennial grassland community at Bibury should also be so responsive. Carry-over effects 'buffering' from the preceding year are limited in part by the annual mowing conducted in the late summer but recent experimental evidence suggests that the main reason why variation in climate from year to year results in marked effects is the presence of an amplification mechanism in the vegetation itself. Following the perspective developed by Boysen-Jensen (1929) it can be envisaged that even quite a small difference in a climatic factor could result in a marked change in the relative proportions of component species. In the finely balanced circumstances where fast-growing perennials are in competition it is not difficult to envisage conditions in which a controlling effect of climate early in the growing season could generate initial differences between plants in the effectiveness of shoot and root foraging which would then become subject to amplification as the main period of growth ensued.

In an attempt to test this hypothesis Dunnett and Grime (1999) examined the responsiveness of five species transplanted from Bibury to a controlled manipulation of temperature applied early in the growing season. The results support the theory that interspecific competition acts as an amplifier in that although all five species exhibited a positive response to warming when grown in isolation a strongly divergent response occurred when the species were combined in additive mixtures.

Disparities in response to-year-to year variation in climate and management may also arise from differences between component species in regenerative characteristics (Grime 1981; Miller 1982; Thompson *et al.* 1996). Some insight into the diversity of regenerative capacities and requirements which may reside within one community can be gained from Figures 100 and 101 which refer to two contrasted types of vegetation in Northern England. In both, it is apparent that within a small area of superficially homogeneous vegetation the component species exhibit a wide range of regenerative strategies, and it may be safely predicted that the identities of the species regenerating most successfully will change from year to year in accordance with fluctuations in climate and in factors such as the timing, distribution, form, and severity of vegetation disturbance. From long-term records (e.g. Fitter *et al.* 1995) there is also abundant evidence that interannual variation in climate is responsible for fluctuations in the reproductive output of plant species. Another source of variety in regenerative response which is conspicuous in

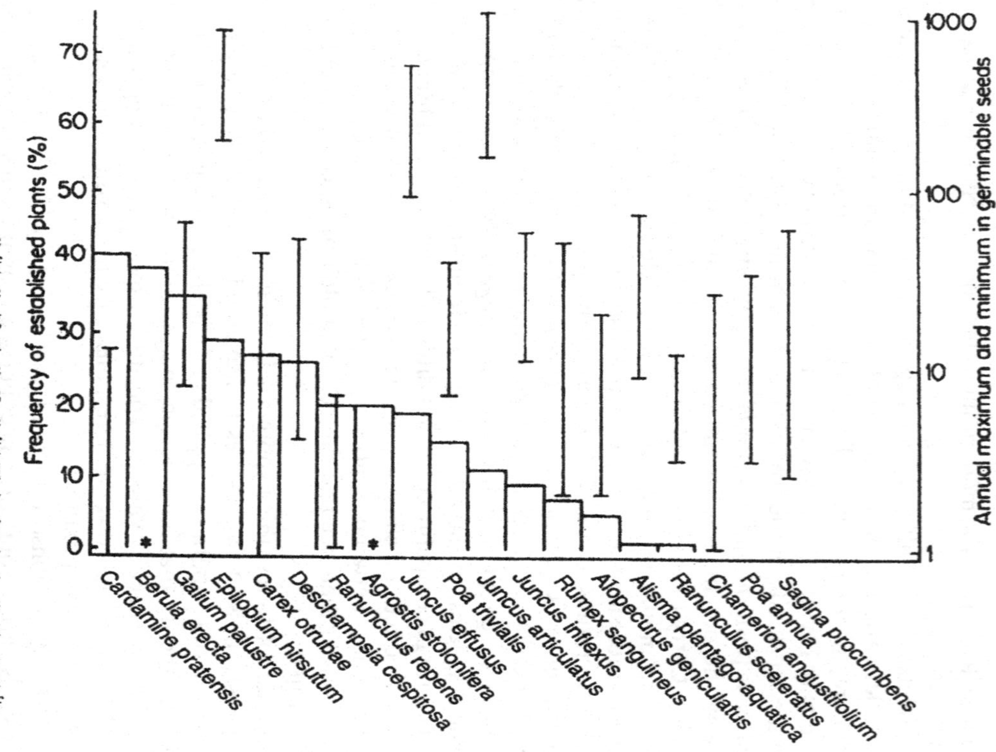

Figure 100. Estimates of the amplitude of seasonal variation in density of germinable seeds recovered from the top 30 mm of the soil profile (including litter) in marshland at the edge of a small lake in northern England. The bars at the upper and lower extremities of the vertical lines indicate the maximum and minimum numbers of seeds detected by a standardised procedure of sampling and laboratory germination applied at regular intervals throughout the year. The histograms describe frequency of occurrence of established plants of the species present at the sites. An asterisk means that no seeds were detected. (Reproduced from Thompson and Grime 1979 by permission of *Journal of Ecology*.)

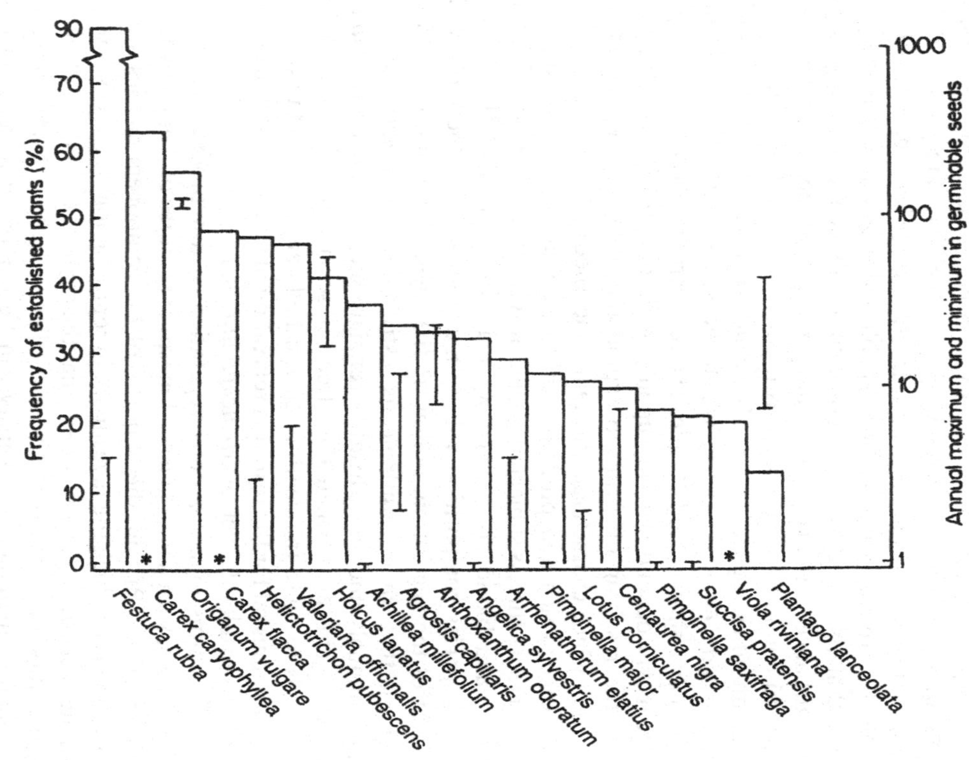

Figure 101. Estimates of the amplitude of seasonal variation in density of germinable seeds recovered from the top 30 mm of the soil profile (including litter) in ungrazed limestone grassland on a north-facing slope in northern England. The bars at the upper and lower extremities of the vertical lines indicate the maximum and minimum numbers of seeds detected by a standardised procedure of sampling and laboratory germination applied at regular intervals throughout the year. The histograms describe frequency of occurrence of established plants of the species present at the sites. An asterisk means that no seeds were detected. (Reproduced from Thompson and Grime 1979 by permission of *Journal of Ecology*.)

the two figures is related to differences in capacity to develop persistent seed banks. In both of the plant communities for which data are presented, large and persistent seed banks are apparent in certain species, a number of which are relatively rare components of the established vegetation. It is reasonable to expect that these plants will become more abundant following years in which disturbance of the vegetation and the soil is exceptionally widespread or severe. In contrast to the plants with persistent seed banks, there are, in both communities, species, e.g. *Arrhenatherum elatius*, *Chamerion angustifolium* and *Festuca rubra*, in which the seeds remain dormant and/or viable for only a short period of time. The regenerative success of these plants in any year will depend crucially upon (1) the output of seeds and (2) the ability of the resulting offspring to exploit the particular opportunities for regeneration which characterise the year concerned.

Where species-rich and predominantly perennial herbaceous vegetation, e.g. limestone pasture, is subjected to year-to-year fluctuations in the intensity of climatic and biotic disturbance we may anticipate that the associated changes in species composition will be most apparent in the more ruderal component of the vegetation. Following the suggestion of Grubb (1976) we may suppose that when disturbance is restricted to the appearance of small gaps tiny annuals and short-lived perennials such *Linum catharticum* will predominate. However, when the effect of severe drought, fire, or biotic disturbance is to cause the death of established perennials and to open large gaps it is not unusual for the resulting relaxation of stress to result in eruptions of competitive ruderals such as *Medicago lupulina*.

In certain forms of herbaceous vegetation of low productivity and high species richness, such as the ancient limestone pastures of damp north-facing slopes in northern Britain, the majority of the component species are long-lived perennials in which the investment in reproduction is small and intermittent. In these circumstances it is clear that the chances of consistently successful regeneration on the part of any particular species are extremely low and the mechanism maintaining diversity can be compared to an interminable game of roulette in which a large number of players place occasional and exceedingly prudent bets (Grime 1979; Chesson 1983; Shmida and Ellner 1984; Petraitis *et al.* 1989; Zobel 1992; van der Maarel and Sykes 1993; Rusch and Fernández-Palacios 1995).

Co-existence and Mycorrhizas

Although the humped-back model originated from field studies in which variation in species richness was related to the above-ground biomass and litter an interesting alternative is to plot species richness against soil properties. A very revealing example using this approach is described by Janssens *et al.* (1998) in which ecologists in five European countries collaborated to examine the relationship between species richness and soil nutrient characteristics in 281 ancient meadows. The most striking result to emerge from this survey was the consistent correlation between high species richness and low extractable phosphorus (Figure 102). None of the meadows yielding more than 5 mg of phosphorus per 100 g of dry soil in an acetate + EDTA extraction contained more than 20 species per 100 m^{-2}.

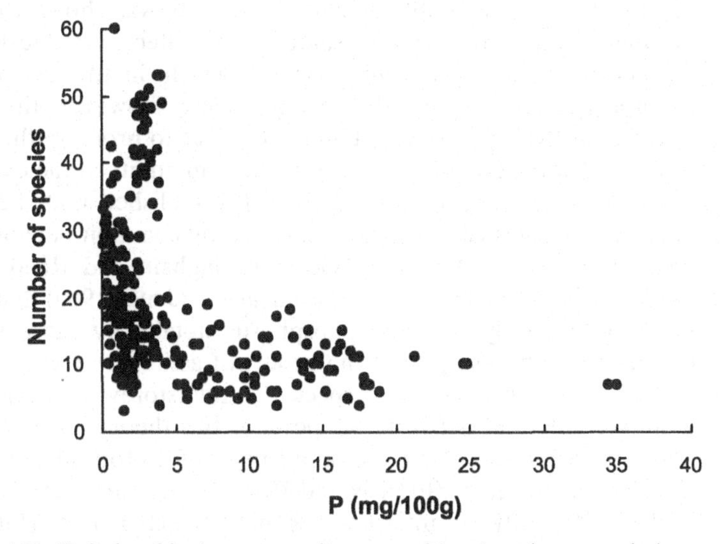

Figure 102. Relationship between soil extractable phosphorus and the number of species per 100 m^2 in meadows sampled at a large number of sites in western Europe. (Modified from Janssens *et al.* 1999 by permission of *Plant and Soil.*)

This correlation between high species richness and low phosphorus status is consistent with the general conclusion reached earlier in this chapter that the dense packing of many different species into a small area of herbaceous vegetation occurs only in conditions of low biomass imposed by soil factors or management. However, the association of species richness with low phosphorus in Janssens *et al.* (1998) points to more specific mechanisms peculiar to the highest levels of diversity. Dense packing of species coincides with the compact morphologies of plants which have been attuned by natural selection over many generations to conditions of chronic mineral nutrient stress. The anatomy and small size of these plants is conducive to efficient internal recycling of captured mineral nutrients and is accentuated by close grazing which may result in exceedingly small phenotypes. However, the influence of mineral nutrient stress upon diversity is not confined to miniaturisation and dense species-packing; low productivity also dictates slow vegetation dynamics which in turn reduce the intensity of competitive interactions, delay the rate at which bare soil in vegetation gaps is recolonised, and thus extend the opportunity for seedling establishment.

The mechanistic basis for the correlation between low phosphorus and high diversity may also involve vesicular-arbuscular (V-A) mycorrhizas which achieve high levels of infection in the roots of many of the vascular plants associated with ancient species-rich grasslands. In a laboratory experiment involving the synthesis of plant communities in microcosms containing phosphorus-deficient soil infected and uninfected by V-A mycorrhizal fungi (Grime *et al.* 1987b), a considerable benefit to diversity was shown to arise from the presence of the mycorrhizas. Some of this effect was due to the survival of species such as *Centaurium erythaea* which, in the absence of infection did not graduate beyond the seedling stage. However, the most pronounced effect of the mycorrhizas (Figure 103) was to promote the yield of the subordinate dicotyledonous members of the community at the expense of the canopy dominant *Festuca ovina*. It is well established that the root systems of different species can become connected by common networks of V-A mycorrhizal fungi (Read *et al.* 1985b; Whittingham and Read 1982; Chiariello *et al.* 1982) and the hypothesis was advanced (Janos 1980) that such connections might not only facilitate phosphorus uptake by those plants joined to the network but also permit the export of assimilate from 'source species' (canopy dominants) to 'sink species' (understorey subordinates) through a common mycelial network. Support for this theory has been derived from experiments recording the movement of isotopically-labelled photosynthate between species (Francis and Read 1984; Francis *et al.* 1986; Grime *et al.* 1987). Recently this interpretation has been challenged (Fitter *et al.* 1998; Robinson and Fitter 1999) and argument continues as to whether the transfers of carbon within the network are restricted to the mycorrhizal component or can move directly from fungal to host cells. However, there can be little doubt concerning the benefit to diversity arising from the V-A infection. It is clearly demonstrated in Figure 103 that the increased yields in understorey species associated with the presence of the fungi were very large but did not occur in *Rumex acetosa*, a species that does not develop associations with V-A mycorrhizas. Further research is required to establish whether the benefits of infection are exclusively concerned with the capture and redistribution of phosphorus or involve transfer of other resources within the network. Additional benefits of V-A mycorrhizas to species richness in plant communities may arise from genetic variation within the mycorrhizal fungi. In a recent experiment (van der Heijden *et al.* 1998) evidence has been obtained of specific stimulation of the yields of particular plant species by individual fungal isolates.

Co-existence Related to Genetic Variation within Populations

Species-rich perennial plant communities are composed mainly of strongly-outbreeding species and it is suspected that genetic variation can influence the persistence and relative abundance of individual populations (Miller and

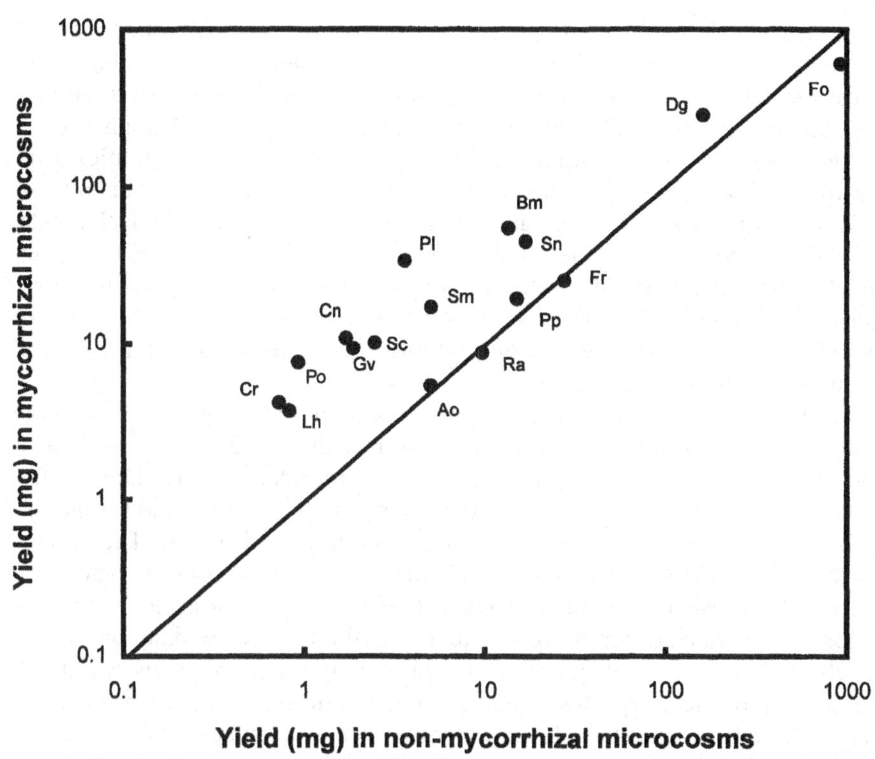

Figure 103. Effects of the presence of mycorrhizal infection on the average shoot weight of individuals of various grassland species grown together for one year in laboratory microcosms. The species are: Ao, *Anthoxanthum odoratum*; Bm, *Briza media*; Cr, *Campanula rotundifolia*; Cn, *Centaurea nigra*; Dg *Dactylis glomerata*; Fo, *Festuca ovina*; Fr, *Festuca rubra*; Gv, *Galium verum*; Lh, *Leontodon hispidus*; Pl, *Plantago lanceolata*; Po, *Pilosella officinarum*; Pp, *Poa pratensis*; Ra, *Rumex acetosa*; Sm, *Sanguisorba minor*; Sc, *Scabiosa columbaria*; Sn, *Silene nutans*. (Reproduced from Grime *et al.* 1987 by permission of *Nature*.)

Fowler 1994; Prentice *et al.* 1995) either by moderating the outcome of interspecific competition (Aarssen 1989) or by reducing vulnerability to pathogen attack (Thompson and Burdon 1992; Burdon 1993; Burdon and Thrall 1999). There are, however, formidable technical and logistical problems in seeking to scale up from the genetics of component populations to the functioning of communities and ecosystems (Antonovics 1976). In consequence we do not yet know whether losses of genetic diversity are a harbinger and cause of losses in species diversity. We also do not know to what extent the frequent failures in attempts to reconstitute species-rich communities in conservation and amenity projects arise from genetic impoverishment of the plant populations employed.

In order to examine experimentally the contribution of plant genetic diversity to the maintenance of species diversity and the functioning of an ecosystem several conditions must be satisfied. It is necessary to identify a vegetation type in which high genetic diversity coincides with high species richness and can be manipulated. To synthesise plant communities with controlled levels of genetic variation it is necessary to propagate individuals vegetatively to produce large numbers of genetically-identical plants. It is also important that the plant populations used should consist of long-lived individuals with low rates of replacement by sexually-derived progeny; fast-growing short-lived species are easier to propagate but do not permit control of the genetic composition of populations over long periods of time.

In 1997 following three years of systematic clonal propagation of large numbers of randomly-selected cuttings from within a 10 m × 10 m area of ancient species-rich limestone pasture at Cressbrookdale in north Derbyshire, UK, a long-term microcosm experiment was initiated (Booth and Grime unpub.) to examine the consequences of genetic impoverishment. The experiment utilises a stock of material derived from cuttings from 16 randomly-selected established individuals from each of 11 species (4 grasses, 3 sedges, 4 forbs), all of which occur at high frequency in the Cressbrookdale turf. Three levels of genetic diversity have been imposed. In one treatment, every individual in each of the 11 species is unique. In a second treatment each species is represented by four each of four randomly-selected biotypes. The third treatment contains no genetic diversity in any of the 11 species; here each replicate consists of a unique combination of biotypes. In this experimental design therefore the vegetation in all microcosms was closely similar in appearance and contained communities with the same initially high level of species richness but with three contrasted levels of genetic diversity. Frequent recording has been conducted to measure the rate of decline in species diversity. The objective is to test the prediction that as a consequence of greater vulnerability to pathogens and lower variation in the outcome of competitive interactions the rate of loss of species evenness and diversity will be greatest in the genetically-impoverished communities. It is interesting to note, however, that it will not be possible to predict the identity of expanding and declining species in particular replicates; this component of the results is predicted to vary idiosyncratically according to the various combinations of biotypes in particular replicates.

By the end of the second growing season, effects of genetic impoverishment began to appear in the experiment. In the communities lacking genetic diversity there has been a significant fall in species diversity and populations of five of the component species have begun to show higher variation in abundance between replicates in comparison with the genetically diverse populations. Repercussions at the community and ecosystem levels are also apparent; the communities without genetic diversity have become more susceptible to disease and more variable in canopy structure (Figure 104). Longer-term recording will be necessary to reveal whether these changes are a precursor to the collapse of the genetically uniform communities.

Co-existence in Woody Vegetation

Although there have been relatively few studies of co-existence between woody species there is reason to suspect that the mechanisms which control the density of trees and shrubs occurring within areas of one hectare or less resemble in many respects those which operate on a smaller scale in herbaceous vegetation.

In temperate regions the highest densities of woody species occur in areas of open woodland and scrub on calcareous soils. In vegetation of this type in southern Britain it is not unusual to find a mixture of shrubs including *Acer campestre*, *Corylus avellana*, *Euonymus europaeus*, *Ligustrum vulgare*, *Prunus spinosa*, *Cornus sanguinea*, and *Viburnum opulus*. On unproductive terrain such as steep, south-facing slopes with a discontinuous soil cover, diverse assemblages of these shrubs are often a persistent feature of the landscape. In these circumstances diversity is maintained, in part, by the low productivity of the habitat which restricts shrubs and trees to a size below that required to exercise major effects of dominance. In addition, it seems likely that spatial heterogeneity arising from variation in factors such as the depth, stability, and mineral nutrient status of the soil encourages the regeneration and persistence of a diversity of species.

In Britain and throughout the temperate zone a marked contrast with the species-rich calcareous scrub is provided by the various types of woodland which develop in areas where the soils are sufficiently stable and fertile to support the growth of large trees such as *Fagus sylvatica*, *Quercus petraea*, *Q. robur*, and *Acer pseudoplatanus*. Some of these species are capable of dominating extensive areas of woodland to the virtual exclusion of other trees and shrubs.

As in the case of herbaceous vegetation, therefore, we may conclude that high species density among trees and shrubs is promoted by moderately severe intensities of environmental stress. Evidence of this effect is not restricted to studies of temperate woodland: ecologists investigating the dynamics of tropical forests have concluded that here also there is an inverse relationship between productivity and species density. A particularly convincing example of this phenomenon is apparent in the work of Holdridge *et al.* (1971) from which it is clear that high densities of species of established trees are correlated with low mineral nutrient (especially phosphorus) status in 40 samples of aboriginal tropical forest examined in Costa Rica.

The parallels between the mechanisms controlling species richness in herbaceous and woody vegetation can be extended further by recognising that high densities of trees and shrubs are not confined to unproductive habitats.

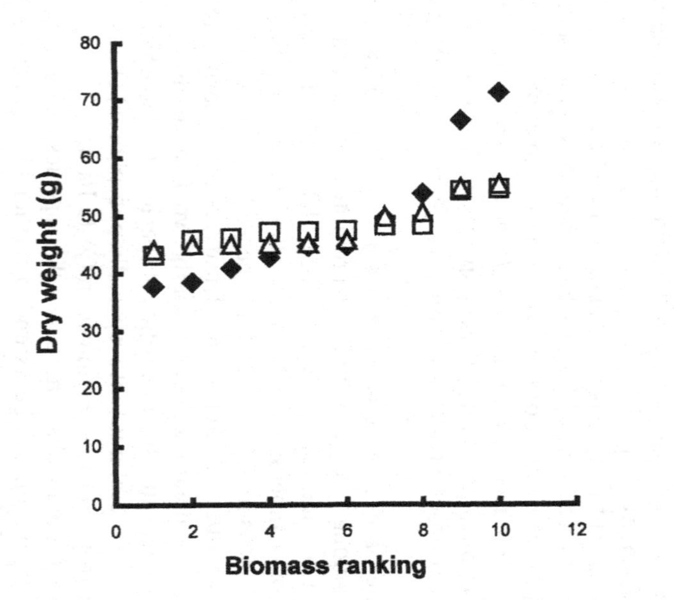

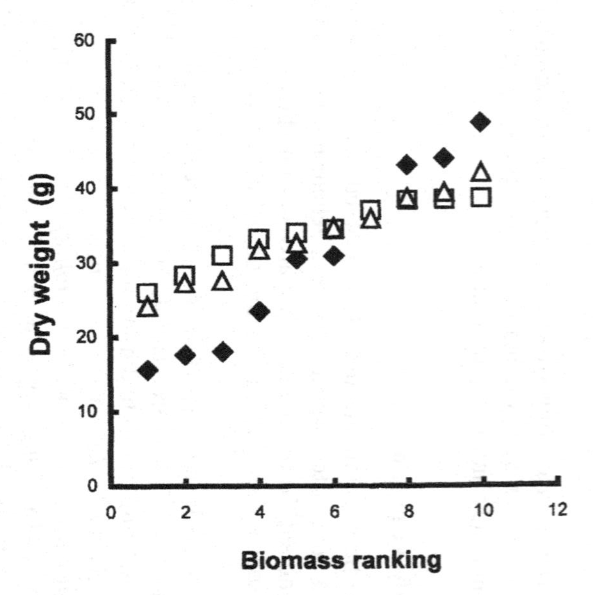

Figure 104. Harvested shoot biomass removed by cutting 25 mm above the soil surface in spring (left) and autumn (right) from microcosms containing calcarous grassland with controlled levels of genetic diversity in all 11 component species. For each level of genetic diversity results from 10 replicate microcosms are ranked in order of increasing biomass, □, all individuals genetically unique; △, 4 genotypes only, ◆, 1 genotype only. (Booth and Grime, unpublished.)

Investigations in permanent plots such as that of Phillips *et al.* (1994) and Burslem and Whitmore (1999) reveal that in productive tropical rainforest high diversity is sustained in conditions where high population turnover is maintained by intermittent disturbance. In coppiced woodlands and hedgerows of the British Isles the effect of management is to so reduce the morphology of the potentially-dominant trees that vegetation types are produced in which 'debilitated' specimens of trees such as oak and beech may be found co-existing with a wide variety of small shrubs.

Differences in regenerative strategy also play a most important part in the vegetation processes which maintain species diversity in natural forests. It has been known for some time (Jones 1945; Watt 1947) that, in temperate deciduous woodland, co-existence by a variety of trees may be related to a continuous cyclic pattern of regeneration whereby openings in the canopy arising from the senescence and death of trees of one species tend to be colonised by seedlings or vegetative sprouts of another. In the Northern oakwoods of the British Isles, for example, gaps arising from the death of oak (*Quercus petraea*) frequently allow a temporary phase of re-colonisation by the wind-dispersed seeds of birch (*Betula pubescens*), which is itself then replaced by oak. Evidence of similar alternations of species has been obtained both for European forests (Nagel 1950; Schaeffer and Moreau 1958) and for mixed deciduous woodland in North America (Auclair and Cottam 1971; Fox 1977).

When attention is turned to the role of regenerative strategies in forests of high species richness in temperate and tropical regions, very close similarities may be observed with the mechanisms of co-existence described for species-rich calcareous grassland (pages 277–284). For tropical forests there is an extensive literature documenting the importance of temporary gaps in providing opportunities for regeneration (Kramer 1933; Richards 1952; van Steenis 1958; Schulz 1960; Knight 1975; Hubbell and Foster 1986). In mature tropical forests where gaps arise mainly as a result of windfall (Jones 1956; Cousens 1965) the most important forms of regeneration are those involving vegetative sprouts, banks of seedlings originating from fruits dispersed by animals, and, more rarely, wind-dispersed seeds. As in the case of the ancient limestone pastures discussed on page 284, many of the species produce seeds intermittently over a long life-history and rates of successful regeneration are exceedingly low.

Two additional features of regeneration in tropical forests provide strong parallels with phenomena occurring in herbaceous vegetation and are therefore relevant to a general model of the mechanism controlling species richness. The first is based upon the observation (Gómez-Pompa 1967; Webb *et al.* 1972) that loss of species richness in severely disturbed tropical forests is often associated with widespread invasion by trees regenerating from buried seed banks (cf. the development of buried seed banks in disturbed areas of herbaceous vegetation and temperate forest (pages 149–151). The second relates to the presence in certain types of mature tropical forest of trees with competitive characteristics. These exceedingly tall, deciduous, fast-growing 'nomads' (van Steenis 1958; Whitmore 1975; Knight 1975) occur as scattered individuals which appear to depend for successful regeneration upon dispersal of wind-dispersed seeds into large openings in the canopy created by windfalls. It would appear that in rainforest it is only in such large gaps that the release from shade and mineral nutrient stress is sustained long enough to allow the establishment of competitive trees. In small gaps establishment is precluded by the rapid rate of canopy closure and the short duration of the pulse of mineral nutrients released by decomposition of fallen trees.

A close parallel may be drawn between the nomads of tropical rainforest and competitive-ruderals such as *Medicago lupulina* which often occur as large isolated individuals in temperate grasslands dominated by slow-growing perennial herbs. As noted on page 284, scrutiny of the distribution of *M. lupulina* in unproductive grassland confirms that, as in the case of the rainforest nomads, this species is confined to sites in which local damage has created exceptionally large gaps in the established vegetation.

A GENERAL MODEL

In Grime (1979) an attempt was made to summarise the relationships between five processes which appear to influence species density in vegetation and this scheme is reproduced here as Figure 105. As in the model already considered on page 263 (Figure 88), three main contingencies are recognised, each corresponding to a different section of the baseline of the figure. Section A includes the species-poor vegetation found in environments subjected to extreme conditions of stress and/or disturbance. The corridor containing vegetation types with the potential for relatively high species richness occupies the intermediate part of the range in biomass (Sections B and C), whilst Section D corresponds to vegetation in which species richness is suppressed by dominance.

Two processes can be identified which determine the species richness attained in particular corridor environments. The first of these has been examined in some detail earlier in this chapter and consists of the degree of spatial and temporal variation and the resulting opportunities for complementary forms of exploitation and regeneration within the environment. The second process, which will be considered under the next heading, is the availability and rate of ingress of potential constituent species from the surrounding landscape.

In the lower part of the figure presented in Grime (1979) an attempt was made to illustrate the relationship of the model to patterns of variation in species richness which can be observed in the field. Horizontal lines have been drawn to indicate those parts of the model which correspond to the range of conditions in particular habitats or vegetation types. It must be emphasised

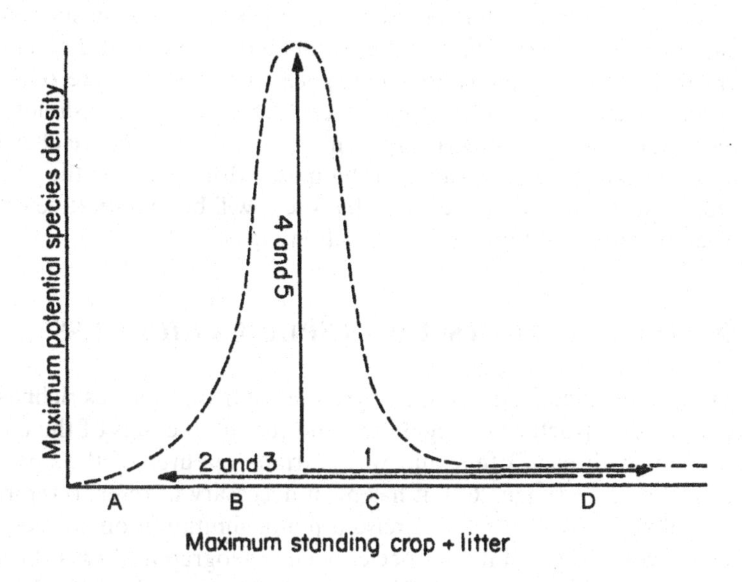

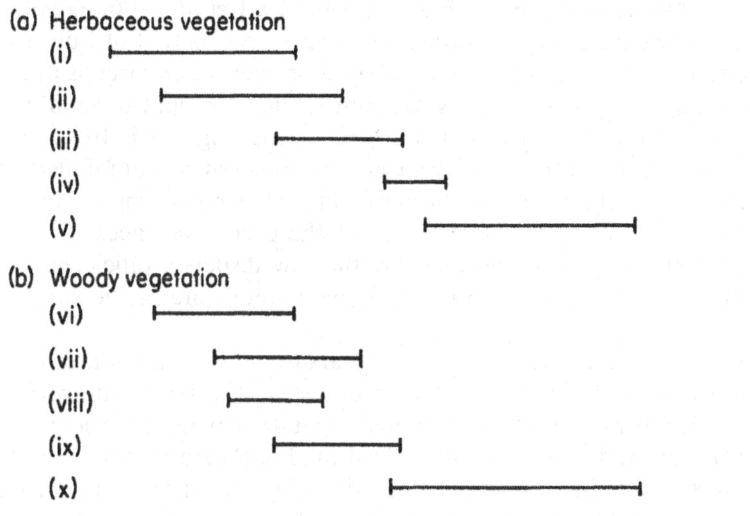

Figure 105. Model summarising the impact of five processes upon species richness in vegetation. Key to processes: 1, dominance; 2, stress; 3, disturbance; 4, niche differentation; 5, ingress of suitable species or genotypes. The horizontal lines describe the range of contingencies encompassed by a number of familiar herbaceous and woody vegetation types. Key to vegetation types: (i) paths, (ii) grazed rock outcrops with discontinuous soil cover, (iii) infertile pastures, (iv) fertilised pastures, (v) derelict fertile pastures, (vi) arctic scrub, (vii) temperate scrub, (viii) temperate hedgerows, (ix) species-rich tropical rainforest, (x) mature temperate or tropical forest on fertile soil. (Reproduced from Grime 1979.)

that although the model may be used to interpret variation in species richness within either herb or tree layers, there will be differences of detail in the way in which the mechanisms controlling species richness operate on herbs and trees. In particular, it is clear that in the tree layer the thresholds at which various phenomena come into play along the baseline of Figure 88 will correspond to very much higher values in biomass. Moreover, for the reasons discussed on page 266, we may expect that there will be quantitative differences between tropical and temperate vegetation types.

'RESERVOIR EFFECTS' UPON SPECIES RICHNESS

So far it has been convenient to analyse the control of species richness mainly in terms of the structure of vegetation and the interactions of the component plants both with each other and with various features of the environment. Already, however (page 265), it has been necessary to refer to the role of an additional factor which may exercise a major limitation on species richness. This is the reservoir of suitable species in each geographical area (Grime 1979; Eriksson 1993; Ricklefs and Schluter 1993; Pärtel *et al.* 1996; Zobel 1997). Particularly in extensively-disturbed landscapes such as that of Lowland Britain, certain habitats such as woodland or unproductive calcareous pasture or marshland have been reduced by agricultural development to small isolated fragments. Where these fragments are of recent origin it is frequently observed that their potential for high species richness remains unfulfilled and we may suppose that this is, in part, because of the slow rate of ingress of suitable plants from the adjacent countryside. In these circumstances, low species richness appears to be determined by the low dispersal efficiencies of the seeds of many plants and also by the depauperate state of the surrounding flora.

A rather different reservoir effect appears to be involved in the consistent differences in species richness which are observed in temperate regions between adjacent and structurally-similar vegetation types on calcareous and acidic soils. From data such as those illustrated in Figure 86, for example, it is apparent that in Britain, regardless of habitat, species richness in herbaceous vegetation is consistently low where the surface soil pH is less than 4.0. A more specific example of the reduction in species richness associated with increasing soil acidity is presented in Figure 106 which is based upon survey data from an area of northern England in which there is an intimate mosaic in which acidic and calcareous grasslands experience comparable effects of climate and land-use. From this figure it is evident that whereas high species richness is attained over the pH range 4.0–8.0, there is an abrupt decline on more acidic soils. When variation in species richness across the soil pH range is compared with the total number of species recorded from all the samples examined in each soil pH category it is apparent that the fall in species richness in individual samples is correlated with a progressive reduction in the size of the species reservoir associated with the transition from calcicoles to calcifuges. Reference to the *Atlas of the British Flora* (Perring and Walters 1962) confirms that the greater abundance of calcicoles over calcifuges in Britain applies not only to grassland herbs but to other ecological groups such as trees, shrubs, and ruderal plants. This suggests that the maximum species richness which can be attained in any particular vegetation type will show a predictable relationship to soil pH. In Figure 107 an attempt has been made to summarise the relationships between maximum species richness, maximum standing crop + litter, and, soil pH in herbaceous vegetation of the British Isles.

We may conclude, therefore, that at latitudes such as that of the British Isles, the higher species richness supported by calcareous soils is related, in part, to the fact that in such regions calcicolous vegetation draws upon a reservoir of species which is considerably larger than that of the calcifuges.

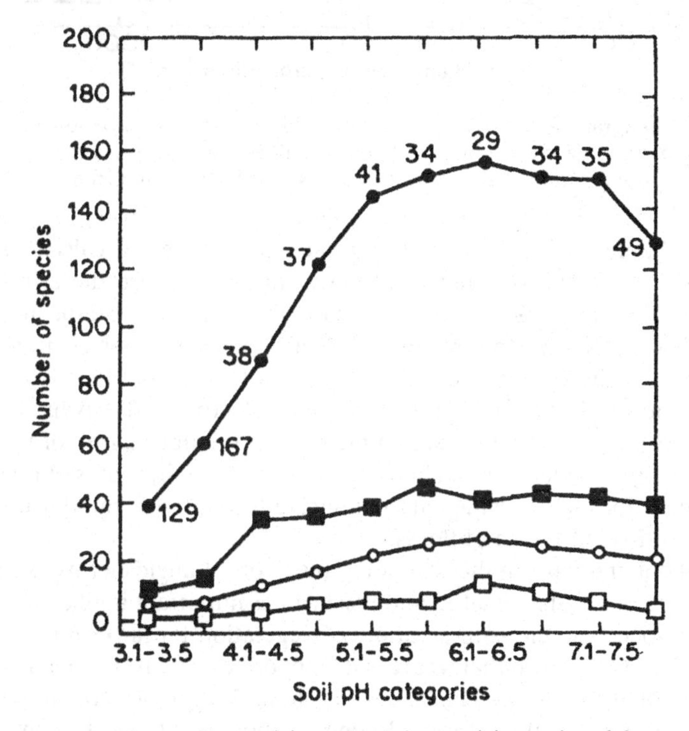

Figure 106. Mean (○), maximum (■), and minimum (□) species richness and total number of species represented (●) in categories of surface soil pH encountered in 593 one square metre quadrats located at random in unmanaged grasslands distributed within an area of 2400 km² in northern England. The values inserted on the topmost curve refer to the number of quadrats falling in each half-unit soil pH category. (Reproduced from Grime 1973c by permission of *Journal of Environmental Management* 1, 151–167. © 1973 Academic Press Inc. (London) Ltd.)

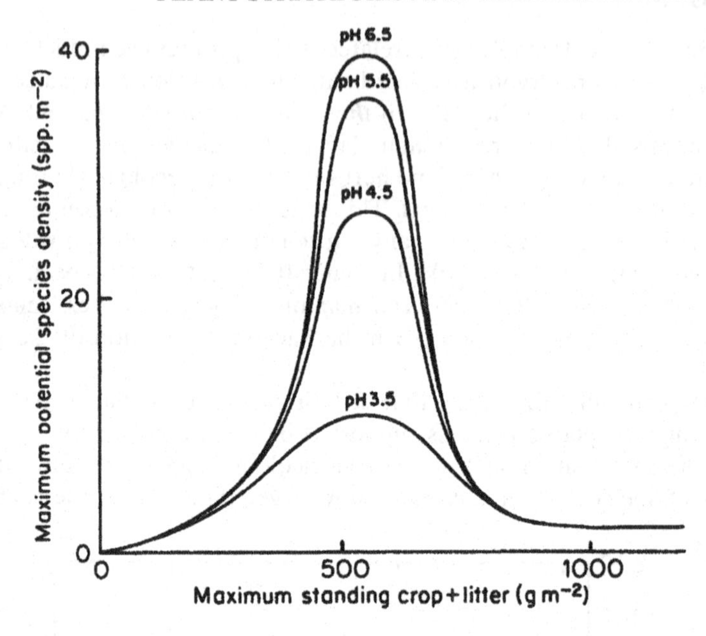

Figure 107. Diagram summarising the relationship between surface soil pH, seasonal maximum in standing crop + litter, and maximum potential species richness in herbaceous vegetation of the British Isles. (Reproduced from Grime 1979.)

The most likely explanation for the greater abundance of calcicoles is that these plants have evolved mainly at lower latitudes where the effect of low precipitation : evaporation ratios is to maintain a high base-status in the soils. This is to suggest that the evolution of calcicoly has occurred in semi-arid environments where, as suggested by Margalef (1968), Stebbins (1952, 1972), Stebbins and Major (1965), and Bartholomew, Eaton, and Raven (1973), the effects of climatic fluctuation are to maintain a greater degree of vegetation disturbance and environmental heterogeneity, with higher rates of turnover in plant populations, all of which are conducive to relatively rapid rates of speciation and diversification of floras.

Latitudinal gradients in the size of the reservoir of angiosperms extend from the polar regions to the equator, and this fact has important implications for the control of species richness in a variety of vegetation types (Grubb 1987). For example, it is well known that whereas in temperate regions the mean density of trees is less than 10 species/Ha, values exceeding 100 species/Ha are frequently encountered in tropical rainforest (Longman and Janik 1974). It seems reasonable to suggest, therefore, that the higher densities of tree species in tropical forests are related to the larger reservoir of trees present at low latitudes and this in turn may be attributed to the long and uninterrupted period over which forest speciation and co-adaptation (Dobzhansky 1950; Janzen 1970; Gilpin 1975) has been able to take place in equatorial habitats.

THE CONTRIBUTION OF EPIPHYTES TO SPECIES-RICH VEGETATION

There are a number of vegetation types in which the exceptionally high species richness depends to a considerable extent upon the presence of a rich epiphytic component. Reference has been made already to the epiphytic habit of certain of the bryophytes in herbaceous vegetation (pages 269 and 273), and in temperate regions many of the most ancient woodlands are characterised by the diverse assemblages of mosses and lichens on the trunks and branches of the older trees (Rose 1974). However, it is in tropical rainforests that epiphytes achieve their greatest biomass and taxonomic variety. In this environment, the contrasted types of micro-habitat which occur at different heights above ground are exploited by a wide range of lichens, bryophytes, pteridophytes, and angiosperms. The abundance of the epiphytic flora may be related in particular to the conditions of moisture and mineral nutrient supply within the rainforest canopy. Because of the frequent rainfall and high humidity, germination, establishment, and survival can take place in the virtual absence of a rooting medium, and in some extreme climates epiphytes colonise not only trunks and branches but also the surfaces of actively-photosynthetic leaves (Pessin 1922). Moreover, it seems reasonable to expect that under such rainfall conditions there will be a constant leaching of solutes from the tree canopy; there is a need to investigate the mechanisms whereby rainforest epiphytes are adapted to exploit this source of mineral nutrients.

CONTROL OF SPECIES RICHNESS BY VEGETATION MANAGEMENT

Management of vegetation, whether for the purposes of agriculture, amenity, landscape restoration, or nature conservation, inevitably brings the practical ecologist into contact with many of the phenomena discussed in this chapter. As an illustration of the relevance of these theoretical concepts to vegetation management two examples in quite different fields of applied ecology will now be considered briefly.

Maintenance of Monocultures in Agricultural Systems

Since much time, energy, and money is expended in the attempt to maintain monocultures of crops and forage plants it is of more than academic interest to attempt to establish where particular agricultural systems lie in relation to the various contingencies summarised in Figure 105. With respect to pastures and meadows in Britain, a valuable clue is provided by the evidence (page 258) of the increasing dominance and reductions in species richness when productivity is stimulated by heavy dressings of mineral fertilisers. When this effect is portrayed diagrammatically (Figure 108), one is prompted to consider whether by further modifications of farming regimes it would be possible to move even closer to the conditions favouring monoculture. As we have seen (page 182), herbaceous monocultures or 'near-monocultures' occur in non-agricultural situations in Britain. However, since these are composed of plants which owe their dominant status to relatively undisturbed conditions it may be unrealistic to expect to be able to create vegetation of this type in cropped systems. It remains to be determined, therefore, whether stable perennial monocultures are attainable and consistent with the objectives of farm management. Nevertheless, it is established that regimes which allow the development of a large standing crop and minimise the frequency of cropping will encourage the trend towards monoculture.

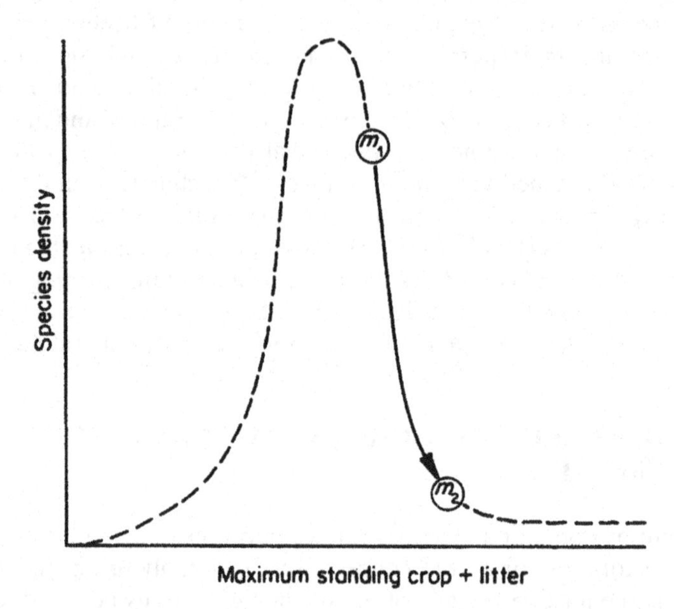

Figure 108. Diagram representing the alteration in species richness associated with the vegetation change ($m_1 \rightarrow m_2$) resulting from the application of a heavy dressing of mineral fertiliser to a species-rich meadow

Management of Vegetation Subject to Trampling

A recurrent problem in managing parkland and areas of natural landscape where herbaceous vegetation is subject to trampling is to estimate the carrying capacity of different habitats and to predict the response of vegetation to increasing intensities of wear. It is interesting to note that in the research reported in this field there is apparently conflicting evidence concerning the impact of trampling in grassland. Some investigators (e.g. Bayfield 1973; Liddle and Greig-Smith 1976) have recorded a fall in species richness with increasing intensity of trampling whilst others (e.g. Westhoff 1967; van der Maarel 1971) have detected a change in the reverse direction. However, as pointed out by Liddle (1975), both of these types of response can be explained by reference to the 'humped-backed' model. This point is illustrated in Figure 109 which depicts the hypothetical responses of a productive and relatively undisturbed grassland community to two intensities of trampling. At the lower intensity the result is a reduction in the vigour of the dominant components of the turf, allowing an ingress of species. The effect of more severe trampling, however, is to create an environment in which only a small number of specialised plants are able to survive.

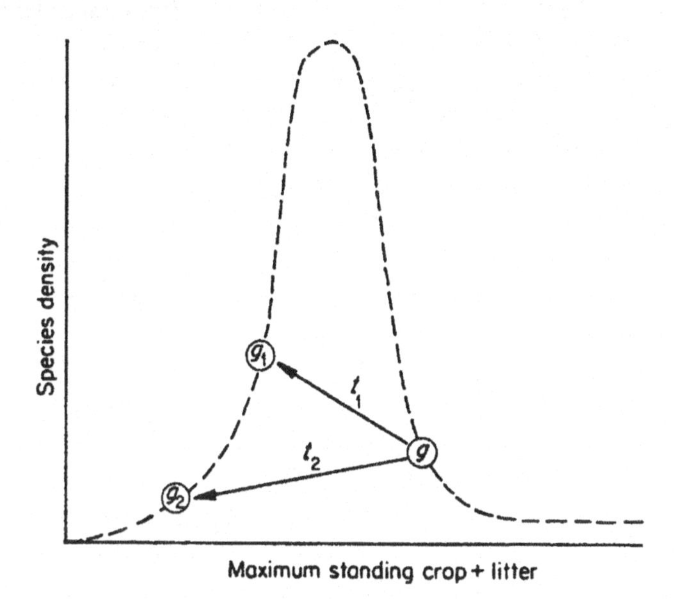

Figure 109. Diagram representing the alteration of species richness which is associated with the changes in vegetation ($g \rightarrow g_1$, $g \rightarrow g_2$) brought about when relatively undisturbed base-rich grassland is subjected to moderate (t1) and severe (t2) intensities of trampling

ANALOGOUS PHENOMENA INVOLVING ANIMALS

Studies of species density and community structure in intertidal (Dayton 1970; Paine 1969, 1974) and benthic communities (Lewis 1914; MacFarlane and Bell 1933; Hehre and Mathieson 1970; Sears and Wilce 1975; Dayton and Hessler 1972) have recognised many of the phenomena included in the general model presented in this chapter. There is also evidence which suggests that many of the same principles may be applied to the control of species density in colonial animals such as corals (Branham *et al.* 1971; Dana *et al.* 1972; Goreau *et al.*

1972), shell-fish (Connell 1961, 1972), and even some noncolonial organisms such as coral reef fish (Sale 1977). In a recent paper, Ritchie and Olff (1999) have recognised that the humped-back model can be extended to account for variation in species richness in larger animals. They have shown that, as in the case of plants examined along a productivity gradient, the species richness of East African mammalian herbivores shows a unimodal distribution in which:

> As resources become more abundant, maximum patch size rapidly increases to allow larger species to exist. However, further increases in resource abundance cause food patches to coalesce, eliminating small, resource-rich patches and requiring greater size separation among smaller species.
>
> Ritchie and Olff (1999)

III Plant Strategies and Ecosystem Properties

10 Trophic Structure, Productivity, and Stability

INTRODUCTION

It has been argued in Part II of this book that the structure and dynamics of the majority of plant communities can be predicted and understood to a major extent from knowledge of the essential biology of the relatively small number of species that dominate the biomass of most plant communities. Further, it has been asserted that much of the variation in this essential biology can be summarised in the form of the CSR theory of primary plant strategies in the established phase of plant life-histories and in theories related to the regenerative strategies of plants. In this final chapter, the argument is extended further to consider the role of plant strategies as controllers of the properties of terrestrial ecosystems. Although, of course, the properties of an ecosystem are influenced by environmental factors, by animals, by micro-organisms, and by management interventions it is an inescapable fact that the capture, processing, and release of energy, carbon, and mineral resources in a terrestrial ecosystem is strongly dependent upon the mass of vegetation and its functional characteristics. On this basis alone we can predict that there will be major differences between ecosystems in their trophic structures, productivities, and stabilities that are strongly correlated with differences in the strategies of the dominant plants.

Although the primary purpose of this chapter is to identify some predictable linkages between features of plants and the properties of the ecosystems they dominate it is necessary to acknowledge other lines of enquiry whereby ecologists are currently seeking to explain the role of plants in the control of ecosystem properties. In recent years a vigorous debate has developed between those attributing ecosystem properties to the *kinds* of plants dominating them (e.g. Grime 1987, 1997; Wardle *et al.* 1997a, 2000; Berendse 1998; Bardgett *et al.* 1999; Aerts and Chapin 2000) and those who have been primarily interested in the effects arising from the *number* of component plant species (e.g. Tilman and Downing 1994; Karieva 1994, 1996; Naeem *et al.* 1994; Hector *et al.* 1999). As we shall see later in this chapter much of the discussion has focussed upon technical issues of experimental design and data analysis. However, there can be little doubt that this debate has a deeper significance which confers upon it benchmark status in the history of ecological research. In part, the importance of the debate arises from the present circumstances which are distinctive—they are driven by concerns about human impacts on global stability and the protection of the planet's fast-declining biodiversity. However, the debate also involves differences of opinion concerning the information required to predict and interpret effectively ecosystem properties from studies of component species. Before proceeding further it may be useful to attempt to identify the origins of the present debate.

Ecological journals and symposium volumes (e.g. Schulze and Mooney 1993b) bear witness to the extent to which during the second half of the 20th century ecologists of many different kinds—botanists, zoologists, microbiologists, theoreticians, and experimenters began to converge on the ecosystem, bringing to bear a diversity of skills, methods, and perspectives. Differences in method and philosophy soon became evident. Many theoreticians and animal ecologists (e.g. Elton 1927, 1958; MacArthur 1955; Hutchinson 1957; May 1972, 1974; Pimm 1991) have placed greatest reliance on insights from population biology. In recent years, there has been a growing concern (Leps *et al.* 1982; Power and Mills 1995; Hurlbert 1997) about the limited potential of numerical approaches to bridge the gap between populations and ecosystems. This is evident in several critiques that have questioned the value of the conceptual devices (e.g. niche, connectance, keystoneness) whereby efforts have been made to scale up to the ecosystem. In parallel with this theoretical school, many plant ecologists and limnologists (e.g. Leps *et al.* 1982; Chapin 1980; Grime 1988c; Carpenter 1988; Reynolds 1998) have focused on resource dynamics.

When scientists from well-defined subdisciplines with contrasted traditions interact within a maturing science the situation often resembles a late stage in the completion of a large jigsaw puzzle at a family party:

> It is at this stage that relationships within the family are tested and begin to determine how quickly the jigsaw is completed. At one extreme is the family in which overall progress is kept under review so that connections between developing islands are established as early as possible allowing the completed picture to be visualised and the remaining gaps to be filled with minimal delay. Not unknown, however, is the family for which the jigsaw provides a longer and more enjoyable diversion in which the construction of each island becomes an absorbing activity in its own right with individual logic, rules and rivalries with neighbouring islands.
>
> JP Grime (1985a)

In keeping with this 'late stage in the jigsaw' scenario, much of the debate about the relative importance of species traits as opposed to species numbers as controllers of ecosystem properties may be explicable as an interaction between plant and animal ecologists. After some early interest in physiology, to an increasing extent animal ecologists have tended to rely heavily on population modelling as their main research method. Despite the strong recent inroads into plant ecology of theoretical, population-based methods derived from animal ecology (e.g. Harper 1977; Silvertown 1982) many plant ecologists have continued to regard physics and chemistry as the pre-eminent parent subjects of ecology with a consequent emphasis upon resource dynamics. As ecosystem research proceeds it will be interesting to compare the insights that result from the two approaches. It is already apparent, however, that resource dynamics have much to contribute to our understanding of the control of productivity, trophic structure, responses to perturbations, and the circulation and storage of organic and inorganic elements.

Before proceeding to a formal review of some of the effects of plant functional types on these different aspects of ecosystem functioning it may be informative briefly to consider two particularly graphic examples where the controlling effects of the resource dynamics of dominant plants on ecosystem properties can be identified. The first example refers to the influence of stress-tolerant plant communities as a whole on the fate of radioactive emissions from the Chernobyl Incident; the second is more subtle and relates to the complementary interaction of co-existing stress-tolerators on the retention of nitrogen in a grassland ecosystem.

TWO EXAMPLES

The Chernobyl Legacy

In 1986, following the major explosion at the nuclear facility at Chernobyl, a large cloud of radioactive material was dispersed westward, and radiocaesium (^{137}Cs) was deposited in rain at various locations in Europe including upland areas of northern England and Wales. Pastures were strongly affected and following the detection of unusually high levels of ^{137}Cs in grasses and sheep carcasses it was necessary to suspend the marketing of animals from more than 100 hill farms in the British Isles. To allay the concerns of farmers and the general public estimates were made of the time that would need to elapse before ^{137}Cs fell to safe levels. To an embarrassing extent these estimates proved to be highly optimistic and some farms have been unable to send animals to market for many years. In 1988, a British Government Inquiry was conducted to determine why there had been such serious underestimation of the persistence of ^{137}Cs. Among the views expressed at this inquiry was the opinion (Grime 1988c) that predictions had been based upon measurements conducted on productive, intensively-managed pastures on fertile soils in Lowland Britain and that there had been a failure to predict the 'slow-dynamics' of unproductive, upland ecosystems that would lead inevitably to sequestration and slow release of ^{137}Cs from both living and dead components. This interpretation was supported by the scheme devised by Chapin (1980) and reproduced in Figure 110 and by more detailed information (Table 24) explaining

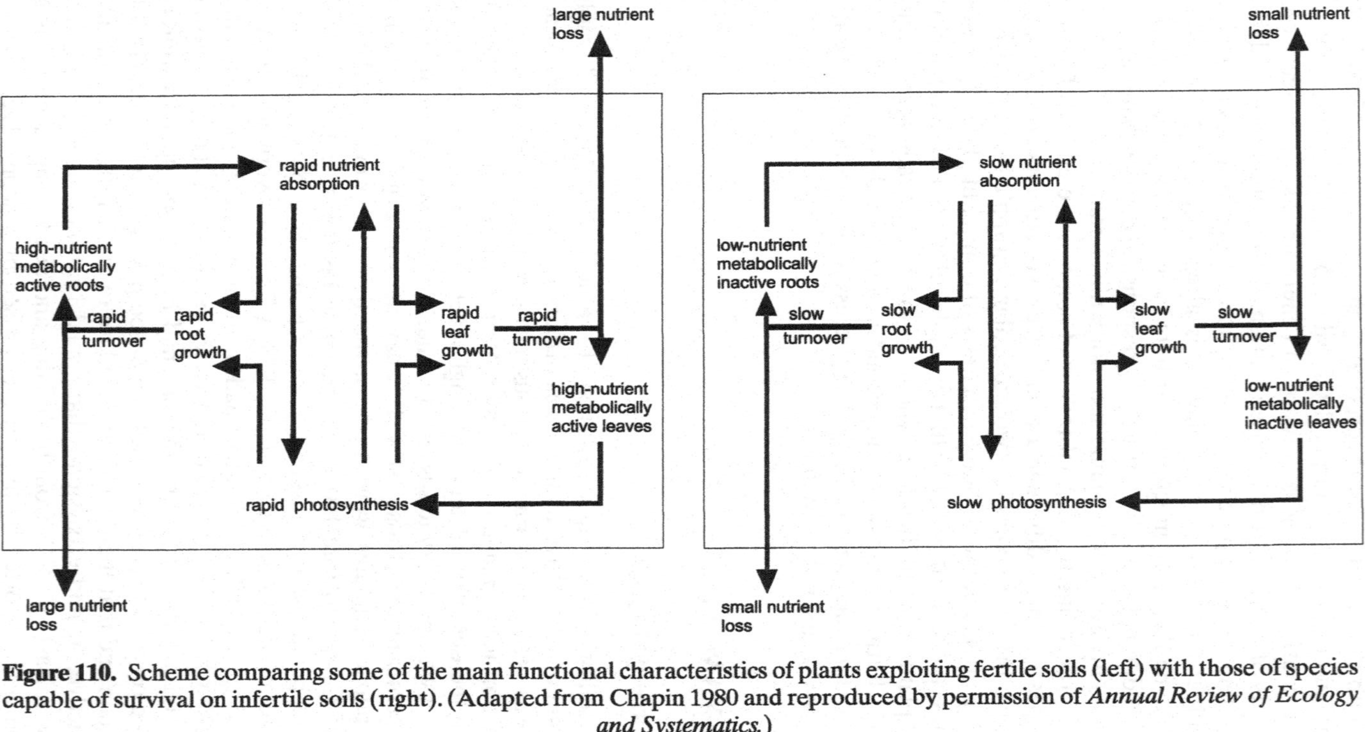

Figure 110. Scheme comparing some of the main functional characteristics of plants exploiting fertile soils (left) with those of species capable of survival on infertile soils (right). (Adapted from Chapin 1980 and reproduced by permission of *Annual Review of Ecology and Systematics.*)

Table 24. Some consistent differences between plants of fertile and infertile soils and their relevance to the persistence of ^{137}Cs in upland areas

Plant characteristics	Species of fertile soils	Species of infertile soils	Implications for persistence on ^{137}Cs in upland areas
1. Life-form	Herbs, shrubs and trees	Lichens, bryophytes, herbs, shrubs, and trees	Lichens and brophytes intercept and absorb ^{137}Cs directly from rainwater through their above-ground surfaces. They also sequester metals for long periods, are unpalatable, and decay slowly.
2 Life-span of whole plant	Often short (<5 years)	Long (>5 years)	^{137}Cs will persist in plants of infertile soils because it is incorporated into potentially long-lived tissues.
3. Life-span of individual leaves and roots	Short (<1 year)	Long (1–3 years)	
4. Leaf phenology	Well-defined peak of growth each spring and summer	Evergreens often showing no seasonal change in biomass	Presence of living foliage throughout the year in species of infertile soils will permit entrapment of ^{137}Cs regardless of the season of deposition, particularly where lichen or bryophyte thallus or higher plant leaves have hairy, rough, or grooved surfaces.
5. Maximum potential relative growth-rate	Rapid	Slow	Slow growth of plants of infertile soils will mean lower rates of tissue turnover (see 2, 3, 4, 8) and tendency to retain ^{137}Cs for long periods.
6. Uptake of mineral nutrients and other ions	Strongly seasonal; and mostly in spring and summer	Opportunistic, capable of occurring at most times of the year	Presence of functional roots throughout the year will allow accumulation of ^{137}Cs from mineralisation pulses regardless of their season of release into the soil solution.
7. Mycorrhizal infection of root system	Light	Heavy	Mycorrhizal fungi are the effective absorbing surface for the root systems of plants of infertile soils and have a very high affinity for metals such as ^{137}Cs. They form an effective network throughout the surface soil and produce residues which are resistant to decay (see 9 and 10).
8. Storage of mineral nutrients	Most mineral nutriants are rapidly incorporated in growth but a proportion is stored and forms the capital for expansion of growth in the following growing season. Little internal recycling from old leaves to new	Storage systems in leaves, stems, and/or roots. Some recycling of minerals from old leaves to new	Because of the weak couplings between mineral uptake and utilisation in growth, ^{137}Cs will tend to be retained in the plant biomass. Some internal recycling of ^{137}Cs is likely to occur, further retaining ^{137}Cs in the living tissues.

(*continued over*)

Table 24. (*continued*)

Plant characteristics	Species of fertile soils	Species of infertile soils	Implications for persistence on ^{137}Cs in upland areas
9. Palatability	High	Low	Low palatablity dictates lower density of sheep in upland pastures and results in a situation where much of the ^{137}Cs is likely to reside in physically repellent plant tissues which tend to be avoided by the animals except in winter when the supply of more palatable leaves is minimal.
10. Rate of litter decomposition	High	Low	The leaf toughness which protects the canopy of slow-growing plants of infertile pastures from heavy defoliation by sheep and invertebrates, remains operational when the leaves die and fall onto the ground surface. In consequence, there is a reduced rate of decay and ^{137}Cs would be expected to be recycled back into the plants at a slow rate. Mycorrhizal roots (especially those associated with ericaceous plants, e.g. heather) tend to be very resistant to decay and will retain ^{137}Cs on the melanised cell walls of fungal residues.

how differences between the functional characteristics of dominant plants of productive and unproductive pastures were likely to dictate differences in the persistence of ^{137}Cs.

Nitrogen Retention at Cressbrookdale

In many heavily-industrialised countries in Western Europe deposition of reactive forms of atmospheric nitrogen on to infertile soils has changed the character of both vegetation and ecosystems and one symptom of such changes is the leakage of soluble forms of nitrogen into the ground water. Figure 111 summarises the results of a lysimeter experiment (Figure 112) designed to investigate the influence of different vegetation components in restricting nitrogen leakage from a species-rich calcareous grassland at Cressbrookdale in North Derbyshire. The results show that three of the most abundant species in the grassland, the grass *Festuca ovina*, the sedge *Carex flacca*, and the forb *Leontodon hispidus* are equally ineffective in retaining nitrogen when grown in isolation. In combination, however, the same three species brought about a very large reduction in nitrogen release. Confirmation that this effect was due to a complementary interaction between the three functional types of plants is available from additional data in Figure 111 which reveal that nitrogen retention was not improved by simply increasing the number of species of grasses, sedges, and forbs in the community.

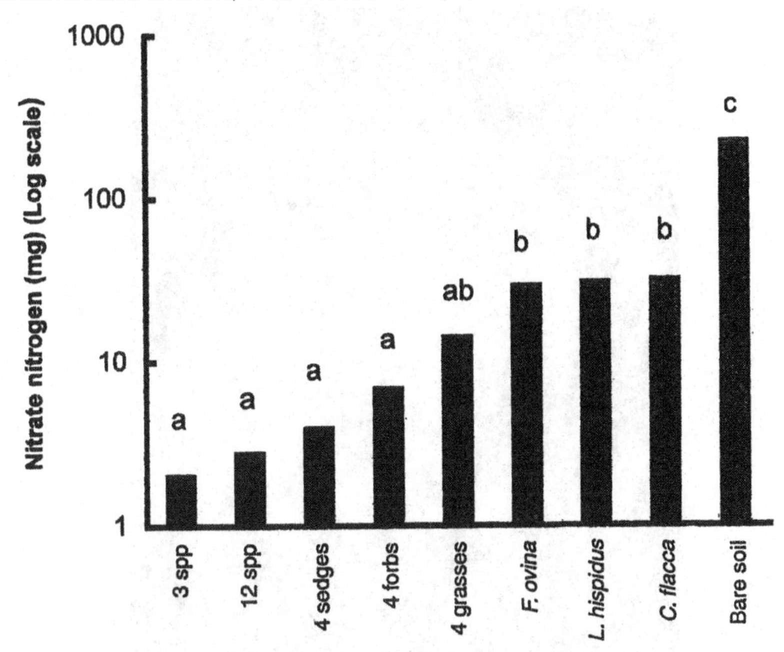

Figure 111. Total nitrate nitrogen leakage over a period of two years from lysimeters containing a natural rendzina soil and nine types of vegetation synthesised from mature transplants and differing in functional composition and species richness. 3 spp. = *Festuca ovina*, *Carex flacca*, and *Leontodon hispidus*; 4 sedges = *Carex flacca*, *C. panicea*, *C. caryophyllea*, and *C. pulicaris*; 4 forbs = *Leontodon hispidus*, *Succisa pratensis*, *Viola riviniana*, *Campanula rotundifolia*; 4 grasses = *Festuca ovina*, *Koeleria macrantha*, *Helictotrichon pratense*, and *Briza media*. Columns that do not share the same lettering are statistically different ($P<0.05$). (Reproduced by permission of Booth and Grime)

Implications of the two Examples

Although the Chernobyl incident was an unusual event it illustrates a general principle that is explored in various ways later in this chapter. This principle, encapsulated in Figure 110, concerns the repercussions upon ecosystems of the functional shift from the dominant plants of fertile soils (ruderals or competitors) to the dominant plants of infertile soils (stress-tolerators). As explained in Chapter 1 this shift coincides, in the sequence R<C<S, with parallel changes increasing the life-spans of individual plants, decreasing the rate of tissue turnover, increasing resistance to herbivory, decreasing rates of decomposition, and slowing the circulation of mineral elements. Differences between upland and lowland sheep pastures in the persistence of ^{137}Cs is therefore only one rather exotic example of the ecosystem effects cascading from the differences in the functional traits of the dominant plants. The remainder of this chapter explores some of these other ecosystem effects with

Figure 112. Lysimeters containing natural rendzina soil and plant communities differing in functional composition and species richness synthesised from transplants from Cressbrookdale, North Derbyshire, England

particular attention to trophic structure, productivity, and ecosystem stability and sustainability.

The message emerging from the investigations of nitrogen leakage from the Cressbrookdale lysimeters is that although the dominant plants occupying a particular ecosystem are likely to be broadly similar in terms of CSR functional type and, in comparison with species in contrasted ecosystems, will exert broadly similar effects on ecosystem processes, fine-tuning of co-existing ecologies will occur and can have significant specific benefits. However, this should not be translated into the uncritical proposition that benefits will accrue in direct proportion to the number of plant species present. There are two main reasons why we should view with considerable caution the notion that species richness delivers ecosystem properties:

1 The admission and persistence of organisms in plant communities and ecosystems depends upon their individual fitness rather than their contribution to group properties. The presence of parasitic plants and pathogens is a scarcely-necessary reminder of this principle.
2 Species traits conferring particular ecosystem effects are often incompatible with and tradeoff against traits affecting other aspects of ecosystem functioning (see discussion of resistance and resilience on page 336).

Having established by preliminary examples and discussion that dominant plants can influence ecosystems through identifiable aspects of their resource dynamics it is now possible to review more broadly the linkages between plant functional types and ecosystem properties. It is appropriate to begin this review by considering the effects of plant strategies on the trophic structure of ecosystems

EFFECTS OF PLANT STRATEGIES ON TROPHIC STRUCTURE

It is not surprising that one of the key steps towards our current understanding of ecosystems (i.e. the recognition of trophic structure) can be traced to an investigation of a small freshwater lake (Lindeman 1942). Within the ecosystems of water bodies there is the opportunity to conduct quantitative studies of the main component populations, to investigate their interactions, to measure the fluxes of energy and resources between trophic components (Carpenter 1988), and where necessary to document fluctuations in organisms and resources over long periods of time (George and Harris 1985). By comparison, investigations of terrestrial ecosystems are beset with difficulties some of which can be identified as follows:

1 The long-lived trees and herbs that dominate many terrestrial ecosystems provide much more difficult subjects for study and have more complex and delayed effects upon ecosystem structure and dynamics than the largely planktonic primary producers of aquatic systems.
2 Although seasonal variation in resource concentrations and environmental conditions occurs within aquatic systems, mixing allows equilibration over a large volume of habitat. Within the leaf canopies and rhizosphere of terrestrial ecosystems, however, the environment is characterised by a complex, fine-scale dynamic mosaic, largely imposed by resource uptake and release by the organisms themselves (Astrom *et al.* 1990; Canham *et al.* 1994; Coomes and Grubb 2000).
3 Many ecosystem processes are located within the soil and depend upon complex interactions involving root systems, soil fauna, and microorganisms.

4 In comparison with small lakes and ponds, it is often extremely difficult to recognise the boundaries of a terrestrial ecosystem. This can be a severe problem in circumstances where the territories of herbivores or carnivores are large and variable and include very different vegetation and soil and hydrological conditions (Weins 1976; McNaughton 1985; Naiman *et al.* 1986; Jeffries 1988).

As a consequence of these difficulties progress has been relatively slow in developing an understanding of how terrestrial ecosystems function and vary in structure and complexity according to location and management. When we consider the methods that are currently used to investigate them a close parallel can be recognised with those applied in studies of plant communities. In reviewing the insights into ecosystem structure and function available from various approaches it is convenient therefore to follow the tripartite classification of methods (direct observation, manipulation, and synthesis) already used for communities in Chapter 5.

Direct Observations in the Field

Forty years ago Hairston *et al.* (1960) asked the simple but provocative question 'Why is the world green?' and proposed that the reason why vegetation survives despite the attentions of herbivores is that populations of herbivores are restricted in size and activity by the controlling effects of carnivores and parasitoids. This theory of 'top-down control' initiated a vigorous debate (e.g. Murdoch 1966; Ehrlich and Birch 1967; Menge and Sutherland 1976; Power 1992) which continues to the present day.

A significant elaboration of the theory has been associated with recognition (Chapter 1, pages 30 and 71) of consistent differences in the ability of primary plant functional types to defend themselves. A number of researchers have shown a negative correlation between plant defence and the maximum relative growth rate of plants (e.g. Grime *et al.* 1968, 1997; Mooney and Gulmon 1982; Coley 1983; Southwood *et al.* 1986; Herms and Mattson 1992; Cebrián and Duarte 1994). Maximum relative growth rate is also related to the productivity of the environment (fast-growing species found at high resource supply rates, slow-growing species found at low resource supply rates) (Mooney 1972; Grime and Hunt 1975; Bazzaz 1979; Chapin 1980; Bryant *et al.* 1983; Coley 1988). Consistent with these relationships SD Fretwell and L Oksanen have proposed a hypothesis that links trophic structure and dynamics to productivity (Fretwell 1977, 1987; Oksanen *et al.* 1981; Oksanen 1990). The essential feature of the Fretwell–Oksanen hypothesis is that the influence of animals on vegetation depends upon productivity. At the lowest productivity the plant cover is too small or unpalatable to support large populations of herbivores. At intermediate productivity, herbivore numbers rise and have important controlling effects on vegetation quantity and quality.

However, as productivity rises still further it is envisaged that herbivores become a consistent food source for their predators which become sufficiently numerous to afford 'top-down protection' to a large and relatively-palatable plant biomass. A classic example of 'top-down' effects at high productivity is that described for guano-enriched marine environments in South Africa (Bosman and Hockey 1988). Here rocky shores are colonised by dense algal mats that are protected from extinction by limpets through the predatory activities of birds such as the African black oystercatcher (*Haematopus moquini*).

It is a short and obvious step from the theories of Fretwell and Oksanen to a general model (Figure 113) in which the trophic structure of ecosystems is predicted as a simple consequence of the primary functional type (*sensu* Grime 1974) of the dominant plants. At extremely low productivity where the vegetation is composed exclusively of stress-tolerators (Figure 113a) the defences against generalist herbivores are so effective that the herbivores are restricted to a small biomass of specialist feeders with predators that are also likely to be extremely sparse and specialised. Where productivity is higher

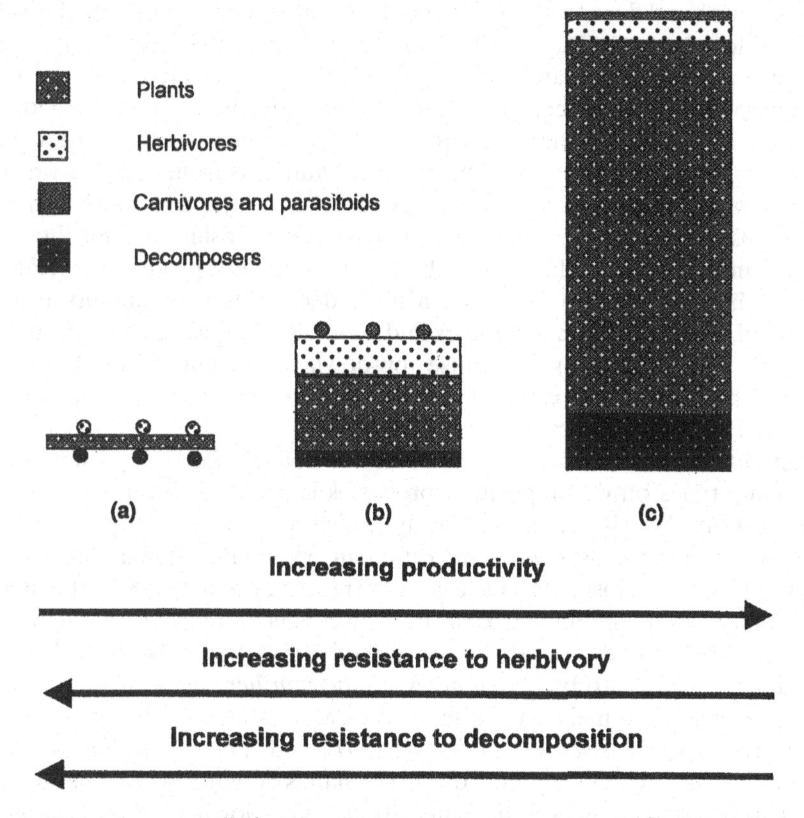

Figure 113. Scheme predicting the changes in abundance of trophic elements along a gradient from low (a) to high (c) productivity

(Figure 113b) and defences are less effective, generalist herbivores may be expected in greater abundance provided that they are insufficient to constitute a large and reliable food source for predators. However, this situation would not be expected at the highest levels of productivity (Figure 113c); here high rates of population growth by herbivores are possible, attracting the constant attention of predators which effectively suppress herbivore numbers (e.g. Bohan *et al.* 2000) and protect the palatable vegetation.

On the same logical basis predictions can be made with regard to the decomposer communities likely to be supported in the circumstances depicted in Figure 113. At the lowest productivity (Figure 113a) the mass of litter present in the ecosystem is likely to be large relative to the living component but tough in texture, high in lignin, and low in nutrient concentrations, making it resistant to decay (Daubenmire and Prusso 1963; Meentemyer 1978) and likely to support only a comparatively small biomass of decomposers. This is because, as suggested by Grime and Anderson (1986), the defence mechanisms that protect the living leaves of stress-tolerators against herbivory are likely to persist in effectiveness against decomposing organisms, restricting the soil fauna and flora to specialist organisms capable of exploiting relatively-intractable substrates (Pugh 1980; Cooke and Rayner 1984). Particularly important organisms here are the mycorrhizal fungi associated with the root systems of trees and ericoid shrubs exploiting infertile soils. A burgeoning literature reviewed by Smith and Read (1997) supports the hypothesis that these fungi have the ability to obtain mineral nutrients from complex organic sources within relatively resistant types of litter. Larger and more dynamic decomposer communities may be anticipated with rising soil fertility and ecosystem productivity (Figures 113b, c). At the highest productivity (Figure 113c) it is interesting to note that a large decomposer community can be predicted, first, because competitors and ruderals produce large quantities of palatable litter and second, because top-down protection against herbivory dictates that a high proportion of the plant tissues survives long enough to be exploited by decomposers.

Support for the existence of predictable controlling effects of primary plant functional types on decomposition processes is available from a large-scale investigation (Wardle *et al.* 1997a) in which vegetation composition and decomposition processes were measured in 50, small, uninhabited islands (Figure 114). Larger islands have a greater frequency of fires due to the higher probability of lightening strikes. In consequence, there is a consistent tendency for smaller islands to be occupied by late successional, slow-growing trees and shrubs (*Picea abies*, *Empetrum hermaphroditum*) and for larger islands to remain in earlier successional states exploited by more productive species such as *Pinus sylvestris* and *Vaccinium intermedium*. Various measurements conducted on the islands (Figure 114) reveal a clear association between rates of decomposition, successional state of the vegetation, and island size. Although this strongly suggests controlling effects of the dominant plants on decomposition through the quality of their litter it is important to note that, as succession proceeds, a positive feed-back is likely in which declining litter quality increasingly dictates the development of stress-tolerant vegetation.

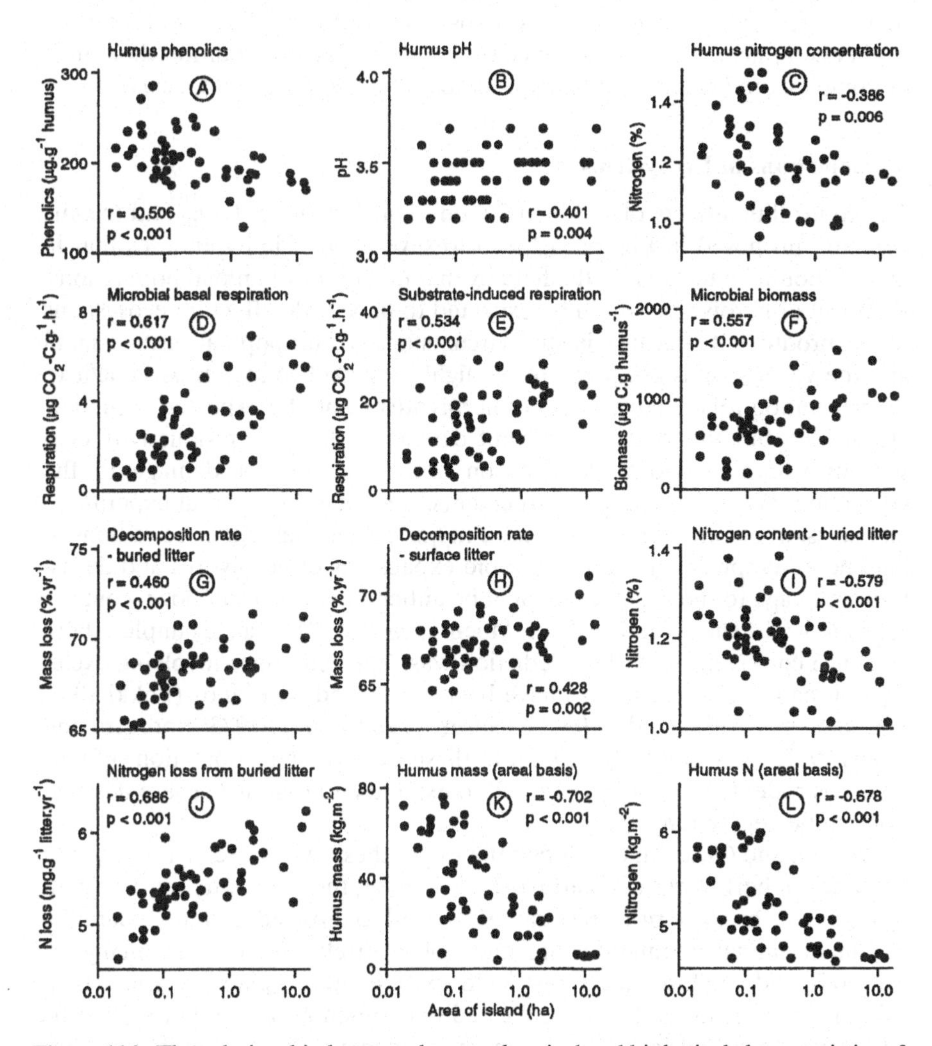

Figure 114. The relationship between humus chemical and biological characteristics of the soil and the area of 50 uninhabited islands in the Baltic Sea. A–F, chemical and biological properties of humus samples collected from each island. Concentrations of phenolics are expressed as mg gallic acid equivalent per g humus. G–J, properties of leaf litter of *Vaccinium myrtillus* placed in litter bags and left to decompose for one year. K, L, total mass of humus and humus nitrogen determined on an areal basis. (Reproduced from Wardle *et al.* 1997a by permission of *Science.*)

The scheme reproduced in Figure 113 is a theoretical construction based upon patterns of ecosystem assembly observed in natural environments and laboratory conditions. Even if the principles upon which it is based are true we should not expect to see all ecosystems conforming closely to this model. As noted earlier, the ecology of herbivores and carnivores is strongly affected by their territories and mobilities and their impacts will be influenced by the spatial distributions of ecosystems across the landscape. Food webs can be further complicated by the presence of omnivores, and in addition there may be complex food chains involving predators that exploit decomposers.

Manipulations of Ecosystems

The controlling effects of productivity on plant functional types and trophic structure proposed in Figure 113 lead to several specific hypotheses that in theory should be testable in the field by manipulations of animal populations. Removal of herbivores would be expected to have little effect on ecosystems of low productivity because in such circumstances their populations are small and the vegetation is predicted to be highly resistant to attack. A benefit to the vegetation following removal of herbivores would be expected at moderate fertility but this effect would not extend to the high productivity ecosystems where top-down controls on herbivores restrict damage to the vegetation. Following the same arguments, we would expect that experimental exclusion of carnivores would have major effects only under conditions of high potential productivity where rapid expansions of herbivores and devastating damage to the vegetation might be anticipated when top-down controls are removed. The experiment of Moen *et al.* (1993) is an example where evidence consistent with this prediction was reported. Tests involving exclusion of mammalian carnivores have been conducted on islands (Pokki 1981; Oksanen *et al.* 1987) and in field exclosures (Krebs *et al.* 1973; Boonstra and Krebs 1977; Desy and Batzli 1989). In these experiments populations of herbivorous rodents have been observed to expand and to inflict severe destruction on the vegetation.

Fretwell and Oksanen developed their hypothesis with specific reference to mammalian herbivores and carnivores. However, the possibility must be considered that the same principles can be applied to those ecosystems where the dominant animal populations are those of invertebrates. Because many arthropods and molluscs have restricted mobility the possibility arises that, using these organisms, highly replicated experimental manipulations can be conducted on a local scale. The recent experiments at three separate locations by Carson and Root (1999, 2000) in which above-ground insect herbivores were suppressed by frequent application of a broad-spectrum herbicide confirm that at moderate productivity herbivores are capable of substantial reductions in the plant biomass and can change the identity of the dominant plant species. It is particularly interesting that these effects took place without obvious signs of damage to the leaf canopy and appeared to be attributable to the xylem-feeding spittle-bug (*Philaenus spumarius*).

The investigations of Carson and Root (1999, 2000) clearly demonstrate the involvement of herbivores in plant community productivity and species composition. In order to accommodate these effects within the broader subject of ecosystem trophic dynamics, however, we need to establish how the role of herbivores varies across sites differing in productivity. Recently an attempt has been made to pursue this objective by herbivore removal along a natural productivity gradient. A conspicuous feature of many of the limestone areas of North Derbyshire in northern England is a mosaic of herbaceous communities varying in productivity from sparsely vegetated outcrops to tall herb communities on deep valley bottom soils (Balme 1953; Pigott and Taylor 1964; Grime and Blythe 1968; Lloyd *et al.* 1971; Grime and Curtis 1976). Laboratory screening experiments have revealed that these local gradients of increasing productivity coincide with increasing vegetation height and potential growth rates in component plant species (Grime and Hunt 1975). They are also associated with a marked decline in the effectiveness of plant defences against generalist invertebrate herbivores (Grime *et al.* 1968, 1996). These results suggest that different mechanisms may be involved in the persistence of the plant communities at opposite ends of the productivity gradient. A manipulative field experiment was therefore conducted (Fraser and Grime 1998a) to test the hypothesis of Fretwell and Oksanen that with increasing productivity there is a shift from direct plant defence (low palatability) to indirect protection of plants by carnivores (top-down control).

Unfortunately, it is not possible to devise experimental treatments that differentiate consistently between herbivorous and carnivorous invertebrates. The method applied by Fraser and Grime therefore involved removal of all invertebrates above and below ground by repeated application of insecticides and molluscicides in field plots corresponding to six positions along a productivity gradient. The whole experiment was contained within a small area of landscape and at each position on the gradient comparisons of vegetation responses were made between treated and untreated plots. The results (Figure 115a) strongly support the Fretwell–Oksanen model of trophic dynamics in that positive responses of the vegetation to the removal of invertebrates by pesticide treatments were confined to the sites of intermediate productivity. This suggests that in the middle of the productivity gradient the amount of palatable plant material was sufficient to support generalist herbivores but did not reach the level required to sustain a density of carnivores sufficient strongly to suppress herbivory. Support for this interpretation was obtained from bioassays (Figure 115b) in which discs of palatable leaf material (lettuce) and prey items (blowfly maggots) were introduced overnight to untreated plots along the productivity gradient. Again evidence of herbivory obtained using this technique was mainly confined to sites of intermediate productivity. The consumption of prey items, however, showed a continuous increase in

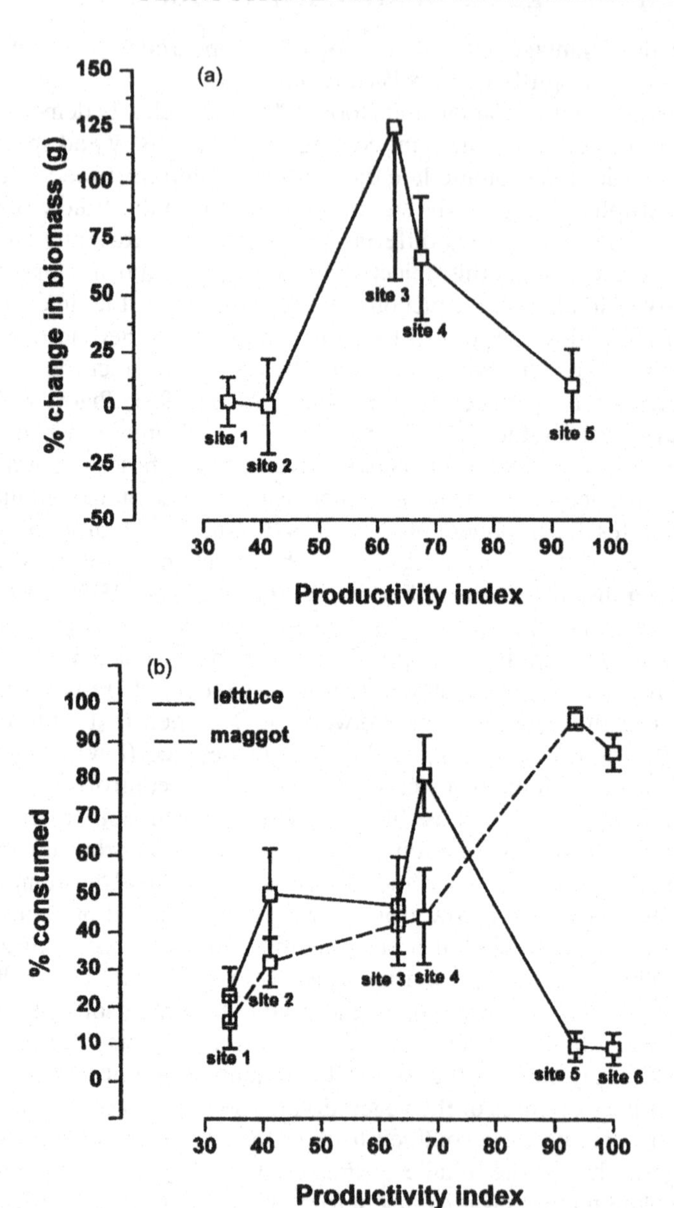

Figure 115. (a) Effects of pesticide treatment on the mean percentages change in shoot biomass for five sites investigated at Tideswell Dale, North Derbyshire, England. (b) Mean percentage of lettuce discs and maggots consumed at each of the six sites at Tideswell Dale. Vertical lines represent 95% confidence limits. (Reproduced from Fraser and Grime 1988 by permission of *Oikos*.)

parallel with rising productivity, a pattern consistent with the prediction of a maximal effect of top-down control at sites of high soil fertility and occupied by productive vegetation.

In theory it should be possible to use manipulative experiments to test predictions of the consequences of variation in the identity and number of plant functional types on decomposer communities. This can be attempted either by removing particular vegetation components (Mikola and Setalä 1998; Wardle *et al.* 1999) or by adding living specimens or their litter (Wardle *et al.* 1997a) to existing communities. So far it has not been possible to demonstrate strong controlling effects of plants on decomposer food webs; in the investigation of Wardle *et al.* (1999), for example, gross manipulations of the species composition of the vegetation, over a three-year period, induced remarkably little response in populations of earthworms, collembola, mites, and microbe-feeding nematodes. This suggests that below-ground food webs may be the product of formative processes operating over long periods of time and buffered against fluctuations in litter quantity and quality. These results also cast further doubts (see page 209) on the efficacy of removal experiments that involve unknown consequences of incomplete removal of roots from soil.

Synthesis of Ecosystems

Experimental results such as those of Fraser and Grime (1998b), when combined with those from similar field manipulations and bioassays (e.g. Coley 1983; Fraser 1998; Hulme 1994), provide circumstantial support for a theory in which productivity and functional characteristics of the dominant plants act as the primary determinants of trophic structure. However, such investigations cannot by themselves provide definitive proofs of the model. This is because in these field studies mechanisms are inferred and often there is limited knowledge of the animals involved and quantitative measurements of the herbivore and carnivore populations are incomplete and difficult to interpret. Particular difficulties arise from the fact that important effects of herbivores, carnivores, and parasitoids may be discontinuous, occurring as a consequence of occasional visits of organisms that achieve higher and more constant densities in neighbouring habitats.

In an attempt to introduce greater precision of measurement and a surer basis for mechanistic interpretation, several ecologists (e.g. Beyers and Odum 1993; Heal and Grime 1991; Lawton 1995; Fraser and Keddy 1997) have advocated the use of microcosms to investigate trophic interactions in terrestrial ecosystems. When ecosystems are allowed to assemble in closed containers there are opportunities to control the trophic structure and the input of species and to manipulate the environmental conditions. It is also possible to monitor more precisely the fate of component populations and to examine the effects of experimental variables on the trophic structure and dynamics of the developing ecosystems. In particular there are excellent opportunities to test the validity of the Fretwell–Oksanen model relating trophic structure to productivity.

Fraser and Grime (1998b) conducted an experiment in which insect-proof outdoor microcosms were used to examine the effect of productivity (controlled by manipulating soil fertility) on the development of trophic structure in an extremely simplified ecosystem synthesised from three grass species, the grass aphid (*Sitobion avenae*), and its predator the seven spot ladybird (*Coccinella septempunctata*). The results supported the Fretwell–Oksanen model in that introduction of the predator brought about a reduction in aphid numbers and an increase in the yield of the most palatable grass (*Poa annua*). Again in accordance with the model this effect was confined to the high fertility treatment; at low productivity plant biomass was low, aphid numbers were small, and ladybird activity was minimal.

The grass/aphid/ladybird experiment provides support for the Fretwell–Oksanen hypothesis by reducing the ecosystem to its bare essentials and allowing a direct test of the underlying principles. However, it must be recognised that tests of this form represent one extremity in the tradeoff in experimental design between precision and reality. In order to investigate more subtle consequences of the mechanisms and interactions implicit in the Fretwell–Oksanen model it may be necessary to develop the microcosm approach to accommodate more complex ecosystems with a greater diversity of organisms. In particular there is considerable scope not only to demonstrate the impacts of different types of invertebrate herbivores on vegetation structure and composition (Fenner 1992; Hanley *et al.* 1995; Hulme 1994, 1996a, 1996b) but also to investigate the possibility that the foraging behaviour of particular predators influences vegetation development by modifying the density, age-structure, and behaviour of herbivore populations. It could be argued that some of these subtleties are more appropriately studied by direct observations in natural vegetation. However, the probability of recognising trophic interactions of general occurrence and significance is likely to be higher where ecosystems are allowed to assemble and are closely studied in well-replicated experiments involving a large initial pool of plant species and controlled manipulations of the trophic structure of the experimental ecosystem. An experiment of this kind using the herbivorous slug, *Deroceras reticulatum*, and its predator the ground beetle, *Pterostichus melanarius*, was conducted over a two-year period on a species-rich turf allowed to assemble from seed in outdoor microcosms (Buckland and Grime 2000). Figure 116, describes the changes that were recorded in the abundance and size structure of populations of four plant species as a consequence of the presence of these herbivores and carnivores. The most conspicuous effect in all four species was the reduction in population size in the presence of herbivores. However, it is also apparent that losses to herbivory were reduced in the microcosms containing beetles. It is interesting to observe that although this

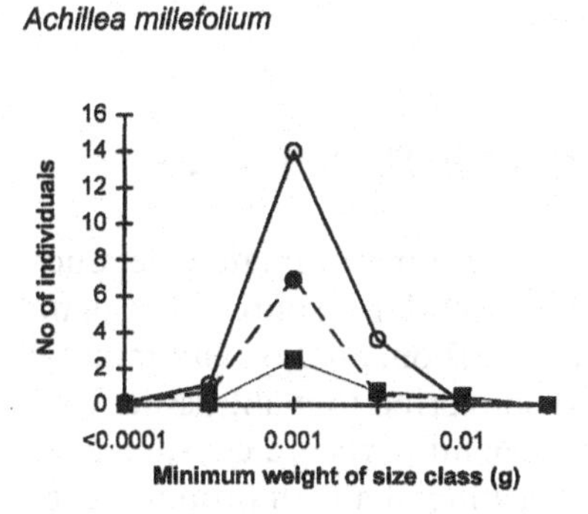

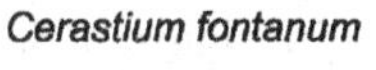

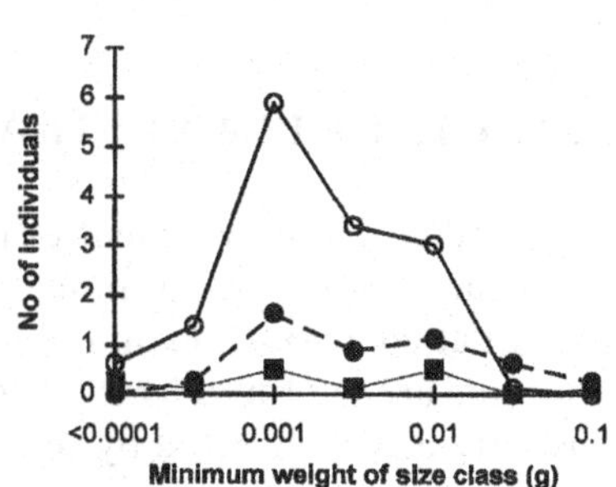

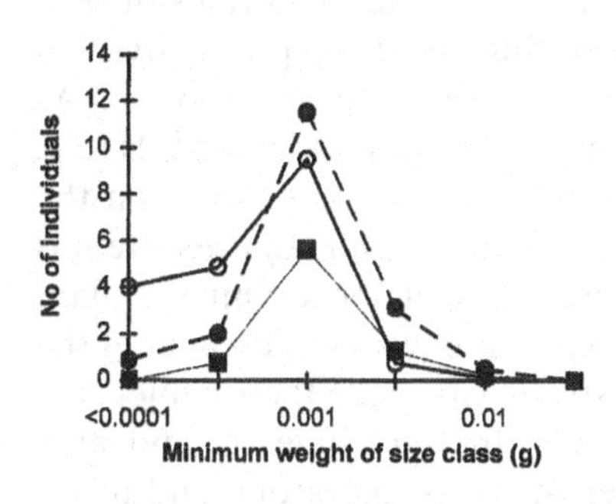

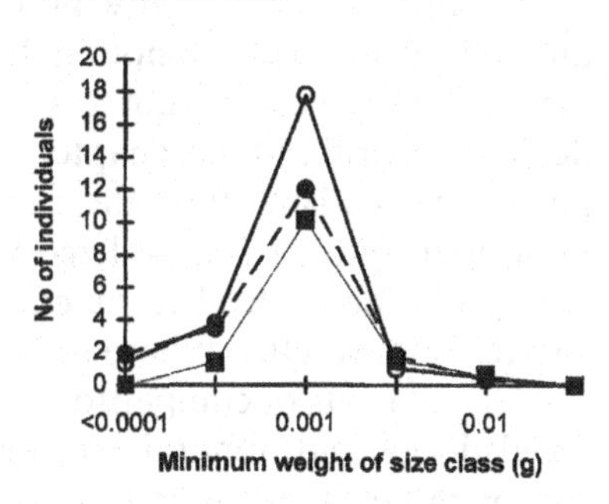

Figure 116. The effects of the presence of the herbivorous slug, *Deroceras reticulatum*, and its predator, the ground beetle, *Pterostichus melanarius*, on the abundance and size distribution of individuals in populations of four dicotyledons in vegetation synthesised from seed and allowed to develop for two years at low soil fertility in ventilated outdoor microcosms. ○ vegetation only; ■, slugs present; and ●, slugs and beetles present. (Reproduced from Buckland and Grime 2000 by permission of *Oikos*.)

protective effect of the beetle was strongly developed in *Bellis perennis* the benefit did not extend to the smaller size classes which remained vulnerable to slug attack.

The slow responses of decomposer food webs to field manipulations have prompted some ecologists to employ the alternative tactic of ecosystem synthesis in microcosms as a method by which to investigate the functional interplay between plants and decomposers. In a comprehensive study of the effect of simple and more complex food webs on plant productivity, for example, Laakso and Setala (1999) were able to identify specific benefits to mineralisation and plant growth arising from the inclusion of fungivores and microbi-detrivores in the decomposer community. This encourages the view that ecosystem synthesis could also provide the basis for tests of the effects of soil nutrient supply and plant functional types on decomposer biomass and composition predicted in Figure 113.

EFFECTS OF PLANT STRATEGIES ON PRODUCTIVITY

The production of living matter in terrestrial ecosystems ultimately depends upon the supply, on an annual basis, of the essential resources for plant growth at each location and the rates at which carbon, energy, and mineral nutrients are captured by the vegetation and converted to plant tissue. Productivity is also affected by the rates at which plant parts are consumed by herbivores and pathogens or removed by harvesting; if a high proportion of the plant biomass is continuously destroyed there is a threshold of damage above which even fast-growing vegetation cannot sustain high rates of production. It is also worth noting that the controlling effects on productivity of plants themselves are not restricted to resource capture and growth. As described earlier in this chapter and demonstrated in terrestrial models (e.g. Pastor and Cohen 1997) the rate at which mineral nutrients (so often the limiting resource in natural ecosystems) are recycled within an ecosystem is strongly affected by the functional characteristics of the dominant plant species (Figure 110). Whereas the short life-span and weaker defences of the leaves and roots of competitor and ruderal strategists make them subjects of rapid decay and mineral nutrient release, the greater longevity and high carbon:mineral nutrients ratio of the tissues of stress-tolerators tend to reduce the rate of recycling by imposing long residence times for mineral nutrients in both living plant parts and litter (Aber and Melillo 1982; Bosatta and Staaf 1982; Berendse *et al.* 1989; Wedin and Tilman 1990; Hobbie 1996).

What are the implications of these fundamental facts for our understanding of how vegetation and productivity relate to each other? Where and when does productivity control vegetation structure and composition and under what conditions does vegetation control productivity? What determines the balance between the two processes? Here it will be suggested that:

1 In most ecosystems both vegetation and productivity and many other ecosystems properties are determined to an overwhelming extent by environmental conditions and management.
2 In some ecosystems a small number of dominant plants can exert modifying effects, increasing or decreasing productivity, sometimes to a considerable extent.

The evidence leading to these conclusions will now be reviewed under two headings. The first summarises information from direct study of natural ecosystems and manipulative experiments and the second examines insights from the synthesis of ecosystems in field plots or microcosms.

Field Observations and Manipulations

There can be no doubt that in many natural plant communities circumstances arise that permit co-existence between species of different functional types. In Chapter 9 numerous examples are cited in which plants with complementary morphologies, physiologies, phenologies, or regenerative strategies can be found within the same communities. Few ecologists will dispute the hypothesis that within such communities co-existence involves exploitation of spatial or temporal niches. It is also tempting to propose that, in some of these communities, the effect of species richness will be to allow a close matching of functional types with niches and a more exhaustive capture of resources leading to higher ecosystem productivity.

Later in this chapter recent experimental evidence of beneficial effects of species richness on productivity will be subjected to critical scrutiny (pages 329–334). First, however, it is necessary to examine the hypothesis linking species richness, niche exploitation, and productivity against the background provided by earlier chapters and, in particular, by reference to the review of mechanisms controlling species richness in vegetation presented in Chapter 9.

Perhaps the most important point to be made here is that although the composition and species richness of plant communities is dependent upon the fitness of component plant populations fitness itself does not invariably involve the assumption of traits that maximise dry matter production. In conditions of high potential productivity and low incidence of vegetation disturbance (pages 14–37) fitness is likely to depend upon high competitive ability and the struggle for dominance between contending species, a process that will tend to drive up productivity. However, there is abundant evidence that on less fertile soils fitness may depend upon attributes that, at least in the short-term, can have the effect of driving down ecosystem productivity. A convenient way in which to examine this evidence and its implications is to return to the humped-back model of species richness in herbaceous vegetation described on pages 292–294. In Figure 117 this model had been redrawn to focus attention, in the following discussion, on the very different relationships between species richness and productivity that can be predicted for the different kinds of vegetation corresponding to sections of the model.

In Zone A, which refers to vegetation of extremely low productivity and species richness both restricted by the extreme conditions (scarce resources, severe disturbance, or some combination of the two), we would not expect to encounter circumstances where differences in species richness are acting as the primary determinant of productivity. It is much more likely that plant characteristics with little direct bearing on productivity will be the focus of natural selection in these circumstances. In the case of resource-poor skeletal

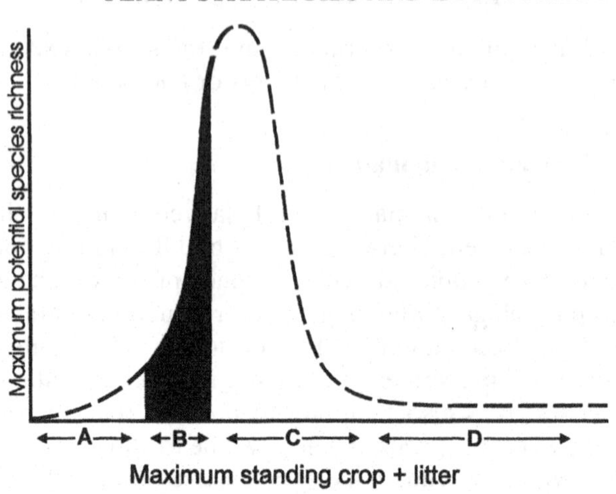

Figure 117. The humped-back model (see Figure 88) redrawn to define four zones (A–D) considered in the text. The shaded area indicates the zone in which rising species richness coincides with increasing biomass + litter

habitats, sequestration of carbon and mineral nutrients and tissue tolerance will take precedence over growth and where frequent and severe disturbance prevails productivity will be restricted by the brevity of life-histories and the early diversion of captured resources into reproduction that characterises success in such unstable conditions. In both of these extreme circumstances characteristic of Zone A it would appear that the scope for complementary exploitation of the environment by a diversity of species is strongly restricted by the severity of the filters operating on community assembly. Vegetation development and species interactions may be so inhibited in Zone A that we may even question whether it is appropriate to apply the terms community and ecosystem here.

In Zone B where the constraints on plant biomass production and/or survival progressively relax we see a positive correlation between species richness and productivity. A challenge of the most profound and practical significance to ecologists, agricultural scientists, and to mankind as a whole is to determine whether the rising curve in Zone B signifies that species richness drives up productivity or alternatively that higher productivity merely allows the admission of more species. The most likely explanation is that both processes can occur, sometimes simultaneously, but that the latter is much more important. Even where beneficial effects of plants are associated with increasing species richness it seems most likely that these are attributable to the presence of particular functional types and species rather than a general effect of species richness. Following the argument developed in Chapter 5 impacts of plant species on ecosystem properties such as productivity depend upon achieving a critical mass (i.e. attaining dominant-status in the community).

Any contribution of beneficial effects of plants to the rising curve in Zone B is therefore more likely to arise from the presence of particular species than from the development of a richer mixture. This is easier to envisage in the case of vegetation on infertile soil where, for example, incursion and expansion to a position of dominance by particular nitrogen-fixing lichens or leguminous plants might bring about an amelioration of conditions to a point where a diversity of more nutrient-demanding species can survive. However, it would be a mistake to overemphasise beneficial effects of dominant plant species in Zone B. It is also possible to recognise well-documented circumstances (Muller 1884; Handley 1954; Grime 1963; Grubb *et al.* 1969; Miller *et al.* 1979; Berendse *et al.* 1989a, b; Aerts 1989; Wardle *et al.* 1997b; Berendse 1998) where establishment and consolidation by stress-tolerant dominants, particularly conifers and Ericaceae, may have a retrogressive effect in which productivity and species richness decline as a consequence of the deposition of poor-quality litter and the reduced accessibility of mineral nutrients. Quite clearly, several interactions and feedbacks may be implicated in the rising curve of Zone B. As explained on page 334, this complexity is highly relevant to the interpretation of experiments in which productivity has been examined in communities synthesised from species-poor and species-rich seed mixtures.

Zone C in Figure 117, corresponds to the very wide range of circumstances, familiar to agricultural researchers, nature reserve managers, and conservationists where rising productivity or reductions in biomass removal have coincided with decreasing species richness. This phenomenon has been documented by observations in natural habitats following changes in management (Watt 1957; Thomas 1960; Duffey *et al.* 1974; Ratcliffe 1984) and in numerous experimental manipulations involving either the application of mineral fertiliser to grasslands (Thurston 1969; Willis 1963, 1989; Jeffrey 1971; Tilman 1987a) or exclusion of grazing animals (Pickworth-Farrow 1916; Tansley and Adamson 1925). In all these recorded observations and experiments the loss of species richness is associated not only with a rising mass of living material and litter but also with increasing dominance of the community by particular plant species. From the evidence assembled in Chapters 4 and 9 there appears to be a wide consensus among ecologists that the descending arm of the humped-back model, contained in Zone C, arises as a consequence of the increasing monopoly of vegetation of high biomass by dominant species.

Finally, in Zone D of the model we encounter vegetation in which diversity has been reduced to a very low level by the dominant effect of one or a few robust species. Chapter 4 contains a review of the various circumstances where such dominance occurs as a consequence either of high rates of resource supply (e.g. intensively fertilised meadows or riverine vegetation) or of dereliction of productive farmland allowing the development of robust plant phenotypes and dense litter accumulation.

Synthesis of Ecosystems

Against a theoretical background in which the traits of dominant plants were widely suspected to be acting as the overriding controllers of ecosystem properties, considerable interest and controversy was generated when a paper appeared (Naeem *et al.* 1994) purporting to demonstrate an immediate benefit to ecosystem productivity arising from high species richness in assemblages of arable weeds that had been created by sowing seed mixtures containing different numbers of species on to bare soil. It was suggested that benefits arose in the species-rich mixtures from the presence of a wider range of morphologies and physiologies, generating complementary and more complete exploitation of resources. Interest in this publication extending beyond the realm of ecology was stimulated by a commentary (Karieva 1994) suggesting that studies of this kind could provide a justification for the conservation of species-rich ecosystems.

Darwin, Diversity and Productivity

Before examining the validity of the conclusions drawn by Naeem *et al.* (1994) it is necessary to draw attention to the apparent conflict between the notion of species diversity promoting productivity and the abundant evidence from direct observation of natural plant communities, reviewed in Chapter 9, that with modest increases in plant biomass across the productivity range which is most relevant to agriculture there is a negative relationship between species richness and biomass in herbaceous vegetation (i.e. in the descending section (Zone C) of the widely-validated humped-back model, see page 262). Naeem *et al.* (1994) and many authors who have recently published similar results and interpretations (e.g. Tilman *et al.* 1996; Hector *et al.* 1999) make no reference to this apparent conflict. Why have these researchers ignored such a large volume of empirical data accumulated by several generations of plant ecologists? In seeking an answer to this question it is instructive to examine the introductory sections of these recent papers. A consistent feature is the reference to Darwin's observations (Darwin 1859) on niche differentiation and productivity in grassland.

> Darwin observed that multi-species plant assemblages are more likely to be more productive than monocultures of the species within the assemblage . . .
>
> Naeem *et al.* (1999)

> Darwin (1872) suggested that greater plant diversity would lead to greater productivity, but his thoughts lay dormant for over a century.
>
> Tilman (1999)

> For example, Darwin (1859) first noted that monocultures of grasses were less productive than ecologically diverse mixtures of plants. Not surprisingly, much of the recent research has also focussed on the productivity of plant communities in relation to the diversity of species and functional groups.
>
> Hector (2000)

These statements reveal that an influential factor prompting recent experimental study of the influences of diversity on vegetation productivity can be traced to the writings of Charles Darwin. It is important, therefore, to examine Darwin's ideas and the evidence upon which they were based. The critical passage appears in Chapter 4 of *Origin of Species* (Darwin 1859) under the subheading 'Divergence of Character':

> It has been experimentally proved, that if a plot of ground be sown with one species of grass, and a similar plot be sown with several distinct genera of grasses, a greater number of plants and a greater weight of dry herbage can thus be raised. The same has been found to hold good when first one variety and then several mixed varieties of wheat have been sown on equal spaces of ground. Hence, if any one species of grass were to go on varying, and those varieties were continually selected which differed from each other in at all the same manner as distinct species and genera of grasses differ from each other, a greater number of individual plants of this species of grass, including its modified descendants, would succeed in living on the same piece of ground. And we well know that each species, and each variety of grass is annually sowing almost countless seeds; and thus, as it may be said, is striving its utmost to increase its numbers. Consequently, I cannot doubt that in the course of many thousands of generations, the most distinct varieties of any one species of grass would always have the best chance of succeeding and of increasing in numbers, and thus of supplanting the less distinct varieties; and varieties, when rendered very distinct from each other, take the rank of species.
>
> The truth of the principle, that the greatest amount of life can be supported by great diversification of structure, is seen under many natural circumstances. In an extremely small area especially if freely open to immigration, and where the contest between individual and individual must be severe, we always find great diversity in its inhabitants. For instance, I found that a piece of turf, three feet by four in size, which had been exposed for many years to exactly the same conditions, supported twenty species of plants, and these belonged to eighteen genera and to eight orders, which shows how much these plants differed from each other. So it is with the plants and insects on small and uniform islets; and so in small ponds of fresh water. Farmers find that they can raise most food by a rotation of plants belonging to the most different orders: nature follows what may be called a simultaneous rotation. Most of the animals and plants which live close round any small piece of ground, could live on it (supposing it not to be in any way peculiar in its nature), and may be said to be striving to the utmost to live there; but, it is seen, that where they come into the closest competition with each other, the advantages of diversification of structure, with the accompanying differences of habit and constitution, determine that the inhabitants, which thus jostle each other most closely, shall, as a general rule, belong to what we call different genera and orders.

These two paragraphs refer to quite different circumstances and kinds of information. In the first, Darwin reports on the work of anonymous researchers who achieved higher yields of herbage from plots sown with a mixture of grasses as compared with monocultures. From the information provided we cannot judge whether the superior yield of the mixtures arose from complementary exploitation of resources, from the inclusion of particular productive species in the mixtures, or from a higher total density of individual plants in the mixtures.

In the second paragraph, Darwin reports his own observations of stable co-existence over many years between a large number of taxonomically diverse plant species in a small area of turf. There is circumstantial evidence here of complementary exploitation of the habitat but no evidence is presented to prove that this was promoting the yield of the turf. There can be little doubt, however, that Darwin believed that the struggle for existence generated diversity in plant species and, anticipating Connell's 'ghost of competition past' (Connell 1980), he interpreted diversity in a plant community as a sign that competition was driving up productivity by more complete exploitation of resources. This raises several important questions. Did Darwin recognise the extent to which diversity was maintained in his species-rich turf by nutritional factors and defoliation (grazing or mowing) that act as powerful constraints (see pages 257–266) restricting competition and permitting co-existence? More fascinating still, how did Darwin explain the existence of vegetation of low diversity? As a keen observer of the British countryside and as a world traveller he must have been closely familiar with circumstances (beech woods, blackthorn thickets, bracken patches, bramble patches, nettle patches, heather moors, cotton sedge bogs, willow carr, reed beds, papyrus swamps, bamboo stands, and sugar-cane fields, to name but a few) where extensive monocultures arise naturally and, in some cases, coincide with exceptionally high productivity. Darwin does not refer to such monocultures and so we do not know whether he recognised circumstances where lack of diversity indicated that competition had been resolved in favour of one or a few monopolistic species.

The towering status of Darwin in biology brings with it the danger that his extensive writings can be selectively and tenuously allied to many current theories. His pre-eminent status must not be allowed to invest his brief comments on productivity and diversity with a decisive significance. To state that 'Darwin suggested that greater plant diversity would lead to greater primary productivity, but his thoughts lay dormant for over a century' (Tilman 1999) is to overstate Darwin's observations and to disregard the very considerable research achievements of several subsequent generations of agricultural and ecological research scientists in many parts of the world. As explained in Chapter 9 and in the opening section of this chapter, for herbaceous vegetation, there was already in existence by 1990 a rich literature defining the circumstances where productivity controls diversity and where diversity *may* control productivity. It is unfortunate that this literature has not played a more influential role in the planning and interpretation of many recent biodiversity/productivity studies.

Experimental Design and Data Interpretation: the Current Debate

Returning now to the specific case of the results published by Naeem *et al.* (1994) considerable doubts have been cast on the validity of the conclusions drawn by the authors of this paper and these have been reviewed in detail by André *et al.* (1994), Garnier *et al.* (1997), Huston (1997), Hodgson *et al.* (1998), Grime (1998), and Wardle (1998, 1999). It appears that, in this study a property attributed to high species richness was in reality mainly due to the presence in the more diverse communities only of species (e.g. *Chenopodium album*) with traits (large size and rapid growth rates) attuned to high productivity. More recent experiments (e.g. Hooper and Vitousek 1997; Tilman *et al.* 1997a; Hector *et al.* 1999) have failed to provide convincing support for effects of high species richness on ecosystem productivity.

On first inspection synthesis of ecosystems from seed mixtures varying in richness appears to provide a direct and informative method by which to determine whether productivity is controlled by the characteristics of dominant species or by the species richness of vegetation. There are two main reasons why this approach has not so far resulted in definitive tests:

1. A crucial flaw in the experiment by Naeem *et al.* (1994) was the inclusion of species in the species-rich assemblages that were not represented in the species-poor mixtures. In an attempt to rectify this design fault, experiments such as those of Tilman *et al.* (1997) and Hector *et al.* (1999) have used species-mixtures in which the composition of the seed mixture sown into individual replicate plots is based upon a random draw of species. However, as pointed out by Huston (1997), this procedure does not avoid the risk that the species-rich mixtures more often than the species-poor mixtures will contain species of high potential productivity. This, of course, arises as a simple consequence of the greater chance of representation of all types of species, including the most productive ones, in the species-rich mixtures. This 'selection probability' effect leads to a confounding of the two possible explanations for any rise in productivity that is found to be associated with increasing the number of species in the seed mixture. Higher yield might be the result of greater species richness but it is also possible that this effect could be the result of including particular species of high potential productivity.
2. In order to differentiate between the two possible explanations (described under 1 above) for any rise in productivity associated with species richness of the seed mixture it is imperative that data analysis and interpretation involves measurements of the numbers of species surviving and their relative abundance in the synthesised communities. In the experiments reported by Naeem *et al.* (1994), Tilman *et al.* (1996, 1997), and Hector *et al.* (1999) estimates of productivity are plotted against the numbers of species sown in the seed mixtures. In the absence of data on the actual

membership of the synthesised communities and the composition of harvested biomass it is not possible to validate the claims in all four of these published studies that benefits to productivity of species richness *per se* have been demonstrated.

The Limitations of Synthesised Ecosystems

In an attempt to circumvent some of the difficulties outlined in the preceding section and, in particular to avoid problems with the creation of communities from seed, Booth and Grime (2001) measured productivity in microcosms (Figure 112, page 310) in which assemblages were created from vegetative transplants of 12 species removed from a small area of ancient calcareous pasture in North Derbyshire. By this method, the vagaries of seedling establishment were avoided and it was possible to exercise close control of both the species-richness and functional composition of the synthesised communities. Several other features of the experimental design are worthy of comment:

1 By using as transplants individuals randomly selected from plant populations of common species co-existing in the same small area of ancient pasture the experiment employed genotypes that had already survived the 'selective sieves' operating in a natural ecosystem.
2 Previous studies at the site from which the transplants originated (Pearce 1987) had shown that species were mainly represented by compact long-lived individuals, small in stature and, in most cases, randomly distributed in the turf. This allowed a convenient planting arrangement in which a standard total number of similar-sized transplants could be introduced to each microcosm using a planting grid in which each species was distributed at random.
3 Confirmation that the manipulations of species richness and functional composition had been realised was obtained by recording the abundance of each species in the vegetation by point-analysis at frequent intervals throughout the experiment.
4 Measures were taken to ensure that the conditions prevailing in the microcosms approximated as far as possible to those of the field site from which the transplants were obtained. The natural, unfertilised soil used in the experiment was removed from a pasture closely-similar to the source site and defoliation and trampling treatments were applied to simulate the effects of sheep grazing.

An important consequence of provisions 1 to 4 was that a turf structure resembling closely that from which the transplants had been removed was established quickly with clearly-defined differences in species composition (Figure 118). Perhaps most important of all, however, was the fact that the

Figure 118. Plant communities synthesised in lysimeters using transplants from Cressbrookdale, North Derbyshire, England. (Left), a monoculture of *Leontodon hispidus*; (right), a species-rich mixture

created assemblages had productivities distributed within the range corresponding to section B of the humped-back model (Figure 117). Location of the experimental conditions in this zone ensured that the manipulations of species richness were conducted in circumstances where, in natural communities, rising species richness is correlated with increasing productivity. By choosing to experiment in circumstances corresponding to section B it can be argued, therefore, that this experiment was biased in favour of species richness and against dominance effects as controllers of ecosystem productivity.

The initial results of this experiment (Figure 119) have provided no evidence of beneficial effects of increasing species richness on productivity. However, interesting insights became available when the data were rearranged to examine the relationship between productivity and the abundance of grasses, sedges, and forbs in the experimental assemblages. Whereas no effect of increasing abundance on productivity was detected in the grasses and forbs, a substantial and statistically-significant benefit was associated with the presence of sedges (Figure 120). In physiological terms, this effect is not hard to interpret; grassland sedges possess dauciform roots (Davies *et al.* 1973), specialised structures (Figure 32, page 78) which are suspected to be capable of facilitating the mineralisation and capture of phosphorus, the limiting element for plant growth in many calcareous ecosystems.

These experimental results have two main implications concerning the use of synthesised ecosystems to investigate the control of productivity. The first

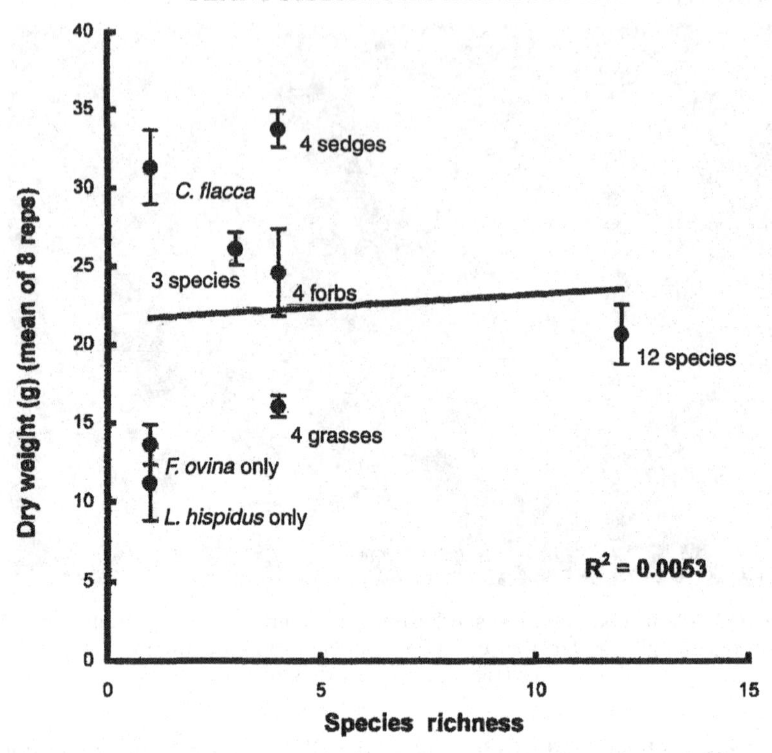

Figure 119. The relationship between total harvested biomass over a two-year period and species richness in grassland communities synthesised from mature transplants from Cressbrookdale, North Derbyshire, England. (Booth and Grime, unpublished.)

becomes obvious if we consider what would have happened if communities drawing upon the same 12 species but varying in species richness had been allowed to assemble from seed mixtures following the experimental protocols of Naeem *et al.* (1994) or Tilman *et al.* (1996). It seems most likely that sedges would have prospered under the conditions of the experiment and their more frequent inclusion in the species-rich seed mixtures would have led inevitably to confounding of a beneficial 'sedge effect' with increasing species richness and rising productivity. Only careful analysis taking account of the species composition of harvested material would have prevented false attribution of this sedge effect as one arising from species richness.

The second implication of the results of the Booth and Grime experiment is that vegetation development in ancient unproductive calcareous pastures involves processes that, at least in the short-term (two years), were not expressed in the microcosms. A conclusion that can be drawn from Figure 120 is that productivity in these ecosystems is strongly affected by

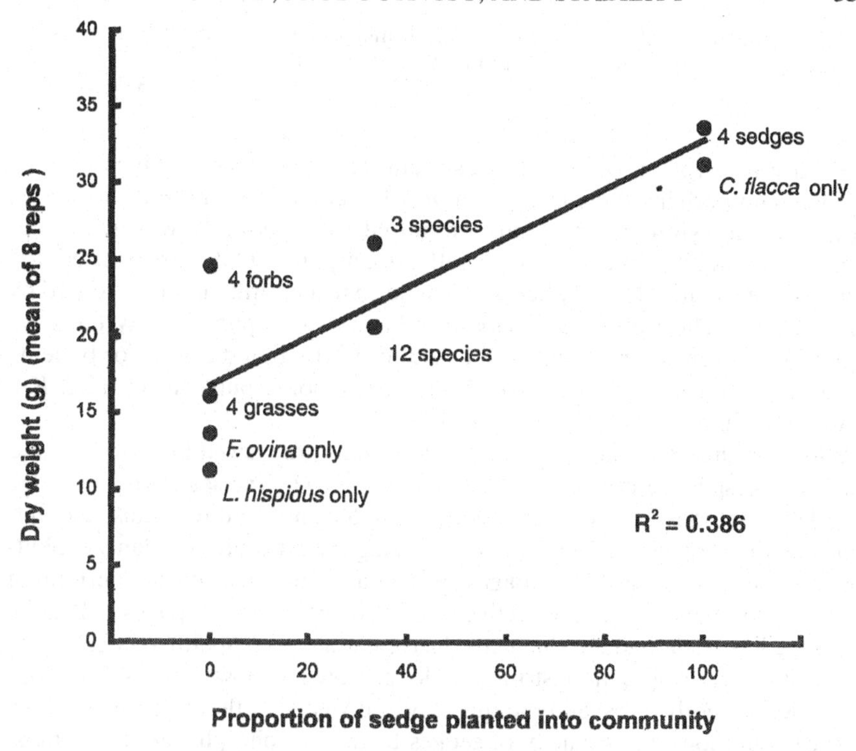

Figure 120. The relationship between total harvested biomass over a two-year period and the abundance of sedges in grassland communities synthesised from mature transplants from Cressbrookdale, North Derbyshire, England. (Booth and Grime, unpublished.)

the capacity of sedges to capture phosphorus and if we rely exclusively upon inferences based upon the microcosm assemblages we might suppose that natural selection in the field would drive the composition of such infertile grasslands towards a turf consisting exclusively of sedges. In reality ancient infertile calcareous pastures contain a large sedge component (Willis 1989) but also present is a rich diversity of grasses, forbs, small shrubs, and bryophytes. How can these insights from microcosm experiments and field observations be reconciled?

In addressing this paradox perhaps the most important first step we can take is to acknowledge the limitations of model experiments in microcosms, garden or field plots. Here it is instructive to consider a comment from the world of quantum mechanics:

> In the beginning natural philosophers tried to understand the world around them. . . . They hit upon the idea of contriving artificially simple situations. . . . Experimental science was born. But experiment is a tool. The aim remains to

understand the world. To restrict quantum mechanics to be exclusively about piddling laboratory operations is to betray the great enterprise.

Bell (1990)

The message from physics for ecosystem ecology is clear. Much can be learned, and perhaps some things can *only* be learned, from the tactic associated with the synthesis and study of simplified systems. However, the approach must not become an end in itself. As Carpenter (1996) recognised, the insights from synthetic ecosystems will need to be continuously confronted by field realities. The dangers in exclusive reliance on simplified ecosystems are only too apparent in the exaggerated and unsubstantiated claims for benefits of species richness to productivity contained in some publications (e.g. Karieva 1994; Tilman 1999; Hector *et al.* 1999).

When insights from the synthesis of ecosystems are viewed in a broader and balanced perspective it is not difficult to envisage why the selective advantage of sedges in a phosphorus-stressed ecosystem does not lead to exclusive dominance by these species. Already, in considering the attributes of plants exploiting infertile soils (Chapter 1, pages 55–63) and the mechanisms permitting species co-existence in unproductive vegetation (Chapter 9, pages 272–292), reference has been made to the inconclusive nature of competitive interactions in circumstances where life-histories are long, growth is slow, biomass is continuously lost to herbivores, and the quantity and vigour of the vegetation is low. In such conditions, the capacity of sedges to capture phosphorus is not translated into exclusive dominance of the vegetation because the mineral nutritional, and grazing constraints on biomass accumulation prevent the dynamic foraging of shoots and roots and the accumulation of the density of vegetation necessary for competitive dominance. It is also likely that, as in the case of legumes in nitrogen-limited pastures, the effect of grazing and dunging by sheep will be to redistribute mineral nutrients (including phosphorus captured by sedges) and to make them accessible to other species. The fitness of the sedges may be reduced by the metabolic costs of phosphorus foraging. More specifically, the inability of sedges to dominate unproductive grassland may be further explained by the seasonal delay in their germination (Schûtz 1998) and in the late seasonal leaf expansion of established plants exhibited by sedges (Grime *et al.* 1985); this dependence upon warmer temperatures (pages 271–278) is likely to provide numerous opportunities for incursions by less thermophilous grasses, forbs, and bryophytes.

This glimpse into the complex realities of a natural, species-rich ecosystem not only reminds us of the necessity to consult natural situations but also raises fundamental questions about the origins and consequences of diversity. Diversity depends upon the fitness of component individuals and populations and not their capacity to drive up ecosystem productivity. This is most obvious in the case of parasitic plants but it illustrates a basic principle that had been undervalued in this research field.

EFFECTS OF PLANT STRATEGIES ON STABILITY AND SUSTAINABILITY

All ecosystems are subject to impacts that affect their functioning and, in extreme form, may threaten their viability and survival. The forces responsible may be climatic (droughts, frost, fires, floods, windstorms) or biotic (grazing episodes, disease outbreaks, cultivation, pollution, urbanisation). Building on Levitt's (1956, 1975, 1980) pioneering and largely physiological broad surveys of plant responses to environmental stresses, Westman (1978) identified two key aspects of the response of an individual plant or a stand of vegetation to an extreme event. Resistance was defined as the ability of the plant biomass to resist displacement from control levels. Resilience was recognised as the speed and completeness of the subsequent return to control levels. When an ecosystem is exposed to an increasing severity or frequency of extreme events due to climate change (Mearns *et al.* 1984; Wigley 1985) or the various insidious consequences of rising human population density (Ratcliffe 1984; Thompson and Jones 1999) the focus of our concern is likely to shift from ecosystem reactions following specific events to the question of ecosystem sustainability. At what point with a rising frequency of extreme events or continuous but less obvious attrition is the essential character of an ecosystem lost and its value to man degraded irreparably? The answers to this question will be strongly affected by our individual subjective judgements about the value and usefulness of particular properties of pristine, severely-impacted or man-made ecosystems. Here in keeping with the main purpose of this book the objective will be to recognise the opportunities for objective analysis and prediction of ecosystem resistance, resilience, and sustainability that arise from recognition of primary plant strategies. A more specific objective is to consider the hypothesis that sustainability is most usefully analysed as the long-term consequence of ecosystem resistance and resilience. At the end of this chapter this hypothesis is accepted but, following the arguments developed in Chapter 5, it is also concluded that a full understanding of sustainability requires an appreciation of the impact of declining plant diversity at the landscape scale on the reassembly of ecosystems.

Resistance and Resilience

Ecologists have often commented informally on the sensitivity of vegetation to extreme events such as droughts, late frosts, windstorms, fires, and herbicide applications. However, in order to develop and test theories of community and ecosystem responses it is necessary to observe a more rigorous protocol in which quantitative measurements are made to record both the immediate consequences of natural or experimental extreme events (resistance) and the speed and completeness of recovery (resilience). Although it would be extremely interesting to conduct these measurements simultaneously on both plant and animal components, attention has so far tended to focus exclusively on the above-ground component of the vegetation, first, because such measurements are relatively easy to make and, second, because natural extreme events, such as drought, are highly unpredictable and the most common source of data is that arising as a fortuitous by-product of vegetation monitoring programmes designed for other purposes. It is also desirable that studies of extreme events should be comparative; only when measurements are made simultaneously in contrasted types of ecosystem is there the opportunity to test predictions identifying particular environmental factors or plant traits as controllers of resistance or resilience.

There are at least three main hypotheses that can be put forward to devise predictions of resistance and resilience in particular ecosystems. The first arises from an historical perspective; it is argued that responses to an extreme event will be strongly affected by past exposures to similar events (e.g. Sankaran and McNaughton 1999). This theory clearly depends upon the assumption that extreme events select species or genotypes with traits that confer resistance and/or resilience. Whether this occurs or not is likely to be affected by the frequency and severity of events and the nature and potency of the selection forces operating on component populations in the periods between extreme events.

The second hypothesis (MacArthur 1955; Elton 1958; Ehrlich and Ehrlich 1981; Tilman and Downing 1994) proposes that resistance and resilience will be higher in species-rich communities. This theory is based on the assumption that species-rich vegetation may contain a greater diversity in the genetic traits conferring tolerance and recovery. The weakness of this theory it that the same diversity that may ensure the presence of advantageous traits within a subset of the community of plants is likely to dictate that there are other subsets which lack resistance or resilience. Clearly, the highest resistance or resilience is likely to be observed where the requisite traits occur in all of the vegetation regardless of whether it is composed of many or a few species. Some monocultures are capable of remarkable resistance; an example is provided by the flood and fire history of the coastal redwoods of California (Stone and Vasey 1968).

A third set of predictions arise directly from the triangular model of primary plant strategies (Chapter 1). Here the predictions are distinctive and specific in proposing (1) a tradeoff between the two very different sets of traits associated with resistance and resilience (Leps *et al.* 1982; MacGillivray *et al.* 1995) and (2) a predictable relationship between resistance and the particular set of traits identified with stress-tolerance (pages 63–80) and between resilience and the sets of traits associated with ruderals (pages 85–87), and to a lesser extent, competitors (pages 46–48). The mechanistic and specific character of the predictions arising from C-S-R theory make them relatively easy to test by observation and experiment. However, in devising tests of these implications of the triangular model four points of clarification are necessary:

1 The prediction of greater resistances of stress-tolerators to extreme events is founded on the assumption that during the evolution of the species exploiting chronically-unproductive, mineral nutrient-limited conditions selection has led to a long life-span both in the individuals and in their vegetative tissues. An inevitable consequence of this longevity is increased experience of extreme events, particularly those arising from climatic fluctuations. However, the mechanisms promoting resistance in stress-tolerators are not confined to those related to greater past exposure to extreme events. Powerful selection forces are also likely to arise from the greater ecological and evolutionary penalties that would follow if plants exploiting unproductive habitats had low resistance to damage; this is because the capacity for replacement of lost tissues and lost individuals is severely limited in these slow-growing plants. Following these arguments it may be expected that many stress-tolerators will exhibit resistance to several different kinds of extreme events, a prediction supported by the investigations of co-tolerance of different climatic stresses documented by Levitt (1975). It does not follow, however, that all stress-tolerators will be highly resistant to all types of extreme events; the spectrum of resistances is likely to be strongly dependent upon the evolutionary history of the species or population concerned.
2 Qualifications are also necessary with respect to predictions of the resistances of stress-tolerators to mechanical damage. As explained on pages 71–73, the majority of plants of unproductive habitats have tough foliage and are resistant to attack by generalist herbivores. It is also evident that the physical structure of the shoots of many stress-tolerators confers resistance to other forms of mechanical damage including trampling (Liddle 1975). However, circumstances arise in which the greater toughness of the foliage of stress-tolerators cannot be expected to confer a significant degree of differential resistance; examples include damage by more catastrophic forces such as fire, flood, wind-damage, the attentions of large herbivores such as large herbivorous dinosaurs, elephant, and giraffe, or the operation of harvesting and mowing machines.
3 In comparison with the rather complex arguments surrounding the concept of resistance, the phenomenon of resilience is relatively straightforward both in theory (De Angelis 1980) and operationally. Regardless of the agency responsible for an extreme event, greater resilience is to be expected in ruderal and competitive strategists on the basis of their faster rates of resource capture and growth. It is also predictable that resilience will be most readily achieved by ruderals because they are composed of species with shorter life-histories than competitors and their communities are relatively unstructured and rapidly reassembled.
4 Although predictions of resistance and resilience using CSR theory can be applied rapidly across ecosystems of widely-contrasted productivities and species composition, there is no justification for exclusive reliance on this

approach. For the study of responses to particular types or intensities of perturbation there are opportunities for more refined analysis and prediction utilising additional plant attributes. Reference to the regenerative strategies of plants may be particularly helpful; here an excellent example is the capacity to accumulate large banks of persistent seeds. Where this trait is allied to the ruderal strategy, as in ephemeral vegetation situated near to the fluctuating water levels of lakes (van der Valk and Davis 1978; Furness and Hall 1981; Singer *et al.* 1996), very precise predictions of resilience can be made that refer not only to the restoration of biomass but also to the degree to which the original species composition of the community is reconstituted.

An early account of an attempt to examine the use of specified criteria as predictors of resistance and resilience is that of Leps *et al.* (1982), who examined the effects of the 1976 drought on two neighbouring grasslands similar in diversity but differing in productivity at a study site in the Czech Republic. One of the grasslands was of recent origin and consisted of fast-growing ruderal and competitive strategists growing on a fertile soil. The second grassland was of greater antiquity and lower productivity and was composed of relatively slow-growing, stress-tolerant species. The results of the investigation are consistent with predictions of resistance derived from CSR theory in that the depression of yield caused by the drought was greater in the more productive vegetation on fertile soil. Also in accordance with plant strategy theory, the rate of recovery (resilience) following the drought was higher in the vegetation composed of populations of ruderal and competitive strategists.

A similar study was reported by Tilman and Downing (1994), who compared the impacts of a summer drought on areas of grassland of contrasted productivity and species composition. In this case, however, the monitored vegetation consisted of experimental plots, some of which had been subjected to applications of nitrogenous fertiliser over the decade preceding the drought, converting the original unproductive species-rich prairie community to a species-poor assemblage of robust fast-growing perennials. In the initial account of this study (Tilman and Downing 1994) greater resistance and resilience under drought was reported in the unfertilised plots and this was interpreted as a benefit from the species richness of the vegetation. Following reanalysis of the data the inference of greater resilience was withdrawn (Tilman 1996) and it has been pointed out (Givnish 1994; Aarssen 1997; Huston 1997) that the most likely explanation for the greater resistance to drought of the vegetation in the unfertilised plots was the lower water use and greater resistance to moisture stress that would be expected in a natural prairie community as compared to a stand of productive vegetation on nitrogen-rich soil. On this basis, there appears to be a close correspondence between the study of Tilman and Downing (1994) and that of Leps *et al.* (1982); in both studies the impact of drought was strongly correlated with site productivity and vegetation responses were predictable from CSR theory.

In 1995 an attempt was made (MacGillivray *et al.* 1995) to examine resistance and resilience on a broader comparative basis by simultaneous experimental applications of three types of extreme events to five neighbouring ecosystems at Buxton in North Derbyshire in England. The treatments were drought, late frost, and fire, the latter involving destruction of all aboveground vegetation using a flame-gun. Resistance, measurable only in the drought and frost treatments, was assessed as the proportion of undamaged shoot biomass immediately after treatment and resilience was estimated from the shoot biomass measured one year after treatment and compared to that of untreated vegetation.

As a first step in interpreting resistances to drought and frost in the five ecosystems, calculations were made of the resistances exhibited by individual plant species. In Figure 121, patterns of resistance to drought and frost are compared in graphs that relate the resistances of species drawn from all the sites to their position on Axis 1 of a principal component analysis conducted on a large database of species traits derived from standardised screening tests in the laboratory. A more detailed account of the results from this screening programme has been presented already in Chapter 1 (page 101); here it is necessary only to recall that Axis 1 is associated with a consistent shift in a set of traits (growth rate, mineral nutrient content, leaf life-span, resistance to herbivory, decomposition rate) predicted to vary in relation to ecosystem productivity and to reflect the transition from stress-tolerator at one extreme to competitors and ruderals at the other.

The results in Figure 121 reveal a relationship between species positions on Axis 1 and resistance to drought and frost. This provides strong support for the hypothesis that ecosystem productivity and the functional composition of the vegetation are controlling factors in resistance to extreme climatic events. Further confirmation of these relationships in the data collected by MacGillivray *et al.* (1995) is presented in Figure 122 where the information for each of the species weighted according to its abundance in the vegetation has been aggregated to allow comparisons of resistances to drought and frost at the level of the plant community. In order to draw upon the majority of the component species at each site when making these calculations of community resistance use of positions on Axis 1 (available for 43 species only) was abandoned in favour of a single attribute, relative growth rate in the seedling phase, which was strongly correlated with Axis 1 and had been measured on many more species (Grime and Hunt 1975). The results show a consistent difference across the sites in community resistances to both drought and frost.

Calculations of resilience in this investigation were complicated by wide variations in the damage caused to species by the drought and frost treatments and this was in part related to the fact that the experimental treatments had less effect and resilience was higher in species with substantial

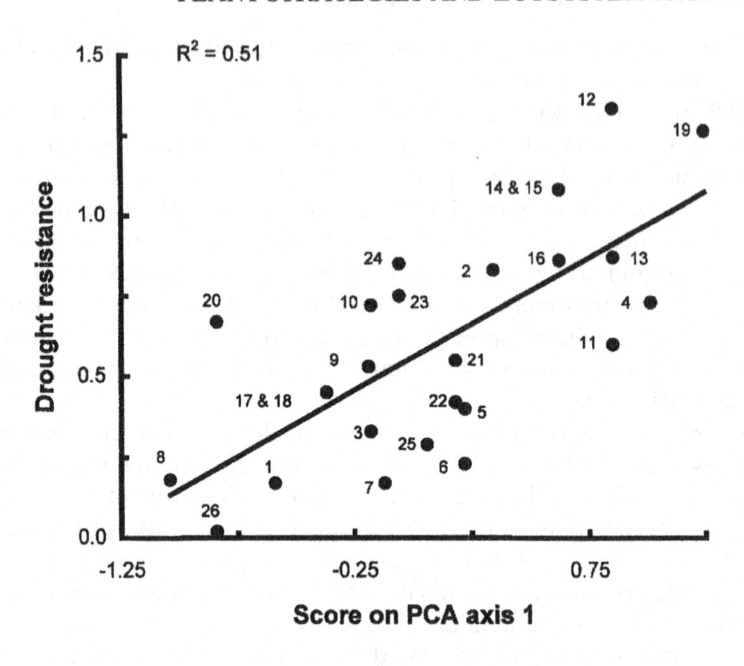

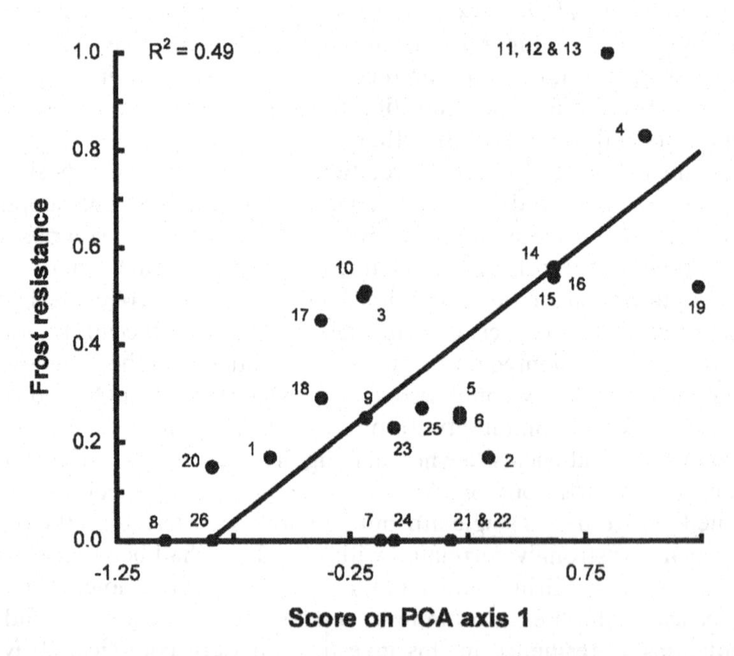

Figure 121. Relationship between estimates of (a) drought resistance and (b) frost resistance of above-ground biomass and the position of species on Axis 1 of a principal component analysis conducted on the results of the Integrated Screening Programme

underground shoots and storage organs. This complexity provides a good example of the necessity discussed earlier (page 338) whenever possible to augment predictions arising from CSR theory classifications with refinements based on plant attributes relevant to specific extreme events. However, despite these complications, resilience was found to be broadly correlated with position on Axis 1. This was most clearly shown in the responses to the burning treatment (Figure 123) where the rather drastic tactic of using a flame-gun to destroy all above-ground vegetation had conveniently created a more standardised starting point for the subsequent assessments of resilience.

A major current concern is to develop and test predictions of the resistances of plant communities and ecosystems to changes in climate and to forecast the way in which resistances will be modified by the large-scale changes in land-use that are occurring as a consequence of human population pressures and intensification of agriculture. Recently data have become available (Grime *et al.* 2000) from the first five years of a long-term experiment in which climate manipulations using non-invasive techniques (Figure 124) have been applied to two contrasted types of calcareous grassland in England.

The first grassland is situated at Wytham in Oxfordshire, England, is representative of the rapidly expanding proportion of the landscape that has been impacted by eutrophication and disturbance. It is of recent origin following abandonment of arable cultivation. The soil contains residues of mineral fertiliser and the flora consists mainly of competitive or ruderal strategists. The second site, at Buxton in North Derbyshire, England, is an ancient pasture that has never been fertilised and is an isolated relic of the unproductive grasslands, dominated by long-lived, slow-growing perennial plants, that extended over large areas of countryside prior to the introduction of intensive agriculture. The results based upon a principal components analysis (PCA, see Figure 125) summarise the extent of changes in plant species composition at both sites over a five-year period (1994–98) in which treatments were applied to elevate winter temperature (+3°C), inflict summer drought (no

---

(see page 338). Both relationships are significant at $P < 0.001$. Resistances were calculated as the ratio of the mean number of point quadrat contacts with the species in the treated turf immediately after the application of the treatments, to that in control plots measured at the same time. In the key to species, the letters in brackets refer to the communities (A–E) in which the observations were made. The species are: 1. *Agrostis capillaris* (D); 2. *Anthoxanthum odoratum* (D); 3. *Arrenatherum elatius* (C); 4. *Helictotrichon pratensis* (D); 5. *Carex flacca* (A); 6. *Carex flacca* (D); 7. *Cerastium fontanum* (B); 8. *Chamerion angustifolium* (C); 9. *Dactylis glomerata* (A); 10. *Dactylis glomerata* (C); 11. *Festuca ovina* (A); 12. *Festuca ovina* (C); 13. *Festuca ovina* (D); 14. *Festuca rubra* (A); 15. *Festuca rubra* (C); 16. *Festuca rubra* (D); 17. *Holcus lanatus* (A); 18. *Holcus lanatus* (C); 19. *Koeleria macrantha* (D); 20. *Leontodon hispidus* (D); 21. *Lotus corniculatus* (A); 22. *Lotus corniculatus* (D); 23. *Poa annua* (E); 24. *Poa annua* (C); 25. *Poa trivialis* (B); 26. *Urtica dioica* (B). (Reproduced from MacGillivray and Grime 1995 by permission of *Functional Ecology*.)

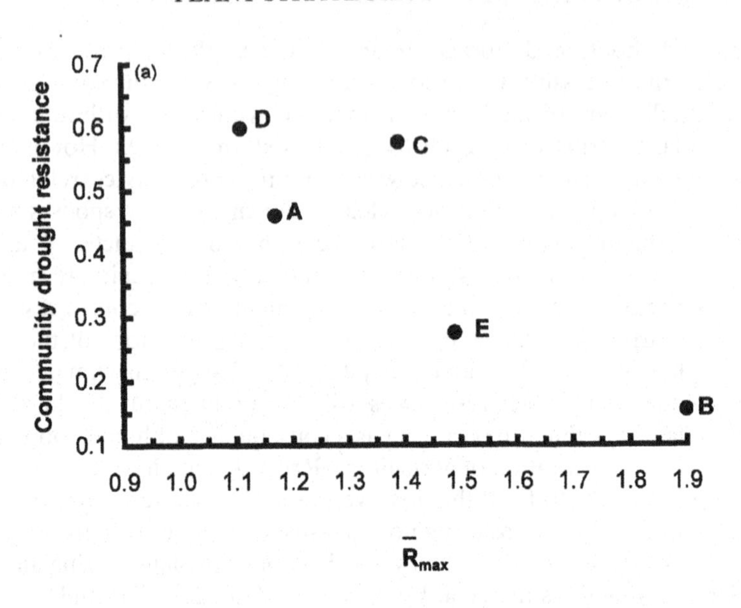

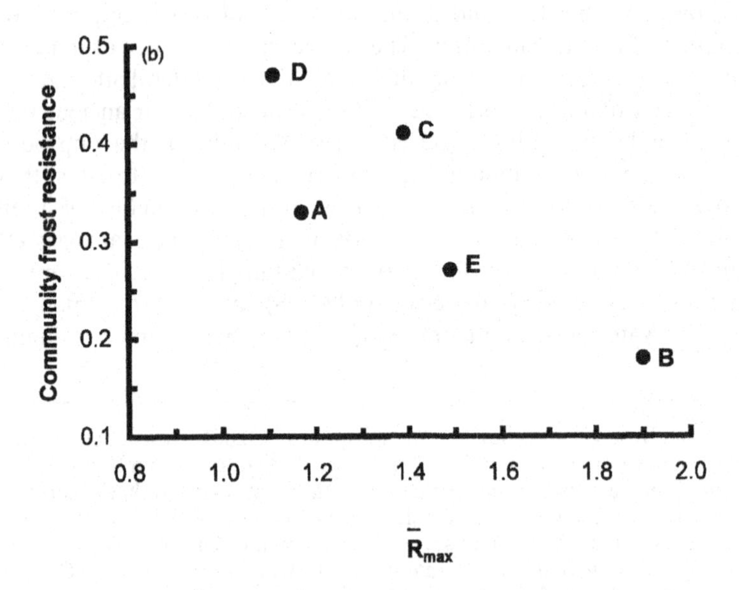

Figure 122. Relationships of R_{max} to (a) community drought resistance and (b) community frost resistance. Community resistances are calculated as the ratio of the estimated total above-ground biomass in the perturbed community to that in control turves examined at the same time. The letter codes (A–E) refer to the five communities in the study, as described in Figure 121. (Reproduced from MacGillivray and Grime 1995 by permission of *Functional Ecology*.)

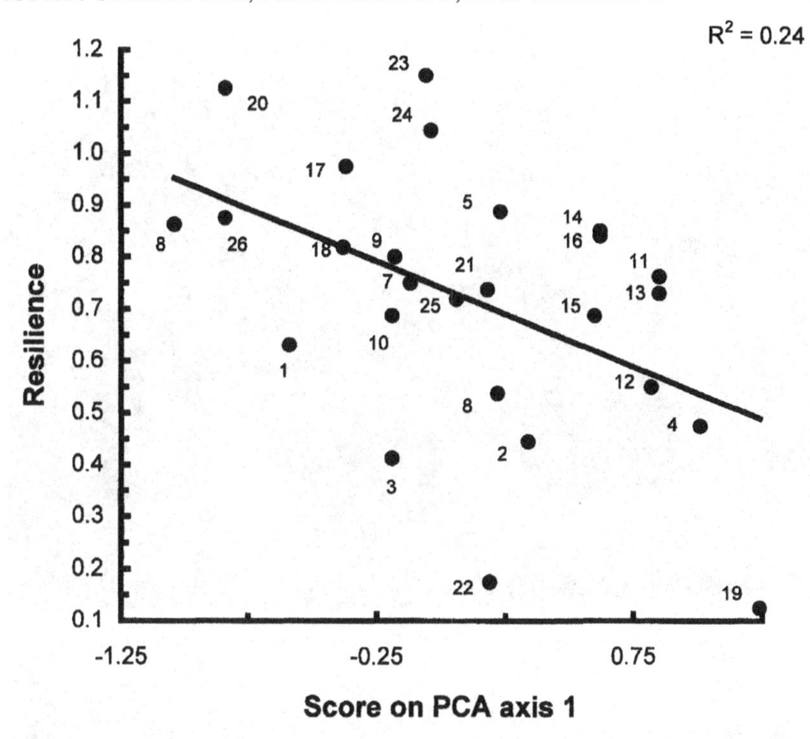

Figure 123. Relationship between the scores on the first principal component analysis axis and fire resilience measured after 12 months ($P < 0.02$; df = 24). Resilience is calculated as the ratio of the mean number of point quadrat contacts in the fire treatment to that in the control plots. (Reproduced from MacGillivray and Grime 1995 by permission of *Functional Ecology*.)

rainfall in July and August), or supplement summer rainfall (to 20% above the long-term average from June to September). Because Wytham and Buxton have almost no plant species in common it was possible by conducting the PCA analysis on the combined floristic data from the two sites to allow the first Axis of the PCA analysis effectively to separate Wytham and Buxton, thus permitting a direct comparison of treatment trajectories at the two sites using PCA axes 2 and 3. The results (Figure 125) expose clearly the very contrasted responses of the vegetation at the two sites. Over five years of climate manipulation the composition of the stress-tolerant vegetation in the treated plots at Buxton varied hardly at all. At Wytham large changes were detected; PCA Axis 3 reveals a successional path followed regardless of treatment and Axis 2 reflects marked treatment effects superimposed on the successional trend.

The results of the Wytham–Buxton comparison therefore provide support for two of the hypotheses outlined earlier. As expected from its strategic composition, the plant community at Wytham was much more labile and

Figure 124. Manipulation of rainfall and temperature at the Buxton Climate Change Impacts Laboratory in North Derbyshire, England. Top: the imposition of summer drought by automatic rainshelters; these are activated by a rain-sensor to slide over the plots during rain showers and conduct water downslope through pipework. Bottom: elevation of winter temperatures by 3°C using heating cables fastened to the soil surface

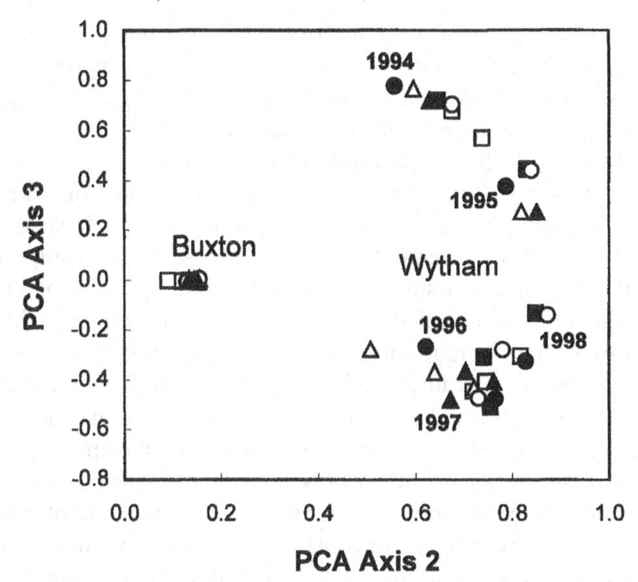

Figure 125. Principal components analysis of the combined floristic data collected from two contrasted grasslands subjected to identical climate manipulations over a five-year period (1994–98). Proportion of variance accounted for by the PCA axes: Axis 1 (not shown), 46%; Axis 2, 27%; Axis 3, 12%. Treatments: ■, control; ▲, winter warming; ●, summer drought; □, supplementary summer rainfall; Δ, winter warming + summer drought; ○, winter warming + supplementary summer rainfall. (Reproduced from Grime *et al.* 2000 by permission of *Science*.)

responsive to the experimental drought and warming treatments. However, there is also some evidence to support the idea that resistance to climate change was partly conditioned by climatic experience. Although the species composition changed substantially at Wytham, the *types* of plants changed least in response to the treatment (summer drought) that most closely approached the experience at the site. Similarly, the plant community at the high-rainfall Buxton site was least affected by the addition of even more rain. In contrast, the data provide little evidence for or against the hypothesis that biodiversity is an important regulator of ecosystem function. Both sites were species rich, so if diversity were a key variable, we might have expected little difference in the response of the two floras to climate change. In fact the sites exhibited very large differences, suggesting that species richness is unlikely to be a major determinant of immediate responses to climate change events.

Sustainability

In the introduction to this section it was pointed out that the term 'sustainability' has many different meanings according to the differing objectives of those engaged in agriculture, forestry, and wildlife management or exploitation. These objectives vary enormously to include products as diverse as total annual crop or livestock biomass, yields from particular timber trees, berried shrubs, game birds, and edible fungi, or the degree of enjoyment of naturalists, hikers, bikers, photographers, or tourists.

If we are to treat such a potentially diffuse subject scientifically it will be necessary as a first step to produce a functional classification of ecosystems that reflects variable attributes such as productivity, physical structure, trophic structure, and the resistances and resiliences of main components. The main benefit of such a classification would be recognition of ecosystems, which although often situated in very different locations and with few taxonomic affinities in dominant organisms, are functionally equivalent. This opens the possibility of transfer of essential knowledge and management techniques from well-studied ecosystems to those that have not been thoroughly investigated. Even where the objectives of management are confined to sustainable exploitation of one or a few species a functional classification of ecosystems would be valuable in allowing specific management objectives to be reviewed in a context where the constraints imposed by the ecosystem and the repercussions of management can be predicted and examined.

Where is such a functional classification of ecosystems to be found? Readers who have persisted to this point in the book will not be surprised to find a recommendation that ecosystems should be classified by reference to the adaptive strategies of the established and regenerative phases in the life-histories of the dominant plants! In various preceding chapters evidence has been presented to support the contention that ecosystem properties as widely divergent as species richness (Chapter 9), trophic structure (Chapter 10), productivity (Chapter 10), resistance, and resilience (Chapter 10) are broadly predictable from CSR theory. It has also been concluded that where required much more specific predictions can be formulated by reference to additional traits of component plant species (e.g. seed banks, dispersal mode, ratio of above- to below-ground biomass, photosynthetic mechanisms, genome size, leaf form, rooting depth, root nodules, mycorrhizas). Despite the strong controlling effects of dominant plants on ecosystem properties, it is also necessary, wherever possible, to develop predictions further by incorporating into models information relating to the functional biology of important herbivores, carnivores, decomposers, symbionts, and pathogens. However, conservation and management guidelines are needed without delay in many thousands of geographical locations and there is much that can be achieved by concentrating initially on the dominant plants. The notion sometimes applied to ecosystems that 'in order to know any of it we must know all of it', whilst intellectually satisfying, can become a barrier to rapid progress and communication between research scientists and ecosystem managers.

Of the many ecosystem properties that can be predicted from CSR theory, the estimates of resistance and resilience appear to be those most directly relevant on a world-wide basis to sustainability. This is because these two properties allow prediction of the very large differences in response that are often observed between neighbouring ecosystems when they are exposed to the changes in land-use and climate that are currently the most potent drivers of global environmental change. In Chapter 6 (pages 218–223) losses of biodiversity in heavily-populated countries have been explained as a consequence of eutrophication and increased frequency of habitat disturbance associated with various human impacts including intensive agriculture, urbanisation, and recreation pressures. The result of these changes which ultimately are a consequence of expanding human numbers or high *per capita* effects is a shift in land cover from ecosystems of high resistance and low resilience to those of low resistance and high resilience (*sensu* Westman 1978). This scenario has profound implications for the sustainability of ecosystems and the future course of development of the terrestrial environment in heavily-populated or exploited landscapes.

Some of these implications can be identified as follows:

1 Remnants of older landscape, containing ecosystems of lower productivity and greater antiquity, will suffer diminution in size and progressive isolation from similar landscape fragments. Following classical island biogeographical theory (MacArthur and Wilson 1967) the floras and faunas of such relictual ecosystems will decline in diversity as species extinctions fail to be compensated by immigrations from increasingly remote 'nearest neighbour' populations.
2 The deterioration of ecosystems situated in fragments of older, relatively unproductive and less disturbed landscape is also a predictable consequence of the life-histories and functional characteristics of many of the plants (Hodgson 1986a, b) and some of the animals (Myers 1996) that exploit such ecosystems. As we have seen from the experiments described earlier in this chapter, resistance to change is considerable in ecosystems dominated by slow-growing, long-lived plants. However, the low reproductive output, limited dispersal range, and long juvenile phase of many stress-tolerant species reduce the prospects for successful recovery when resistance is eventually broken and population reductions and extinctions occur.
3 In marked contrast to the circumstances prevailing in fragments of the more ancient landscape a much more dynamic temporal and spatial flux of plant and animal populations with 'weedy' (i.e. R or C traits) can be predicted for the areas of landscape associated with human habitation and intensive exploitation. In both agricultural and industrial areas vegetation processes tend to proceed rapidly and to be constantly returned to an early successional stage by eutrophication and by various forms of disturbance.
4 A significant additional factor accounting for the rapid and unpredictable changes in flora and fauna often observed in modern landscapes is their susceptibility to colonisations by species already present at low frequency

or new to a locality. The increasing incidence of such expansions or invasions in areas subject to eutrophication and disturbance is consistent with the model of invasibility reviewed in Chapter 7 and can be interpreted as a result of the probability of a coincidence between increased resource supply or release and incursions by invader propagules in heavily-impacted landscapes.

There is little doubt that a consistent effect of the landscape changes summarised in points 1 to 4 is a continuing decline in biodiversity. This implies obvious damage in terms of the unrealised potential benefits to man of extinct species and the cultural impoverishment and losses in local, regional, and national heritage. However, it is important to establish whether losses in biodiversity have effects on ecosystem sustainability. Ecologists tend to agree that damaging effects on ecosystems are likely to follow from losses of plant and animal species but there is wide disagreement concerning the mechanism whereby this damage occurs. Some researchers suggest that significant damage to ecosystem properties follows as an immediate consequence of declining species richness in plant, animal, or microbial communities. Others suspect that properties will be mainly affected in the longer-term and will be associated with the progressive failure of the most appropriate dominant plants and other key organisms to be recruited during the continuous reassembly of ecosystems that is a normal feature of the long-term dynamics of landscapes. On the basis of the evidence reviewed in Chapters 4, 5, 8, 9, and 10 it is concluded that the balance of evidence in this debate currently rests in favour of the second hypothesis.

References

Aarssen LW (1989) Competitive ability and species coexistence: a 'plant's-eye' view. *Oikos* **56**: 386–401.

Aarssen LW (1997) High productivity in grassland ecosystems: effected by species diversity or productive species? *Oikos* **80**: 183–184.

Aber JD and Melillo JM (1982) Nitrogen immobilization in decaying hard wood leaf litter as a function of initial nitrogen and lignin content. *Canadian Journal of Botany* **60**: 2263–2269.

Abrahamson WG and Gadgil M (1973) Growth form and reproductive effort in goldenrods *(Solidago,* Compositae). *American Naturalist* **107**: 651–661.

Aerts R (1989) Above-ground biomass and nutrient dynamics of *Calluna vulgaris* and *Molinia caerulea* in a dry heathland. *Oikos* **56**: 31–38.

Aerts R (1995) The advantages of being evergreen. *Trends in Ecology and Evolution* **10**: 402–407.

Aerts R (1996) Nutrient resorption from senescing leaves of perennials: are there general patterns? *Journal of Ecology* **84**: 597–608.

Aerts R *et al.* (1990) Competition in heathland along an experimental gradient of nutrient availability. *Oikos* **57**: 310–318.

Aerts R and Chapin III FS (2000) The mineral nutrition of wild plants revisited: a re-evaluation of processes and patterns. *Advances in Ecological Research* **30**: 1–67.

Aerts R and van der Peijl (1993) A simple model to explain the dominance of low-productive perennials in nutrient-poor habitats. *Oikos* **66**: 144–147.

Aguiar, MR and Sala OE (1994) Competition, facilitation, seed distribution and the origin of patches in a Patagonian steppe. *Oikos* **70**: 26–34.

Aguilera MO and Lauenroth WK (1993) Seedling establishment in adult neighbourhoods—intraspecific constraints in the regeneration of the bunchgrass *Bouteloua gracilis*. *Journal of Ecology* **81**: 253–261.

Akinola MO, Thompson K, and Hillier SH (1998) Development of soil seed banks beneath synthesised meadow communities after seven years of climate manipulations. *Seed Science Research* **8 (4)**: 493–500.

Allee WC (1938) *The Social Life of Animals.* Norton, New York.

Allessio ML and Tieszen LL (1975) Patterns of carbon allocation in an arctic tundra grass *Dupontia fischeri* (Gramineae) at Barrow, Alaska. *American Journal of Botany* **62**: 797–807.

Al-Mufti MM *et al.* (1977) A quantitative analysis of shoot phenology and dominance in herbaceous vegetation. *Journal of Ecology* **65**: 759–791.

Alpert P (1991) Nitrogen sharing among ramets increases clonal growth in *Fragaria chiloensis*. *Ecology* **72**: 69–80.

Ampofo ST, Moore KG, and Lovell PH (1976) The role of the cotyledons in four *Acer* species and in *Fagus sylvatica* during early seedling development. *New Phytologist* **76**: 31–39.

Andersen AN (1991) Parallels between ants and plants: implications for community ecology. In *Ant-Plant Interactions*, eds. CR Huxely and DF Cutler, pp. 539–558. Oxford University Press, Oxford.

Andersen AN (1995) A classification of Australian ant communities based on functional groups which parallel plant life-forms in relation to stress and disturbance. *Journal of Biogeography* **22**: 15–29.

Andersen MC (1991) Mechanistic models for the seed shadows of wind-dispersed plants. *American Naturalist* **137**: 476–497.

André M, Bréchignac F, and Thibault P (1994) Biodiversity in model ecosystems. *Nature* **371**: 565.

Andreev AV (1991) Winter adaptations in the willow ptarmigan. *Arctic* **44**: 106–114.

Andrewartha HG and Birch LC (1954) *The Distribution and Abundance of Animals*. University of Chicago Press, Chicago.

Antonovics J (1976) The population genetics of mixtures. In *Plant Relations in Pastures*, ed. JR Wilson, pp. 233–252. CSIRO, Melbourne.

Arendt J (1997) Adaptive intrinsic growth rates: an integration across taxa. *Quarterly Review of Biology* **72**: 149–177.

Arno SE and Habeck JR (1972) Ecology of Alpine Larch (*Larix lyalli*, Parl.) in the Pacific Northwest. *Ecological Monographs* **42**: 417–450.

Aronson J *et al.* (1992) Adaptive phenology of desert and Mediterranean populations of annual plants grown with or without water stress. *Oecologia* **89**: 17–26.

Ashenden TW *et al.* (1996) Responses to SO_2 pollution in 41 British herbaceous species. *Functional Ecology* **10**: 483–490.

Ashton DH (1958) The ecology of *Eucalyptus regnans* F. Muell: the species and its frost resistance. *Australian Journal of Botany* **6**: 154–176.

Ashton DH and Macauley BJ (1972) Winter leaf spot disease of seedlings of *Eucalyptus regnans* and its relation to forest litter. *Transactions of the British Mycological Society* **58**: 377–386.

Askew AP *et al.* (1997) A new apparatus to measure the rate of fall of seeds. *Functional Ecology* **11**: 121–125.

Astrom M, Lundberg P, and Danell K (1990) Partial prey consumption by browsers: trees as patches. *Journal of Animal Ecology* **59**: 287–300.

Auclair AN and Cottam G (1971) Dynamics of black cherry (*Prunus serotina* Erhr.) in Southern Wisconsin oak forests. *Ecological Monographs* **41**: 153–177.

Austin MP (1982) Use of physiological performance value in the prediction of performance in multispecies mixtures from monoculture performance. *Journal of Ecology* **70**: 559–570.

Austin MP and Austin BO (1980) Behaviour of experimental plant communities along a nutrient gradient. *Journal of Ecology* **68**: 891–918.

Austin MP *et al.* (1988) Competition and relative yield estimation and interpretation at different densities and under various nutrient concentrations using *Silybum marianum* and *Cirsium vulgare*. *Journal of Ecology* **49**:157–171.

Austin MP and Gaywood MJ (1994) Current populations on environmental gradients and species response curves in relation to continuum theory. *Journal of Vegetation Science* **5**: 473–487.

Baker HG (1965) Characteristics and modes of origin of weeds. In *The Genetics of Colonising Species*, eds. HG Baker and GL Stebbins, pp. 147–168. Academic Press, New York.

Baker HG (1972) Seed weight in relation to environmental conditions in California. *Ecology* **53**: 997–1010.

Bakker D (1960) A comparative life-history of *Cirsium arvense* (L) Scop. and *Tussilago farfara* (L), the most troublesome weeds in the newly reclaimed polders of the former Zuiderzee. In *The Biology of Weeds*, ed. JL Harper, pp. 205–222.

Bakker JP (1989) *Nature Management by Cutting and Grazing*. Dr W Junk Publishers, Dordrecht.

Ballaré CL *et al.* (1987) Early detection of neighbour plants by phytochrome perception of spectral changes in reflected sunlight. *Plant, Cell and Environment* **10**: 551–557

Balme OE (1953) Edaphic and vegetational zoning on the carboniferous limestone of the Derbyshire Dales. *Journal of Ecology* **41**: 331–344.

Band SR and Grime JP (1981) Chemical composition of leaves. *Annual Report*, pp. 6–8. Unit of Comparative Plant Ecology (NERC), University of Sheffield.

Bardgett RD *et al.* (1999) Plant species and nitrogen effects on soil biological properties of temperate upland grasslands. *Functional Ecology* **13**: 1–11.

Barrera *et al.* (2000) Structural and functional changes in *Nothofagus pumilo* forests along an altitudinal gradient in Tierra del Fuego, Argentina. *Journal of Vegetation Science* **11**: 178–188.

Barrett GW (1968) The effects of an acute insecticide stress on a semi-enclosed grassland ecosystem. *Ecology* **49**: 1019–1035.

Bartholomew B, Eaton LC, and Raven PV (1973) *Clarkia rubicunda*: a model of plant evolution in semi-arid regions. *Evolution* **27**: 505–517.

Barton LV (1961) *Seed Preservation and Longevity*. Hill, London.

Baskin CC and Baskin JM (1998) *Seeds; Ecology, Biogeography and Evolution of Dormancy and Germination*. Academic Press, San Diego.

Batzli GO and Pitelka FA (1970) Influence of meadow mouse populations on California grassland. *Ecology* **51**: 1027–1035.

Bayfield NG (1973) Use and deterioration of some Scottish hill paths. *Journal of Applied Ecology* **10**: 639–648.

Bazzaz FA (1979) The physiological ecology of plant succession. *Annual Review of Ecology and Systematics* **10**: 351–371.

Beadle NCW (1954) Soil phosphate and the delimitation of plant communities in eastern Australia, I. *Ecology* **35**: 370–375.

Beadle NCW (1962) Soil phosphate and the delimitation of plant communities in eastern Australia, II. *Ecology* **43**: 281–288.

Beatley JC (1967) Survival of winter annuals in the northern Mojave Desert. *Ecology* **48**: 745–750.

Beaton MJ and Cavalier-Smith T (1999) Eukaryotic non-coding DNA is functional: evidence from the differential scaling of cryptomonad genomes. *Proceedings of the Royal Society of London* **B266**: 2053–2059.

Begon M, Harper JL, and Townsend CR (1996) *Ecology*. Blackwell, Oxford.

Bekker *et al.* (1998a) Seed size, shape and vertical distribution in the soil: indicators of seed longevity. *Functional Ecology* **12**: 834–842.

Bekker RM *et al.* (1998b) Seed bank characteristics of Dutch plant communities. *Acta Botanica Neerlandica* **47**: 15–26.

Bell JS (1990) Against 'measurement'. *Physics World* **3**: 33–40.

Bender EA, Case TJ, and Gilpin ME (1984) Perturbation experiments in community ecology: theory and practice. *Ecology* **65**: 1–13.

Bengtsson J *et al.* (1995) Food webs in soil: an interface between population and ecosystem ecology. In *Linking Species and Ecosystems*, eds. CG Jones and JH Lawton, pp. 159–165. Chapman and Hall, New York.

Bennett MD (1971) The duration of meiosis. *Proceedings of the Royal Society of London* **B178**: 277–299.

Bennett MD (1972) Nuclear DNA content and minimum generation time. *Proceedings of the Royal Society of London* **B181**: 109–135.

Bennett MD (1976) DNA amount, latitude and crop plant distribution. *Environmental and Experimental Botany* **16**: 93–108.

Bennett MD and Leitch IJ (1995) Nuclear DNA amounts in angiosperms. *Annals of Botany* **76**: 113–176.

Bennett MD and Leitch IJ (1997) Nuclear DNA amounts in angiosperms—583 new estimates. *Annals of Botany* **80**: 169–196.

Bennett MD and Smith JP (1976) Nuclear DNA amounts in angiosperms. *Philosophical Transactions of the Royal Society* **B274**: 227–274.

Bennett MD, Smith JB, and Heslop-Harrison JS (1982) Nuclear DNA amounts in angiosperms. *Proceedings of the Royal Society of London* **B216**: 179–199.

Bentley PJ (1966) Adaptations of amphibia to arid environments. *Science* **152**: 619–623.

Berendse F (1998) Effects of dominant plant species on soils during succession in nutrient poor ecosystems. *Biogeochemistry* **42**: 73–88.

Berendse F, Bobbink R, and Rouwenhorst G (1989) A comparative study on nutrient cycling in wet heathland ecosystems. II Litter decomposition and nutrient mineralisation. *Oecologia* **78**: 338–348.

Berendse F and Elberse WT (1989) Competition and nutrient losses from the plant. In *Causes and Consequences of Variation in Growth Rate and Productivity of Higher Plants*, eds. H Lambers *et al.* SPB Academic Publishing, The Hague.

Berendse F *et al.* (1992) Experiments on the restoration of species-rich meadows in the Netherlands. *Biological Conservation* **62**: 59–65.

Berendse F, Schmitz M, and Devisser W (1994) Experimental manipulations of succession in heathland ecosystems. *Oecologia* **100**: 38–44.

Berg RY (1954) Development and dispersal of the seed of *Pedicularis silvatica. Nytt. magasin for botanikk* **2**: 1–58.

Bernard JM and McDonald JG (1974) Primary production and life-history of *Carex lacustris. Canadian Journal of Botany* **52**: 117–123.

Bernard JM and Solsky BA (1977) Nutrient cycling in a *Carex lacustris* wetland. *Canadian Journal of Botany* **55**: 630–638.

Bernhard F (1970) Etude de la litière et de sa contribution au cycle des éléments mineraux en forêt ombrophile de Cote-d'lvoire. *Oecologia Plantarum* **5**: 247–266.

Bertalanaffy L von (1950) The theory of open systems in physics and biology. *Science* **111**: 23–29.

Beyers RJ and Odum HT (1993) *Ecological Microcosms*. Springer-Verlag, New York.

Bhat KKS and Nye PH (1973) Diffusion of phosphate to plant roots in soil. I. Quantitative autoradiography of the depletion zone. *Plant and Soil* **38**: 161–175.

Bibbey RO (1948) Physiological studies of weed seed germination. *Plant Physiology Lancaster* **23**: 467–484.

Billings WD and Godfrey PJ (1968) Acclimation effects on metabolic rates of arctic and alpine *Oxyria* populations subjected to temperature stress (Abstr.). *Bulletin of the Ecological Society of America* **49**: 68–69.

Billings WD *et al.* (1971) Metabolic acclimation to temperature in arctic and alpine ecotypes of *Oxyria digyna. Arctic Alpine Research* **3**: 277–289.

Billings WD and Mooney HA (1968) The ecology of arctic and alpine plants. *Biological Reviews* **43**: 481–529.

Birrel KS and Wright ACS (1945) A serpentine soil in New Caledonia. *New Zealand Journal of Science and Technology* **27A**: 72–76.

Björkman O (1968a) Carboxydismutase activity in shade-adapted and sun-adapted species of higher plants. *Physiologia Plantarum* **21**: 1–10.

Björkman O (1968b) Further studies on differentiation of photosynthetic properties in sun and shade ecotypes of *Solidago virgaurea. Physiologia Plantarum* **21**: 84–99.

Björkman O and Holmgren P (1963) Adaptability of the photosynthetic apparatus to light intensity in ecotypes from exposed and shaded habitats. *Physiologia Plantarum* **16**: 889–914.

Björkman O and Holmgren P (1966) Adaptation to light intensity in plants native to shaded and exposed habitats. *Physiologia Plantarum* **19**: 854–859.

Black JM (1958) Competition between plants of different initial seed sizes in swards of subterranean clover (*Trifolium subterraneum L.)* with particular reference to leaf area and the light micro-climate. *Australian Journal of Agricultural Research* **9**: 299–312.

Blackman GE and Black JN (1959) Physiological and ecological studies in the analysis of plant environment: XI. A further assessment of the influence of shading on the growth of different species in the vegetative phase. *Annals of Botany NS* **23**: 51–63.

Blackman GE and Rutter AJ (1948) Physiological and ecological studies in the analysis of plant environment: III. The interaction between light intensity and mineral nutrient supply in leaf development and in the net assimilation rate of the bluebell *(Scilla non-scripta). Annals of Botany NS* **12**: 1–6.

Blackman GE and Wilson GL (1951a) Physiological and ecological studies in the analysis of plant environment: VI. The constancy for different species of a logarithmic relationship between net assimilation rate and light intensity and its ecological significance. *Annals of Botany NS* **15**: 63–94.

Blackman GE and Wilson AJ (1951b) Physiological and ecological studies in the analysis of plant environment: VII. An analysis of the differential effects of light intensity on the net assimilation rate, leaf-area ratio and relative growth rate of different species. *Annals of Botany NS* **15**: 373–408.

Bliss LC (1962) Adaptations of arctic and alpine plants to environmental conditions. *Arctic* **15**: 117–144.

Bliss LC (1971) Arctic and alpine life cycles. *Annual Review of Ecology and Systematics* **2**: 405–438.

Bliss LC (1985) Alpine. In *Physiological Ecology of North American Plant Communities*, eds. BF Chabot and HA Mooney, pp. 41–65. Chapman and Hall, London.

Bloom AJ, Chapin FS, and Mooney HA (1985) Resource limitation in plants—an economic analogy. *Annual Review of Ecology and Systematics* **16**: 363–393.

Böcher TW (1949) Racial divergencies in *Prunella vulgaris* in relation to habitat and climate. *New Phytologist* **48**: 285–314.

Böcher TW (1961) Experimental and cytological studies in plant species: VI. *Dactylis glomerata* and *Anthoxanthum odoratum. Botaniska Tidsskrift* **56**: 314–355.

Böcher TW and Larsen K (1958) Geographical distribution of initiation of flowering, growth habit, and other characters in *Holcus lanatus. Botaniska Notiser* **3**: 289–300.

Bohan DA *et al.* (2000) Spatial dynamics of predation by carabid beetles on slugs. *Journal of Animal Ecology* **69**: 367–379.

Bolker BM and Pacala SW (1999) Spatial moment equations for plant competition: understanding spatial strategies and the advantages of short dispersal. *American Naturalist* 153: 575–602.

Bond WJ (1989) The tortoise and the hare: ecology of angiosperm dominance and gymnosperm persistence. *Biological Journal of the Linnaen Society* **36**: 227–249.

Boonstra R and Krebs CJ (1977) A fencing experiment on a high density population of *Microtus townsendii. Canadian Journal of Zoology* **55**: 1166–1175.

Boot R, Raynal DJ, and Grime JP (1986) A comparative study of the influence of moisture stress on flowering in *Urtica dioica* and *Urtica urens. Journal of Ecology* **74**: 485–495.

Booth RE and Grime JP (2001) Plant diversity and ecosystem productivity: new evidence from synthesised communities. *Journal of Ecology* (in review).

Bordeau PF and Laverick ML (1958) Tolerance and photosynthetic adaptability to light intensity in white pine, red pine, hemlock, and ailanthus seedlings. *Forest Science* **4**: 196–207.

Bormann FH and Likens GE (1979) *Patterns and Processes in a Forested Ecosystem.* Springer-Verlag, Berlin.

Bosatta E and Staaf H (1982) The control of nitrogen turnover in forest litter. *Oikos* **39**: 143–151.

Bosman AL and Hockey PAR (1988) Life history patterns of populations of the limpet *Patella granularis*: the dominant roles of food supply and mortality rate. *Oecologia* **75**: 412–419.

Boutin, C, and Keddy, P.A. (1992) A functional classification of wetland plants. *Journal of Vegetation Science* **4**: 591–600.

Bowen BJ and Pate JS (1991) Adaptations of S.W. Australian members of the Proteaceae: allocation of resources during early growth. In *Proceedings of the International Protea Association Sixth Biennial Conference*, pp. 289–301. Promaco Conventions, Perth.

Box EO (1981) *Macroclimate and Plant Forms: An Introduction to Predictive Modelling in Phytogeography*. Kluwer, The Hague.

Box EO (1996) Plant functional types and climate at the global scale. *Journal of Vegetation Science* **7**: 309–320.

Boysen-Jensen P (1929) Studier over Skovtracerres Forhold til Lyset. *Dansk. Skovforeningens Tidsskrift* **14**: 5–31.

Boysen-Jensen P (1932) *Die Stoffproduktion der Pflanzen*, Jena.

Bradbury IK and Hofstra G (1976) The partitioning of net energy resources in two populations of *Solidago canadensis* during a single developmental cycle in southern Ontario. *Canadian Journal of Botany* **54**: 2449–2456.

Bradshaw AD (1959) Population differentiation in *Agrostis tenuis* Sibth: 1. Morphological differentiation. *New Phytologist* **58**: 208–227.

Bradshaw AD (1977) Conservation problems in the future. In *Scientific Aspects of Nature Conservation in Great Britain, Proceedings of the Royal Society of London* **B197**: 77–96.

Bradshaw AD *et al.* (1964) Experimental investigations into the mineral nutrition of several grass species: IV. Nitrogen level. *Journal of Ecology* **52**: 665–676.

Bragg AN (1945) The spadefoot toads in Oklahoma with a summary of our knowledge of the group. *American Naturalist* **79**: 52–72.

Branham JM *et al.* (1971) Coral-eating sea stars *Acanthaster planci* in Hawaii. *Science* **172**: 1155–1157.

Bratton SP (1976) Resource division in an understory herb community: responses to temporal and microtopographic gradients. *American Naturalist* **110**: 679–693.

Braun-Blanquet J (1932) *Plant Sociology: the Study of Plant Communities*. McGraw Hill, New York.

Bray JR (1958) Notes toward an ecologic theory. *Ecology* **39**: 770–776.

Brenchley WE (1918) Buried weed seeds. *Journal of Agricultural Science* **9**: 1–31.

Brenchley WE and Warington K (1930) The weed seed population of arable soil: I. Numerical estimation of viable seeds and observations on their natural dormancy. *Journal of Ecology* **18**: 235–272.

Brenchley WE and Warington K (1958) *The Park Grass Plots at Rothamsted 1856–1949*. Rothamsted Experimental Station, Harpenden.

Brewer SW and Rejmánek M (1999) Small rodents as significant dispersers of tree seeds in a neotropical forest. *Journal of Vegetation Science* **10**: 165–174.

Brinck P (1980) Theories in population and community ecology. *Oikos* **35**: 129–130.

Britten RJ and Davidson EH (1971) Repetitive and non-repetitive DNA sequences and a speculation on the origins of evolutionary novelty. *Quarterly Review of Biology* **46**: 111–137.

Bronson FW (1964) Agonistic behaviour in wood-chucks. *Animal Behaviour* **12**: 270–278.

Brooker RW and Callaghan TV (1998) The balance between positive and negative plant interactions and its relationship to environmental gradients: a model. *Oikos* **81**: 196–207.

Brouwer R (1962a) Distribution of dry matter in the plant. *Netherlands Journal of Agricultural Science* **10**: 361–376.

Brouwer R (1962b) Nutritive influences on the distribution of dry matter in the plant. *Netherlands Journal of Agricultural Science* **10**: 399–408.

Brown ES (1962) The African army worm, *Spodoptera exempta* (Walker) *(I. epidoptera, Noctuidae)*: a review of the literature. Commonwealth Institute of Entomology, London.

Brown JS (1989) Desert rodent community structure: a test of four mechanisms of coexistence. *Ecological Monographs* **20**: 1–20.

Brown MJ and Parker GG (1994) Canopy light transmittance in a chronosequence of mixed-species deciduous forests. *Canadian Journal of Forest Research* **24**: 1694–1703.

Brown VK and Gange AC (1992) Secondary plant succession: how is it modified by insect herbivory? *Vegetatio* **101**: 3–13.

Brown VK, Jepson M, and Gibson CWD (1998) Insect herbivory: effects on early old field succession demonstrated by chemical exclusion methods. *Oikos* **52**: 293–302.

Bryant JP, Chapin FS, and Klein DR (1983) Carbon/nutrient balance of boreal plants in relation to vertebrate herbivory. *Oikos* **40**: 357–368.

Brzeziek B and Kienast F (1994) Classifying the life-history strategies of trees on the basis of the Grimian model. *Forest Ecology and Management* **69**: 167–187.

Buckland SM and Grime JP (2000) The effects of trophic structure and soil fertility on the assembly of plant communities: a microcosm experiment. *Oikos* **91**: 336–352.

Buckland SM *et al.* (1997) A comparison of plant responses to the extreme drought of 1995 in Northern England. *Journal of Ecology* **85**: 875–882.

Bunce RGH *et al.* (1998) *Vegetation of the British Countryside; the Countryside Vegetation System. ECOFACT Volume 1*, London DETR.

Burdon JJ (1993) The structure of pathogen populations in natural plant communities. *Annual Review of Phytopathology* **31**: 305–323.

Burdon JJ and Thrall PH (1999) Spatial and temporal patterns in coevolving plant and pathogen associations. *American Naturalist* **153**: 515–533.

Burgeff H (1961) *Mikrobiologie des Hochmoores*, Gustav Fischer, Stuttgart. p. 197.

Burgess PF (1972) Studies on the regeneration of the hill forests of the Malay Peninsula. *Malay Forest* **35**: 103–123.

Burgess TL, Bowers JE, and Turner RM (1991) Exotic plants of the desert laboratory, Tuscon, Arizona. *Madroño* **38**: 96–114.

Burke MJW and Grime JP (1996) An experimental study of plant community invasibility. *Ecology* **77**: 776–790.

Burns GP (1923) Studies in tolerance of New England forest trees: IV. Minimum light requirement referred to a definite standard. *Bulletin Vermont Agricultural Experimental Station* **235**: 15–25.

Burslem DFRP and Whitmore TC (1999) Species diversity, susceptibility to disturbance and tree population dynamics in tropical rain forest. *Journal of Vegetation Science* **10**: 767–776.

Burt-Smith G, Tilman D, and Grime JP (2001) Seedling traits as predictors of the role of herbivory in a prairie grassland. *Journal of Ecology* (submitted).

Caldwell MM and Pearcy RW (1994) *Exploitation of Environmental Heterogeneity by Plants.* Academic Press, San Diego.

Callaghan TV (1976) Growth and population dynamics of *Carex bigelowii* in an alpine environment. Strategies of growth and population dynamics of tundra plants, 3. *Oikos* **27**: 402–413.

Callaghan TV and Emanuelsson U (1985) Population structure and processes of tundra plants and vegetation. In *The Population Structure of Vegetation*, ed. J White, pp. 399–439. Dr. W Junk Publishers, Dordrecht.

Callaway RM and Walker LR (1997) Competition and facilitation: a synthetic approach to interactions in plant communities. *Ecology* **78**: 1958–1965.

Calow P (1977) The joint effect of temperature and starvation on the metabolism of triclads. *Oikos* **29**: 87–92.

Calow P and Woolhead AS (1977) The relationship between ration, reproductive effort, and age-specific mortality in the evolution of life-history strategies—some observations on freshwater triclads. *Journal of Animal Ecology* **46**: 765–781.

Campbell BD and Grime JP (1989a) A comparative study of plant responsiveness to the duration of episodes of mineral nutrient enrichment. *New Phytologist* **112**: 261–267.

Campbell BD and Grime JP (1989a) A new method of exposing developing root systems to controlled patchiness in mineral nutrient supply. *Annals of Botany* **63**: 395–400.

Campbell BD and Grime JP (1992) An experimental test of plant strategy theory. *Ecology* **73**: 15–29.

Campbell BD, Grime JP, and Mackey JML (1991a) A trade-off between scale and precision in resource foraging. *Oecologia* **87**: 532–538.

Campbell BD, Grime JP, and Mackey JML (1992) Shoot thrust and its role in plant competition. *Journal of Ecology* **80**: 633–641.

Campbell BD *et al.* (1991b) The quest for a mechanistic understanding of resource competition in plant communities: the role of experiments. *Functional Ecology* **5**: 241–253.

Canham CD *et al.* (1994) Causes and consequences of resource heterogeneity in forests: interspecific variation in light transmission by canopy trees. *Canadian Journal of Forestry Research* **24**: 337–349.

Cannell MGR and Sheppard LJ (1982) Seasonal changes in the frost hardiness of provenances of *Picea sitchensis* in Scotland. *Forestry* **55**: 137–153.

Caraco T and Kelly CK (1991) On the adaptive value of physiological integration in clonal plants. *Ecology* **72**: 81–93.

Carlsson BA and Callaghan TV (1991) Positive plant interactions in tundra vegetation and the importance of shelter. *Journal of Ecology* **79**: 973–983.

Carpenter AT *et al.* (1990) Plant community dynamics in relation to nutrient addition following a major disturbance. *Plant and Soil* **126**: 91–100.

Carpenter SR (1988) *Complex Interactions in Lake Communities*. Springer-Verlag, New York.

Carpenter SR (1996) Microcosm experiments have limited relevance for community and ecosystem ecology. *Ecology* **77**: 677–680.

Carson WP and Barrett GW (1988) Succession in old-field plant communities: effects of contrasting types of nutrient enrichment. *Ecology* **69**: 984–994.

Carson WP and Root RB (1999) Top-down effects of insect herbivores during early succession: influence on biomass and plant dominance. *Oecologia* **121**: 260–272.

Carson WP and Root RB (2000) Herbivory and plant species coexistence: community regulation by an outbreaking phytophagous insect. *Ecological Monographs* **70**: 73–99.

Carter RN and Prince SD (1981) Epidemic models used to explain biogeographical distribution limits. *Nature* **293**: 644–645.

Case TJ (1990) Invasion resistance arises in strongly interacting species-rich model competition communities. *Proceedings of the National Academy of Sciences* **87**: 9610–9614.

Castro-Diez P *et al.* (1998) Stem anatomy and relative growth rate in seedlings of a wide range of woody plant species and types. *Oecologia* **116**: 57–66.
Caswell H (1978) Predator-mediated coexistence: a non-equilibrium world. *American Naturalist* **112**: 127–154.
Caswell H (1989) *Matrix Population Models*. Sinauer, Sunderland.
Caswell H and Reed FC (1976) Plant herbivore interactions: the indigestibility of C4 bundle sheath cells by grasshoppers. *Oecologia* **26**: 151–156.
Caswell H *et al.* (1973) Photosynthetic pathways and selective herbivory: a hypothesis. *American Naturalist* **107**: 465–480.
Cates RG and Orians GH (1975) Successional status and the palatability of plants to generalised herbivores. *Ecology* **56**: 410–418.
Cavalier-Smith T (1978) Nuclear volume control by nucleoskeletal DNA, selection for cell volume and cell growth rate and the solution of the DNA C-value paradox. *Journal of Cell Science* **34**: 247–278.
Cavalier-Smith T and Beaton MJ (1999) The skeletal function of non-genetic nuclear DNA: new evidence from ancient chimeras. *Genetica* **106**: 3–13.
Cavers PB and Harper JL (1966) Germination polymorphism in *Rumex crispus* and *Rumex obtusifolius*. *Journal of Ecology* **54**: 307–382.
Cavers PB and Harper JL (1967) Studies in the dynamics of plant populations. I. The fate of seed and transplants introduced into various habitats. *Journal of Ecology* **55**: 59–71.
Cebrián J and Duarte CM (1994) The dependence of herbivory on growth rate in natural communities. *Functional Ecology* **8**: 518–525.
Chabot BF and Hicks DF (1982) The ecology of leaf life spans. *Annual Review of Ecology and Systematics* **13**: 229–259.
Champness SS and Morris K (1948) Populations of buried viable seeds in relation to contrasting pasture and soil types. *Journal of Ecology* **36**: 149–173.
Chapin FS (1980) The mineral nutrition of wild plants. *Annual Review of Ecology and Systematics* **11**: 233–260.
Chapin FS (1983) Direct and indirect effects of temperature on arctic plants. *Polar Biology* **2**: 47–52.
Chapin FS (1991) Integrated responses of plants to stress: a centralised system of physiological responses. *Bioscience* **41**: 29–36.
Chapin FS and Bloom A (1976) Phosphate absorption: adaptation of tundra graminoids to a low temperature, low phosphorus environment. *Oikos* **26**: 111–121.
Chapin FS and Shaver GR (1985) Arctic. In *Physiological Ecology of North American Plant Communities*, eds. BF Chabot and HA Mooney, pp. 16–40. Chapman and Hall, London.
Chapin FS, Schulze ED, and Mooney HA (1990) The ecology and economics of storage in plants. *Annual Review of Ecology and Systematics* **21**: 423–447.
Charles M, Jones G, and Hodgson JG. (1997) FIBS in archaeobotany: functional interpretation of weed floras in relation to husbandry practices. *Journal of Archaeological Science* **24**: 1151–1161.
Chew RM and Butterworth BB (1964). Ecology of rodents in Indian Cove (Mojave Desert), Joshua Tree National Monument, California. *Journal of Mammalogy* **45**: 203–225.
Chew RM and Chew AE (1970) Energy relationships of the mammals of a desert shrub *(Larrea tridentata)* community. *Ecological Monographs* **40**: 1–21.
Chiariello N, Hickman JC, and Mooney HA (1982) Endomycorrhizal role for interspecific transfer of phosphorus in a community of annual plants. *Science* **217**: 941–943.

Chippindale HG and Milton WEJ (1934) On the viable seeds present in the soil beneath pastures. *Journal of Ecology* **22**: 508–531.
Choong MF *et al.* (1992) Leaf fracture toughness and sclerophylly; their correlations and ecological implications. *New Phytologist* **121**: 597–610.
Clapham AR (1956) Autecological studies and the 'Biological Flora of the British Isles'. *Journal of Ecology* **44**: 1–11.
Clarkson DT (1967) Phosphorus supply and growth rate in species of *Agrostis* L. *Journal of Ecology* **55**: 111–118.
Clausen J, Keck DD, and Hiesey WM (1940) Experimental studies on the nature of species: I. Effect of varied environments on western North American plants. *Carnegie Institute Washington Pub.* 520.
Clay K (1990) Fungal endophytes of grasses. *Annual Review of Ecology and Systematics* **21**: 275–297.
Clay K and Holah J (1999) Fungal endophyte symbiosis and plant diversity in successional fields. *Science* **285**: 1742–1744.
Clay K, Marks S, and Cheplick GP (1995) Effects of insect herbivory and fungal endophyte infection on competitive interactions among grasses. *Ecology* **74**: 1767–1777.
Clements FE (1916) *Plant Succession. An Analysis of the Development of Vegetation.* Carnegie Institute, Washington.
Clutton-Brock TH and Harvey PH (1979) Comparison and adaptation. *Proceedings of the Royal Society of London, Series* **B205**: 547–565.
Cody M (1966) A general theory of clutch size. *Evolution* **20**: 174–184.
Colasanti RL (2000) Individual-based models in plant ecology. PhD Thesis, University of Sheffield.
Colasanti RL and Grime JP (1993) Resource dynamics and vegetation processes: A deterministic model using two dimensional cellular automata. *Functional Ecology* **7**: 169–177.
Colasanti RL and Hunt R (1997) Resource dynamics and plant growth: a self-assembling model for individuals, populations and communities. *Functional Ecology* **11**: 133–145.
Cole LC (1954) The population consequences of life history phenomena. *Quarterly Review of Biology* **29**: 103–137.
Coley PD (1983) Herbivory and defensive characteristics of tree species in a lowland tropical forest. *Ecological Monographs* **53**: 209–233.
Coley PD (1988) Effects of plant growth rate and leaf life-time on the amount and type of anti-herbivore defence. *Oecologia* **74**: 531–536.
Coley PD, Bryant JP, and Chapin III FS (1985) Resource availability and plant anti-herbivore defence. *Science* **230**: 895–899
Commoner B (1964) Roles of deoxyribonucleic acid in inheritance. *Nature* **202**: 277–298.
Connell JH (1961) Effect of competition, predation by *Thais lapillus* and other factors on natural populations of the barnacle *Balanus balanoides. Ecological Monographs* **31**: 61–104.
Connell JH (1972) Community interactions on marine intertidal shores. *Annual Review of Ecology and Systematics* **3**: 169–192.
Connell JH (1980) Diversity and the coevolution of competitors, or the ghost of competition past. *Oikos* **35**: 131–138.
Connell J and Slatyer RO (1977) Mechanisms of succession in natural community stability and organisation. *American Naturalist* **111**: 1119–1144.
Conolly AP (1977) The distribution and history in the British Isles of some alien species of *Polygonum* and *Reynoutria. Watsonia* **11**: 291–311.

Convey P (1997) How are the life history strategies of Antarctic terrestrial invertebrates influenced by extreme environmental conditions? *Journal of Thermal Biology* **22**: 429–440.

Cook R (1980) The biology of seeds in the soil. In *Demography and Evolution in Plant Populations*, ed. OT Solbrig, pp. 107–129. Blackwell, New York.

Cook RE (1976) Photoperiod and the determination of potential seed number in *Chenopodium rubrum* L. *Annals of Botany* **40**: 1085–1099.

Cook SCA, Lefebvre C and McNeilly T (1972) Competition between metal tolerant and normal plant populations on normal soil. *Evolution* **26**: 366–372.

Cooke RC and Rayner ADM (1984) *Ecology of Saprotrophic Fungi*. Longman, London.

Coomes DA and Grubb PJ (2000) Impacts of root competition in forests and woodlands: a theoretical framework and review of experiments. *Ecological Monographs* **70**: 171–207.

Cooper JP and Tainton NM (1968) Light and temperature requirements for the growth of tropical and temperate grasses. *Herbage Abstracts* **38**: 167–176.

Cooper-Driver G (1985) Anti-predation strategies in pteridophytes: a biochemical approach. *Proceedings of the Royal Society of Edinburgh* **86B**: 397–402.

Cornelissen JHC (1996) An experimental comparison of leaf decomposition rates in a wide range of temperate plant species and types. *Journal of Ecology* **84**: 573–582.

Cornelissen JHC, Castro-Diaz P, and Carnelli AL (1998) Variation in relative growth rate among woody species. In *Inherent Variation in Plant Growth; Physiological Mechanisms and Ecological Consequences*, eds. H Lambers, H Poorter, and MMI Van Vuuren, pp. 363–392. Backhuys Publishers, Leiden.

Cornelissen JHC, Diez Castro P, and Hunt R (1996) Seedling growth, allocation and leaf attributes in a wide range of woody plant species and types. *Journal of Ecology* **84**: 755–765.

Cornelissen JHC *et al.* (1999) Leaf structure and defence control litter decomposition rate across species, life forms and continents. *New Phytologist* **143**: 191–200.

Cornelissen JHC *et al.* (1997) Foliar nutrients in relation to growth, allocation and leaf traits in seedlings of a wide range of woody plant species and types. *Oecologia* **111**: 460–469.

Corré WJ (1983) Growth and morphogenesis of sun and shade plants. II The influence of light quality. *Acta Botanica Neerlandica* **32**, 185–202.

Cottam WP, Tucker JM and Drobnick R (1959) Some clues to Great Basin postpluvial climates provided by oak distributions. *Ecology* **40**: 361–377.

Cousens JE (1965) Some reflections on the nature of Malayan lowland rain forest. *Malay Forest.* **28**: 122–128.

Crawley MJ (1987) What makes a community invasible? In *Colonization, Succession and Stability*, eds. AJ Gray, MJ Crawley, and PJ Edwards, pp 424–453. Blackwell, Oxford.

Crawley MJ and May RM (1987) Population dynamics and plant community structure: competition between annuals and perennials. *Journal of Theoretical Biology* **125**: 475–489.

Crawley MJ *et al.* (1999) Invasion-resistance in experimental grassland communities: species richness or species identity? *Ecology Letters* **2**: 140–148.

Cresswell E and Grime JP (1981) Induction of a light requirement during seed development and its ecological consequences. *Nature* **291**: 583–585.

Crick JC and Grime JP (1987) Morphological plasticity and mineral nutrient capture in two herbaceous species of contrasted ecology. *New Phytologist* **107**: 403–414.

Crisp MD and Lange RT (1976) Age structure, distribution and survival under grazing of the arid-zone shrub *Acacia burkittii. Oikos* **27**: 86–92.

Crocker RL and Major J (1955) Soil development in relation to vegetation and surface age at Glacier Bay, Alaska. *Journal of Ecology* **43**: 427–448.

Cunningham SA, Summerhays B, and Westoby M (1999) Evolutionary divergences in leaf structure and chemistry, comparing rainfall and soil nutrient gradients. *Ecology* 69: 569–588.

Currey DR (1965) An ancient bristlecone pine stand in Eastern Nevada. *Ecology* **46**: 564–566.

Curtis JT (1959) The vegetation of Wisconsin. *An Ordination of Plant Communities.* University of Wisconsin Press, Madison.

Curtis JT and Cottam G (1950) Antibiotic and autotoxic effects in prairie sunflower. *Bulletin of Torrey Botany Club,* **77**: 187–191.

D'Antonio CM (1993) Mechanisms controlling invasions of coastal plant communities by the alien succulent *Carpobrotus edulis. Ecology* **74**: 83–95.

Dana F, Newman WA, and Fager EW (1972) *Acanthaster* aggregations: Interpreted as primarily responses to natural phenomena. *Pacific Science* **26**: 355–372.

Daniel CP and Platt RB (1968) Direct and indirect effects of short term ionising radiation on old field succession. *Ecological Monographs* **38**: 1–29.

Darwin C (1859) *The Origin of Species by Means of Natural Selection or the Preservation of Favoured Races in the Struggle for Life.* Murray, London.

Datta SC, Evenari M, and Gutterman Y (1970) The heteroblasty of *Aegilops ovata* L. *Israeli Journal of Botany* **19**: 463–483.

Daubenmire R and Prusso DC (1963) Studies of the decomposition rates of tree litter. *Ecology* **44**: 589–593.

Davey AJ (1961) Biological flora of the British Isles—*Epilobium nerterioides* A.Cnn.. *Journal of Ecology* **49**: 753–759.

Davidson EH and Britten RJ (1973) Organisation, transcription and regulation in the animal genome. *Quarterly Review of Biology* **48**: 565–613.

Davies J, Briarty LG, and Rieley JO (1973) Observations on the swollen lateral roots of the Cyperaceae. *New Phytologist* **72**: 167–174.

Davis MA, Grime JP, and Thompson K (2000) Fluctuating resources in plant communities : a general theory of invasibility. *Journal of Ecology* **99** (in press).

Davis MA, Thompson K, and Grime JP (2001) Charles S Elton and the dissociation of invasion ecology from the rest of ecology. *Distributions and Diversity* (in press).

Davis MA, Wrage KJ, and Reich PB (1998) Competition between tree seedlings and herbaceous vegetation: support for a theory of resource supply and demand. *Journal of Ecology* **86**: 652–661.

Davis MA *et al.* (1999) Survival, growth and photosynthesis of tree seedlings competing with herbaceous vegetation along a water-light-nitrogen gradient. *Plant Ecology* **145**: 341–350.

Davis RF (1928) The toxic principle *of Juglans nigra* as identified with synthetic juglone and its toxic effects on tomato and alfalfa plants. *American Journal of Botany* **15**: 620.

Davison AW (1977) The ecology of *Hordeum murinum* L: III. Some effects of adverse climate. *Journal of Ecology* **65**: 523–530.

Dawson JH (1970) *Proceedings of the 23rd Southern Weed Conference*, Time and duration of weed infestations in relation to weed crop competition, pp. 13–35.

Day RT, Keddy PA, and McNeil J (1988) Fertility and disturbance gradients: a summary model for riverine marsh vegetation. *Ecology* **69**: 1044–1054.

Dayton PK (1970) Competition, disturbance and community organisation: The provision and subsequent utilisation of space in a rocky inter-tidal community. *Ecological Monographs* **41**: 351–389.

Dayton PK and Hessler RR (1972). Role of biological disturbance in maintaining diversity in the deep sea. *Deep-sea Research* **19**: 199–208.

De Angelis DL (1980) Energy flow, nutrient cycling and ecosystem resilience. *Ecology* **61**: 764–771.

del Moral R and Muller CH (1969) Fog drip: a mechanism of toxin transport from *Eucalyptus globulus. Bulletin of the Torrey Botany Club* **96**: 467–475.

del Moral R and Wood DM (1993) Early primary succession on the volcano Mount St-Helens. *Journal of Vegetation Science* **4**: 23–234.

Denslow JS, Ellison AM, and Sanford RE (1998) Tree-fall gap size effects on above- and below-ground processes in a tropical wet forest. *Journal of Ecology* **86**: 597–609.

Desy EA and Batzli GO (1989) Effects of food availability and predation on prairie vole demography: A field experiment. *Ecology* **70**: 411–421.

DeWit CT (1960) On competition. *Verslagen Landbouwkundige Onderzoekingen* **66**: 1–82.

DeWit CT (1961) Space relationships within populations of one or more species. In *Mechanisms in Biological Competition*, ed. FL Milthorpe, pp. 314–329. Cambridge University Press, London.

Diaz S *et al.* (2001) Plant traits and terrestrial ecosystem functions: a cross-continental comparison. *Ecology* (in review).

Diaz S and Cabido M (1997) Plant functional types and ecosystem function in relation to global change. *Journal of Vegetation Science* **8**: 463–474.

Diaz S, Cabido M, and Casanoves F (1999) Functional implications of trait-environmental linkages in plant communities. In *Ecological Assembly Rules*, eds. E Weiher and PA Keddy, pp. 338–362. Cambridge University Press, Cambridge.

Diaz S *et al.* (1999) Plant functional traits, ecosystem structure and land-use history along a climatic gradient in central western Argentina. *Journal of Vegetation Science* **10**: 651–660.

Diaz S *et al.* (1993) Evidence of a feedback mechanism limiting plant response to elevated carbon dioxide. *Nature* **364**: 616–617.

Diaz Barrados MC *et al.* (1999) Plant functional types and ecosystem function in Mediterranean shrubland. *Journal of Vegetation Science* **10**: 709–716.

Dixon KW, Pate JS, and Bailey WJ (1980) Nitrogen nutrition of the tuberous sundew *Drosera erythrorhiza* Lindl. with special reference to catch of arthropod fauna by its glandular leaves. *Australian Journal of Botany* **28**: 283–297.

Dixon S (1892) The effects of settlement and pastoral occupation in Australia upon the indigenous vegetation. *Transactions of the Royal Society of South Australia* **15**: 195–206.

Dobzhansky T (1950) Evolution in the tropics. *American Scientist* **38**: 209–221.

Dodd M *et al.* (1995) Community stability: a 60-year record of trends and outbreaks in the occurrence of species in the Park Grass Experiment. *Journal of Ecology* **83**: 277–285.

Donald CM (1958) The interaction of competition for light and for nutrients. *Australian Journal of Agricultural Research* **9**: 421–432.

Donnegan JA and Rebertus AJ (1999) Rates and mechanisms of subalpine forest succession along an environmental gradient. *Ecology* **80**: 1370–1384.

Doolittle WF and Sapienza C (1980) Selfish genes, the phenotypic paradigm and genome evolution. *Nature* **284**: 601–603.

Downes JA (1964) Arctic insects and their environment. *Canadian Entomologist* **96**: 279–307.

Dowson CG, Rayner ADM, and Boddy L (1986) Outgrowth patterns of mycelial cord-forming Basidiomycetes from and between woody resource units in soil. *Journal of General Microbiology* **132**: 203–211.

Drew MC (1975) Comparison of the effects of a localised supply of phosphate, nitrate, ammonium and potassium on the growth of the seminal root system, and the shoot, in barley. *New Phytologist* **75**: 479–490.

Drew MC, Saker LR, and Ashley TW (1973) Nutrient supply and the growth of the seminal root system. *Journal of Applied Botany* **24**: 1189–1202.

Dring MJ (1982) *The Biology of Marine Plants*. Arnold, London.

Duffey E *et al.* (1974) *Grassland Ecology and Wildlife Management,* Chapman and Hall, London.

Dukes JS and Mooney HA (1999) Does global change increase the success of biological invaders? *Trends in Ecology and Evolution* **14**: 135–139.

Dunham RJ and Nye PH (1976) The influence of soil water content on the uptake of ions by roots. III Phosphate, potassium, calcium and magnesium uptake and concentration gradients in soil. *Journal of Applied Ecology* **13**: 967–984.

Dunnett NP and Grime JP (1999) Competition as an amplifier of short-term vegetation responses to climate: an experimental test. *Functional Ecology* **13**: 388–395.

Dunnett *et al.* (1998) A 38-year study of relations between weather and vegetation dynamics in road verges near Bibury, Gloucestershire. *Journal of Ecology* **86**: 610–623.

During HJ and ter Horst B (1983) The diaspore bank of bryophytes and ferns in chalk grassland. *Lindbergia* **9**: 57–64.

During HJ and Willems JH (1984) Diversity models applied to chalk grassland. *Vegetatio* **57**: 103–114.

Egler FE (1954) Vegetation science concepts. I. Initial floristic composition, a factor in old-field development. *Vegetatio* **4**: 412–417.

Egunjobi JK (1974a) Dry matter, nitrogen and mineral element distribution in an unburnt savannah during the year. *Oecologia Plantarum* **9**: 1–10.

Egunjobi JK (1974b) Litter fall and mineralisation in a teak *(Tectona grandis)* stand. *Oikos* **25**: 222–226.

Ehleringer J (1985) Annuals and perennials of warm deserts. In *Physiological Ecology of North American Plant Communities*, eds. BF Chabot and HA Mooney, pp. 162–180. Chapman and Hall, London.

Ehleringer JR (1993) Carbon and water relations in desert plants: an isotopic perspective. In *Stable Isotopes and Plant Carbon-water Relations*, eds. JR Ehleringer, AE Hall and GD Farquahar, pp. 155–172. Academic Press, San Diego.

Ehrlen J and Groenendael JM van (1998) The trade-off between dispersibility and longevity: an important aspect of plant species diversity. *Applied Vegetation Science* **1**: 29–36.

Ehrlich PR and Birch LC (1967) The 'Balance of Nature' and 'Population Control'. *American Naturalist* **101**: 97–108.

Ehrlich PR and Ehrlich AH (1981) *Extinction. The Causes and Consequences of the Disappearance of Species*. Random House, London

Eissenstat DM and Yanai R (1997) The ecology of root life-span. *Advances in Ecological Research* **27**: 1–60.

Elberse WT, Van den Bergh JP and Dirven JGP (1983) Effects of use and mineral supply on the botanical composition and yield of old grassland on heavy clay soil. *Netherlands Journal of Agricultural Science* **31**: 63–88.

Ellenberg H (1963) *Vegetation Mitteleuropas mit den Alpen.* Eugen Ulmer, Stuttgart.

Ellenberg H (1974) *Zeigwerte der Gefässpflanzen Mitteleeuropas*. Scripta Geobotanica 9 Goeltze, Gottingen.

Ellenberg H and Mueller-Dombois D (1974). *Aims and methods of Vegetation ecology.* Wiley, New York.

Elton C (1927) *Animal Ecology*. Sidgwick and Jackson, London.

Elton CS (1958) *The Ecology of Invasions by Animals and Plants*. Methuen, London.

Elton CS and Miller RS (1954). The ecological survey of animal communities: with a practical system of classifying habitats by structural characteristics. *Journal of Ecology* **42**: 460–496.

Eriksson O (1993) The species-pool hypothesis and plant community diversity. *Oikos* **68**: 371–374.

Evans EM (1953) Studies in bryophyte ecology. PhD Thesis, University of Sheffield.

Evans JP (1988) Nitrogen translocation in a clonal dune perennial, *Hydrocotyle bonariensis*. *Oecologia* **77**: 64–68.

Facelli JM and Pickett STA (1991) Plant litter: light interception and effects on an old-field plant community. *Ecology* **72**: 1024–1031.

Farrar JF (1976a). Ecological physiology of the lichen *Hypogymnia physodes*: II. Effects of wetting and drying cycles and the concept of physiological buffering. *New Phytologist* **77**:105–113.

Farrar JF (1976b). Ecological physiology of the lichen *Hypogymnia physodes*: III The importance of the rewetting phase. *New Phytologist* **77**: 115–125.

Farrar JF (1976c) The uptake and metabolism of phosphate by the lichen *Hypogymnia physodes. New Phytologist* **77**: 127–134.

Feeny PP (1968) Effects of oak leaf tannins on larval growth of the winter moth *Operophtera brumata. Journal of Insect Physiology* **14**: 805–817.

Feeny PP (1969) Inhibitory effect of oak leaf tannins on the hydrolysis of proteins by trypsin. *Phytochemistry* **8**: 2119–2126.

Feeny PP (1970). Seasonal changes in oak leaf tannins and nutrients as a cause of spring feeding by winter moth caterpillars. *Ecology* **5**: 565–581.

Feeny PP (1975) Biochemical coevolution between plants and their insect herbivores. In *Coevolution of Plants and Animals*, eds. LE Gilbert and PH Raven, pp. 3–19. University of Texas Press.

Fenner M J (1978) A comparison of the abilities of colonisers and closed-turf species to establish from seed in artificial swards. *Journal of Ecology* **66**: 953–964.

Fenner M (1992) *Seeds: the Ecology of Regeneration in Plant Communities*. CAB International, Wallingford.

Ferguson CW (1968) Bristlecone Pine: science and esthetics. *Science* **159**: 839–846.

Field C and Mooney HA (1986) The photosynthesis-nitrogen relationship in world plants. In *On the Economy of Plant Form and Function*, ed. TV Givnish, pp. 25–55. Cambridge University Press, Cambridge.

Finlay RD, Frostegard A, and Sonnerfeldt AM (1992) Utilization of organic and inorganic nitrogen sources by ectomycorrhizal fungi in pure culture and in symbiosis with *Pinus contorta* Doug. Ex. Loud. *New Phytologist* **120**: 105–115.

Fisher HJ, Myers LF, and Williams JD (1974) Nutrient responses of an indigenous *Poa* Tussock and *Lolium perenne* L. grown separately and together in pot culture. *Australian Journal of Agricultural Research* **25**: 863–874.

Fitter AH *et al.* (1995) Relationships between first flowering date and temperature in the flora of a locality in Central England. *Functional Ecology* **9**: 55–60.

Fitter AH *et al.* (1998) Carbon transfer between plants and its control in networks of arbuscular mycorrhizas. *Functional Ecology* **12**: 406–412.

Ford ED and Newbould PJ (1977) The biomass and production of ground vegetation and its relation to tree cover through a deciduous woodland cycle. *Journal of Ecology* **65**: 201–212.

Foulds W (1993) Nutrient concentrations of foliage and soil in south-western Australia. *New Phytologist* **125**: 529–546.

Fowler N (1981) Competition and coexistence in a North Carolina grassland. II The effects of the experimental removal of species. *Journal of Ecology* **69**: 843–854.

Fowler N (1986) The role of competition in plant communities in arid and semi-arid regions. *Annual Review of Ecology and Systematics* **17**: 89–105.

Fox JF (1977) Alternation and coexistence of tree species. *American Naturalist* **111**: 69–89.

Fox JF (1981) Intermediate levels of soil disturbance maximise alpine plant diversity. *Nature* **293**: 564–565.

Francis R, Finlay RD, and Read DJ (1986) Vesicular-arbuscular mycorrhizas in natural vegetation systems. IV. Transfer of nutrients in inter- and intra-specific combinations of host plants. *New Phytologist* **102**: 103–111.

Francis R and Read DJ (1984) Direct transfer of carbon between plants connected by vesicular-arbuscular mycorrhizal mycelium. *Nature* **307**: 53–56.

Francis R and Read DJ (1995) Mutualism and antagonism in the mycorrhizal symbiosis, with special reference to impacts on plant community structure. *Canadian Journal of Botany* **73**: S1301-S1309.

Franco AC and Nobel PS (1989) Effect of nurse plants on the microhabitat and growth of cacti. *Journal of Ecology* **77**: 870–886.

Fraser LH (1998) Top-down vs. bottom-up control influenced by productivity in a North Derbyshire dale. *Oikos* **81**: 99–108.

Fraser LH and Grime JP (1998a) Primary productivity and trophic dynamics investigated in a North Derbyshire dale. *Oikos* **80**: 499–508.

Fraser LH and Grime JP (1998b) Top-down control and its effect on the biomass and composition of three grasses at high and low soil fertility in outdoor microcosms. *Oecologia* **113**: 239–246.

Fraser L and Grime JP (1999a) Aphid fitness on 13 grass species: a test of plant defence theory. *Canadian Journal of Botany* **77**: 1783–1789.

Fraser LH and Grime JP (1999b) Interacting effects of herbivory and fertility on a synthesised plant community. *Journal of Ecology* **87**: 514–525.

Fraser LH and Keddy P (1997) The role of experimental microcosms in ecological research. *Trends in Ecology and Evolution* **12**: 478–481.

Fredrickson AG and Stephanopoulos G (1981) Microbial competition. *Science* **213**: 972–979.

French NR (1967) Life spans of *Dipodomys* and *Perognathus* in the desert. *Journal of Mammalogy* **48**: 537–548.

French NR, Maza BG, and Aschwanden AP (1966) Periodicity of desert rodent activity. *Science* **154**: 1194–1195.

French NR *et al.* (1974) A population study of irradiated desert rodents. *Ecological Monographs* **44**: 45–72.

Fretwell D (1977) The regulation of plant communities by food chains exploiting them. *Perspectives in Biology and Medicine* **20**: 169–185.

Fretwell D (1987) Food chain dynamics: the central theory of ecology? *Oikos* **50**: 291–301.

Friedman D and Alpert P (1991) Reciprocal transport between ramets increases growth in *Fragaria chiloensis* when light and nitrogen occur in separate patches but only if patches are rich. *Oecologia* **86**: 76–80.

Frith HJ (1973). Discussion: role of environment in reproduction as a source of 'predictive' information. In *Breeding Biology of Birds*, ed. DS Farner, pp. 147–154, National Academy of Sciences, Washington DC.

Furness SB (1978) Ecological investigations of growth and temperature responses in bryophytes. PhD Thesis, University of Sheffield.

Furness SB and Grime JP (1982a) Growth rate and temperature responses in bryophytes. I. An investigation of *Brachythecium rutabulum*. *Journal of Ecology* **70**: 513–523.

Furness SB and Grime JP (1982b) Growth rate and temperature responses in bryophytes. II. A comparative study of species of contrasted ecology. *Journal of Ecology* **70**: 525–536.

Furness SB and Hall RH (1981) An explanation for the intermittent occurrence of *Physcomitrium sphaericum*. *Journal of Bryology* **11**: 733–742.

Gadallah FL and Jeffries RL (1995a) Comparison of the nutrient contents of the principal forage plants utilised by lesser snow geese on summer breeding grounds. *Journal of Applied Ecology* **32**: 263–275.

Gadallah FL and Jeffries RL (1995b) Forage quality in brood rearing areas of the lesser snow goose and the growth of captive goslings. *Journal of Applied Ecology* **32**: 276–287.

Gadgil M and Solbrig OT (1972) The concept of r- and K-selection: evidence from wild flowers and some theoretical considerations. *American Naturalist* **106**: 14–31.

Gardner G (1977) The reproductive capacity of *Fraxinus excelsior* on the Derbyshire limestone. *Journal of Ecology* **65**: 107–118.

Gardner RA and Bradshaw KE (1954) Characteristics and vegetative relationships of some podzolic soils near the coast of northern California. *Soil Science Society of America Proceedings*. **18**: 320–325.

Garnier E and Laurent G (1994) Leaf anatomy, specific mass and water content in congeneric annual and perennial grass species. *New Phytologist* **128**: 725–736.

Garnier E *et al.* (1997) A problem for biodiversity-productivity studies: how to compare the productivity of multispecific plant mixtures to that of monocultures. *Acta Ecològica* **18**: 657–670.

Gates DM (1968) Transpiration and leaf temperature. *Annual Review Plant Physiology* **19**: 211–238.

Gaudet CL and Keddy PA (1988) A comparative approach to predicting competitive ability from plant traits. *Nature* **334**: 242–243.

George DG and Harris GP (1985) The effect of climate on the long-term changes in the crustacean zooplankton biomass of Lake Windermere, UK. *Nature* **316**: 536–539.

George DG and Taylor AH (1995) UK lake plankton and the Gulf Stream. *Nature* **378**: 139.

George LO and Bazzaz FA (1999a) The fern understorey as an ecological filter: emergence and establishment of canopy-tree seedlings. *Ecology* **80**: 833–845.

George LO and Bazzaz FA (1999b) The fern understorey as an ecological filter: growth and survival of canopy-tree seedlings. *Ecology* **80**: 846–856.

Getz LL (1960) Standing crop of herbaceous vegetation in southern Michigan. *Ecology* **41**: 393–395.

Gilbert L (1977) The role of insect-plant co-evolution in the organisation of ecosystems. In *Comportement des Insects et Milieu Tropique*, ed. V Labyrie, pp. 399–413. CNRS, Paris.

Gilbert OL (1984) *The Ecology of Urban Habitats*. Chapman and Hall, London.

Gilpin ME (1975) Limit cycles in competition communities. *American Naturalist* **109**: 51–60.

Ginzo HD and Lovell PH (1973) Aspects of the comparative physiology of *Ranunculus bulbosus* L. and *Ranunculus repens* L. II Carbon dioxide assimilation and distribution of photosynthates. *Annals of Botany* **37**: 765–776.

Gitay H, Noble IR, and Connell JH (1999) Deriving functional types for rainforest trees. *Journal of Vegetation Science* **10**: 641–650.

Givnish TJ (1982) On the adaptive significance of leaf height in forest herbs. *American Naturalist* **120**: 353–381.

Givnish TJ (1994) Does diversity beget stability? *Nature* **371**: 113–114.

Gleeson S and Tilman D (1990) Allocation and the transient dynamics of succession on poor soils. *Ecology* **71**: 1144–1155.

Godwin H (1956) *The History of the British flora.* Cambridge University Press, London.

Goldberg DE (1987) Neighbourhood competition in an old field plant community. *Ecology* **68**: 1211–1223.

Goldberg DE (1997) Competitive ability: definitions, contingency and correlated traits. In *Plant Life Histories; Ecology, Phylogeny and Evolution*, eds. J Silvertown, M Franco, and JL Harper, pp. 283–301. Cambridge University Press, Cambridge.

Goldberg DE and Barton AM (1992) Patterns and consequences of interspecific competition in natural communities: a review of field experiments with plants. *American Naturalist* **139**: 771–801.

Goldberg DE and Fleetwood L (1987) Competitive effect and response in four annual plants. *Journal of Ecology* **75**: 1131–1143.

Goldberg DE and Werner PA (1983) Equivalence of competitors in plant communities: a null hypothesis and a field experimental approach. *American Journal of Botany* **70**: 1098–1104.

Golley FB (1960) Energy dynamics of a food chain of an old-field community. *Ecological Monographs* **30**: 187–206.

Golley FB (1965) Structure and function of an old-field broomsedge community. *Ecological Monographs* **35**: 113–137.

Gómez-Pompa A (1967) Some problems of tropical forest plant ecology. *J Arnold Arboretum* **48**: 105–121.

Gordon WS and Jackson RB (2000) Nutrient concentrations in fine roots. *Ecology* **81**: 275–280.

Goreau NI and Yonge CM (1971) Reef corals: autotrophs or heterotrophs? *Biological Bulletin Marine Biol. Lab. Woods Hole* **141**: 247–260.

Goreau TF *et al.* (1972) Structure and ecology of the Saipan reefs in relation to predation by *Acanthaster planci* (L). *Bulletin of Marine Science* **22**: 113–152.

Gorski T (1975) Germination of seeds in the shadow of plants. *Physiologia Plantarum* **34**: 342–346.

Gough L, Grace JB, and Taylor KL (1994) The relationship between species richness and community biomass: the importance of environmental variables. *Oikos* **70**: 271–279.

Gough L *et al.* (2000) Fertilization effects on species density and primary productivity in herbaceous plant communities. *Oikos* **89**: 428–439.

Grace J and Woolhouse HW (1970) A physiological and mathematical study of the growth and productivity of a *Calluna–Sphagnum* community: I. Net photosynthesis of *Calluna vulgaris* (L.) Hall. *Journal of Applied Ecology* **7**: 363–381.

Grace JB (1991) A clarification of the debate between Grime and Tilman. *Functional Ecology* **5**: 583–587.

Grace JB (1999) The factor controlling species density in herbaceous plant communities. *Perspectives in Plant Ecology, Evolution and Systematics* **3**: 1–28.

Grace JB and Jutila H (1999) The relationship between species density and community biomass in grazed and ungrazed coastal meadows. *Oikos* **85**: 398–408.

Grace JB and Wetzel RG (1981) Habitat partitioning and competitive displacement in cattails (Typha): experimental field studies. *American Naturalist* **118**: 463–474.

Granato TC and Raper CD (1989) Proliferation of maize (*Zea mays* L.) roots in response to localised supply of nitrate. *Journal of Experimental Botany* **40**: 263–275.

Grant PR (1970) Experimental studies of competitive interactions in a two-species system: II. The behaviour of *Microtus*, *Peromyscus* and *Clethrionomys* species. *Animal Behaviour* **18**: 411–426.

Green DG (1989) Simulated effects of fire, dispersal and spatial pattern on competition within forest mosaics. *Vegetatio* **82**: 139–153.

Greene DF and Johnson EA (1996) Wind dispersal of seeds from a forest into a clearing. *Ecology* **77**: 595–609.

Greenslade PJM (1983) Adversity selection and the habitat templet. *American Naturalist* **122**: 352–365.

Grieve BJ (1956) Studies in the water relations of plants: I. Transpiration of western Australian (Swan Plain) sclerophylls. *Royal Society of Western Australia* **40**: 15–20.

Grime JP (1963) An ecological investigation at a junction between two plant communities in Coombsdale on the Derbyshire limestone. *Journal of Ecology* **51**: 391–402.

Grime JP (1965) Comparative experiments as a key to the ecology of flowering plants. *Ecology* **45**: 513–515.

Grime JP (1966) Shade avoidance and tolerance in flowering plants. In *Light as an Ecological Factor.* eds. R Bainbridge, GC Evans, and O Rackman, pp. 281–301. Blackwell, Oxford.

Grime JP (1972) The creative approach to nature conservation. In *The Future of Man*, eds. FJ Ebling and GW Heath, pp. 47–54. Academic Press, London.

Grime JP (1973a) Competitive exclusion in herbaceous vegetation. *Nature* **242**: 344–347.

Grime JP (1973b) Competition and diversity in herbaceous vegetation – a reply. *Nature* **244**: 310–311.

Grime JP (1973c) Control of species density in herbaceous vegetation. *Journal of Environmental Management* **1**: 151–167.

Grime JP (1974) Vegetation classification by reference to strategies. *Nature* **250**: 26–31.

Grime JP (1977) Evidence for the existence of three primary strategies in plants and its relevance to ecological and evolutionary theory. *American Naturalist* **111**: 1169–1194.

Grime JP (1979) *Plant Strategies and Vegetation Processes.* Wiley, Chichester.

Grime JP (1981) The role of seed dormancy in vegetation dynamics. *Annals of Applied Biology* **98**: 555–558.

Grime JP (1983) Prediction of weed and crop response to climate based upon measurements of nuclear DNA content. *Aspects of Applied Biology* **4**: 87–98.

Grime JP (1984) The ecology of species, families and communities of the contemporary British flora. *New Phytologist* **98**: 15–33.

Grime JP (1985a) Towards a functional classification of vegetation In *The Population Structure of Vegetation*, ed. J White, pp. 503–514. Dr W Junk Publishers, Dordrecht.

Grime JP (1985b) Factors limiting the contribution of pteridophytes to a local flora. In: *The biology of pteridophytes*, eds. AF Dyer and CM Page. *Proceedings of the Royal Society of Edinburgh* **B86**: 403–421.

Grime JP (1986) The circumstances and characteristics of spoil colonisation within a local flora. *Phil. Transactions of the. Royal Society London.* **B314**, 637–654.

Grime JP (1987) Dominant and subordinant components of plant communities implications for succession, stability and diversity. In *Colonization, Succession and stability*, eds. A Gray, P Edwards, and M Crawley, pp. 413–428. Blackwell Scientific Publications, Oxford.

Grime JP (1988a) The CSR model of primary plant strategies—origins, implications and tests. In *Evolutionary Plant Biology*, eds. LD Gottlieb and S Jain, pp. 371–393. Chapman and Hall, London.

Grime JP (1988b) Fungal strategies in ecological perspective. *Proceedings of the Royal Society of Edinburgh* **B94**: 167–169.

Grime JP (1988c) Appendix 39. Memorandum submitted by Professor JP Grime, NERC, Unit of Comparative Plant Ecology, University of Sheffield. Agriculture Committee. *Second Report. Chernobyl: The Government's reaction. Volume II.* Minutes of Evidence and Appendices, pp. 399–403. HMSO; London.

Grime JP (1989a) The stress debate: symptom of impending synthesis? In: *Evolution, Ecology and Environmental Stress*, ed. P Calow. *Special issue of the Biological Journal of the Linnean Society* **37**: Proceedings of the Linnean Society Bicentenary Symposium, June 1988, 3–17.

Grime JP (1989b) Mechanisms promoting floristic diversity in calcareous grassland. In *Calcareous Grasslands—Ecology and Management*, eds. SH Hillier, DWH Walton, and DA Wells, pp. 51–56. Bluntisham Books, Huntingdon.

Grime JP (1991) Nutrition, environment and plant ecology: an overview. In *Plant Growth: Interactions with Nutrition and Environment*, eds. JR Porter and DW Lawlor, pp. 249–267. Cambridge University Press, Cambridge.

Grime JP (1993) Ecology sans frontières. *Oikos* **68**: 385–392.

Grime JP (1994) The role of plasticity in exploiting environmental heterogeneity In *Exploitation of Environmental Heterogeneity in Plants*, eds. M Caldwell and R Pearcy, pp. 1–18. Academic Press, San Diego.

Grime JP (1997) Biodiversity and ecosystem function: the debate deepens. *Science* **277**: 1260–1261.

Grime JP (1998) Benefits of plant diversity to ecosystems: immediate, filter and founder effects. *Journal of Ecology* **86**: 902–910.

Grime JP (1999) The Magnesian Limestone ecosystems of Northern England: What is their future in the 21st Century? *The Naturalist* **124**: 12–16.

Grime JP and Anderson JM (1986) Environmental controls over organism activity. In *Forest Ecosystems in the Alaskan Taiga; a Synthesis of Structure and Function*, eds. K van Cleve *et al.*, pp. 89–95. Springer-Verlag, Berlin.

Grime JP and Blythe GM (1968) An investigation of the relationships between snails and vegetation at the Winnats Pass. *Journal of Ecology* **57**: 45–66.

Grime JP *et al.* (2000) The response of two contrasting limestone grasslands to simulated climate change. *Science* **289**: 762–765.

Grime JP *et al.* (1996) Evidence of a causal connection between anti-herbivore defence and the decomposition rate of leaves. *Oikos* **77**: 489–494.

Grime JP, Crick JC, and Rincon E (1986) The ecological significance of plasticity. In *Plasticity in Plants*, eds. DH Jennings and AJ Trewavas, pp. 5–29, Company of Biologists, Cambridge.

Grime JP and Curtis AV (1976) The interaction of drought and mineral nutrient stress in calcareous grassland. *Journal of Ecology* **64**: 976–998.

Grime JP, Hodgson JG, and Hunt, R (1988) *Comparative Plant Ecology: a Functional Approach to Common British Species*. Unwin Hyman, London.

Grime JP *et al.* (1997) Functional types: testing the concept in northern England. In *Plant Functional Types, their Relevance to Ecosystem Properties and Global change*, eds. TM Smith, HH Shugart, and FI Woodward, pp. 122–150. International Geosphere-Biosphere programme book series 1. Cambridge University Press, Cambridge.

Grime JP and Hunt R (1975) Relative growth rate: its range and adaptive significance in a local flora. *Journal of Ecology* **63**: 393–422.

Grime JP, Hunt R, and Krzanowski WJ (1987a) Evolutionary physiological ecology of plants. In *Evolutionary Physiological Ecology*, ed. P Calow, pp. 105–126. Cambridge University Press, Cambridge.

Grime JP and Jarvis BC (1975) Shade avoidance and shade tolerance in flowering plants: II. Effects of light on the germination of species of contrasted ecology. In *Light as an Ecological Factor*, eds. R Bainbridge, GC Evans, and O Rackman, pp. 525–532. Blackwell, Oxford.

Grime JP and Jeffrey DW (1965) Seedling establishment in vertical gradients of sunlight. *Journal of Ecology* **53**: 621–642.

Grime JP and PS Lloyd (1973) *An Ecological Atlas of Grassland Plants.* Arnold, London.

Grime JP *et al.* (1987b) Floristic diversity in a model system using experimental microcosms. *Nature* **328**: 420–422.

Grime JP, MacPherson-Stewart SF, and Dearman RS (1968) An investigation of leaf palatablility using the snail *Cepaea nemoralis* L. *Journal of Ecology* **56**: 405–420.

Grime JP *et al.* (1981) A comparative study of germination characteristics in a local flora. *Journal of Ecology* **69**: 1017–1059.

Grime JP and Mowforth MA (1982) Variation in genome size—an ecological interpretation. *Nature* **299**: 151–153.

Grime JP, Shacklock JML, and Band SR (1985) Nuclear DNA contents, shoot phenology and species coexistence in a limestone grassland community. *New Phytologist* **100**: 435–445.

Grime JP *et al.* (1997) Integrated screening validates primary axes of specialisation in plants. *Oikos* **79**: 259–281.

Grime JP *et al.* (1994) Climate–vegetation relationships in the Bibury road verge experiments. In *Insight from Foresight: Long-term Experiments in Agricultural and Ecological Sciences*, eds. RA Leigh and AE Johnston, pp. 271–285. CAB International, Wallingford.

Gross KL (1984) Effects of seed size and growth form on seedling establishment of six monocarpic perennial plants. *Journal of Ecology* **72**: 369–387.

Gross KL and Werner PA (1982) Colonising abilities of 'biennial' plant species in relation to ground cover; implications for their distributions in a successional sere. *Ecology* **63**: 921–931.

Gross KL *et al.* (2000) Patterns of species density and productivity at different spatial scales in herbaceous plant communities. *Oikos* **89**: 417–427.

Grouzis M, Berger A, and Heim G (1976) Polymorphisme et germination des graines chez trois espécès anuelles du genre *Salicornia. Oecologia Plantarum* **11**: 41–52.

Grubb PJ (1976) A theoretical background to the conservation of ecologically distinct groups of annuals and biennials in the chalk grassland ecosystem. *Biological Conservation* **10**: 53–76.

Grubb PJ (1977) The maintenance of species-richness in plant communities: the importance of the regeneration niche. *Biological Reviews* **52**: 107–145.

Grubb PJ (1980) Review of Grime (1979) *Plant Strategies and Vegetation Processes. New Phytologist* **86**: 123–124.

Grubb PJ (1985) Plant populations and vegetation in relation to habitat, disturbance and competition: problems of generalisation. In *The Population Structure of Vegetation*, ed. J White, pp. 595–621. Dr W Junk Publishers, Dordrecht.

Grubb PJ (1987) Global trends in species richness in terrestrial vegetation: a view from the northern hemisphere. In *Organisation of Communities Past and Present*, eds. JHR Gee and PS Giller, pp. 99–118. Blackwell, Oxford.

Grubb PJ (1992) A positive distrust in simplicity—lessons from plant defences and from competition among plants and among animals. *Journal of Ecology* **80**: 585–610.

Grubb PJ (1994) Root competition in soils of different fertility: a paradox resolved? *Phytocoenologia* **24**: 494–505.

Grubb PJ (1998) A reassessment of the strategies of plants which cope with shortages of resources. *Perspectives in Plant Ecology, Evolution and Systematics* **1**: 3–31.

Grubb PJ, Green HE, and Merrifield RCJ (1969) The ecology of chalk heath; its relevance to the calcicole-calcifuge and soil acidification problems. *Journal of Ecology* **57**: 175–212.

Grubb PJ, Kelly D, and Mitchley J (1982) The control of relative abundance in communities of herbaceous plants. The Plant Community as a Working Mechanism, British Ecological Society No.1 (ed. Newman E), pp. 77–97. Blackwell, Oxford.

Gutterman Y (1980) Influences on seed germinability: phenotypic maternal effects during seed maturation. *Israeli Journal of Botany* **29**: 105–117.

Gutterman Y (1993) *Seed Germination in Desert Plants*. Springer-Verlag, Berlin.

Haag RW (1974) Nutrient limitations to plant production in two tundra communities. *Canadian Journal of Botany* **52**: 103–116.

Hackett C (1965) Ecological aspects of the nutrition of *Deschampsia flexuosa* (L.) Trin: II. The effects of Al, Ca, Fe, K, Mn, N, P, and pH on the growth of seedlings and established plants. *Journal of Ecology* **53**: 315–333.

Hadley EB and LC Bliss (1964) Energy relationships of alpine plants on Mt. Washington, New Hampshire. *Ecological Monographs* **34**: 331–357.

Hairston NG, Smith FE, and Slobodkin LD (1960) Community structure, population control and competition. *American Naturalist* **94**: 421–425.

Hald AB and Vintner E (2000) Restoration of a species-rich fen-meadow after abandonment: response of 64 plant species to management. *Applied Vegetation Science* **3**: 15–24.

Hallè F and Oldeman RAA (1975) Essay on the architecture and dynamics of growth of tropical trees. Penerbit University, Kuala Lumpur.

Hancock JM (1996) Simple sequences and the expanding genome. *BioEsssays* **18**: 421–425.

Handel SN (1978) New ant-dispersal species in the genera Carex, Luzula and Claytonia. *Canadian Journal of Botany* **56**: 2925–2927.

Hanes TL (1971). Succession after fire in the chaparral of southern California. *Ecological Monographs* **41**: 27–52.

Hanley ME, Fenner M, and Edwards PJ (1995) An experimental study of the effects of molluscan grazing on seedling recruitment and survival in grassland. *Journal of Ecology* **83**: 621–627.

Hansen AP and Pate JS (1987) Comparative growth and symbiotic performance of seedlings of *Acacia pulchella* and *A. alata* in defined pot culture or as natural understorey components of a eucalyptus forest ecosystsem in S.W. Australia. *Journal of Experimental Botany* **38**: 13–25.

Hansen K (1976) Ecological studies in Danish heath vegetation. *Dansk Botanisk Arkv* **41**: 1–118.

Harley JL (1969). *The Biology of Mycorrhiza.* 2nd ed. Hill, London.

Harley JL (1970) Mycorrhiza and nutrient uptake in forest trees. In *Physiology of Tree Crops*, eds. L Luckwcll and C Cutting, pp. 163–178. Academic Press, New York.

Harley JL (1971) Fungi in ecosystems. *Journal of Ecology* **59**: 653–686.

Harley JL and Harley EL (1987) A checklist of mycorrhizas in the British Flora. *New Phytologist* (Suppl.) **105**: 1–102.

Harper JL (1957) Biological Flora of the British Isles: *Ranunculus acris* L., *Ranunculus repens* L., and *Ranunculus bulbosus* L. *Journal of Ecology* **45**: 289–342.

Harper JL (1961) Approaches to the study of plant competition. In *Mechanisms in Biological Competition*, ed. FL Milthorpe, pp. 1–39. *Symp. Soc. Exp. Biol.* **15**. Cambridge University Press.

Harper JL (1977) *The Population Biology of Plants.* Academic Press, London.

Harper JL (1982) After description. In *The Plant Community as a Working Mechanism*, ed. EI Newman, pp. 11–25. Special Publication No.1. BES Blackwell, Oxford.
Harper JL and White J (1974) The demography of plants. *Annual Review of Ecology and Systematics* **5**: 419–163.
Harper JL, Williams JT, and Sagar GR (1965) The behaviour of seeds in soil. Part 1. The heterogeneity of soil surfaces and its role in determining the establishment of plants from seed. *Journal of Ecology* **53**: 273–286.
Harrington GN (1991) Effects of soil moisture on shrub seedling survival in a semi-arid grassland. *Ecology* **72**: 1138–1149.
Harrington JF and Thompson RC (1952) Effect of variety and area of production on subsequent germination of lettuce seed at high temperatures. *Proceedings of the American Society for Horticultural Science* **597**: 445–450.
Harris P (1960) Production of pine resin and its effect on survival of *Ryhacionia beolina. Canadian Journal of Zoology* **38**: 121–130.
Hartsema AM (1961) Influence of temperature on flower formation and flowering of bulbous and tuberous plants. *Handbuch der Pflanzenphysiologie* **16**: 123–167.
Hastings A (1980) Disturbance, coexistence, history and competition for space. *Theoretical Population Biology* **18**: 363–373.
Heal OW and Grime JP (1991) Comparative analysis of ecosystems: past lessons and future directions. In *Comparative Analyses of Ecosystems, Patterns, Mechanisms and Theories*, eds. J Cole, G Lovett, and S Findlay, pp. 7–23. Springer-Verlag.
Hector A (2000) Biodiversity and ecosystem functioning. *Progress in Environmental Science* (in press).
Hector A *et al.* (1999) Plant diversity and productivity in European grasslands. *Science* **286**: 1123–1127.
Hehre EJ and Mathieson AC (1970) Investigations of New England marine algae: III. Composition, seasonal occurrence and reproductive periodicity of the marine Rhodophyceae in New Hampshire. *Rhodora* **72**: 194–239.
Hellmers H (1964) An evaluation of the photosynthetic efficiency of forests. *Quarterly Review of Biology* **39**: 249–257.
Hendry GAF and Grime JP (1993) *Comparative Plant Ecology—A Laboratory Manual.* Chapman and Hall, London.
Hendry GAF *et al.* (1994) Seed persistence: a correlation between seed longevity in the soil and ortho-dihydroxyphenol concentration. *Functional Ecology* **8**: 658–664.
Herben T *et al.* (1997) Fine scale species interactions of clonal plants in a mountain grassland: a removal experiment. *Oikos* **78**: 299–310.
Herms DA and Mattson WJ (1992) The dilemma of plants: to grow or defend? *Quarterly Review of Biology* **67**: 283–335.
Herrera CM (1998) Long-term dynamics of Mediterranean frugivorous birds and fleshy fruits: a 12-year study. *Ecological Monographs* **68**: 511–538.
Hewitt LJ (1952) A biological approach to the problems of soil acidity. *Transactions of the International Society of Soil Science Joint Meeting Dublin* **1**: 107–118.
Hicklenton PR and Oechel WC (1976) Physiological aspects of the ecology of *Dicranum fuscescens* in the subarctic: I. Acclimation and acclimation potential of CO_2 exchange in relation to habitat, light and temperature. *Canadian Journal of Botany* **54**: 1104–1119.
Hickman JC (1975). Environmental unpredictability and plastic energy allocation strategies in the annual *Polygonum cascadense* (Polygonaceae). *Journal of Ecology* **63**: 689–701.
Higgs DEB and James DB (1969) Comparative studies on the biology of upland grasses: I. Rate of dry matter production and its control in four grass species. *Journal of Ecology* **57**: 553–563.

Hik DS and Jefferies RL (1990) Increase in net above-ground primary production of a salt marsh forage grass: a test of the predictions of the herbivore optimisation model. *Journal of Ecology* **78**: 180–195.

Hik DS, Jefferies RL, and Sinclair ARE (1992) Foraging by geese, isostatic uplift and asymmetry in the development of salt-marsh plant-communties. *Journal of Ecology* **80**: 395–406.

Hill MO *et al.* (1999) *Ellenberg's Indicator Values for British Plants. Technical Annexe to Volume 2, ECOFACT*. ITE Monks Wood, Abbots Ripton, Huntingdon.

Hillier SH, Sutton F, and Grime JP (1994) A new technique for the experimental manipulation of temperature in plant communities. *Functional Ecology* **8**: 755–762.

Hillier SH (1990) Gaps, seed banks and plant species diversity in calcareous grasslands. In *Calcareous Grasslands—Ecology and Management*, eds. SH Hillier, DWH Walton, and DH Wells, pp. 57–66. Bluntisham Books, Huntingdon.

Hills JM *et al.* (1994) A method for classifying European riverine wetland ecosystems using functional vegetation groups. *Functional Ecology* **8**: 242–252.

Hobbie SE (1996) Temperature and plant species control over litter decomposition in Alaskan tundra. *Ecological Monographs* **66**: 503–522.

Hobbs RJ (1989) The nature and effects of disturbance relative to invasions. In *Biological Invasion: a Global Perspective*, eds. JA Drake *et al.*, pp. 389–405. Wiley & Sons, New York.

Hobbs RJ and Atkins L (1988) Effect of disturbance and nutrient addition on native and introduced annuals in plant communities in the Western Australia wheatbelt. *Australian Journal of Ecology* **13**: 171–179

Hobbs RJ and Mooney HA (1985) Community and population dynamics of serpentine grassland annuals in relation to gopher disturbance. *Oecologia* **67**: 342–351.

Hobbs RJ and Mooney HA (1991) Effects of rainfall variability and gopher disturbance on serpentine annual grassland dynamics. *Ecology* **72**: 59–68.

Hodgson GL and Blackman GE (1956) An analysis of the influence of plant density on the growth of *Vicia faba*. *Journal of Experimental Botany* **7**: 146–165.

Hodgson JG (1986a) Commonness and rarity in plants with special reference to the Sheffield Flora. Part I. The identity, distribution and habitat characteristics of the common and rare species. *Biological Conservation* **36**: 199–252.

Hodgson JG (1986b) Commonness and rarity in plants with special reference to the Sheffield Flora. Part II. The relative importance of climate, soils and land use. *Biological Conservation* **36**: 253–274.

Hodgson JG (1986c) Commonness and rarity in plants with special reference to the Sheffield Flora. Part III. Taxonomic and evolutionary aspects. *Biological Conservation* **36**: 275–296.

Hodgson JG (1989) What is happening to the British flora? An investigation of commonness and rarity. *Plants Today* **2**: 26–32.

Hodgson JG (1991) The use of ecological theory and autecological datasets in studies of endangered plant and animal species and communities. *Pirineos* **138**: 3–28.

Hodgson JG (1993) Commonness and rarity in British butterflies. *Journal of Applied Ecology* **30**: 407–427.

Hodgson JG and Mackey JML (1986) The ecological specialisation of dicotyledonous families within a local flora: some factors constraining optimisation of seed size and their possible evolutionary significance. *New Phytologist* **104**: 497–513.

Hodgson JG *et al.* (1998) Does biodiversity determine ecosystem function? The Ecotron experiment revisited. *Functional Ecology* **12**: 843–848.

Hodgson JG *et al.* (1999) Allocating CSR plant functional types: a soft approach to a hard problem. *Oikos* **85**: 282–294.

Hodkinson DJ and Thompson K (1997) Plant dispersal: the role of man. *Journal of Applied Ecology* **34**: 1484–1496.

Hoffman GR (1966) Ecological studies of *Funaria hygrometrica* (L.) Hebw. in Eastern Washington and Northern Idaho. *Ecological Monographs* **36**: 157–180.

Holdridge LR (1947) Determination of world plant formations from simple climatic data. *Science* **105**: 367–368.

Holdridge LR *et al.* (1971) *Forest environment in Tropical Life Zone: a pilot study.* Pergamon Press, Oxford.

Holmes MG and Smith H (1975) The function of phytochrome in plants growing in the natural environment. *Nature* **254**: 512–514.

Holmgren M, Scheffer M, and Huston MA (1997) The interplay of facilitation and competition in plant communities. *Ecology* **78**: 1966–1975.

Holt BR (1972) Effect of arrival time on recruitment, mortality and reproduction in successional plant populations. *Ecology* **53**: 668–673.

Holt RD (1977) Predation, apparent competition and the structure of prey communities. *Theoretical Population Biology* **12**: 197–229.

Hooper DU and Vitousek PM (1997) The effects of plant composition and diversity on ecosystem processes. *Science* **277**: 1302–1305.

Hopkins B (1966) Vegetation of the Olokemeji Forest Reserve, Nigeria: IV. The litter and soil with special reference to their seasonal changes. *Journal of Ecology* **54**: 687–703.

Hopkins B (1978) The effects of the 1976 drought on chalk grassland in Sussex, England. *Biological Conservation* **14**: 1–12.

Horn HS (1971) *The Adaptive Geometry of Trees.* Princeton University Press, NJ.

Horn HS (1974) The ecology of secondary succession. *Annual Review of Ecology and Systematics* **5**: 25–37.

Hosakawa T, Odani N, and Tagawa H (1964) Causality of the distribution of corticolous species in forests with special reference to the physiological approach. *Bryologist* **67**: 396–411.

Howe HF (1977) Bird activity and seed dispersal of a tropical wet forest tree. *Ecology* **58**: 539–550.

Howe HF and Smallwood J (1982) Ecology of seed dispersal. *Annual Review of Ecology and Systematics* **13**: 201–228.

Huante P, Rincon E, and Acosta I (1995) Nutrient availability and growth rate of 34 woody species from a tropical deciduous forest in Mexico. *Functional Ecology* **9**: 849–858.

Hubbell SP and Foster RB (1986) Biology, chance and history and the structure of the tropical rain forest tree communities. In *Community Ecology*, eds. J Diamond and TJ Close, pp. 314–329. Harper and Row, New York.

Huenneke LK *et al.* (1990) Effects of soil resources on plant invasions and community structure in Californian serpentine grassland. *Ecology* **71**: 478–491.

Hughes MK (1975) Ground vegetation net production in a Danish beech wood. *Oecologia* (*Berl.*), **18**: 251–258.

Hughes RF, Kauffman JB, and Jaramillo VJ (1999) Biomass, carbon, and nutrient dynamics of secondary forests in a humid tropical region of Mexico. *Ecology* **80**: 1892–1907.

Hulme PE (1994) Seedling herbivory in grassland: relative impact of vertebrate and invertebrate herbivores. *Journal of Ecology* **84**: 43–51.

Hulme PE (1996a) Herbivores and the performance of grassland plants: a comparison of arthropod, mollusc and rodent activity. *Journal of Ecology* **84**: 43–51.

Hulme PE (1996b) Herbivory, plant regeneration and species coexistence. *Journal of Ecology* **84**: 609–615.

Hunt R *et al.* (1991) Response to CO_2 enrichment in 27 herbaceous species. *Functional Ecology* **5**: 410–421.

Hunt R and Nicholls AO (1986) Stress and the coarse control of growth and root-shoot partitioning in herbaceous plants. *Oikos* **47**: 149–158.
Hurlbert SH (1997) Functional importance vs keystoneness: reformulating some questions in theoretical biocenology. *Australian Journal of Ecology* **22**: 369–382.
Huston M (1979) A general hypothesis of species diversity. *American Naturalist* **113**: 81–101.
Huston MA (1994) *Biological Diversity*. Cambridge University Press, Cambridge.
Huston MA (1997) Hidden treatments in ecological experiments: re-evaluating the ecosystem function of biodiversity. *Oecologia* **110**: 449–460.
Huston MA and De Angelis DL (1994) Competition and coexistence: the effects of resource transport and supply rates. *American Naturalist* **144**: 954–977.
Huston MA and Gilbert L (1996) Consumer diversity and secondary production. In *Biodiversity and Ecosystem processes in Tropical Forests*, eds. G Orians, R Dirzo, and JH Cushman, pp. 33–47. Springer-Verlag, Berlin.
Huston MA and Smith T (1987) Plant succession: life history and competition. *American Naturalist* **130**: 168–198.
Hutchings MJ and de Kroon H (1994) Foraging in plants; the role of morphological plasticity in resource acquisition. *Advances in Ecological Research* **25**: 159–238.
Hutchinson GE (1951) Copepodology for the ornithologist. *Ecology* **32**: 571–577.
Hutchinson GE (1957) Concluding remarks. Cold Spring Harbour Symposium. *Quarterly Review of Biology* **22**: 415–427.
Hutchinson GE (1959) Homage to Santa Rosalia or why are there so many kinds of animals? *American Naturalist* **93**: 145–159.
Hutchinson TC (1967) Comparative studies of the ability of species to withstand prolonged periods of darkness. *Journal of Ecology* **55**: 291–299.
Idso SB (1992) Shrubland expansion in the American southwest. *Climate Change* **22**: 85–86.
Ingestad T (1973) Mineral nutrient requirements of *Vaccinium vitis-idaea* and *V. myrtillus*. *Physiologia Plantarum* **29**: 239–246.
Inghe O and Tamm CO (1985) Survival and flowering of perennial herbs IV. *Oikos* **45**: 400–420.
Inouye DW *et al.* (1987) Old field succession on a Minnesota sand plain. *Ecology* **68**: 12–26.
Iwasa Y and Roughgarden J (1984) Shoot/root balance of plants: optimal growth of a system with many vegetative organs. *Theoretical Population Biology* **25**: 78–104.
Jackson RB and Caldwell MM (1996) Integrating resource heterogeneity and plant plasticity: modelling nitrate and phosphate uptake in a patchy soil environment. *Journal of Ecology* **84**: 891–903.
Jackson RB, Manwaring JH, and Caldwell MM (1990) Rapid physiological adjustment of roots to localised soil enrichment. *Nature* **344**: 58–60.
Jalili A (1991) An investigation of the influences of drought and other soil factors on the structure of calcareous grassland. PhD Thesis, University of Sheffield.
James DW and Jurinak JJ (1978) Nitrogen fertilisation of dominant plants in the northeastern Great Basin Desert. In *Nitrogen in Desert Ecosystems*, eds. NE West and JJ Skujins, pp. 219–231. Dowden, Hutchinson and Ross, Stroudsberg, Pennsylvania.
Janos DP (1980) Vesicular-arbuscular mycorrhizae effect lowland tropical rain forest plant growth. *Ecology* **61**: 151–162.
Janssens F *et al.* (1998) Relationship between soil chemical factors and grassland diversity. *Plant and Soil* **202**: 69–78.
Janzen DH (1970) Herbivores and the number of tree species in tropical forests. *American Naturalist* **104**: 501–528.

Janzen DH (1971) Seed predation by animals. *Annual Review of Ecology and Systematics* **2**: 265–292.

Janzen DH (1973) Community structure of secondary compounds in plants. In *Chemistry in Evolution and Sytematics*, ed. T Swain, pp. 529–538. Russak, New York.

Janzen DH (1976) The depression of reptile biomass by large herbivores. *American Naturalist* **110**: 371–400.

Janzen DH (1983) The dispersal of small seeds by vertebrate guts. In *Coevolution*, eds. DJ Futuyman and M Slatkin, pp. 232–262. Sinacier Association, Sunderland.

Janzen DH (1986) Lost plants. *Oikos* **46**: 129–131.

Jarvis PG and McNaughton KG (1986) Stomatal control of transpiration; scaling up from leaf to region. *Advances in Ecological Research* **15**: 1–49.

Jarvis PG and Jarvis MS (1964) Growth rates of woody plants. *Physiologia Plantarum* **17**: 654–666.

Jeffrey DW (1967) Phosphate nutrition of Australian heath plants: I. The importance of proteoid roots in *Banksia* (Proteaceac). *Australian Journal of Botany* **15**: 403–411.

Jeffrey DW (1971) The experimental alteration of a *Kobresia*-rich sward in Upper Teesdale. In *The Scientific Management of Animal and Plant Communities for Conservation*, eds. E Dufley and AS Watt, pp. 78–89. Blackwell, Oxford.

Jeffries RA, Bradshaw AD, and Putwain PD (1981) Growth, nitrogen accumulation and nitrogen transfer by legume species established on mine spoil. *Journal of Applied Ecology* **18**: 945–956.

Jeffries RL (1988) Vegetation mosaics, plant-animal interactions and resources for plant growth. In *Plant Evolutionary Biology*, eds. LD Gottlieb and SK Jain, pp. 341–369. Chapman and Hall, New York.

Jeffries RL (1999) Herbivores, nutrients and trophic cascades in terrestrial environments. In *Herbivores: Between Plants and Predators*, eds. H Olff, VK Brown, and RH Drent, pp. 301–330. Blackwell Science, Oxford.

Jeffries RL and Willis AJ (1964) Studies on the calcicole-calcifuge habit: II. The influence of calcium on the growth and establishment of four species in soil and sand cultures. *Journal of Ecology* **52**: 691–707.

Jenkowska-Blaszezuk M and Grubb PJ (1997) Soil seed banks in primary and secondary deciduous forest in Bialowieza, Poland. *Seed Science Research* **7**: 281–292.

Jenny H, Gessel SP, and Bingham FT (1949) Comparative study of decomposition rates of organic matter in temperate and tropical regimes. *Soil Science* **6**: 419–432.

Johnson HB, Polley HW, and Mayeux HS (1993) Increasing CO_2 and plant–plant interactions: effects on natural vegetation. *Vegetatio* **104–105**: 157–170.

Johnstone IM (1986) Plant invasion windows: a time-based classification of invasion potential. *Biological Review* **61**: 369–394.

Jonasson S (1989) Implications of leaf longevity, leaf nutrient re-absorption and translocation for the resource economy of five evergreen plant species. *Oikos* **55**: 121–131.

Jonasson S and Chapin FS (1991) Seasonal uptake and allocation of phosphorus in *Eriophorum vaginatum* L. measured by labelling with ^{32}P. *New Phytologist* **118**: 349–357.

Jones EW (1956) Ecological studies on the rainforest of southern Nigeria: IV. The plateau forest of the Okurnu Forest Reserve. *Journal of Ecology* **44**: 83–117.

Jones FW (1945) Structure and reproduction of the virgin forest of the north temperate zone. *New Phytologist* **44**: 130–148.

Jones RK (1974) A study the phosphorus responses of a wide range of accessions from the genus *Styolanthes. Australian Journal of Agricultural Research* **25**: 847–862.

Jordan CF and Kline JR (1972) Mineral cycling: some basic concepts and their application in a tropical rain forest. *Annual Review of Ecology and Systematics* **3**: 33–50.

Jordano P (1982) Migrant birds are the main seed dispersers of blackberries in southern Spain. *Oikos* **38**: 183–193.

Jordano P (1992) Traits and frugivory. In *Seeds; the Ecology of Regeneration in Plant Communities*, ed. M Fenner, pp. 105–156. CAB International, Wallingford.

Jowett D (1964) Population studies on lead-tolerant *Agrostis tenuis*. *Evolution* **18**: 70–80.

Jurado E, Westoby M, and Nelson D (1991) Diaspore weight, dispersal, growth form and perenniality of central Australian plants. *Journal of Ecology* **79**: 811–828.

Kadmon R and Shmida A (1990) Competition in a variable environment: an experimental study in a desert annual population. *Israel Journal of Botany* **39**: 403–412.

Karieva P (1994) Diversity begets productivity. *Nature* **368**: 686–287.

Karieva P (1996) Diversity and sustainability on the prairie. *Nature* **379**: 673–674.

Karlson PS and Pate JS (1992) Contrasting effects of supplementary feeding of insects or mineral nutrients on the growth and nitrogen and phosphorus economy of pygmy species of *Drosera*. *Oecologia* **92**: 8–13.

Kautsky L (1988) Life strategies of aquatic soft bottom macrophytes. *Oikos* **53**: 126–135.

Keast A (1959) Australian birds: their zoogeography and adaptation to an arid continent. In *Biogeography and ecology in Australia*, eds. A Keast, RL Crocker, and CS Christian, pp. 89–114. Dr W Junk Publishers, The Hague.

Keay RWJ (1957) Wind-dispersed species in a Nigerian forest. *Journal of Ecology* **45**: 471–478.

Keddy PA (1989) *Competition*. Chapman and Hall, London

Keddy PA (1992a) A pragmatic approach to functional ecology. *Functional Ecology* **6**: 621–626.

Keddy PA (1992b) Assembly and response rules: two goals for predictive community ecology. *Journal of Vegetation Science* **3**: 157–164.

Keddy PA (1993) Reflections on the 21st birthday of geographical ecology and on the applications of the Hertzsprung-Russell star-chart to ecology. *Trends in Ecology and Evolution* **9**: 231–234.

Keddy P and Fraser LH (1999) On the diversity of land plants. *Ecoscience* **6**: 366–380.

Keddy PA, Gaudet C, and Fraser LH (2000) Effects of low and high nutrients on the competitive heirarchy of 26 shoreline plants. *Journal of Ecology* **88**: 413–423.

Keddy PA, Twolan-Strutt L, and Wisheu IC (1994) Competitive effect and response rankings in 20 wetland plants: are they consistent across three environments? *Journal of Ecology* **82**: 635–643.

Keeley JE and Fotheringham CJ (1988) Mechanism of smoke-induced seed germination in a postfire chaparral annual. *Journal of Ecology* **86**: 27–36.

Keever C (1950) Causes of succession on old fields of the Piedmont, North Carolina. *Ecological Monographs* **20**: 229–250.

Keever C (1973) Distribution of major forest species in southeastern Pennsylvania. *Ecological Monographs* **43**: 303–327.

Kent DH (1956) *Senecio squalidus* L. in the British Isles. I. Early records (to 1877). *Proceedings of the Botanical Society of the British Isles* **2**: 115–118.

Kent DH (1960) *Senecio squalidus* L. in the British Isles. II. The spread from Oxford 1879–1939. *Proceedings of the Botanical Society of the British Isles* **3**: 375–379.

Kerbes R, Kotanen PM, and Jeffries RL (1990) Destruction of wetland habitats by Lesser Snow Geese: a keystone species in the west coast of the Hudson Bay. *Journal of Applied Ecology* **27**: 242–258.

Kershaw KA (1977a) Physiological–environmental interactions in lichens: II. The pattern of net photosynthetic acclimation in *Peltigera canina* (L.) Willd var *praetextata* (Floerke in Somm.) Hue, and *P. polydactyla* (Neck.) Hoffm. *New Phytologist* **79**: 377–390.

Kershaw KA (1977b). Physiological–environmental interactions in lichens: III. The rate of net photosynthetic acclimation *in Peltigera canina* (L.) Willd. var *praetextata*

(Floerke in Somm.) Hue, and *P. polydactyla* (Neck.) Hoffm. *New Phytologist* **79**: 391–402.
King TJ (1975) Inhibition of seed germination under leaf canopies in *Arenaria serpyllifolia, Veronica arvensis* and *Cerastium holosteoides. New Phytologist* **75**: 87–90.
King TJ (1976) The viable seed content of ant-hill and pasture soil. *New Phytologist* **77**: 143–147.
King TJ (1977a) The plant ecology of ant-hills in calcareous grasslands: I. Patterns of species in relation to ant-hills in southern England. *Journal of Ecology* **65**: 235–256.
King TJ (1977b). The plant ecology of ant-hills in calcareous grasslands: II. Succession on the mounds. *Journal of Ecology* **65**: 257–278.
King TJ (1977c) The plant ecology of ant-hills in calcareous grasslands: III. Factors affecting the population sizes of selected species. *Journal of Ecology* **65**: 279–316.
Kingsbury RW *et al.* (1976) Salt stress responses in *Lasthenia glabrata*, a winter annual composite endemic to saline soils. *Journal of Botany* **54**: 1377–1385.
Kinsman EA *et al.* (1996) Effects of temperature and elevated CO_2 on cell division in shoot meristems: differential responses of two natural populations of *Dactylis glomerata* L. *Plant, Cell and Environment* **19**: 775–780.
Kleyer M (1999) Distribution of plant functional types along gradients of disturbance intensity and resource supply in an agricultural landscape. *Journal of Vegetation Science* **10**: 697–708.
Knight DH (1975) A phytosociological analysis of species-rich tropical forest on Barro Colorado Island, Panama. *Ecological Monographs* **45**: 259–284.
Koford CB (1958) Prairie dogs, white faces and blue grama. *Wildlife Monographs* **3**: 1–78.
Koller D (1969) The physiology of dormancy and survival of plants in desert environments. In *Dormancy and Survival*, ed. HW Woolhouse, pp. 449–469.
Körner C and Larcher W (1988) Plant life in cold climates. In *Plants and Temperature*, eds. SP Long and FI Woodward, pp. 25–57. Company of Biologists, Cambridge.
Körner C *et al.* (1989) Functional morphology of mountain plants. *Flora* **182**: 353–383.
Kramer F (1933) De natuurlijke verjonging in het Goenoeng-Gedehcomplex. *Tectona* **26**: 156–185.
Krebs CJ *et al.* (1973) Population cycles in small rodents. *Science* **179**, 35–41.
Kropac Z (1966) Estimation of weed seeds in arable soil. *Pedobiologia* **6**: 105–128.
Kruckeberg AR (1954) The ecology of serpentine soils: III. Plant species in relation to serpentine soils. *Ecology* **35**: 267–274.
Kubicek F and Brechtl J (1970) Production and phenology of the herb layer in an oak-hornbeam forest. *Biologia* **25**: 651–666.
Kubiena WL (1953) *The Soils of Europe; Illustrated Diagnosis and Systematics.* Allen and Unwin, London.
Kuhn TS (1962) *The Structure of Scientific Revolutions.* Chicago University Press, Chicago.
Laakso J and Setala H (1999) Sensitivity of primary production to changes in the architecture of below-ground food webs. *Oikos* **87**: 57–84.
Lack D (1947) The significance of clutch size. Parts I and II. *Ibis* **89**: 302–352.
Lack D (1948) The significance of clutch size. Part III. *Ibis* **90**: 25–45.
Lamb HH (1964) *The English Climate*. The English Universities Press, London.
Lambers H and Poorter H (1992) Inherent variation in growth rate between higher plants: a search for physiological causes and ecological consequences. *Advances in Ecological Research* **23**: 187–261.
Lamont BB (1984) Specialized modes of nutrition. In *Kwongan—Plant Life of the Sandplain*, eds. JS Pate and JS Beard, pp. 236–245. University of Western Australia Press, Nedlands.

Landsberg J, Lavoral S, and Stol J (1999) Grazing response groups among understorey plants in arid rangelands. *Journal of Vegetation Science* **10**: 683–696.
Larsen DW and Kershaw KA (1975) Acclimation in arctic lichens. *Nature* **254**: 421–423.
Lavorel S, McIntyre S, and Grigulis K (1999) Plant response to disturbance in a Mediterranean grassland: how many functional groups? *Journal of Vegetation Science* **10**: 661–672.
Law R (1979) The cost of reproduction in annual meadow grass. *American Naturalist* **112**: 3–16.
Law R, Bradshaw AD, and Putwain PD (1977) Life history variation in *Poa annua*. *Evolution* **31**: 233–246.
Lawrence DB and Hulbert L (1960) Growth stimulation of adjacent plants by lupin and alder on recent glacier deposits in south-east Alaska. *Bulletin of the Ecological Society of America* **31**: 58.
Lawrence JM (1990) The effect of stress and disturbance on echinoderms. *Zoological Science* **7**: 17–28.
Lawrence JM and Baghim A (1998) Life-history strategies and the potential of sea-urchins for aquaculture. *Journal of Shellfish Research* **17**: 1515–1522.
Lawton JG *et al.* (1993) The Ecotron: a controlled environmental facility for the investigation of population and ecosystem processes. *Philosophical Transactions of the Royal Society of London* **B341**: 181–194.
Lawton JL (1995) Ecological experiments with model systems. *Science* **269**: 328–331.
Leake JR (1994) The biology of myco-heterotrophic 'saprotrophic' plants. *New Phytologist* **127**: 171–216.
Leigh EG (1975) Structure and climate in tropical rain forest. *Annual Review of Ecology and Systematics* **6**: 67–86.
Leishman MR and Westoby M (1992) Classifying plants into groups on the basis of associations of individual traits—evidence from Australian semi-arid woodlands. *Journal of Ecology* **80**: 417–425.
Leishman MR and Westoby M (1998) Seed size and shape are not related to persistence in soil in Australia in the same way as in Britain. *Functional Ecology* **12**: 480–485.
Leith H (1960) Patterns of change within grassland communities. In *The Biology of Weeds*, ed. JL Harper. British Ecological Society Symposium No. 1, Blackwell, Oxford.
Leps J, Osbornova-Kosinova J, and Rejmanek K (1982) Community stability, complexity and species life-history strategies. *Vegetatio* **511**: 53–63.
Levin DA (1971). Plant phenolics: an ecological perspective. *American Naturalist* **105**: 157–181.
Levin DA and Funderburg SW (1979) Genome size in angiosperms; temperate versus tropical species. *American Naturalist* **114**: 784–795.
Levin SA (1974) Dispersion and population interactions. *American Naturalist* **108**: 207–228.
Levins R and Culver D (1971) Regional coexistence of species and competition between rare species. *Proceedings of the National Academy of Sciences of the USA* **6**: 1246–1248.
Levitt J (1956) *The Hardiness of Plants*. Academic Press, New York.
Levitt J (1975) *Responses of Plants To Environmental Stresses*. Academic Press, New York.
Levitt J (1980) *Responses of Plants to Environmental Stresses. Volume II. Water, Radiation, Salt and Other Stresses*. Academic Press, New York.
Lewis IF (1914) The seasonal life-cycle of some red algae at Woods Hole. *Plant World* **17**: 31–35.

Lewontin RC (1971) The effect of genetic linkage on the mean fitness of a population. *Proceedings of the National Academy of Science* **68**: 984–986.
Li X and Wilson SD (1998) Facilitation among woody plants establishing in an old field. *Ecology* **79**: 2694–2705.
Liddle M (1975) A selective review of the ecological effects of human trampling on natural ecosystems. *Biological Conservation* **7**: 17–36.
Liddle MJ and Greig-Smith P (1976) A survey of tracks and paths in a sand dune ecosystem: II. Vegetation. *Journal of Applied Ecology* **12**: 909–930.
Liebig J (1840) *Chemistry and its Application to Agriculture and Physiology*. Taylor and Walton, London.
Lindeman RI (1942) The trophic-dynamic aspect of ecology. *Ecology* **23**: 399–418.
Livingstone RB and Allessio ML (1968) Buried viable seed in successional field and forest stands, Harvard Forest, Massachusetts. *Bull. Torrey Bot. Club* **95**: 58–69.
Lloyd PS (1968) The ecological significance of fire in limestone grassland communities of the Derbyshire Dales. *Journal of Ecology* **56**: 811–826.
Lloyd PS, Grime JP, and Rorison IH (1971) The grassland vegetation of the Sheffield Region. I. General features. *Journal of Ecology* **59**: 863–886.
Lloyd PS and Pigott CD (1967) The influence of soil conditions on the course of succession on the Chalk of Southern England. *Journal of Ecology* **55**: 137–146.
Loach K (1967) Shade tolerance in tree seedlings: I. Leaf photosynthesis and respiration in plants raised under artificial shade. *New Phytologist* **66**: 607–621.
Loach K (1970) Shade tolerance in tree seedlings: II. Growth analysis of plants raised under artificial shade. *New Phytologist* **69**: 273–286.
Lodge DM (1993) Biological invasions: lessons from ecology. *Trends in Ecology and Evolution* **8**: 133–136.
Longman KA and Janik J (1974) *Tropical Forest and its Environment.* Longman, London.
Lonsdale WM (1999) Global patterns of plant invasions and the concept of invasibility. *Ecology* **80**: 1522–1536.
Loomis WD (1967) Biosynthesis and metabolism of monoterpenes. In *Terpenoids in Plants*, ed. JB Pridham, pp. 59–82. Academic Press, London.
Loveless AR (1961) A nutritional interpretation of sclerophylly based on differences in chemical composition of sclerophyllous and mesophytic leaves. *Annals of Botany NS* **25**: 168–176.
Lovett-Doust L (1981) Population dynamics and local specialisation in a clonal perennial (*Ranunculus repens*). I The dynamics of ramets in contrasting habitats. *Journal of Ecology* **69**: 743–755.
Lutz HJ (1928) Trends and silvicultural significance of upland forest successions in southern New England. *Yale University. School of Forestry Bulletin* **22**: 1–68.
Mabry C, Ackerly D, and Gerhardt F (2000) Landscape and species-level distribution of morphological and life-history traits in a temperate woodland flora. *Journal of Vegetation Science* **11**: 213–224.
MacArthur RH (1955) Fluctuation of animal populations and a measure of community stability. *Ecology* **36**: 533–536.
MacArthur RH (1968) The theory of the niche. In *Population Biology and Evolution*, ed. RC Lewontin, pp. 159–176. Syracuse University Press, Syracuse.
MacArthur RH (1972) Geographical Ecology: Patterns in the distribution of species. Harper and Row, London.
MacArthur RH and Wilson ED (1967) *The Theory of Island Biogeography*. Princeton University Press, Princeton, NJ.
MacFarlane C and Bell HP (1933) Observations of the seasonal changes in the marine algae in the vicinity of Halifax, with particular reference to winter conditions. *Proceedings of NS Institute of Science* **18**: 134–176.

MacFarlane JD and Kershaw KA (1977) Physiological–environmental interactions in lichens: IV. Seasonal changes in the nitrogenase activity of *Peltigera canina* (L.) Willd. var. *praetextata* (Floerke in Somm.) Hue, and *P. canina* (L.) Willd. var. *rufescens* (Weiss) Mudd. *New Phytologist* **79**: 403–408.

MacGillivray CW, Grime JP, and the ISP team (1995) Testing predictions of resistance and resilience of vegetation subjected to extreme events. *Functional Ecology* **9**: 640–649.

Mack RN (1996) Predicting the identity and fate of plant invaders: emergent and emerging approaches. *Biological Conservation* **78**: 107–121.

Mackay AD and Barber SA (1985) Soil moisture effects on root growth and phosphorus by corn. *Agronomy Journal* **77**: 519–523.

Mackie-Dawson LA (1999) Nitrogen uptake and root morphological responses of defoliated *Lolium perenne* (L.) to a heterogeneous nitrogen supply. *Plant and Soil* **209**: 111–118.

MacLeod J (1894) Over de bevruchting der bloemen in het Kempisch gedeelte van Vlaanderen. Deel II. *Botanische Jaarboek* **6**: 119–511.

Madge DS (1965) Leaf fall and litter disappearance in a tropical forest. *Pedobiologia* **5**: 272–288.

Mahmoud A (1973) A laboratory approach to ecological studies of the grasses *Arrhenatherum elatius* (L.) Beauv. Ex. J. and C. Presl., *Agrostis tenuis* Sibth and *Festuca ovina* L. PhD Thesis, University of Sheffield.

Mahmoud A and Grime JP (1974) A comparison of negative relative growth rates in shaded seedlings. *New Phytologist* **73**: 1215–1219.

Mahmoud A and Grime JP (1976) An analysis of competitive ability in three perennial grasses. *New Phytologist* **77**: 431–435.

Mahmoud A, Grime JP, and Furness SB (1975) Polymorphism in *Arrhenatherum elatius* (L.) Beauv. Ex J. and C. Presl. *New Phytologist* **75**: 269–276.

Main AR, Littlejohn MJ, and Lee AK (1959) Ecology of Australian frogs. In *Biogeography and Ecology in Australia*, eds. A Keast, RL Crocker, and CS Christian, pp. 396–411. Dr W Junk Publishers, The Hague.

Major J and Pyott WT (1966) Buried viable seeds in two California bunch grass sites and their bearing on the definition of a flora. *Vegetatio* **13**: 253–282.

Margalef R (1963a) On certain unifying principles in ecology. *American Naturalist* **97**: 357–374.

Margalef R (1963b) Successions of populations. *Adv. Frontiers of Plant Sci., Inst. Adv. Sci. Cult., New Delhi, India* **2**: 137–188.

Margalef R (1968) *Perspectives in Ecological Theory*. Chicago Press, Chicago.

Marimoto T, Anderson JPE, and Domisch KH (1982a) Mineralisation of nutrients from soil microbial biomass. *Soil Biology and Biochemistry* **14**: 469–475.

Marimoto T, Anderson JPE, and Domisch KH (1982b) Decomposition of ^{14}C and ^{15}N-labelled microbial cells in soil. *Soil Biology and Biochemistry* **14**: 461–467.

Marks PL (1974) The role of pin cherry (*Prunus pensylvanica* L.) in the maintenance of stability in northern hardwood ecosystems. *Ecological Monographs* **44**: 73–88.

Marshall R (1927) The growth of hemlock before and after release from suppression. *Harvest For. Bulletin*, **11**: 66.

Masaki T *et al.* (1998) The seed bank dynamics of *Cornus controversa* and their role in regeneration. *Seed Science Research* **8**: 53–63.

Mason HL (1946) The edaphic factor in narrow endemism: I. The nature of environmental influences. *Madrono* **8**: 209–226.

May RM (1972) Will a large complex system be stable? *Nature* **238**: 413–414.

May RM (1974) *Stability and Complexity in Model Ecosystems*, 2nd edition. Princeton University Press, NJ.

May RM and Seger J (1986) Ideas in ecology. *American Scientist* **74**: 256–267.

McClure FA (1966) Flowering, fruiting and animals in the canopy of a tropical rain forest. *Malay Forest* **29**: 182–203.
McEvoy PB and Cox CS (1987) Wind dispersal distances in dimorphic achenes of ragwort, *Senecio jacobaea. Ecology* **68**: 2006–2014.
McLendon T and Redente EF (1991) Nitrogen and phosphorus effects on secondary succession dynamics on a semi-arid sagebrush site. *Ecology* **72**: 2016–2024.
McLachlan KD (1976) Comparative phosphorus responses in plants to a range of available phosphorus situations. *Australian Journal of Agricultural Research* **27**: 323–341.
McMillan C (1956) The edaphic restriction of *Cupressus* and *Pinus* in the Coast Ranges of central California. *Ecological Monographs* **26**: 177–212.
McNaughton SJ (1975) r- and K-selection in *Typha. American Naturalist* **109**: 251–261.
McNaughton SJ (1978a) Stability and diversity of ecological communities. *Nature* **274**: 251–253.
McNaughton SJ (1978b) Serengeti ungulates: feeding selectivity influences the effectiveness of plant defence guilds. *Science* **99**: 806–807.
McNaughton SJ (1983) Compensatory plant growth as a response to herbivory. *Oikos* **40**: 329–336.
McNaughton SJ (1985) Ecology of a grazing system: the Serengeti. *Ecological Monographs* **55**: 259–294.
McPeek MA and Kalisz S (1998) The joint evolution of dispersal and dormancy in metapopulations. *Archive für Hydrobiologie* **52**: 33–51.
McPherson JK and Muller CH (1969) Allelopathic effects of *Adenostoma fasciculatum,* 'Chamise', in the California Chaparral. *Ecological Monographs* **39**: 177–198.
McRill M (1974) The ingestion of weed seed by earthworms. *Proceedings of the 12th British Weed Control Conference* **2**: 519–525.
McRill M and Sagar GR (1973) Earthworms and seeds. *Nature* **243**: 482.
Mearns LO, Kaatz RW, and Schneider SH (1984) Changes in the probabilities of extreme high temperature events with changes in global mean temperature. *Journal of Climate and Applied Meterology* **23**: 1601–1613.
Meentemyer V (1978) Macroclimate and lignin control of litter decomposition rates. *Ecology* **59**: 465–472.
Mellinger MV and McNaughton SJ (1975) Structure and function of successional vascular plant communities in Central New York. *Ecological Monographs* **45**: 161–182.
Menge BA and Sutherland JP (1976) Species diversity gradients: synthesis of the roles of predation, competition and temporal heterogeneity. *American Naturalist* **110**: 351–369.
Menges ER and Waller DM (1983) Plant strategies in relation to elevation and light in floodplain herbs. *American Naturalist* **122**: 454–473.
Menhinick EF (1967) Structure, stability and energy flow in plants and arthropods in a *Sericea lesedeza* stand. *Ecological Monographs* **37**: 255–272.
Metcalfe DJ and Grubb PJ (1995) Seed mass and light requirement for regeneration in south east Asian rain forest. *Canadian Journal of Botany* **73**: 817–826.
Mikola J and Setälä H (1998) Relating species diversity to ecosystem functioning: mechanistic backgrounds and an experimental approach with a decomposer food web. *Oikos* **83**: 180–194.
Milberg P and Andersson L (1997) Seasonal variation in dormancy and light sensitivity in buried seeds of eight annual weed species. *Canadian Journal of Botany* **75**: 1998–2004.
Miller HG *et al.* (1979) Nutrient cycles in pine and their adaptations to poor soils. *Canadian Journal of Forestry Research* **9**: 19–26.

Miller RE and Fowler NL (1994) Life history variation and local adaptation within two populations of *Bouteloua rigideseta* (Texas grama). *Journal of Ecology* **82**: 855–854.
Miller RS (1967) Pattern and process in competition. *Advances in Ecological Resarch* **4**: 1–74.
Miller RS (1968) Conditions of competition between redwings and yellow-headed blackbirds. *Journal of Animal Ecology* **37**: 43–62.
Miller TE (1982) Community diversity and interactions between size and frequency of disturbance. *American Naturalist* **120**: 533–536.
Milne A (1961) Definition of competition among animals. In *Mechanisms in Biological Competition*, ed. FL Milnthorpe, pp. 40–61. Cambridge University Press, London.
Milton WEJ (1939) The occurrence of buried viable seeds in soils at different elevations and in a salt marsh. *Journal of Ecology* **27**: 149–159.
Milton WEJ (1940) The effect of manuring, grazing and cutting on the yield, botanical and chemical composition of natural hill pastures. *Journal of Ecology* **28**: 328–356.
Milton WEJ (1943) The yields of ribwort plantain (ribgrass) when sown in pure plots and with grass and clover species. *Welsh Journal of. Agriculture* **27**: 109–116.
Miyata I and Hosakawa T (1961) Seasonal variations of the photosynthetic efficiency and chlorophyll content of epiphytic mosses. *Ecology* **42**: 766–775.
Moen J *et al.* (1993) Grazing by food-limited microtine rodents on a productive experimental community: does the 'green desert' exist? *Oikos* **68**, 401–41.
Moles AT, Hodgson DW, and Webb CJ (2000) Seed size and persistence in the soil in the New Zealand flora. *Oikos* **89**: 541–545.
Monk CD (1966) An ecological significance of evergreenness. *Ecology* **47**: 504–505.
Monk CD (1971) Leaf decomposition and loss of ^{45}Ca from deciduous and evergreen trees. *American Midland Naturalist* **86**: 379–385.
Monk D, Pate JS, and Loneragan WA (1981) Biology of *Acacia pulchella* R.B. with special reference to symbiotic nitrogen fixation. *Australian Journal of Botany* **29**: 579–592.
Monsi M and Saeki T (1953) Über den liehfaktor in den Pflanzengesell-schaften und seine Bedentung für die Stoffproduktion. *Japanese Journal of Botany* **14**: 22–52.
Monteith JL (1973) Principles of Environmental Physics. Edward Arnold, London.
Montgomery GG and Sunquist ME (1974) Impact of sloths on neotropical energy flow and nutrient cycling. In *Trends in Tropical Ecology: Ecological Studies IV*, eds. E Medina and FB Golley. Springer, New York.
Mooney HA (1972) The carbon balance of plants. *Annual Review of Ecology and Systematics* **3**: 315–346.
Mooney HA and Billings WD (1961) The physiological ecology of arctic and alpine populations of *Oxyria digyna*. *Ecological Monographs* **31**: 1–29.
Mooney HA and Gulmon SL (1982) Constraints on leaf structure and function in reference to herbivory. *Bioscience* **32**: 198.
Mooney HA and Harrison AT (1970) The influence of conditioning temperature on subsequent temperature-related photosynthetic capacity in higher plants. In *Prediction and Measurement of Photosynthetic Productivity*, ed. CT de Wit, pp. 411–417. Centre for Agricultural Publishing and Documentation, Wageningen.
Mooney HA and Shropshire F (1967) Population variability in temperature related photosynthetic acclimation. *Oecologia Plantarum* **2**: 1–13.
Mooney HA and West M (1964) Photosynthetic acclimation of plants of diverse origin. *American Journal of Botany* **51**: 825–827.
Mooney HA, Winner WE, and Pell EJ, eds. (1991) Response of plants to multiple stresses. Academic Press, San Diego.
Moore DRJ and Keddy PA (1989) The relationship between species richness and standing crop in wetlands: the importance of scale. *Vegetatio* **79**: 99–106.
Moore RJ (1978) Is *Acanthaster planci* an r-strategist? *Nature* **271**: 56–57.

Moreno-Casasola P, Grime JP, and Martinez ML (1994) A comparative study of the effects of fluctuations in temperature and moisture supply on hard coat dormancy in seeds of coastal tropical legumes in Mexico. *Journal of Tropical Ecology* **10**: 67–86.

Morhardt SS and Gates DM (1974) Energy-exchange analysis of the belding ground squirrel and its habitat. *Ecological Monographs* **44**: 17–44.

Morse DH (1971) The insectivorous birds as an adaptive strategy. *Annual Review of Ecology and Systematics* **2**: 177–200.

Morse DH (1974) Niche breadth as a function of social dominance. *American Naturalist* **108**: 818–830.

Mortimer AM (1974) Studies of germination and establishment of selected species with special reference to the fate of seeds. PhD Thesis, University College of North Wales.

Mott JJ (1972) Germination studies on some annual species from an arid region of Western Australia. *Journal of Ecology* **60**: 293–304.

Mueller-Dombois D and Sims HP (1966) Response of three grasses to two soils and a water table depth gradient. *Ecology* **47**: 644–648.

Muller CH (1940) Plant succession in the *Larrea–Flourensia* climax. *Ecology* **21**: 206–212.

Muller CH and Chou CH (1972) Phyto-toxins: an ecological phase of phytochemistry. In *Phytochemical Ecology*, ed. JB Harborne. Academic Press, London.

Muller G and Foerster E (1974) Entwicklung von Weideansaaten im Überflutungsbereich das Rheines bei Kleve. Z. *Acker und Pflanzenban* **140**: 61–174.

Muller PE (1884) Studier over skovjord, som bidrag til skovdyrkningens theori. II Om muld og mor i egeskove og paa heder. *Tidsskrift for Skovbrug* **7**: 232.

Muller WH and Muller CH (1956) Association patterns involving desert plants that contain toxic products. *American Journal of Botany* **43**: 354–361.

Munz PA (1959). *A California Flora*. University of California Press, Berkeley, California.

Murdoch WW (1966) Community structure, population control and competition—a critique. *American Naturalist* **100**: 219–226.

Murphy BF and Nier AO (1941) Variations in the relative abundance of the carbon isotopes. *Physics Reviews* **59**: 771–772.

Murray KG (1988) Avian seed dispersal of 3 neotropical gap-dependent plants. *Ecological Monographs* **58**: 271–298.

Murray KG *et al.* (1994) Fruit laxatives and seed passage rates in frugivores: consequences for plant reproductive success. *Ecology* **75**, 989–994.

Murtagh GJ, Dyer PS, and Crittenden PD (2000) Sex and the single lichen. *Nature* **404**: 564.

Muscatine L and Cernichiari E (1969) Assimilation of photosynthetic products of zooxanthellae by a reef coral. *Biology Bulletin* **137**: 506–523.

Myers N (1996) The biodiversity crisis and the future of evolution. *The Environmentalist* **16**: 1614–1174.

Myster RW (1993) Tree invasion and establishment in old fields at Hutcheson Memorial Forest. *Botanical Review* **59**: 251–572.

Naeem S *et al.* (1994) Declining biodiversity can alter the performance of ecosystems. *Nature* **368**: 734–737.

Naeem S *et al.* (1999) Plant neighborhood diversity and production. *Ecoscience* **6**: 355–365.

Nagel JL (1950) Changement d'essences. *J Forest Suisse* (*Schwei. Z. Fortswissenschaften*), **101**: 95–104.

Naiman RJ, Mellilo JM, and Hobbie JE (1986) Ecosystem alteration of boreal forest streams by beaver (*Castor canadensis*). *Ecology* **67**: 1254–1269.

Nakashizuka T *et al.* (1993) Seed dispersal and vegetation development on a debris avalanche on the Ontake volcano, Central Japan. *Journal of Vegetation Science* **4**: 537–542.

Naveh Z (1961). Toxic effects *of Adenostoma fasciculatum* (Chamise) in the Californian chaparral. *Proceedings of the 4th Congress Sci. Soc. (Rehovot, Israel),* 1 page.

Nemani R and Running SW (1996) Implementation of hierarchical global vegetation classification in ecosystem function models. *Journal of Vegetation Science* **7**: 337–346.

Nemeth K *et al.* (1987) Organic compounds extracted from arable and forest soils by electro-ultra filtration and recovery rates of amino acids. *Biology and Fertility of Soils* **5**: 271–275.

New JK (1958) A population study of *Spergula arvensis*. *Annals of Botany* **22**: 457–477.

Newman EI (1963) Factors controlling the germination date of winter annuals. *Journal of Ecology* **51**: 625–638.

Newman EI (1973) Competition and diversity in herbaceous vegetation. *Nature* **244**: 310.

Newman EI and Rovira AD (1975) Allelopathy among some British grassland species. *Journal of Ecology* 63: 727–737.

Newton I (1964) Bud-eating by bullfinches in relation to the natural food supply. *Journal of Applied Ecology* **1**: 265–279.

Nichols JD *et al.* (1976) Temporally dynamic reproductive strategies and the concept of r- and K-selection. *American Naturalist* **110**: 995–1005.

Nicholson IA, Paterson IS, and Currie A (1970) A study of vegetational dynamics: selection by sheep and cattle in *Nardus* pasture. In *Animal Populations in Relation to their Food Resources*, ed. A Watson, pp. 73–98. Blackwell, London.

Nier AO and Gulbransen EA (1939) Variation in the relative abundance of the carbon isotopes. *Journal of the American Chemical Society* **61**: 697–698.

Niering WA and Goodwin RH (1962) Ecological studies in the Connecticut Arboretum Natural Area. I. Introduction and survey of vegetation types. *Ecology* **43**: 41–54.

Nieto JH, Brondo MH, and Gonzalez JT (1968) Critical periods of the crop growth cycle for competition from weeds. *Pesticide Article and News Summaries C* **14**: 159–166.

Nitsan Z, Dvorin A, and Nir I (1981) Composition and amino acid content of the carcass, skin and feathers of the growing gosling. *British Poultry Science* **22**: 79–84.

Noble IR and Slatyer RO (1979) The use of vital attributes to predict successional changes in plant communities subject to recurrent disturbances. *Vegetatio* **43**: 5–21.

Norberg RA (1977) An ecological theory on foraging time and energetics and choice of optimal food searching method. *Journal of Animal Ecology* **46**: 511–530.

Novoplansky A, Cohen D, and Sachs T (1990) How Portulaca seedlings avoid their neighbours. *Oecologia* **82**: 138–140.

Noy-Meir I (1973) Desert ecosystems: environment and producers. *Annual Review of Ecology and Systematics* **4**: 25–51.

Numata M (ed.) (1979) *Ecology of Grasslands and Bamboo Lands of the World*. Dr W Junk Publishers, The Hague.

Nye PH (1961) Organic matter and nutrient cycles under moist tropical forest. *Plant and Soil* **13**: 333–346.

Odum EP (1963) *Ecology*. Holt, Rinehart, and Winston, New York.

Odum EP (1969) The strategy of ecosystem development. *Science* **164**: 262–270.

Odum EP (1971) *Fundamentals of Ecology*, 3rd ed. Saunders, Philadelphia.

Odum HT and Odum EP (1955) Trophic structure and productivity of a windward coral reef community on Eniwetok Atoll. *Ecological Monographs* **25**: 291–320.

Odum HT and Pinkerton RC (1955) Time's speed regulator: the optimum efficiency for maximum power output in physical and biological systems. *American Scientist* **43**: 331–343.

Oechel WC and Collins NJ (1973) Seasonal patterns of CO_2 exchange in bryophytes at Barrow, Alaska. In *Primary Production and Production Processes. Tundra Biome, Proceedings of the Conference, Dublin, Ireland*, eds. C Bliss and FE Wielogolaski, pp. 197–203. Swedish IBP Committee, Wenner-Gren Center, Stockholm, Sweden.

Ogden J (1974) The reproductive strategy of higher plants: II. The reproductive strategy of *Tussilago farfara* L. *Journal of Ecology* **62**: 291–324.

Ohtonen R, Aikio S, and Vare H (1997) Ecological theories in soil biology. *Soil Biology and Biochemistry* **29**: 1613–1619.

Oksanen L (1990) Exploitation ecosystems in seasonal environments. *Oikos* **57**: 14–24.

Oksanen L *et al.* (1981) Exploitation ecosystems in gradients of primary productivity. *American Naturalist* **118**: 240–261.

Oksanen L *et al.* (1987) The role of phenol-based inducible defense in the interaction between tundra populations of the vole *Clethrionomys rufocannus* and the dwarf shrub *Vaccinium myrtillus*. *Oikos* **50**: 371–380.

Olff H (1992) *On the Mechanisms of Vegetation Succession.* PhD Thesis, University of Groningen.

Olff H and Ritchie ME (1998) Effects of herbivores on grassland plant diversity. *Trends in Ecology and Evolution* **13**: 261–265.

Olff H, van Andel J, and Bakker JP (1990) Biomass and shoot:root allocation of five species from a grassland succession series at different combinations of light and nutrient availability. *Functional Ecology* **4**: 193–200.

Olmo E (1983) Nucleotype and cell size in vertebrates: a review. *Basic and Applied Histochemistry* **27**: 227–256.

Olmo E (1987) DNA variation, phylogenesis and speciation in reptiles. *Bolletino de Zoologia* **54**: 49–54.

Olmsted NW and Curtis JD (1947) Seeds of the forest floor. *Ecology* **28**: 49–52.

Olsen JS (1963) Energy storage and the balance of producers and decomposers in ecological systems. *Ecology* **44**: 322–331.

Onipchenko VG, Semenova GV, and van der Maarel E (1998) Population strategies in severe environments: alpine plants in the northwestern Caucasus. *Journal of Vegetation Science* **9**: 27–40.

Oomes MJM (1992) Yield and species density of grasslands during restoration management. *Journal of Vegetation Science* **3**: 271–274.

Oosting HJ (1942) An ecological analysis of the plant communities of Piedmont, North Carolina. *American Midland Naturalist* **28**: 1–126.

Oosting HJ and Kramer PJ (1946) Water and light in relation to pine reproduction. *Ecology* 28: 47–53.

Orgel LE and Crick FHC (1980) Selfish DNA: the ultimate parasite. *Nature* **284**: 604–607.

Orians GH and Collier G (1963) Competition and blackbird social systems. *Evolution* **17**: 449–459.

Osbornova J *et al.* (1989) *Succession in Abandoned Fields: Studies in Central Bohemia, Czechoslovakia.* Kluwer, Dordrecht.

Paine RT (1969) The *Pisaster-Tegula* interaction: Prey patches, predator food preference and intertidal community structure. *Ecology* **50**: 950–961.

Paine RT (1974) Intertidal community structure: Experimental studies on the relationship between a dominant competitor and its principal predator. *Oecologia (Berl.)* **15**: 93–120.

Paine RT (1984) Ecological determinism in the competition for space. *Ecology* **65**: 1339–1348.

Park T (1954) Experimental studies of interspecific competition: II. Temperature, humidity and competition in two species of *Tribolium. Physiological Zoology* **27**: 177–238.

Parrish JAD and Bazzaz FA (1982) Competitive interactions in plant communities of different successional ages. *Ecology* **62**: 314–320.

Parsons RF (1968a) The significance of growth rate comparisons for plant ecology. *American Naturalist* **102**: 595–597.

Parsons RF (1968b) Ecological aspects of the growth and mineral nutrition of three mallee species of Eucalyptus. *Oecologia Plantarum* **3**: 121–136.

Pärtel M *et al.* (1996) The species pool and its relation to species richness: evidence from Estonian plant communities. *Oikos* **75**: 111–117.

Pastor J and Cohen Y (1997) Herbivores, the functional diversity of plant species and the cycling of nutrients in ecosystems. *Theoretical Population Biology* **51**: 1–15.

Pate JS (1993) Structural and functional responses to fire and nutrient stress: case studies from the sandplains of south-west Australia. In *Plant Adaptation to Environmental Stress*, eds. L Fowden, TA Mansfield, and JL Stoddart, pp. 189–205. Chapman and Hall, London.

Pate JS *et al.* (1985) Biology of fire ephemerals of the sandplains of the Kwongan of south-western Australia. *Australian Journal of Plant Physiology* **12**: 641–655.

Pate JS and Dell B (1984) Economy of mineral nutrients in sandplain species. In *Kwongan—Plant Life of the Sandplain*, eds. JS Pate and JS Beard, pp. 227–252. University of Western Australia Press, Nedlands.

Peace WJH and Grubb PJ (1982) Interaction of light and mineral nutrient supply in the growth of *Impatiens parviflora. New Phytologist* **90**: 127–150.

Pearce AK (1987) An investigation of phenotypic-plasticity in coexisting populations of four grasses in a calcareous pasture in North Derbyshire. PhD Thesis, University of Sheffield.

Pearcy RW *et al.* (1994) Photosynthetic ultilization of sunflecks: a temporally patchy resource on a time scale of seconds to minutes. In *Exploitation of Environmental Heterogeneity by Plants*, eds. MM Caldwell and RW Pearcy, pp. 175–208. Academic Press, San Diego.

Pearsall WH (1950) *Mountains and Moorlands*. Bloomsbury Books, London.

Peart MH (1984) The effects of morphology, orientation and position of grass diaspores on seedling survival. *Journal of Ecology* **72**: 437–453.

Pemadesa MA and Lovell PH (1974) Some factors affecting the distribution of some annuals in the dune system at Aberffray, Anglesey. *Journal of Ecology* **62**: 403–416.

Pennings SC and Callaway RM (2000) The advantages of clonal integration under different ecological conditions: a community-wide test. *Ecology* **81**: 709–710.

Perring FH (1959) Topographical gradients of chalk grassland. *Journal of Ecology* **47**: 447–481.

Perring FH (1968) *Critical Supplement to the Atlas of the British Flora.* Nelson, Edinburgh.

Perring FH and Walters SM (1962) *Atlas of the British Flora*. Nelson, Edinburgh.

Perry TO (1971) Winter-season photosynthesis and respiration by twigs and seedlings of deciduous and evergreen trees. *Forest Science* **17**: 41–43.

Pessin LJ (1922) Epiphyllous plants of certain regions in Jamaica. *Bulletin of the Torrey Botany Club* **49**: 1–14.

Peterken GF and Lloyd PS (1967) Biological Flora of the British Isles: *Ilex aquifolium. Journal of Ecology* **51**: 841–858.

Petraitis PS, Latham RE, and Niesenbaum RA (1989) The maintenance of species diversity by disturbance. *The Quarterly Review of Biology* **64**: 393–418.

Phillips OL *et al.* (1994) Dynamics and species richness of tropical rainforests. *Proceedings of the National Academy of Science, USA*. **91**: 2805–2809.

Pianka ER (1966) Convexity, desert lizards and spatial heterogeneity. *Ecology* **47**: 1055–1059.

Pianka ER (1970) On r- and K-selection. *American Naturalist* **104**: 592–597.

Pickett STA and Bazzaz FA (1978) Organisation of an assemblage of early successional species on a soil moisture gradient. *Ecology* **59**: 1248–1255.

Pickett STA and White PS (eds.) (1985) *The Ecology of Natural Disturbance and Patch Dynamics*. Academic Press, New York.

Pickworth-Farrow E (1916) On the ecology of the vegetation of Breckland. II. Factors relating to the relative distributions of Calluna-heath and grass-heath in Breckland. *Journal of Ecology* **4**: 57–64.

Pigott CD (1968) Biological Flora of the British Isles: *Cirsium acaulon. Journal of Ecology* **56**: 597–612.

Pigott CD (1971) Analysis of the response of *Urtica dioica* to phosphate. *New Phytologist* **70**: 953–966.

Pigott CD (1983) Regeneration of oak birch woodland following exclusion of sheep. *Journal of Ecology* **71**: 629–646.

Pigott CD and Huntley JP (1981) Factors controlling the distribution of *Tilia cordata* at the northern limit of its geographical range. 3. Nature and cause of seed sterility. *New Phytologist* **87**: 817–839.

Pigott CD and Taylor K (1964) The distribution of some woodland herbs in relation to the supply of nitrogen and phosphorus in the soil. *Journal of Ecology* **52**: 175–185.

Pimm SL (1991) *The Balance of Nature? Ecological Issues in the Conservation of Species and Communities*. University of Chicago Press, Chicago.

Piroznikow E (1983) Seed bank in the soil of stablized ecosystem of a deciduous forest (Tilio-Carpinetum) in the Bialowieza National Park. *Ekilogia Polska* **31**: 145–172.

Platt WJ (1975) The colonisation and formation of equilibrium plant species associations in badger disturbances in a tall-grass prairie. *Ecological Monographs* **45**: 285–305.

Pokki J (1981) Distribution, demography and dispersal of the field vole, *Microtus agrestis* (L.), in the Tvarminne archipelago, Finland. *Acta Zoologica. Fennoscandia* **164**: 1–48.

Polunin N (1948) Botany of the Canadian Eastern Arctic, Part III. Vegetation and ecology. *National Museum of Canada Bulletin* **104**: 304.

Pons TL (1989a) Dormancy, germination and mortality of seeds in heathland and inland sand dunes. *Acta Botanica Neerlandica* **38**: 327–335.

Pons TL (1989b) Breaking of seed dormancy by nitrate as a gap detection mechanism. *Annals of Botany* **63**: 139–143.

Pons TL (1991a) Dormancy, germination and mortality of seeds in a chalk-grassland flora. *Journal of Ecology* **79**: 765–780.

Pons TL (1991b) Induction of dark dormancy in seeds: its importance for the seed bank in the soil. *Functional Ecology* **5**: 669–675.

Poorter H and Bergkotte A (1992) Chemical composition of 24 wild species differing in relative growth rate. *Plant Cell and Environment* **15**: 221–229.

Poorter H *et al.* (1991) Respiratory energy requirements of roots vary with the potential growth rate of a plant species. *Physiologia Plantarum* **83**: 469–475.

Portnoy S and Willson MF (1993) Seed dispersal curves: the behaviour of the tail of the distribution. *Evolutionary Ecology* **7**: 25–44.

Poschlod P and Bonn S (1998) Changing dispersal processes in the central European landscape since the last ice age: an explanation for the actual decrease of plant species richness in different habitats? *Acta Botanica Neerlandica* **47**: 27–44.

Power ME (1992) Top-down and bottom-up forces in food webs: do plants have primacy? *Ecology* **73**: 733–746.

Power ME and Mills LS (1995) The keystone cops meet in Hilo. *Trends in Ecology and Evolution* **10**: 182–184.
Prach K and Pysek P (1994) Clonal plants – what is their role in succession? *Folia Geobotanica and Phytaxa.*, Praha **29**: 307–320.
Precsenyi I (1969) Analysis of the primary production (phytobiomass) in an *Artemisio-Festucetum pseudovinae. Acta Botanica Acad. Hungary* **15**: 309–325.
Prentice HC *et al.* (1995) Associations between allele frequencies in *Festuca ovina* and habitat variation in the alvar grasslands on the Baltic island of Oland. *Journal of Ecology* **83**: 391–402.
Priestley DA (1986) *Seed Aging: Implications for Seed Storage and Persistence in the Soil.* Cornell University Press, Ithaca, NY.
Prince SD, Carter RN, and Dancy KJ (1985) The geographical distribution of prickly lettuce (*Lactuca serriola*). II. Characteristics of populations near its distribution limit in Britain. *Journal of Ecology* **73**: 39–48.
Promislow DEL and Harvey PH (1990) Living fast and dying young: a comparative analysis of life history variation among mammals. *Journal of Zoology* **220**: 417–437.
Proulx M and Mazumder A (1998) Reversal of grazing impact on plant species richness in nutrient-poor vs. nutrient-rich ecosystems. *Ecology* **79**: 2581–2592.
Pugh GJF (1980) Strategies in fungal ecology. *Transactions of the British Mycological Society* **75**: 1–14.
Rabotnov TA (1969) On coenopopulations of perennial herbaceous plants in natural coenoses. *Vegetatio* **19**: 87–95.
Rabotnov TA (1983) Types of plant strategies. *Ékologiya* **3**: 3–12.
Rabotnov TA (1985) Dynamics of plant coenotic populations. In *The Population Structure of Vegetation*, ed. J White, pp.121–142. Dr W Junk Publishers, Dordrecht.
Rainey RC *et al.* (1990) *Migrant 'pests': Progress, Problems and Potentialities.* The Royal Society, London.
Ramenskii LG (1938) *Introduction to the Geobotanical Study of Complex Vegetations.* Selkozgiz, Moscow.
Rapson GL, Thompson K, and Hodgson JG (1997) The humped relationship between species richness and biomass—testing its sensitivity to sample quadrat size. *Journal of Ecology* **85**: 99–100.
Ratcliffe D (1961) Adaptation to habitat in a group of annual plants. *Journal of Ecology* **49**: 187–203.
Ratcliffe DA (1984) Post-mediaeval and recent changes in British vegetation: the culmination of human influence. *New Phytologist* **98**: 73–100.
Raunkiaer C (1913) Formationsstätistiske Underğelser paa Skagens Odde. *Botanisk Tidsskraft, Kobenhaven* **33**: 197–228.
Raunkiaer C (1934) *The Life Forms of Plants and Statistical Plant Geography, being the collected papers of C. Raunkiaer.* Clarendon Press, Oxford
Raven JA (1981) Nutritional strategies of submerged benthic plants: the acquisition of C, N and P by rhizophytes and haptophytes. *New Phytologist* **88**: 1–30.
Raynal DJ and FS Bazzaz (1975) The contrasting life-cycle strategies of three summer annuals found in abandoned fields in Illinois. *Journal of Ecology* **63**: 587–596.
Rayner ADM, Boddy L, and Dowson CG (1987) Genetic interactions and developmental versatility during establishment of decomposer Basidiomycetes in wood and tree litter. In *Ecology of Microbial Communities*, eds. TRG Gray, M Fletcher, and G Jones, pp. 83–123. Cambridge, Cambridge University Press.
Read DJ, Francis R, and Finlay RD (1985) Mycorrhizal mycelia and nutrient cycling in plant communities. In *Ecological Interactions in Soil*, eds. AH Fitter *et al.* pp. 193–217. Blackwell, Oxford.
Reader PM and Southwood TRE (1981) The relationship between palatability to invertebrates and the successional status of a plant. *Oecologia* **51**: 271–275.

Reader R *et al.* (1992) A comparative study of plasticity in seedling rooting depth in drying soil. *Journal of Ecology* **81**: 543–550.

Reader RJ (1993) Control of seedling emergence by ground cover and seed predation in relation to seed size for some old-field species. *Journal of Ecology* **81**: 69–175.

Reader RJ and Best BJ (1989) Variation in competition along an environmental gradient: *Hieracium floribundum* in an abandoned pasture. *Journal of Ecology* **77**: 673–684.

Redente EF, Friedlander JE, and McLendon T (1992) Response of early and late semiarid seral species to nitrogen and phosphorus gradients. *Plant and Soil* **140**: 127–135.

Redmann RE (1975) Production ecology of grassland plant communities in western and north Dakota. *Ecological Monographs* **45**: 83–106.

Reed ML (1987) Ericoid mycorrhizas of Epacridaceae in Australia. In *Mycorrhizae in the Next Decade: Practical Applications and Research Priorities*, eds. DM Silva, LL Hung, and JH Graham, p. 335. University of Florida, Gainsville.

Rees M (1993) Trade-offs among dispersal strategies in the British flora. *Nature* **366**: 150–152.

Rees M, Grubb PJ, and Kelly D (1996) Quantifying the impact of competition and spatial heterogeneity on the structure and dynamics of a four-species guild of winter annuals. *American Naturalist* **147**: 1–32.

Reich PB *et al.* (1991) Leaf lifespan as a determinant of leaf structure and function among 23 tree species in Amazonian forest communities. *Oecologia* **86**: 16–24.

Reich PB, Walters MB, and Ellesworth DS (1992) Leaf life-span in relation to leaf, plant and stand characteristics among diverse ecosystems. *Ecological Monographs* **62**: 365–392.

Reiling K and Davison AW (1992) The response of native herbaceous species to ozone: growth and fluorescence screening. *New Phytologist* **120**: 29–33.

Rejmánek M (1989) Invasibility of plant communities. In *Ecology of Biological Invasion: a Global Perspective*, eds. JA Drake *et al.* pp. 369–388. Wiley & Sons, New York.

Rejmánek M (1996) A theory of seed plant invasiveness: the first sketch. *Biological Conservation* **78**: 171–181.

Reynolds CS (1998) *Vegetation Processes in the Pelagic: A Model for Ecosystem Theory*. Oldendorf/Lutie, Germany.

Reynolds HG (1958) The ecology of the Merriam kangaroo rat *(Dipodomys merriami* Mearns) on the grazing lands of southern Arizona. *Ecological Monographs* **28**: 111–127.

Reynoldson TB (1961) Environment and reproduction in freshwater triclads. *Nature* **189**: 329–330.

Reynoldson TB (1968) Shrinkage thresholds in freshwater triclads. *Ecology* **49**: 584–586.

Rhoades DF (1976) The anti-herbivore defences of Larrea. In *The Biology and Chemistry of the Creosote Bush. A Desert Shrub*, eds. TJ Mabry, JH Hunziker, and D R DiFeo. Dowden, Hutchinson, and Ross, Stroudsburg, Pennsylvania.

Rhoades DF and Cates RG (1976) Toward a general theory of plant anti-herbivore chemistry. In *Recent Advances in Phytochemistry, Vol. 10: Biochemical Interactions between Plants and Insects*, ed. J Wallace. Plenum, New York.

Rice EL (1974) *Allelopathy*. Academic Press, New York.

Richards PW (1952) *The Tropical Rainforest,* Cambridge University Press, London.

Ricklefs R and Schluter D (1993) Species diversity, regional and historical influences. In *Species Diversity in Ecological Communities*, eds. RE Ricklefs and D Schluter, pp. 350–363. University of Chicago Press, Chicago.

Ricklefs RE (1977) On the evolution of reproductive strategies in birds: reproductive effort. *American Naturalist* **111**: 453–478.

Ridley HN (1930) *The Dispersal of Plants Throughout the World.* Reeve, Ashford.

Rincon E (1990) Growth responses of *Brachythecium rutabulum* to different litter arrangements. *Journal of Bryology* **16**: 120–122.

Rincon E and Grime JP (1989) An analysis of seasonal patterns of bryophyte growth in a natural habitat. *Journal of Ecology* **77**: 447–455.

Rincon E and Huante P (1994) Influence of mineral nutrient availability on growth of tree seedlings from the tropical deciduous forest. *Trees* **9**: 93–97.

Risser PG (1985) Grasslands. In *Physiological Ecology of North American Plant Communities*, eds. BF Chabot and HA Mooney, pp. 232–256. Chapman and Hall, London.

Ritchie ME and Olff H (1999) Spatial scaling laws yield a synthetic theory of biodiversity. *Nature* **400**: 557–560.

Ritchie ME and Tilman D (1995) Responses of legumes to herbivores and nutrients during succession on a nitrogen-poor soil. *Ecology* **76**: 2648–2655.

Roberts HA, Bond W, and Hewson RT (1976) Weed competition in drilled summer cabbage. *Annals of Applied Biology* **84**: 91–95.

Roberts, HA and Stokes FG (1966) Studies on the weeds of vegetable crops: VI. Seed populations of soil under commercial cropping. *Journal of Applied Ecology* **3**: 181–190.

Robinson D (1994) The responses of plants to non-uniform supplies of nutrients. *New Phytologist* **127**: 635–674.

Robinson D and Fitter A (1999) The magnitude and control of carbon transfer between plants linked by a common mycellial network. *Journal of Experimental Botany* **50**: 9–13.

Robinson D *et al.* (1998) Plant root proliferation in nitrogen-rich patches confers competitive advantage. *Proceedings of the Royal Society of London* **B266**: 431–435

Robinson D and Van Vuuren MMI (1998) Responses of wild plants to nutrient patches in relation to growth rate and life-form. In *Inherent Variation in Plant Growth: Physiological Mechanisms and Ecological Consequences*, eds. H Lambert, H Poorter, and MMI Van Vuuren, pp. 237–257. Backhuys, Leiden.

Robinson T (1974) Metabolism and function of alkaloids in plants. *Science* **184**: 430–435.

Rodwell JS (1991) *British Plant Communities. Vol. 1. Woodlands and Scrub.* Cambridge University Press, Cambridge.

Rogers RW (1988) Succession and survival strategies in lichen populations on a palm trunk. *Journal of Ecology* **76**: 759–776.

Rogers RW (1990) Ecological strategies of lichens. *Lichenologist* **22**: 149–162.

Rogers RW and Clifford HT (1993) The taxonomic and evolutionary significance of leaf longevity. *New Phytologist* **123**: 811–821.

Rorison IH (1960) Some experimental aspects of the calcicole–calcifuge problem: I. The effects of competition and mineral nutrition upon seedling growth in the field. *Journal of Ecology* **48**: 585–599.

Rorison IH (1968) The response to phosphorus of some ecologically distinct plant species: I. Growth rates and phosphorus absorption. *New Phytologist* **67**: 913–923.

Rose F (1974) The epiphytes of oak. In *The British Oak, its History and Natural History*, eds. MG Morris and FH Perring, pp. 250–273. Classey, Faringdon.

Ross BA, Bray JR, and Marshall WH (1970) Effects of long-term deer exclusion on a *Pinus resinosa* forest in north-central Minnesota. *Ecology* **51**: 1088–1093.

Ross MA and Harper JL (1972) Occupation of biological space during seedling establishment. *Journal of Ecology* **60**: 77–88.

Roughton RD (1972) Shrub age structures on a mule deer range in Colorado. *Ecology* **53**: 615–625.

Rusch G and Fernández-Palacios JM (1995) The influence of spatial heterogeneity on regeneration by seed in a limestone grassland. *Journal of Vegetation Science* **6**: 417–426.

Ryel RJ, Caldwell MM, and Manwaring JH (1996) Temporal dynamics of soil spatial heterogeneity in sage-brush-wheatgrass steppe during a growing season. *Plant and Soil* **184**: 299–309.

Ryser P and Lambers H (1995) Root and leaf attributes accounting for the performance of fast- and slow-growing grasses at different nutrient supply. *Plant and Soil* **170**: 251–265.

Sagar GR and Harper J (1961) Controlled interference with natural populations of *Plantago lanceolata*, *P. major* and *P. media*. *Weed Research* **1**: 163–176.

Sale PF (1977) Maintenance of high diversity in coral reef fish communities. *American Naturalist* **11**: 337–359.

Salisbury EJ (1932) The East Anglian flora. *Transactions of the Norfolk Naturalists Society* **8**: 191–263.

Salisbury EJ (1942) *The Reproductive Capacity of Plants*. Bell, London.

Salisbury EJ (1953) A changing flora as shown in the study of weeds of arable land and waste places. In *The Changing Flora of Britain*, ed. JE Lousely, pp. 130–139. Buncle, Arbroath.

Salisbury EJ (1964) *Weeds and Aliens*. 2nd edn. Collins, London.

Sanders FE *et al.* (1977) The development of endomycorrhizal root systems. I. Spread of infection and growth promoting effects with four species of vesicular-arbuscular mycorrhizas. *New Phytologist* **78**: 257–268.

Sankaran M and McNaughton SM (1999) Determinants of biodiversity regulate compositional stability of communities *Nature* **401**: 691–693.

Sarukhan J (1974) Studies in plant demography: *Ranunculus repens* L. *R. bulbosus* L. and *R. acris* I.: II. Reproductive strategies and seed population dynamics. *Journal of Ecology* **62**: 151–177.

Saverimutta T and Westoby M (1996) Components of variation in seedling potential growth rate: phylogenetically independent contrasts. *Oecologia* **105**: 281–285.

Savile DBO (1972) *Arctic Adaptations in Plants*. Information Canada, Ottawa.

Savory CJ and Gentle MJ (1976) Changes in food intake and gut size in Japanese quail in response to manipulation of dietary fibre content. *British Poultry Science* **17**: 571–580.

Scaife MA (1976) The use of a simple dynamic model to interpret the phosphate response of plants grown in solution culture. *Annals of Botany* **40**: 1217–1229.

Schaeffer K and Moreau R (1958) L'alternance des essences. *Soc. Forest. France-Compte Bull.*, **29**: 1–12, 76–84, 277–298.

Schmitt RJ (1996) Exploitative competition in mobile grazers: trade-offs in use of a limited resource. *Ecology* **77**: 408–425.

Schouw JF (1823) Grundzüge einer allgemeinen Pflanzengeographie. Berlin (in Danish, Copenhagen 1822).

Schulz JP (1960) Ecological studies on rainforest in northern Surinam. *Verh. K. Ned. Akad. Wet.*, **53**: 1–367.

Schulze ED *et al.* (1991) The utilization of nitrogen from insect capture by different growth forms of *Drosera* from south-west Australia. *Oecologia* **87**: 240–246.

Schulze ED and Mooney HA (eds.) (1993a) *Design and execution of experiments on CO_2 enrichment.* Commission of the European Communities, Brussels.

Schulze ED and Mooney HA (eds.) (1993b) *Biodiversity and Ecosystem Function.* Springer-Verlag, Berlin.

Schütz W (1998) Dormancy cycles and germination phenology in sedges (*Carex*) of various habitats. *Wetlands* **18**: 288–297.

Scurfield G (1953) Ecological observations in southern Pennine woodlands. *Journal of Ecology* **41**: 1–12.

Sears JR and Wilce RT (1975) Sublittoral, benthic marine algae of southern Cape Cod and adjacent islands: seasonal periodicity, associations, diversity and floristic composition. *Ecological Monographs* **45**: 337–365.

Sharpe DM and Fields DE (1982) Integrating effects of climate and seed fall velocities on seed dispersal by wind: a model and application. *Ecological Modelling* **17**: 297–310.

Shen-Miller J *et al.* (1995) Exceptional seed longevity and robust growth: ancient sacred lotus from China. *American Journal of Botany* **82**: 1367–1380.

Shepherd SA (1981) Ecological strategies in a deep water red algal community. *Botanica Marina* **24**: 457–463

Shields JA, Paul EA, and Lowe WE (1973) Turnover of microbial tissue in soil under field conditions. *Soil Biology and Biochemistry* **5**: 753–764.

Shipley B and Keddy PA (1988) The relationship between relative growth rate and sensitivity to nutrient stress in twenty-eight species of emergent macrophytes. *Journal of Ecology* **76**: 1101–1110.

Shipley B and Keddy PA (1994) Evaluating the evidence for competitive hierarchies in plant communities. *Oikos* **69**: 340–345.

Shipley B *et al.* (1991) A model of species density in shoreline vegetation. *Ecology* **72**: 1658–1667.

Shipley B and Parent M (1991) Germination responses of 64 wetland species in relation to seed size, minimum time to reproduction and seedling relative growth rate. *Functional Ecology* **5**: 111–118.

Shipley B and Peters RH (1990) A test of the Tilman model of plant strategies: relative growth rates and biomass partitioning. *American Naturalist* **136**: 139–153.

Shipley B and Peters RH (1991) The seduction by mechanism: a reply to Tilman. *American Naturalist* **138**: 1276–1282.

Shmida A and Ellner S (1984) Coexistence of plant species with similar niches. *Vegetatio* **58**: 29–55.

Shreve F (1942) The desert vegetation of North America. *Botanical Reviews* **8**: 195–246.

Shure DJ (1971) Insecticide effects on early succession in an old-field ecosystem. *Ecology* **52**: 271–279.

Sibly RM and Calow P (1983) An integrated approach to life-cycle evolution using selective landscapes. *Journal of Theoretical Ecology* **102**: 527–547.

Sibly RM and Grime JP (1986) Strategies of resource capture by plants: evidence for adversity selection. *Journal of Theoretical Biology* **118**: 247–250.

Siccama TG, Bormann FH, and Likens GE (1970) The Hubbard Brook ecosystem study: productivity, nutrients and phytosociology of the herbaceous layer. *Ecological Monographs* **40**: 389–402.

Siegler D and Price PW (1976) Secondary compounds in plants: primary functions. *American Naturalist* **110**: 101–105.

Silander JA and Antonovics J (1982) Analysis of interspecific interactions in a coastal plant community: a perturbation approach. *Nature* **298**: 557–560.

Silvertown J (1980) The dynamics of a grassland ecosystem: botanical equilibrium in the Park Grass Experiment. *Journal of Applied Ecology* **17**: 491–504.

Silvertown J *et al.* (1999) Hydrologically-defined niches reveal a basis for species richness in plant communities. *Nature* **400**: 61–63.
Silvertown J *et al.* (1994) Rainfall, biomass variation and community composition in the Park Grass Experiment. *Ecology* **75**: 2430–2437.
Silvertown J, Franco M, and McConway . (1992) A demographic interpretation of Grime's triangle. *Functional Ecology* **6**: 130–136.
Silvertown JW (1982) *Introduction to Plant Population Ecology*. Longman, London.
Simard SW *et al.* (1997) Net transfer of carbon between ectomycorrhizal tree species in the field. *Nature* **388**: 579–582.
Simpson DA (1984) A short history of the introduction and spread of Elodea Michx in the British Isles. *Watsonia* **15**: 1–9.
Singer DK *et al.* (1996) Differentiating climatic and successional influence in long-term developments of a marsh. *Ecology* **77**: 1765–1778.
Singh JS and Misra R (1969) Diversity, dominance, stability and net production in the grasslands at Varanasi, India. *Canadian Journal of Botany* **47**: 425–427.
Skutch AF (1929) Early stages of plant succession following forest fire. *Ecology* **10**: 177–190.
Slade AJ and Hutchings MJ (1987) An analysis of the costs and benefits of physiological integration between ramets in the clonal perennial herb *Glechoma hederacea*. *Oecologia* **73**: 425–431.
Slatkin M (1974) Competition and regional coexistence. *Ecology* **55**: 128–134.
Slatyer RO (1967) *Plant–Water Relationships.* Academic Press, London.
Slavikova J (1958) Einfluss der Buche (*Fagus silvatica* L.) als Edificator auf die Entwicklung der Krautschicht in den Buchenphytozonosen. *Presalia* **30**: 19–42.
Smida A and Wilson MV (1985) Biological determinants of species diversity. *Journal of Biogeography* **12**: 1–20.
Smith CC (1970) The coevolution of pine squirrels *(Tamiasciurus)* and conifers. *Ecological Monographs* **40**: 349–371.
Smith CJ, Elston J, and Bunting AH (1971) The effects of cutting and fertilizer treatments on the yield and botanical composition of chalk turf. *Journal of the British Grasslands Society* **26**: 213–223.
Smith H (1982) Light quality, photoperception and plant strategy. *Annual Review of Plant Physiology* **33**: 481–518.
Smith RH (1966) Resin quality as a factor in the resistance of pines to bark beetles. In *Breeding Pest-Resistant Trees*, eds. HD Gorhold *et al.* pp. 189–196. Pergamon, New York.
Smith SE and Read DJ (1997) *Mycorrhizal Symbiosis*. Academic Press, San Diego.
Smith SE *et al.* (1994) Transport of phosphate from fungus to plant in VA mycorrhizas: calculations of the area of symbiotic interface and of fluxes from two different fungi to *Allium porrum* L. *New Phytologist* **127**: 93–97.
Smith TM, Shugart HH, and Woodward FI (eds) (1996) *Plant Functional Types*. Cambridge University Press, Cambridge.
Snaydon RW (1962) Microdistribution of *Trifolium repens* L. and its relation to soil factors. *Journal of Ecology* **50**: 133–143.
Snow AA and Vince SW (1984) Plant zonation in an Alaskan salt marsh. II. An experimental study of the role of edaphic conditions. *Journal of Ecology* **72**: 669–684.
Sommer U (ed.) (1989) *Plankton Ecology: Succession in Plankton Communities*. Springer-Verlag, Berlin.
Sonesson M and Callaghan TV (1991) Strategies of survival in plants of the Fennoscandian tundra. *Arctic* **44**: 95–105.
Southwood TRE (1977) Habitat, the templet for ecological strategies? *Journal of Animal Ecology* **46**: 337–365.

Southwood TRE (1988) Tactics, stategies and templets. *Oikos* **52**: 13–18.
Southwood TRE, Brown VK, and Reader PM (1986) Leaf palatability, life expectancy and herbivore damage. *Oecologia* **70**: 544–548.
Southwood TRE *et al.* (1974) Ecological strategies and population parameters. *American Naturalist* **108**: 791–804.
Sporne KR (1982) The advancement index vindicated. *New Phytologist* **91**: 137–145.
Sprent JI (1979) *The Biology of Nitrogen-Fixing Organisms*. McGraw-Hill, London.
Stampfli A (1992) Year-to-year changes in unfertilised meadows of great species richness detected by point quadrat analysis. *Vegetatio* **103**: 125–132.
Stainforth RJ and PB Cavers (1977) The importance of cottontail rabbits in the dispersal of *Polygonum* spp. *Journal of Applied Ecology* **14**: 261–267.
Stearns SC (1976) Life-history tactics; a review of the ideas. *Quarterly Review of Biology* **51**: 3–47.
Stebbins GL (1952) Aridity as a stimulus to plant evolution. *American Naturalist* **86**: 33–48.
Stebbins GL (1971) Adaptive radiation of reproductive characteristics in angiosperms: II. Seeds and seedlings. *Annual Review of Ecology and Systematics* **2**: 237–260.
Stebbins GL (1972) Ecological distribution of centers of major adaptive radiation in angiosperms. In *Taxonomy, Phytogeography and Evolution*, ed. DH Valentine, pp. 7–34. Academic Press, London.
Stebbins GL and Major J (1965) Endemism and speciation in the Californian flora. *Ecological Monographs* **35**: 1–36.
Stockey A and Hunt R (1994) Predicting secondary succession in wetland mesocosms on the basis of autecological information on seeds and seedlings. *Journal of Applied Ecology* **31**: 543–559.
Stöcklin J and Bäumler E (1996) Seed rain, seedling establishment and clonal growth strategies on a glacier foreland. *Journal of Vegetation Science* **7**: 45–56.
Stohlgren TJ *et al.* (1999) Exotic plant species invade hot spots of native plant diversity. *Ecological Monographs* **69**: 25–46.
Stone EC and Vasey RB (1968) Preservation of coastal redwoods on alluvial flats. *Science* **159**: 157–161.
Strain KR and Chase VC (1966) Effect of past and prevailing temperatures on the carbon dioxide exchange capacities of some woody desert perennials. *Ecology* **47**: 1043–1045.
Swingland IR (1977) Reproductive effort and life-history strategy of the Aldabran giant tortoise. *Nature* **269**: 402–404.
Sydes C (1981) Investigations into the effects of tree litter on herbaceous vegetation. PhD Thesis, University of Sheffield.
Sydes CL (1984) A comparative study of leaf demography in limestone grassland. *Journal of Ecology* **72**: 331–345.
Sydes C and Grime JP (1981a) Effects of tree leaf litter on herbaceous vegetation in deciduous woodland. I Field investigations. *Journal of Ecology* **69**: 237–248.
Sydes C and Grime JP (1981b) Effects of tree leaf litter on herbaceous vegetation in deciduous woodland. II An experimental investigation. *Journal of Ecology* **69**: 249–262.
Sydes CL and Grime JP (1984) A comparative study of root development using a simulated rock crevice. *Journal of Ecology* **72**: 937–946.
Symstad AJ (2000) A test of the effects of functional group richness and composition on grassland invasibility. *Ecology* **81**: 99–109.
Tallis JH (1958) Studies in the biology and ecology of *Racomitrium lanuginosum* Brid: I. Distribution and ecology. *Journal of Ecology* **46**: 271–288.
Tallis JH (1959) Studies in the biology, and ecology of *Racomitrium lanuginosum* Brid: II. Growth, reproduction and physiology. *Journal of Ecology* **47**: 325–350.

Tallis JH (1964) Growth studies on *Racomitrium lanuginosum*. *Bryologist* **67**: 417–422.
Tamm CO (1956) Further observations on the survival and flowering of some perennial herbs: I. *Oikos* **7**: 273–292.
Tamm CO (1972) Further observations on the survival and flowering of some perennial herbs: II and III. *Oikos* **23**: 23–28 and 159–166.
Tansley AG (1939). *The British Islands and their Vegetation.* Cambridge University Press, London.
Tansley AG and Adamson RS (1925) Studies of the vegetation of the English Chalk. III. The chalk grasslands of the Hampshire–Sussex Border. *Journal of Ecology* **13**: 177–223.
Taylor AA, De-Felice J, and Havill DC (1982) Seasonal variation in nitrogen availability and utilisation in an acidic and calcareous soil. *New Phytologist* **92**: 141–152.
Taylor RJ and Pearcy RW (1976) Seasonal patterns in the CO_2 exchange characteristics of understory plants from a deciduous forest. *Canadian Journal of Botany* **54**: 1094–1103.
Taylor DR, Aarssen LW, and Loehle C (1990) On the relationship between r/K selection and environmental carrying capacity: a new habitat templet for plant life-history strategies. *Oikos* **58**: 239–250.
Taylorson RB and HA Borthwick (1969). Light filtration by foliar canopies; significance for light-controlled weed seed germination. *Weed Science* **17**: 48–51.
Ter Heerdt GNJ *et al.* (1996) An improved method for seedbank analysis: seedling emergence after removing the soil by sieving. *Functional Ecology* **10**: 144–151.
Tessier AJ, Leibold MA, and Tsao J (2000) A fundamental tradeoff in resource exploitation by *Daphnia* and consequences to plankton communities. *Ecology* **81**: 826–841.
Tevis L (1958) A population of desert ephemerals germinated by less than one inch of rain. *Ecology* **39**: 688–695.
Thanos CA and Rundel PW (1995) Fire-followers in chaparral: nitrogenous compounds trigger seed germination. *Journal of Ecology* **83**: 207–216.
Thomas AS (1960) Changes in vegetation since the advent of myxomatosis. *Journal of Ecology* **48**: 287–306.
Thomas M (1949) Physiological studies in acid metabolism in green plants: I. CO_2 fixation and CO_2 liberation in crassulacean acid metabolism. *New Phytologist* **48**: 390–420.
Thomas WA and Grigal DF (1976) Phosphorus conservation by evergreenness of mountain laurel. *Oikos* **27**: 19–26.
Thompson JD (1991) The biology of an invasive plant. *Bioscience* **41**: 393–401.
Thompson JW and Burdon JJ (1992) Gene for gene coevolution between plants and parasites. *Nature* **360**: 121–126.
Thompson K (1977) An ecological investigation of germination responses to diurnal fluctuations in temperature. PhD Thesis, University of Sheffield.
Thompson K (1987) Seeds and seed banks. *New Phytologist* **106**: 23–24.
Thompson K (1992) The functional ecology of seed banks. In *Seeds: the Ecology of Regeneration in Plant Communities*, ed. M Fenner, pp. 231–258. CAB International, Wallingford.
Thompson K (1994) Predicting the fate of temperate species in response to human disturbance and global change. In *Biodiversity, temperate ecosystems and global change*, eds. TJB Boyle and CEB Boyle. Springer-Verlag, Berlin.
Thompson K (1998) Weed seed banks; evidence from the north-west European seed bank database. *Aspects of Applied Biology* **51**: 105–112.
Thompson K and Band SR (1997) Survival of a lowland heathland seed bank after a 33-year burial. *Seed Science Research* **7**: 409–411.

Thompson K, Band SR, and Hodgson JG (1993a) Seed size and shape predict persistence in soil. *Functional Ecology* **7**: 236–241.
Thompson K *et al.* (1998) Ecological correlates of seed persistence in soil in the N.W. European flora. *Journal of Ecology* **86**: 163–169.
Thompson K and Baster K (1992) Establishment from seed of selected *Umbelliferae* in unmanaged grassland. *Functional Ecology* **6**: 346–352.
Thompson K and Grime JP (1979) Seasonal variation in the seed banks of herbaceous species in ten contrasting habitats. *Journal of Ecology* **66**: 893–921.
Thompson K and Grime JP (1983) A comparative study of germination responses to diurnally-fluctuating temperatures. *Journal of Applied Ecology* **20**: 141–156.
Thompson, K and Grime, JP (1988) Competition reconsidered—a reply to Tilman. *Functional Ecology*, **2**.
Thompson K, Grime JP, and Mason G (1977) Seed germination in response to diurnal fluctuations of temperature. *Nature* **67**: 147–149.
Thompson K *et al.* (1996) A functional analysis of a limestone grassland community. *Journal of Vegetation Science* **7**: 371–380.
Thompson K *et al.* (1993b) Ellenberg numbers revisited. *Phytocoenologia* **23**: 277–289.
Thompson K, Hodgson JG, and Rich TCG (1995) Native and alien invasive plants: more of the same? *Ecography* **18:** 390–402.
Thompson K and Jones A (1999) Human population density and prediction of local plant extinction in Britain. *Conservation Biology* **15**: 1–6.
Thompson K *et al.* (1997) A comparative study of leaf nutrient concentrations in a regional herbaceous flora. *New Phytologist* **136**: 679–689.
Thompson K and Whatley JC (1983) Germination responses of naturally-buried weed seeds to diurnal temperature fluctuations. In *Aspects of Applied Biology 4*, pp. 71–77. Association of Applied Biologists, HRI, Wellesbourne.
Thompson L *et al.* (1993) The effects of earthworms and snails in a simple plant community. *Oecologia* **95**: 171–178.
Thorpe PC, MacGillivray CW, and Priestman GH (1993) A portable device for the simulation of air frosts at remote field locations. *Functional Ecology* **7**: 503–505.
Thurston JM (1969)The effects of liming and fertilizers on the botanical composition of permanent grassland, and on the yield of hay. In *Ecological Aspects of the Mineral Nutrition of Plants*, ed. IH Rorison, pp. 3–10. Blackwell, Oxford.
Tielborger K and Kadmon R (2000) Temporal environmental variation tips the balance between facilitation and interference in desert plants. *Ecology* **81**: 1544–1553.
Tilman D (1981) Tests of resource competition theory using four species of Lake Michigan Algae. *Ecology* **62**: 802–815.
Tilman D (1982) *Resource Competition and Community Structure*. Princeton University Press, NJ.
Tilman D (1985) The resource-ratio hypothesis of plant succession. *American Naturalist* **125**: 827–852.
Tilman D (1987a) Secondary succession and the patterns of plant dominance along experimental nitrogen gradients. *Ecological Monographs* **57**: 189–214.
Tilman D (1988) *Plant Strategies and the Dynamics and Structure of Plant Communities*. Princeton University Press, NJ.
Tilman D (1990) Constraints and tradeoffs: toward a predictive theory of competition and succession. *Oikos* **58**: 3–15.
Tilman D (1994) Competition and biodiversity in spatially structured habitats. *Ecology* **78**: 81–92.
Tilman D (1996) Biodiversity: population versus ecosystem stability. *Ecology* **77**: 97–106.
Tilman D (1999) Ecological consequences of biodiversity: a search for general principles. *Ecology* **80**: 1455–1474.
Tilman D and Downing JA (1994) Biodiversity and stability in grasslands. *Nature* **367**: 363–365.

Tilman D *et al.* (1997) The influence of functional diversity and composition on ecosystem processes. *Science* **277**: 1300–1302.
Tilman D, Wedin D, and Knops J (1996) Productivity and sustainability influenced by biodiversity in grassland ecosystems. *Nature* **379**: 718–720.
Tinkle DW (1969) The concept of reproductive effort and its relation to the evolution of life histories of lizards. *American Naturalist* **103**: 501–515.
Tinoco-Ojanguren C and Pearcy RW (1992) Dynamic stomatal behaviour and its role in carbon gain during lightflecks of a gap phase and an understorey *Piper* species acclimated to high and low light. *Oecologia* **92**: 222–228.
Tissue DT and Nobel PS (1988) Parent-ramet connections in *Agave desert*: influences of carbohydrates on growth. *Oecoligia* 75, 266–271.
Titman D (1976) Ecological competition between algae: experimental confirmation of resource-based competition theory. *Science* **192**: 463–465.
Topham PB (1997) Colonisation, growth, succession and competition. In *Lichen Ecology*, ed. MRD Seaward, pp. 31–68. Academic Press, London.
Traveset A (1998) Effect of seed passage through vertebrate frugivores' guts on germination: a review. *Perspectives in Plant Ecology, Evolution and Systematics*, **1**, 151–190
Tribe DE (1950) The behaviour of the grazing animal—a critical review of present knowledge. *Journal of the British Grassland Society* **5**: 200–214.
Troughton JH (1972) Carbon isotope fractionation by plants. In *Proceedings of the Eight International Radiocarbon Dating Conference*, eds. TA Rafter and T Grant-Taylor, pp. E40-E57. Royal Society of New Zealand, Lower Hutt, New Zealand.
Turkington R and Klein E (1991) Integration among ramets of *Trifolium repens*. *Canadian Journal of Botany* **69**: 226–228.
Turner FB *et al.* (1970) The demography of the lizard, *Uta stansburiana* Baird and Girard, in southern Nevada. *Journal of Animal Ecology* **39**: 505–519.
Turner MG *et al.* (1997) Effects of fire size and pattern on early succession in Yellowstone National Park. *Ecological Monographs* **67**: 411–433.
Turpin DH (1988) Physiological mechanisms in phytoplankton resource competition. In *Growth and Reproduction Strategies of Freshwater Phytoplankton*, ed. CD Sandgren, pp. 316–368. Cambridge University Press, Cambridge.
Tyler G (1971) Studies in the ecology of Baltic seashore meadows: IV. Distribution and turnover of organic matter and minerals in a shore meadow ecosystem. *Oikos* **22**: 265–291.
Underwood AJ (1986) The analysis of competition by field experiments. In *Community Ecology: Pattern and Process*, eds. DJ Anderson and J Kikkawa, pp. 240–260. Blackwell, Melbourne.
Vaartaja O (1952) Forest humus quality and light conditions as factors influencing damping-off. *Phytopathology* **42**: 501–506.
Vaartaja O and Cran HW (1956) Damping-off pathogens of conifers and of caragana in Saskatchewan. *Phytopathology* **46**: 391–397.
Valerio DePatta P (1999) On the identification of optimal plant functional types. *Journal of Vegetation Science* **10**: 631–640.
Van Andel J, Bakker JP, and Grootjans AP (1993) Mechanisms of vegetation succession: a review of concepts and perspectives. *Acta Botanica Neerlandica* **42**: 413–433.
Van Andel J and Vera F (1977) Reproductive allocation in *Senecio sylvaticus* and *Chamaenerion angustifolium* in relation to mineral nutrition. *Journal of Ecology* **65**: 747–758.
van der Heijden MGA *et al.* (1998) Mycorrhizal fungal diversity determines plant biodiversity, ecosystem variability and productivity. *Nature* **396**: 69–72.
van der Maarel E (1971) Plant species diversity in relation to management. In *The Scientific Management of Animal and Plant Communities for Conservation*, eds. E Duffey and AS Watt, pp. 45 64. Blackwell, Oxford.

van der Maarel E and Sykes MT (1993) Small-scale plant species turnover in grasslands: the carousel model and a new niche concept. *Journal of Vegetation Science* **4**: 179–188.

van der Peijl L (1972) *Principles of Dispersal in Higher Plants*. Springer-Verlag, Berlin.

van der Steen W and Scholten M (1985) Methodological problems in evolutionary biology. IV. Stress and stress tolerance, an exercise in definition. *Acta Biotheoretica* **34**: 81–90.

van der Valk AG (1981) Succession in wetlands: a Gleasonian approach. *Ecology* **62**: 688–696.

van der Valk AG and Davis CB (1976) The seed banks of prairie glacial marshes. *Canadian Journal of Botany* **54**: 1832–1838.

van der Valk AG and Davis CB (1978) The role of the seed bank in the vegetation dynamics of prairie glacial marshes. *Ecology* **59**: 322–335.

Van der Wall R *et al.* (2000) Effects of resource competition and herbivory on plant performance along a natural productivity gradient. *Journal of Ecology* **88**: 317–330.

Van der Wall SB (1990) *Food Hoarding in Animals*. Chicago University Press, Chicago.

Van der Wall SB and Balda RP (1977) Co-adaptations of the Clark's Nutcracker and the pinon pine for efficient seed harvest and dispersal. *Ecological Monographs* **47**: 89–111.

Van der Werf A *et al.* (1993) Contribution of physiological and morphological plant traits to a species' competitive ability at high and low nitrogen supply. *Oecologia* **94**: 434–440.

van Dobben WH (1967) Physiology of growth in two *Senecio* species in relation to their ecological position. *Jaarboek Instituut voor Biologisch en Scheikundig Onderzoek van Landbouwgewassen*, pp. 75–83.

van Steenis CGGJ (1958) Rejuvenation as a factor for judging the status of vegetation types: the biological nomad theory. In *Study of Tropical Vegetation*, pp. 212–215. Proceedings of the Kandy Symposium, UNESCO.

van Steenis CGGJ (1972) *The Mountain Flora of Java*. EJ Brill, Leiden.

van Tooren BF (1988) The fate of seeds after dispersal in chalk grasslands: the role of the bryophyte layer. *Oikos* **53**: 41–48.

Van't Hof J and Sparrow AH (1963) A relationship between DNA content, nuclear volume and minimum mitotic cycle time. *Proceedings of the National Academy of Sciences USA* **49**: 897–902.

Vàzquez-Yanes C and Orozco-Sergovia A (1990) Ecological significance of light controlled seed germination in two contrasting tropical habitats. *Oecologia* **83**: 171–175.

Vegelin K *et al.* (1997) Wind dispersal of a species-rich fen-meadow (*Polygono-Cirsietum oleracei*) in relation to the restoration perspectives of degraded valley fens. In *Species Dispersal and Land Use Processes*, eds. A Cooper and J Power, pp. 85–92. IALE (UK), Aberdeen.

Venable DL and Brown JS (1988) The selective interactions of dispersal, dormancy, and seed size as adaptations for reducing risk in variable environments. *American Naturalist* **131**: 360–384.

Venable DL and Lawlor L (1980) Delayed germination and dispersal in desert annuals: escape in space and time. *Oecologia* **46**: 272–282.

Veneklaas EJ and Poorter L (1998) Growth and carbon partitioning of tropical tree seedlings in contrasting light environments. In *Inherent Variation in Plant Growth*, eds. H Lambers, H Poorter, and MMI Van Vauren, pp. 337–361. Backhuys, Leiden.

Veresoglou DS and Fitter AH (1984) Spatial and temporal patterns of growth and nutrient uptake of five coexisting grasses. *Journal of Ecology* **72**: 259–272.

Verkaar HJ and Schenkeveld AJ (1984) On the ecology of short-lived forbs in chalk grasslands: life-history characteristics. *New Phytologist* **98**: 659–672.

Vermeer JG and Berendse F (1983) The relationship between nutrient availability, shoot biomass and species richness in grassland and wetland communities. *Vegetatio* **53**: 121–136.

Viereck LA (1966) Plant succession and soil development on gravel outwash of the Muldrow Glacier, Alaska. *Ecological Monographs* **36**: 181–199.

Villiers TA (1974) Seed aging: chromosome stability and extended viability of seeds stored fully imbibed. *Plant Physiology* **53**: 875–878.

Vitousek PM *et al.* (1996) Biological invasions as global environmental change. *American Scientist* **84**: 468–478.

Vogel JC (1980) Fractionation of the carbon isotopes during photosynthesis. In *Sitzungsberichte der Heidelberger Akademie der Wissenschafen, Mathematisch-naturwissenschaftliche Klasse Jahrgang 1980*, pp. 111–135. Springer-Verlag, Berlin.

Vogl RJ (1973) Ecology of Knobcone pine in the Santa Ana Mountains, California. *Ecological Monographs* **43**: 125–143.

Wakamiya I *et al.* (1993) Genome size and environmental factors in the Genus *Pinus*. *American Journal of Botany* **80**: 1235–1241.

Walker RB (1954) Factors affecting plant growth on serpentine soils. *Ecology* **35**: 259–266.

Walter H (1973) *Vegetation of the Earth in relation to Climate and the Ecophysiological Conditions.* English Universities Press, London.

Ward LK (1990) Management of grassland-scrub mosaics. In *Calcareous Grasslands; Ecology and Management*, eds. SH Hillier, DWH Walton, and DA Wells, pp. 134–139. Bluntisham Books, Huntingdon.

Wardle DA (1998) A more reliable design for biodiversity study? *Nature* **394**: 30.

Wardle DA (1999) Is 'sampling effect' a problem for experiments investigating biodiversity ecosystem function relationships? *Oikos* **87**: 403–407.

Wardle DA *et al.* (1999) Plant removals in perennial grassland: vegetation dynamics, decomposers, soil biodiversity and ecosystem properties. *Ecological Monographs* **69**: 535–568.

Wardle DA, Bonner KI, and Nicholson KS (1997b) Biodiversity and plant litter: experimental evidence which does not support the view that enhanced species richness improves ecosystem function. *Oikos* **79**: 247–258.

Wardle DA *et al.* (2000) Biodiversity and ecosystem function: an issue in ecology. *Bulletin of the Ecological Society of America* **81**: 235–239.

Wardle DA *et al.* (1997a) The influence of island area on ecosystem properties. *Science* **277**: 1296–1299.

Wardle P (1959) The regeneration of *Fraxinus excelsior* in woods with a field layer of *Mercurialis perennis. Journal of Ecology* **47**: 483–497.

Waring RH (1988) Ecosystems; fluxes of matter and energy. In *Ecological Concepts; the Contribution of Ecology to an Understanding of the Natural World*, ed. JM Cherrett, pp. 17–41. Blackwell, Oxford.

Washitani I and Masuda M (1990) A comparative study of the germination characteristics of seeds from a moist tall grassland community. *Functional Ecology* **4**: 543–557.

Watt AS (1919) On the causes of failure of the natural regeneration in British oak woods. *Journal of Ecology* **7**: 173–203.

Watt AS (1925) On the ecology of the British beechwoods with special reference to their regeneration. Part II. Sections II and III. The development of structure of beech communities on the Sussex Downs. *Journal of Ecology* **13**: 27–73.

Watt AS (1947) Pattern and process in the plant community. *Journal of Ecology* **35**: 1–22.

Watt AS (1955) Bracken versus heather; a study in plant sociology. *Journal of Ecology* **43**: 490–506.

Watt AS (1957) The effect of excluding rabbits from Grassland B (Mesobrometum) in Breckland. *Journal of Ecology* **45**: 861–878.

Watt, AS (1960) Population changes in acidophilous grass-heath in Breckland 1936–57. *Journal of Ecology* **48**: 605–629.

Watt KEF (1971) Dynamics of populations: a synthesis. In *Dynamics in Population*, eds. PJ DenBoer and GR Gradwell, pp. 568–580. Centre for Agric. Publ. Documentation, Wageningen.

Watt TA (1976) The emergence, growth, flowering and seed production of *Holcus lanatus* L. sown monthly in the field. *Proceedings of the 1976 British Crop Protection Conference—Weeds*; 567–574.

Weaver JE and Albertson FW (1956) *Grasslands of the Great Plains*. Johnsen, Lincoln, Nebraska.

Webb LJ, Tracey JG, and Haydock KP (1967) A factor toxic to seedlings of the same species associated with living roots of the non-gregarious subtropical rainforest tree *Grevillea robusta*. *Journal of Applied Ecology* **4**: 13–25.

Webb LJ, Tracey JG, and Williams WT (1972) Regeneration and pattern in the subtropical rain forest. *Journal of Ecology* **60**: 675–695.

Wedin DA (1995) Species, nitrogen, and grassland dynamics: the constraints of stuff. In *Linking Species and Ecosystems*, eds. CG Jones and JH Lawton, pp. 253–262. Chapman and Hall, New York.

Wedin DA and Tilman D (1990) Species effects on nitrogen cycling: a test with perennial grasses. *Oecologia* **84**: 433–441.

Wedin DA and Tilman D (1993) Competition among grasses along a nitrogen gradient: initial conditions and mechanisms of competition. *Ecological Monographs* **63**:199–229.

Wedin DA and Tilman D (1996) Influence of nitrogen loading and species composition on the carbon balance of grasslands. *Science* **274**: 1720–1723.

Weiher E and Keddy PA (1995) The assembly of experimental wetland plant communities. *Oikos* **73**: 323–335.

Weiher E *et al.* (1999) Challenging Threophrastus: a common core list of plant traits for functional ecology. *Journal of Vegetation Science* **10**: 609–620.

Weins JA (1976) Population responses to patchy environments. *Annual Review of Ecology and Systematics* **7**: 81–120.

Weins JA (1977) On competition and variable environments. *American Scientist* **65**: 590–597.

Welden CW and Slauson WL (1986) The intensity of competition versus its importance: an overlooked distinction and some implications. *The Quarterly Review of Biology* **61**: 23–44.

Wells GJ (1974) The autecology of *Poa annua* L. in perennial ryegrass pastures. PhD Thesis, University of Reading.

Wells TCE (1967) Changes in a population of *Spiranthes spiralis* (L.) Chevall. at Knocking Hoe National Nature Reserve, Bedfordshire, 1962–65. *Journal of Ecology* **55**: 83–99.

Went FW (1948) Ecology of desert plants: I. Observations on germination in the Joshua Tree National Monument, California. *Ecology* **29**: 242–253.

Went FW (1949) Ecology of desert plants: II. The effect of rain and temperature on germination and growth. *Ecology* **30**: 1–13.

Went FW (1955) The ecology of desert plants. *Scientific American* **192**: 68–75.

Weny DG (2000) Seed dispersal, seed predation and seedling recruitment of a neotropical montane tree. *Ecological Monographs* **70**: 331–351.

Werner PA (1975) Predictions of fate from rosette size in teasel *(Dipsacus fullonum* L.). *Oecologia (Berl)* **20**: 197–201.

Wesson G and Wareing PF (1969a) The role of light in the germination of naturally-occurring populations of buried weed seeds. *Journal of Experimental Botany* **20**: 401–413.

Wesson G and Wareing PF (1969b) The induction of light sensitivity in weed seeds by burial. *Journal of Experimental Botany* **20**: 414–425.

Westhoff V (1967) The ecological impact of pedestrian, equestrian and vehicular traffic on vegetation. P-v. *Un. Int. Conserv. Nat.,* **10**: 218–223.

Westman WE (1975) Edaphic climax pattern of the pygmy forest region of California. *Ecological Monographs* **45**: 109–135.

Westman WE (1978) Measuring inertia and resilience of ecosystems. *Bioscience* **28**: 705–710.

Westoby M (1998) A leaf-height-seed (LHS) plant ecology strategy scheme. *Plant and Soil* **199**: 213–227.

Westoby M, Jurado E, and Leishman MR (1992) Comparative evolutionary ecology of seed size. *Trends in Ecology and Evolution* **7**: 368–372.

Wheeler BD and Giller KE (1982) Species richness in herbaceous fen vegetation on Broadland, Norfolk in relation to the quantity of above-ground plant material. *Journal of Ecology* **70**: 179–200.

Wheeler BD and Shaw SC (1991) Above-ground crop mass and species-richness of the principal types of herbaceous rich-fen vegetation of lowland England and Wales. *Journal of Ecology* **79**: 285–301.

Whelan BR and Edwards DG (1975) Uptake of potassium by *Setaria anceps* and *Macroptilium atropurpureum* from the same standard solution culture. *Australian Journal of Agricultural Research* **26**: 819–829.

White TA *et al.* (2000a) Impacts of extreme climatic events on competition during grassland invasions. *Global Change Biology* (in press).

White TA *et al.* (2000b) Sensitivity of three grassland communities to simulated extreme temperature and rainfall events. *Global Change Biology* (in press).

White TRC (1993) *The Inadequate Environment.* Springer-Verlag, Berlin.

Whitmore TC (1975) *Tropical Rainforests of the Far East.* Oxford University Press, London.

Whittaker RH (1965) Dominance and diversity in land plant communities. *Science* **147**: 250–260.

Whittaker RH (1966) Forest dimensions and production in the Great Smoky Mountains. *Ecology* **47**: 103–121.

Whittaker RH (1975) *Communities and Ecosystems*, 2nd ed. Macmillan, New York.

Whittaker RH and Feeny PP (1971) Allelochemics: chemical interactions between species. *Science* **171**: 757–770.

Whittaker RH and Goodman D (1979) Classifying species according to their demographic strategy. I. Population fluctuations and environmental heterogeneity. *American Naturalist* **113**: 185–200.

Whittingham J and Read DJ (1982) Vesicular-arbuscular mycorrhizas in natural vegetation systems. II. Nutrient transfer between plants with mycorrhizal interconnections. *New Phytologist* **90**: 277–284.

Wigley TML (1985) Impact of extreme events. *Nature* **316**: 106–107.

Wilbur HM, Tinkle DW, and Collins JP (1974) Environmental certainty, trophic level, and resource availability in life history evolution. *American Naturalist* 108: 805–817.

Wilde SA and White DP (1939) Damping-off as a factor in the natural distribution of pine species. *Phytopathology* **29**: 367–369.

Williams GC (1966) Natural selection, the costs of reproduction, and a refinement of Lack's principle. *American Naturalist* **100**: 687–692.

Williamson MH (1999) Invasions. *Ecography* **22**: 5–12.

Williamson MH and Fitter A (1996) The characters of successful invaders. *Biological Conservation* **78**: 163–170.

Williamson P (1976) Above-ground primary production of chalk grassland allowing for leaf death. *Journal of Ecology* **64**: 1059–1075.

Willis AJ (1963) Braunton Burrows: the effects on the vegetation of the addition of mineral nutrients to the dune soils. *Journal of Ecology* **51**: 353–374.

Willis AJ (1972) Long-term ecological changes in sward composition following application of Maleic Hydrazide and 2–4-D. In *Proceedings of the 11th British Weed Control Conference*, pp. 360–367. British Crop Protection Council, London.

Willis AJ (1988) The effects of growth retardent and selective herbicide on roadside verges at Bibury, Gloucestershire, over a thirty-year period. *Aspects of Applied Biology* **16**: 19–26.

Willis AJ (1989) Effects of the addition of mineral nutrients on the vegetation of the Avon Gorge, Bristol. *Proceedings of the Bristol Naturalists Society* **49**: 55–68.

Willis AJ *et al.* (1995) Does Gulf Stream position affect vegetation dynamics in Western Europe? *Oikos* **73**: 408–410.

Wilson EO (1971) Competitive and aggressive behaviour. In *Man and Beast*, eds. W Dillon and JF Eisenberg, pp. 183–217. Smithsonian Institution, Washington DC.

Wilson PJ (1998) The causes and consequences of recent vegetation change in Britain. PhD Thesis, University of Sheffield.

Wilson PJ, Thompson K, and Hodgson JG (1999) Specific leaf area and leaf dry matter content as alternative predictors of plant strategies. *New Phytologist* **143**: 155–162.

Wilson SD and Keddy PA (1986) Species competitive ability and position along a natural stress/disturbance gradient. *Ecology* **67**: 1236–1242.

Wilson SD and Shay JM (1990) Competition, fire and nutrients in a mixed-grass prairie. *Ecology* **71**: 1959–1967.

Winemiller KO and Rose KA (1992) Patterns of life-history diversification in North American fishes: implications for population regulation. *Canadian Journal of Fisheries and Aquatic Science* **49**: 2196–2218.

Wisheu IC and Keddy PA (1989) Species richness-standing crop relationships along four lakeshore gradients: constraints on the general model. *Canadian Journal of Botany* **67**: 1609–1617.

Wisheu IC and Keddy PA (1994) The low competitive ability of Canada's Atlantic coastal plain shoreline flora: implications for conservation. *Biological Conservation* **68**: 2247–252.

Wolfram S (1984) Cellular automata as models of complexity. *Nature* **311**: 419–424.

Woods DB and Turner NC (1971) Stomatal response to changing light by four tree species of varying shade tolerance. *New Phytologist* **70**: 77–84.

Woods KD (2000) Dynamics in late successional hemlock hardwood forests over three decades. *Ecology* **81**: 110–126.

Woodward FI (1987) *Climate and Plant Distribution*. Cambridge University Press, Cambridge.

Woodwell G and Rebuck AL (1967) Effects of chronic gamma radiation on the structure and diversity of an oak-pine forest. *Ecological Monographs* **37**: 53–69.

Woolhouse HW (1982) Aspects of the carbon and energy requirements of photosynthesis considered in relation to environmental constraints. In *Physiological Ecology: an Evolutionary Approach to Resource Use*, eds. CR Townsend and P Calow. Blackwell, Oxford.

Wourms JP (1972) The developmental biology of annual fishes: III. Pre-embryonic and embryonic diapause of variable duration in the eggs of annual fishes. *Journal of Experimental Zoology* **182**: 389–414.

Wright IJ and Westoby M (2000) Cross-species relationships between seedling relative growth rate, nitrogen productivity and root vs leaf function in 28 Australian woody species. *Functional Ecology* **14**: 97–107.

Yemm EW and Willis AJ (1962) The effects of maleic hydrazide and 2, 4-dichlorophenoxyacetic acid on roadside vegetation. *Weed Research* **2**: 24–40.

Yoda K *et al.* (1963) Self-thinning in overcrowded pure stands under cultivated and natural conditions. *Journal of Biology, Osaka City University* **14**: 107–129.

Youngman BJ (1951) Germination of old seeds. *Kew Bulletin* **6**: 423–426.
Zobel DB (1969) Factors affecting the distribution of *Pinus pungens*, an Appalachian endemic. *Ecological Monographs* **39**: 303–333.
Zobel K, Zobel M, and Rosen E (1994) An experimental test of diversity maintenance mechanisms by a species removal experiment in a species-rich wooded meadow. *Folia Geobotanica, Phytotax* **29**: 449–457.
Zobel M (1992) Plant species coexistence: the role of historical, evolutionary and ecological factors. *Oikos* **65**: 314–320.
Zobel M (1997) The relative role of species pools in determining plant species richness: an alternative explanation of species coexistence? *Trends in Ecology and Evolution* **12**: 266–269.

Species List

Index